QUANTITATIVE GRAPH THEORY

Mathematical Foundations and Applications

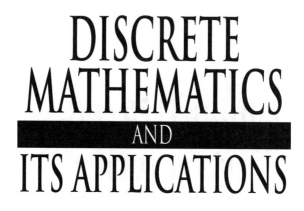

DISCRETE MATHEMATICS AND ITS APPLICATIONS

R. B. J. T. Allenby and Alan Slomson, How to Count: An Introduction to Combinatorics, Third Edition

Craig P. Bauer, Secret History: The Story of Cryptology

Juergen Bierbrauer, Introduction to Coding Theory

Katalin Bimbó, Combinatory Logic: Pure, Applied and Typed

Katalin Bimbó, Proof Theory: Sequent Calculi and Related Formalisms

Donald Bindner and Martin Erickson, A Student's Guide to the Study, Practice, and Tools of Modern Mathematics

Francine Blanchet-Sadri, Algorithmic Combinatorics on Partial Words

Miklós Bóna, Combinatorics of Permutations, Second Edition

Jason I. Brown, Discrete Structures and Their Interactions

Richard A. Brualdi and Dragoš Cvetković, A Combinatorial Approach to Matrix Theory and Its Applications

Kun-Mao Chao and Bang Ye Wu, Spanning Trees and Optimization Problems

Charalambos A. Charalambides, Enumerative Combinatorics

Gary Chartrand and Ping Zhang, Chromatic Graph Theory

Henri Cohen, Gerhard Frey, et al., Handbook of Elliptic and Hyperelliptic Curve Cryptography

Charles J. Colbourn and Jeffrey H. Dinitz, Handbook of Combinatorial Designs, Second Edition

Abhijit Das, Computational Number Theory

Matthias Dehmer and Frank Emmert-Streib, Quantitative Graph Theory: Mathematical Foundations and Applications

Martin Erickson, Pearls of Discrete Mathematics

Martin Erickson and Anthony Vazzana, Introduction to Number Theory

Titles (continued)

Steven Furino, Ying Miao, and Jianxing Yin, Frames and Resolvable Designs: Uses, Constructions, and Existence

Mark S. Gockenbach, Finite-Dimensional Linear Algebra

Randy Goldberg and Lance Riek, A Practical Handbook of Speech Coders

Jacob E. Goodman and Joseph O'Rourke, Handbook of Discrete and Computational Geometry, Second Edition

Jonathan L. Gross, Combinatorial Methods with Computer Applications

Jonathan L. Gross and Jay Yellen, Graph Theory and Its Applications, Second Edition

Jonathan L. Gross, Jay Yellen, and Ping Zhang Handbook of Graph Theory, Second Edition

David S. Gunderson, Handbook of Mathematical Induction: Theory and Applications

Richard Hammack, Wilfried Imrich, and Sandi Klavžar, Handbook of Product Graphs, Second Edition

Darrel R. Hankerson, Greg A. Harris, and Peter D. Johnson, Introduction to Information Theory and Data Compression, Second Edition

Darel W. Hardy, Fred Richman, and Carol L. Walker, Applied Algebra: Codes, Ciphers, and Discrete Algorithms, Second Edition

Daryl D. Harms, Miroslav Kraetzl, Charles J. Colbourn, and John S. Devitt, Network Reliability: Experiments with a Symbolic Algebra Environment

Silvia Heubach and Toufik Mansour, Combinatorics of Compositions and Words

Leslie Hogben, Handbook of Linear Algebra, Second Edition

Derek F. Holt with Bettina Eick and Eamonn A. O'Brien, Handbook of Computational Group Theory

David M. Jackson and Terry I. Visentin, An Atlas of Smaller Maps in Orientable and Nonorientable Surfaces

Richard E. Klima, Neil P. Sigmon, and Ernest L. Stitzinger, Applications of Abstract Algebra with Maple™ and MATLAB®, Second Edition

Richard E. Klima and Neil P. Sigmon, Cryptology: Classical and Modern with Maplets

Patrick Knupp and Kambiz Salari, Verification of Computer Codes in Computational Science and Engineering

William Kocay and Donald L. Kreher, Graphs, Algorithms, and Optimization

Donald L. Kreher and Douglas R. Stinson, Combinatorial Algorithms: Generation Enumeration and Search

Hang T. Lau, A Java Library of Graph Algorithms and Optimization

C. C. Lindner and C. A. Rodger, Design Theory, Second Edition

San Ling, Huaxiong Wang, and Chaoping Xing, Algebraic Curves in Cryptography

Nicholas A. Loehr, Bijective Combinatorics

Toufik Mansour, Combinatorics of Set Partitions

Titles (continued)

Alasdair McAndrew, Introduction to Cryptography with Open-Source Software

Elliott Mendelson, Introduction to Mathematical Logic, Fifth Edition

Alfred J. Menezes, Paul C. van Oorschot, and Scott A. Vanstone, Handbook of Applied Cryptography

Stig F. Mjølsnes, A Multidisciplinary Introduction to Information Security

Jason J. Molitierno, Applications of Combinatorial Matrix Theory to Laplacian Matrices of Graphs

Richard A. Mollin, Advanced Number Theory with Applications

Richard A. Mollin, Algebraic Number Theory, Second Edition

Richard A. Mollin, Codes: The Guide to Secrecy from Ancient to Modern Times

Richard A. Mollin, Fundamental Number Theory with Applications, Second Edition

Richard A. Mollin, An Introduction to Cryptography, Second Edition

Richard A. Mollin, Quadratics

Richard A. Mollin, RSA and Public-Key Cryptography

Carlos J. Moreno and Samuel S. Wagstaff, Jr., Sums of Squares of Integers

Gary L. Mullen and Daniel Panario, Handbook of Finite Fields

Goutam Paul and Subhamoy Maitra, RC4 Stream Cipher and Its Variants

Dingyi Pei, Authentication Codes and Combinatorial Designs

Kenneth H. Rosen, Handbook of Discrete and Combinatorial Mathematics

Douglas R. Shier and K.T. Wallenius, Applied Mathematical Modeling: A Multidisciplinary Approach

Alexander Stanoyevitch, Introduction to Cryptography with Mathematical Foundations and Computer Implementations

Jörn Steuding, Diophantine Analysis

Douglas R. Stinson, Cryptography: Theory and Practice, Third Edition

Roberto Tamassia, Handbook of Graph Drawing and Visualization

Roberto Togneri and Christopher J. deSilva, Fundamentals of Information Theory and Coding Design

W. D. Wallis, Introduction to Combinatorial Designs, Second Edition

W. D. Wallis and J. C. George, Introduction to Combinatorics

Jiacun Wang, Handbook of Finite State Based Models and Applications

Lawrence C. Washington, Elliptic Curves: Number Theory and Cryptography, Second Edition

DISCRETE MATHEMATICS AND ITS APPLICATIONS

QUANTITATIVE GRAPH THEORY

Mathematical Foundations and Applications

Edited by

Matthias Dehmer

Institute for Theoretical Computer Science, Mathematics and Operations Research
Department of Computer Science
Universität der Bundeswehr München
Neubiberg-München
Germany

Frank Emmert-Streib

Computational Biology and Machine Learning Laboratory
Center for Cancer Research and Cell Biology
School of Medicine, Dentistry and Biomedical Sciences
Faculty of Medicine, Health and Life Sciences
Queen's University Belfast, Belfast, UK

CRC Press
Taylor & Francis Group
Boca Raton London New York

CRC Press is an imprint of the
Taylor & Francis Group, an **informa** business

A CHAPMAN & HALL BOOK

CRC Press
Taylor & Francis Group
6000 Broken Sound Parkway NW, Suite 300
Boca Raton, FL 33487-2742

© 2015 by Taylor & Francis Group, LLC
CRC Press is an imprint of Taylor & Francis Group, an Informa business

No claim to original U.S. Government works

Printed on acid-free paper
Version Date: 20140916

International Standard Book Number-13: 978-1-4665-8451-8 (Hardback)

Library of Congress Cataloging-in-Publication Data

Quantitative graph theory : mathematical foundations and applications / [edited by] Matthias Dehmer, Frank Emmert-Streib.
     pages cm. -- (Discrete mathematics and its applications)
     Includes bibliographical references and index.
     ISBN 978-1-4665-8451-8 (hardback)
     1. Graph theory--Data processing. 2. Combinatorial analysis. I. Dehmer, Matthias, 1968- editor. II. Emmert-Streib, Frank, editor.

QA166.Q36 2015
511'.5--dc23   2014033366

Visit the Taylor & Francis Web site at
http://www.taylorandfrancis.com

and the CRC Press Web site at
http://www.crcpress.com

To Marion

Contents

Preface   xi

Editors   xv

Contributors   xvii

1   **What Is Quantitative Graph Theory?**   1
Matthias Dehmer, Veronika Kraus, Frank Emmert-Streib, and Stefan Pickl

2   **Localization of Graph Topological Indices via Majorization Technique**   35
*Monica Bianchi, Alessandra Cornaro, José Luis Palacios,
and Anna Torriero*

3   **Wiener Index of Hexagonal Chains with Segments of Equal Length**   81
Andrey A. Dobrynin

4   **Metric-Extremal Graphs**   111
Ivan Gutman and Boris Furtula

5   **Quantitative Methods for Nowhere-Zero Flows and Edge Colorings**   141
Martin Kochol

6   **Width-Measures for Directed Graphs and Algorithmic Applications**   181
Stephan Kreutzer and Sebastian Ordyniak

7   **Betweenness Centrality in Graphs**   233
Silvia Gago, Jana Coroničová Hurajová, and Tomáš Madaras

8   **On a Variant Szeged and PI Indices of Thorn Graphs**   259
Mojgan Mogharrab and Reza Sharafdini

9   **Wiener Index of Line Graphs**   279
Martin Knor and Riste Škrekovski

10   **Single-Graph Support Measures**   303
Toon Calders, Jan Ramon, and Dries Van Dyck

11 Network Sampling Algorithms and Applications   **325**
Michael Drew LaMar and Rex K. Kincaid

12 Discrimination of Image Textures Using Graph Indices   **355**
Martin Welk

13 Network Analysis Applied to the Political Networks of Mexico   **387**
Philip A. Sinclair

14 Social Network Centrality, Movement Identification, and the
Participation of Individuals in a Social Movement: The Case of the
Canadian Environmental Movement   **407**
David B. Tindall, Joanna L. Robinson, and Mark C.J. Stoddart

15 Graph Kernels in Chemoinformatics   **425**
Benoît Gaüzère, Luc Brun, and Didier Villemin

16 Chemical Compound Complexity in Biological Pathways   **471**
Atsuko Yamaguchi and Kiyoko F. Aoki-Kinoshita

Index   **495**

Preface

Graph-based approaches have been employed extensively in several disciplines such as biology, computer science, chemistry, among others. In the 1990s, exploration of the topology of complex networks became quite popular and was triggered by the breakthrough of the Internet and the examinations of random networks. As a consequence, the structure of random networks has been explored using graph-theoretic methods and stochastic growth models. However, it turned out that besides exploring random graphs, quantitative approaches to analyze networks are crucial as well. This relates to quantifying structural information of complex networks by using a measurement approach. As demonstrated in the scientific literature, graph- and information-theoretic measures, and statistical techniques applied to networks have been used to do this quantification.

It has been found that many real-world networks are composed of network patterns representing nonrandom topologies. Graph- and information-theoretic measures have been proven efficient in quantifying the structural information of such patterns. The study of relevant literature reveals that quantitative graph theory has not yet been considered a branch of graph theory. In fact, most of the existing theories related to analyzing graphs structurally are descriptive in nature, for example, Kuratowski's theorem for characterizing planar graphs and the Eulerian paths. Compared with the number of existing descriptive approaches in graph theory, quantitative techniques have been clearly underrepresented.

So far, quantitative approaches for analyzing graphs have been examined from different perspectives in a variety of disciplines including discrete mathematics, computer science, computational biology, ecology, structural chemistry, medicinal chemistry, and sociology. In biology, computer science, and discrete mathematics, the focus has been on the comparative analysis of graphs, for example, computation of their structural similarity/distance. In addition, determining the structural complexity of graphs using so-called numerical graph invariants has been a growing field for several decades. For instance, graph invariants have been used extensively to characterize molecular structures. Besides applications in drug design such as QSAR and QSPR, their mathematical properties such as their uniqueness and structural interpretation have also been investigated.

The main goal of this book is to present and demonstrate existing novel methods for analyzing graphs quantitatively. The underlying mathematical methods have been developed with the aid of graph theory, information theory, measurement theory, and statistical techniques. The book is intended for researchers and graduate and advanced undergraduate students in the fields of mathematics, computer science, mathematical chemistry, cheminformatics, physics, bioinformatics, and systems biology. As the potential of quantitative graph methods is huge, this list of scientific fields cannot be complete and will hopefully be extended in the future.

The topics covered in this book include a broad range of quantitative graph-theoretical concepts and methods including those pertaining to real and random graphs such as

- Comparative approaches (graph similarity or distance)

- Graph measures to characterize graphs quantitatively

- Applications of graph measures in social network analysis and other disciplines

- Metrical properties of graphs and measures

- Mathematical properties of quantitative methods or measures in graph theory

- Network complexity measures and other topological indices

- Quantitative approaches to graphs using machine learning, for example, clustering.

- Graph measures and statistics

- Information-theoretic methods to analyze graphs quantitatively (e.g., entropy)

By covering this broad range of topics, the book aims to fill a gap in the contemporary literature of discrete and applied mathematics, computer science, systems biology, and related disciplines.

This book is dedicated, with all my heart, to my sister Marion Dehmer who passed away in 2012. She was always helpful and creative when helping me with earlier projects. Her positive spirit would have been highly beneficial for this and future projects.

Many colleagues, knowingly or unknowingly have provided us with input, help, and support before and during the preparation of this book. In particular, we thank Marion Dehmer, Andrey A. Dobrynin, Boris Furtula, Ivan Gutman, Bo Hu, D. D. Lozovanu, Alexei Levitchi, Abbe Mowshowitz, Miriana Moosbrugger, Andrei Perjan, Stefan Pickl, Stefan Shetschew, Yongtang Shi, Ricardo de Matos Simoes, Fred Sobik, Shailesh Tripathi, Kurt Varmuza, and Dongxiao Zhu and apologize for all whose names that have been inadvertently omitted. We also thank acquiring editor Sunil Nair, editorial assistant Sarah Gelson, and project coordinator Jennifer Ahringer from CRC Press who have been always available and helpful. Last but not least, Matthias Dehmer thanks the Austrian Science Funds (Project P26142) and the Universität der Bundeswehr München (Project RiKoV, Grant No. 13N12304) for supporting this work.

To the best of our knowledge, this is the first book devoted exclusively to quantitative graph theory. Existing books dealing with graph theory and complex networks have limited scope as they mainly consider descriptive approaches to explore networks. Therefore, we hope that this book will help establish quantitative graph theory as a modern and independent branch of network sciences and graph theory. It will also broaden the scope of pure and applied researchers who deal with graph-based techniques. Finally, we hope this book conveys the enthusiasm and joy we have for this field and inspires fellow researchers in their own practical or theoretical work.

Matthias Dehmer
Hall/Tyrol, Austria and Munich, Germany

Frank Emmert-Streib
Belfast, United Kingdom

MATLAB® is a registered trademark of The MathWorks, Inc. For product information, please contact:

The MathWorks, Inc.
3 Apple Hill Drive
Natick, MA 01760-2098 USA
Tel: 508-647-7000
Fax: 508-647-7001
E-mail: info@mathworks.com
Web: www.mathworks.com

Editors

Matthias Dehmer, PhD, studied mathematics and computer science at the University of Siegen, Germany, and graduated from there in 1998. His studies focused on discrete mathematics, complex analysis, and statistics. Later, he earned his PhD in computer science from the Darmstadt University of Technology, where his thesis dealt with the problem of measuring the structural similarity of graphs. From 2005 to 2008, he held several research positions at the University of Rostock (Germany), Vienna Bio Center (Austria), Vienna Technical University (Austria), and University of Coimbra (Portugal). He earned his habilitation in applied discrete mathematics from the Vienna University of Technology. His research focuses on investigating network-based methods in the context of systems biology, structural graph theory, operations research, and information theory. In particular, Dr. Dehmer has been extensively exploring graph similarity techniques, information theory, and statistics (e.g., entropy, statistical estimation, and optimization) to investigate complex networks. He has over 165 peer-reviewed publications in mathematics and computer science. Moreover, he is an editor of the book series Quantitative and Network Biology published by Wiley-VCH. He has also organized and co-organized several international scientific conferences and workshops in the United States. He was recently made a member of the editorial board of *Scientific Reports* (Nature Publishing Group) and *PLOS ONE.*

Frank Emmert-Streib, PhD, studied physics at the University of Siegen, Germany, and earned his PhD in theoretical physics from the University of Bremen. After postdoc positions in the United States, he joined the Center for Cancer Research and Cell Biology at the Queen's University Belfast (United Kingdom), where he is currently an associate professor (senior lecturer) leading the Computational Biology and Machine Learning Laboratory. His research interests are in the fields of computational biology, biostatistics, and network medicine focused on the development and application of methods from statistics and machine learning for the analysis of high-dimensional data from genomics experiments.

Contributors

Kiyoko F. Aoki-Kinoshita
Faculty of Engineering
Department of Bioinformatics
Soka University
Tokyo, Japan

Monica Bianchi
Faculty of Economics
Department of Mathematics
  and Econometrics
Catholic University
Milan, Italy

Luc Brun
Research Group on Computer,
  Imaging, Control and Instrumentation
Superior National School of Engineering
  and Research Center
Caen, France

Toon Calders
Faculty of Applied Sciences
Department of Computer & Decision
  Engineering
Free University of Brussels
Bruxelles, Belgium

Alessandra Cornaro
Faculty of Economics
Department of Mathematics
  and Econometrics
Catholic University
Milan, Italy

Matthias Dehmer
Division for Bioinformatics and
Translational Research
UMIT—The Health and Life Sciences
  University
Tyrol, Austria

and

Faculty of Computer Science
Institute of Theoretical Computer
  Science, Mathematics and Operations
  Research
Bundeswehr University Munich
Munich, Germany

Andrey A. Dobrynin
Sobolev Institute of Mathematics
Siberian Branch of the Russian Academy
  of Sciences
Novosibirsk, Russia

Frank Emmert-Streib
Faculty of Medicine, Health and Life
  Sciences
Computational Biology and Machine
  Learning Laboratory
Center for Cancer Research and Cell
  Biology
School of Medicine, Dentistry and
  Biomedical Sciences
Queen's University Belfast
Belfast, United Kingdom

Boris Furtula
Faculty of Science
University of Kragujevac
Kragujevac, Serbia

Contributors

Silvia Gago
Department of Applied Mathematics III
and
Barcelona College of Industrial
  Engineering
Polytechnic University of Catalonia
Barcelona, Spain

Benoît Gaüzère
Departement of Computer Science
Research Group in Computer Science,
  Image, Control and Electronics
and
National Graduate School of Engineering
  and Research Center
and
University of Caen
Caen, France

Ivan Gutman
Faculty of Science
University of Kragujevac
Kragujevac, Serbia

Jana Coroničová Hurajová
Institute of Mathematics
Pavol Jozef Šafárik University in Košice
Košice, Slovakia

Rex K. Kincaid
Department of Mathematics
The College of William and Mary
Williamsburg, Virginia

Martin Knor
Faculty of Civil Engineering
Department of Mathematics
Slovak University of Technology in
  Bratislava
Bratislava, Slovakia

Martin Kochol
Mathematical Institute
Slovak Academy of Sciences
Bratislava, Slovakia

Veronika Kraus
Institute for Bioinformatics and
  Translational Research
UMIT—The Health and Life Sciences
  University
Tyrol, Austria

Stephan Kreutzer
School IV—Electrical Engineering and
  Computer Science
Technical University of Berlin
Berlin, Germany

Michael Drew LaMar
Department of Biology
Integrated Science Center
The College of William and Mary
Williamsburg, Virginia

Tomáš Madaras
Institute of Mathematics
Pavol Jozef Šafárik University in Košice
Košice, Slovakia

Mojgan Mogharrab
Department of Mathematics
Persian Gulf University
Bushehr, Iran

Sebastian Ordyniak
Faculty of Informatics
Masaryk University
Brno, Czech Republic

José Luis Palacios
Department of Electrical and Computer
  Engineering
University of New Mexico
Albuquerque, New Mexico

Stefan Pickl
Faculty of Computer Science
Institute of Theoretical Computer
    Science, Mathematics and Operations
    Research
Bundeswehr University Munich
Munich, Germany

Jan Ramon
Department of Computer
    Science
Katholieke Universiteit Leuven
Leuven, Belgium

Joanna L. Robinson
Glendon College
York University
Toronto, Ontario, Canada

Reza Sharafdini
Department of Mathematics
Persian Gulf University
Bushehr, Iran

Philip A. Sinclair
School of Computing, Mathematics and
    Digital Technology
Manchester Metropolitan University
Manchester, United Kingdom

Riste Škrekovski
Department of Mathematics
University of Ljubljana
Ljubljana, Slovenia

and

Faculty of Information Studies
Novo Mesto, Slovenia

and

Faculty of Mathematics, Natural
    Sciences and Information Technologies
University of Primorska
Koper, Slovenia

Mark C.J. Stoddart
Department of Sociology
Memorial University of Newfoundland
St. John's, Newfoundland, Canada

David B. Tindall
Department of Sociology
University of British Columbia
Vancouver, British Columbia, Canada

Anna Torriero
Faculty of Economics
Department of Mathematics
    and Econometrics
Catholic University
Milan, Italy

Dries Van Dyck
Databases and Theoretical Computer
    Science Research Group
Hasselt University
and
Transnational University of Limburg
Hasselt, Belgium

Didier Villemin
Laboratory of Molecular and
    Thio-Organic Chemistry
National Graduate School of Engineering
    and Research Center
Caen, France

Martin Welk
Biomedical Image Analysis Division
Institute for Biomedical Computer
    Science
UMIT—The Health and Life Sciences
    University
Tyrol, Austria

Atsuko Yamaguchi
Database Center for Life Science
Research Organization of Information
    and Systems
Kashiwa, Japan

Chapter 1

What Is Quantitative Graph Theory?

Matthias Dehmer, Veronika Kraus, Frank Emmert-Streib, and Stefan Pickl

Contents

1.1 Introduction and Concept Formation..1
1.2 Quantitative versus Descriptive Graph Theory....................................3
1.3 Aspects of Quantitative Graph Theory..4
    1.3.1 Comparative Graph Analysis...5
        1.3.1.1 Isomorphism-Based Measures...................................5
        1.3.1.2 Graph Edit Distance...6
        1.3.1.3 Iterative Methods...8
        1.3.1.4 String-Based Measures.......................................8
        1.3.1.5 Graph Kernels...8
        1.3.1.6 Similarity of Document Structures...........................9
        1.3.1.7 Tree Similarity...9
        1.3.1.8 Molecular Similarity..9
        1.3.1.9 Statistical Graph Matching..................................10
    1.3.2 Numerical Graph Invariants...11
        1.3.2.1 Uniqueness of Graph Measures................................13
        1.3.2.2 Information Inequalities.....................................15
        1.3.2.3 Correlation of Graph Measures...............................18
    1.3.3 Software for Quantitative Graph Analysis.................................19
        1.3.3.1 R-Packages..19
        1.3.3.2 Conjecture Engines..19
1.4 Summary and Conclusion...20
References..20

1.1 Introduction and Concept Formation

Graph theory [12,83,86] is a relatively new branch of mathematics and the main body of fundamental results dates back to the seminal contribution due to König [111]. Earlier contributions dealing with graphs are due to Euler [72] and Kirchhoff [104]. Importantly, Euler solved the so-called *Königsberg bridge problem* by introducing graph-theoretical terms that contributed to modern graph theory extensively [14,75].

Later, investigating topological aspects of graphs has led to seminal work due to Kuratowski [117] and Gross et al. [80] and, finally, to the emergence of the branch topological graph theory [80]. This relates to investigate, for example, embeddings of graphs in surfaces and graphs as topological spaces. A highlight in this theory is surely the theorem of Kuratowski [117] for characterizing planar graphs. Other numerous methods to explore structural properties of networks have also contributed and have led to a vast number of journal and book publications (e.g., see [12,14,39,75,83,86,111]). Note that other branches of graph theory such as extremal or random graph theory have also been explored (see [14]). Based on those multifaceted methods, numerous problems by means of graphs have been solved in disciplines such as computer science [2,75], chemistry [61,63,82], ecology [172], and sociology and cognitive sciences [84,85,163,177].

In a wider sense, graph-based approaches have been used extensively in various disciplines. The hype dealing with complex networks also contributed a lot to modern graph or network theory and has been triggered from the breakthrough of the World Wide Web and other physically oriented studies exploring networks (graphs) as complex systems [2,4,178,179]. Besides investigating only random graph models for analyzing complex systems, it turned out that there is a strong need to further develop *quantitative* approaches. A main reason for this was the insight that many real-world networks are composed of nonrandom topologies where quantitative methods such as graph measures [37,39] have been proven essential to quantify structural information of graphs.

When studying the existing literature dealing with classical aspects of graph theory [12,83,86,111], it turns out that most of the existing contributions are *descriptive* approaches for describing graphs structurally. Examples thereof are Kuratowski's theorem mentioned earlier, the description of Eulerian paths, and graph colorings [12,83,86]. This leads us straightforwardly to a definition of *quantitative graph theory*:

> Instead of characterizing graphs only descriptively, Quantitative Graph Theory deals with quantifying structural aspects of graphs.

We see that the aspect of *measurement* is crucial here. To quantify structural information of a graph means to employ any kind of measurement, that is, a local or global one.

Concrete examples of quantitative approaches in the context of graph theory are as follows:

- Graph similarity or distance measures [50,69,155,156,183].

- Graph measures to characterize the topology of a graph [37,39,169]. This group of measurements includes numerical graph invariants that are also referred to as *topological indices* [18,61].

- Exploring metrical properties of graphs and deriving measures thereof [97,153].

- Treating structural graph measures as network complexity measures and exploring properties thereof [16,17,40,54].

- Information theory and statistics of graphs [5,44,110].

- Applications of the mentioned graph measures in social network analysis, chemistry, and other disciplines [18,61,177].

- Using machine learning for deriving quantitative approaches for graphs [33,78].

As known, various graph measures have been defined and investigated [37,39] even in the early 1950s [11,84]. Seminal contributions from this development have been graph centrality measures to investigate aspects in sociology such as group performance [11,84]. Numerous other graph-theoretical measurements have been developed afterward, but to date, there has been no development to consider quantitative graph theory as a graph-theoretical branch. In view of the vast number of existing descriptive methods [12,83,86], quantitative techniques to analyze graphs are clearly underrepresented so far. In fact, quantitative approaches to explore graphs have been examined from different perspectives in a variety of disciplines including discrete mathematics, computer science, biology, chemistry, and related areas [43,101]. In biology, computer science, and discrete mathematics, the focus has tended to be on the comparative analysis of graphs such as graph similarity and graph distances; in chemistry and related fields, the principal aim has been to quantify structural features of graphs leading to numerical graph invariants [63].

1.2 Quantitative versus Descriptive Graph Theory

Based on our concept formation outlined in Section 1.1, it is reasonable to distinguish between descriptive and quantitative graph theory. Typical descriptive approaches involve structural concepts such as graph colorings, graph embeddings and decompositions, and the characterization of graphs. Typical quantitative methods are quantitative graph measures that map graphs to the reals by taking structural features into account [37,39,169]. It is clear that those measures can be further divided into several classes, for instance, information theoretic and noninformation theoretic. An exhaustive overview on the classification of quantitative network measures has been given by Emmert-Streib and Dehmer [68].

We emphasize that some methods may be ambiguous as they belong to both descriptive and quantitative graph theory. An example thereof is the term graph characterization. On the one hand, the term *Euler graph* [83,86] characterizes a graph based on the existence of a special kind of path. On the other hand, a numerical invariant such as the Wiener index [180] or the global clustering coefficient [133] also characterizes a given graph as the invariant maps it to a real number. Then, the characterization of the graph corresponds to the magnitude of this number describing the complexity of the given relational structure. This implies that any structural graph measures has a *structural interpretation* whose exploration has been a challenging undertaking [20,53,128].

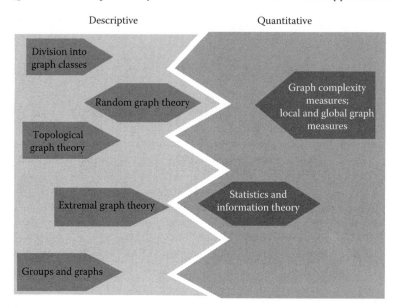

FIGURE 1.1:   Exemplaric classification of descriptive and quantitative graph theory.

Figure 1.1 shows a classification of descriptive and quantitative graph theory by way of example. On the left-hand side, we see typical branches of descriptive graph theory. By contrast, the quantitative graph theory shown on the right-hand side employs measurements and relates to quantify structural information of graphs by using, for example, quantitative network measures [53,68,169]. Another possibility to derive quantitative approaches for graphs is to use tools from statistics and information theory [35] such as entropy and mutual information [53,68]. Figure 1.1 also indicates that hybrid concepts exist, that is, methods that are based on both quantitative and descriptive techniques. It is evident that quantitative measures can be easily derived from descriptive methods. The closer a branch is located to the parting line of Figure 1.1, the more overlap possess the corresponding branches. For instance, statistics and information theory is depicted closely to the parting line as statistical and information-theoretic methods can be used for both describing and quantifying properties of graphs.

1.3   Aspects of Quantitative Graph Theory

Quantitative graph theory can be divided into two subject areas, namely, graph similarity problems and the use and analysis of numerical graph invariants. In the upcoming Sections 1.3.1 and 1.3.2, we will go into detail about the different aspects

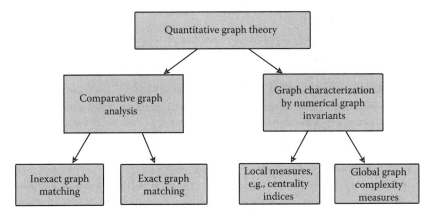

FIGURE 1.2: Classification of quantitative graph theory.

of these areas, giving an overview of existing theories and literature. A graphical overview of the principal aspects of quantitative graph theory is given in Figure 1.2.

1.3.1 Comparative Graph Analysis

In this section, we review the most important methods for performing comparative graph analysis [32]. To start, we note that comparative graph analysis relates to determine the structural similarity or distance between graphs. This problem is often referred to as graph matching and relates to compare graphs either by finding isomorphic relations (*exact* graph matching) [26,155,183] or by using *inexact* (approximative) techniques for matching graphs in an error-tolerant way [26]. Evidently, measuring the structural similarity of large graphs is a critical undertaking because of the often inefficient computational complexity. It is important to emphasize that graph matching techniques [26] have been extensively applied in various scientific disciplines, for example, artificial intelligence and pattern recognition [32], biology [67,101], chemoinformatics [137,138,174], cognitive modeling [159–162], computational linguistics [125], image recognition [95,165], machine learning [21,93], and web mining [50].

1.3.1.1 Isomorphism-Based Measures

The problem of finding an isomorphism between two given graphs G and H is often referred to as exact graph matching [102,155]. In particular, this holds if $n(G) = n(H)$ where $n(G)$ and $n(H)$ denote the number of vertices of G and H, respectively. In case $n(G) \neq n(H)$, there is a need to identify subgraph isomorphic relations between G and H (see, e.g., [26,102,155]). The first graph distance measure based on determining isomorphic relations has been developed by Zelinka [183]:

$$d_Z(G, H) := n - n(G, H), \quad G, H \in \mathcal{G}_n. \tag{1.1}$$

$\mathcal{G}_n$ is the set of graphs having $n = n(G)$ vertices, and $n(G, H)$ stands for the number of vertices of the largest common induced subgraph of both G and H. Note that this metric only holds for the set of all graphs with n vertices without reflexive and parallel edges. After this seminal work [183], this graph metric has been generalized by Sobik [155] to determine the similarity of labeled graphs of different order:

$$d_{S_1}(G, H) := \max \big(n(G), n(H)\big) - n(G, H), \quad G, H \in \mathcal{G}. \tag{1.2}$$

$\mathcal{G}$ denotes the set of all graphs. Also, Kaden [102] defined a related metric:

$$d_K(G, H) := 1 - \frac{n(G, H)}{\max\big(n(G), n(H)\big)}, \quad G, H \in \mathcal{G}. \tag{1.3}$$

Several variants of those graph metrics have been investigated by Sobik [155,157] and Kaden [102]. For more related work, see [24,26,32]. A complementary approach to Zelinka's distance has been introduced by Sobik [155,157]. It is based on the minimal order of a distinguishing subgraph, that is, the minimal order of an induced subgraph contained in exactly one of the considered graphs. An example for a resulting graph metric is

$$d_{S_2}(G, H) := \max \big(n(G), n(H)\big) + 1 - u(G, H), \quad G, H \in \mathcal{G}. \tag{1.4}$$

Here, $u(G, H)$ stands for the number of vertices of the smallest induced subgraph contained in exactly one of the graphs G and H, respectively. Note that Sommerfeld [159–162] applied these methods for modeling cognitive structures.

However, we remark that the subgraph isomorphism problem is known to be NP-complete [77]. By contrast, the complexity of the graph isomorphism problem, the problem is believed to be neither NP-complete nor polynomial [167]. Following the work of Toda [167], it is widely assumed that the complexity of the graph isomorphism problem lies between deterministic polynomial-time algorithms and NP-completeness. Hence, this causes major drawbacks when applying the just mentioned measures to real-world networks. To overcome this problem, Dehmer et al. [47] recently explored highly discriminating graph invariants and found that they are useful for performing isomorphism testing. Note that there is a considerable body of literature dealing with the problem of deriving complete graph invariants [41,120], but until now, such invariants computable with polynomial-time complexity have not yet been found.

1.3.1.2   Graph Edit Distance

The *graph edit distance* (GED) for matching graphs in an error-tolerant way is an important measure developed by Bunke [22]. GED belongs to the inexact graph matching paradigm (see, e.g., [22,26]). An early attempt in the area of pattern recognition was to explore the well-known *string edit distance* (SED) done by Levenstein [118]. Graphs can often be represented as string representations [50,141], and, hence, GED can be considered as an extension of the SED measure. In order to understand the underlying principle of GED, note that this measure is based on defining graph edit operations such as insertions or deletions of edges/vertices or relabelings of vertices along with certain edit costs associated with each such operation. In [22], Bunke defined a sequence of edit operations that transforms a graph G into H by producing minimal transformation costs as an optimal inexact match. Note that these operations in fact map G to a graph isomorphic to H. By assuming $m_1, m_2, \ldots, m_n$ to be all possible transformations that map G to H, then the optimal inexact match m' is defined by [22]

$$c(m') = \min\{c(m_i)|1 \leq i \leq n\}. \tag{1.5}$$

$c(m_i)$ is the cost of m_i. Finally, the GED of two given graphs is the minimum cost associated with a an optimal inexact match, that is, $c(m')$. Also, the optimal error-correcting graph isomorphism is then defined as the resulting isomorphism of H obtained by the optimal sequence of edit operations [22]. Finally, the key result due to Bunke [22] is as follows.

Theorem 1.3.1 *Let $d(H, G)$ be the costs for determining the optimal inexact match between H and G. $d(H, G)$ is a graph metric.*

Note that Theorem 1.3.1 requires that graph edit costs are in some sense symmetric. If G is mapped to H by a sequence of operations m, then H is mapped to G by the inverse operations, where deletion is inverse to addition. Since a metric requires symmetry, that is, $d(G, H) = d(H, G)$, the cost $c(m')$ must be the same for m' and its inverse. This problem has been overcome in computing by Nehaus and Bunke [132].

Importantly, many variants of the GED measure for finding appropriate models to determine matching costs have been investigated (see, e.g., [22,25]). Further, Bunke [23] pointed out there is an interesting relation between the edit distance and the maximum common subgraph (MCS) method. In particular, Bunke [23] demonstrated that the optimal error-correcting graph isomorphism for a certain graph edit cost function is the MCS. As to aspects of computational complexity, it has been shown that the graph matching problem is NP-complete [26,173]. But particularly for graphs possessing unique vertex labels, Dickinson et al. [62] proved that the computational complexity to compute the GED is $O(n^2)$. Moreover, they proved a useful theorem stating that some related computational graph problems require quadratic time complexity [62].

Theorem 1.3.2 *Start from the class of graphs having unique node labels. Then, the following problems possess quadratic time complexity with respect to n:*

- *GED*

- *Graph isomorphism problem*

- *Subgraph isomorphism problem*

- *MCS problem*

Besides GED-like measures, other graph similarity or distance measures based on graph transformations have been discussed by Chartrand et al. [29,30].

1.3.1.3   Iterative Methods

Iterative methods to determine similarity scores between graph elements (e.g., vertices or edges) have been proven useful, for example, for analyzing graph-based patterns in web mining [13,109,182]. When starting from two given graphs G and H, the main idea for deriving such similarity scores is to consider an element (e.g., a vertex or an edge) of G and an element of H as similar if their corresponding neighborhoods are similar (see [182]). Seminal work to explore iterative graph algorithms has been done by Kleinberg [109]. Afterward, other approaches based on the idea of recursively calculating vertex similarity scores using the scores of neighboring vertices have also been developed (see, e.g., [126]). For instance, the approach due to Blondel et al. [13] is based on Kleinberg's method for developing a similarity measure between any two vertices in any two graphs. But note that the limit value of the procedure depends on the initial conditions [13,182]. To overcome this problem, Zager et al. [182] proposed a more sophisticated approach where the resulting similarity scores for edges and vertices do not depend on initial values. For more related work in this area, see [182].

1.3.1.4   String-Based Measures

A powerful alternative for tackling the graph matching problem is based on converting graphs into strings and then applying *string alignment* techniques [118]. But such a technique heavily relies on the procedure to transform graphs into strings.

For instance, Robles-Kelly et al. [141] used the so-called spectral seriation method based on the leading eigenvector of the underlying adjacency matrix for converting graphs to string representations. In order to match two graphs, the corresponding string representations are then matched and evaluated [141]. In order to do so, an important problem is to derive the strings in a way that edge ordering of the vertices is preserved. Finally, Robles-Kelly et al. [141] tackled this problem by using a probabilistic framework.

Dehmer et al. [50] also developed a technique to determine the structural similarity of hierarchical structures based on using string alignment techniques [50]. The graphs under consideration are called *generalized trees* [51] representing rooted-tree-like structures (hierarchical graphs). Note that this technique to measure the structural similarity between generalized trees has been used to determine the similarity of arbitrary undirected graphs (see [69]). Hence, the method is not restricted to hierarchical graphs. Also, several extensions for measuring the similarity between arbitrary graphs have been developed by Dehmer and Emmert-Streib [42].

1.3.1.5   Graph Kernels

Another method for measuring the structural similarity between graphs is based on using the so-called graph kernels [21,93]. In order to compare isomorphic subgraphs contained in two given graphs, graph kernels based on random walks, paths, cyclic patterns, and trees have been employed (see [21,93]). Since the kernel method maps graphs into vectors to be treated in a Hilbert space of graph features [21], machine learning techniques such as support vector machines are applicable.

Early results when investigating mathematical properties of kernel functions have been achieved by Schölkopf et al. [148]. Further related work dealing with graph kernels can be found in [21]. Recent applications using graph kernel approaches are in bioinformatics [21] and chemoinformatics [21,79].

1.3.1.6 Similarity of Document Structures

In the context of web and text mining, graph or tree similarity measures have extensively been used. Many of such measures are based on transforming web-based document structures into special graphs such as rooted trees (DOM-trees) [28] or more complex patterns representing generalized trees [51]. Then, by deriving structural features of the graphs, for example, paths and degree sequences, measures for determining the structural similarity of graph-based document structures can be derived (see, e.g., [27,36,50,100,124]).

1.3.1.7 Tree Similarity

Early work for exploring tree similarity is based on using string alignment techniques [147]. Seminal work was done by Selkow [149] who extended the string alignment problem to undirected and the so-called ordered trees [149]. To use the concept of tree similarity in bioinformatics, Höchstmann et al. [87] developed an approach for measuring the structural similarity of RNA secondary structures. But this approach is mainly based on the well-known tree alignment technique developed by Jiang et al. [99]. In phylogeny [74], tree measures have also been investigated to measure the distance between unrooted trees; the Robinson–Foulds distance [74] is a prominent example thereof.

1.3.1.8 Molecular Similarity

To study molecular similarity measures has been an active research field since many years [123,137,138,154]. Particularly in chemical graph theory, many graph measures have been explored for determining the similarity between molecular structures [154]. Historically seen, an early attempt to measure the structural similarity between chemical graphs is to represent and compare graphs as vectors where the entries are real numbers [123]. Then, based on binary vector comparison (the vectors indicate the absence or presence of predefined subgraphs [174]) using the Tanimoto index [123], further similarity measures have been obtained [174]. Also, Klein [105–108] explored graph metrics in structural chemistry and employed metrical properties of graphs to tackle the graph similarity problem. Topological indices [15,17,169] and the concept of the MCS [154] have also been employed to measure the similarity between chemical graphs. A typical example is the following measure [123]:

$$d(G_i, G_j) := |G_i|_E + |G_j|_E - 2|MCS(G_i, G_j)|_E. \tag{1.6}$$

The subscript E indicates that the cardinality solely refers to the edges. As a final remark, other well-known quantities such as Tanimoto index and Hamming distance have also been used extensively to explore structural similarity between chemical graphs [123,174]. As a conclusive remark, other measures for determining the similarity between chemical graphs have been also developed by Ruiz et al. (see [123,143–146]).

1.3.1.9   Statistical Graph Matching

In this section, we briefly sketch some known approaches in this area. The motivation to use statistical graph matching is often triggered by the major drawback of some classical graph matching techniques (e.g., exact graph matching), namely, the graphs are assumed to be inferred without any uncertainty. Thus, a statistical treatment of the problem is required. Note that statistical graph matching techniques have been employed in various disciplines such as image analysis [165,166], computer and neuroscience [150,158], and computational linguistics [125]. But in view of the number of approaches that has been developed, (1) such techniques are clearly underrepresented, and (2) there is no general theory for interpreting and understanding the existing results mathematically.

In image analysis, graph similarity techniques have often been used (see [32, 166]). A particular contribution using statistical graph matching is due to Theoharatos et al. [166]. In this work, images representing minimum spanning trees (MSTs) have been matched. Note that by using the visual content of images, vector distributions and, finally, minimum spanning trees can be inferred. For the technical comparison of two images, Theoharatos et al. [166] employed a statistical technique using the known Wald–Wolfowitz test. As a result, the outcome of this test is used to define a similarity measure for images. Finally, to measure the similarity between two images, the corresponding MST graphs have been compared using the nonparametric multivariate generalization of the Wald–Wolfowitz two-sample problem leading to a similarity measure for analyzing image contents [166].

Another notation of graph similarity to compare control flow graphs has been found by Sokolsky et al. [158]. In [158], such graphs possess vertex and edge labels and often occur when exploring software artifacts. A crucial feature of this method is that structural similarity is less important; it rather measures similarity between vertex or edge labels. This is typical for methods based on *simulation* or *bisimulation*. Technically, the used procedures are similar to those for analyzing Markov decision processes.

In Section 1.3.1.4, we have sketched a method due to Dehmer et al. [50] for determining the structural similarity of hierarchical structures (generalized trees). A generalization thereof to compare large undirected graphs by using a statistical technique has been proposed by Emmert-Streib et al. [69]. The key step is as follows: instead of comparing two graphs G_1 and G_2 directly, the tree sets resulting from a generalized tree decomposition [69] of G_1 and G_2 have been compared. By employing the method for pairwise calculating generalized trees (see Section 1.3.1.4), three similarity distributions have been inferred and compared based on a chi-square test. Finally, Emmert-Streib et al. [69] defined G_1 and G_2 to be similar if the three similarity distributions obtained from pairwise calculating similarity scores of the different trees sets are similar.

The last approach sketched by this section is due to Mehler et al. [125] who performed a quantitative analysis of special linguistic networks. In order to derive information-theoretic graph metrics, the mutual information between vertices of the given graphs and the local entropy of graphs must be determined. For this, Mehler et al. [125] calculated these quantities [38] by determining vertex probabilities using information functionals [38]. This particular study led to the graph distance measure [125]

$$d(G_1, G_2)|_{(v,w)} = 1 - \frac{I(v; w)}{\max\{H(v), H(w)\}}, \qquad (1.7)$$

and the similarity measure

$$S(G_1, G_2)|_{(v,w)} = 1 - d(G_1, G_2)|_{(v,w)}. \qquad (1.8)$$

$I(v; w)$ is the local mutual information of the pair v, w and $H(v)$ is a local vertex entropy (see [125]).

1.3.2 Numerical Graph Invariants

A large variety of numerical graph invariants has been introduced since the 1940s to describe graphs by a single numerical value (see, e.g., [169]). Such invariants can quantify either local or global properties of graphs. Examples for local measures are vertex eccentricities [153] or local clustering coefficients [179]. Also, vertex entropies have been defined by Konstantinova [114]. Global measures, on the contrary, encode topological information of the entire graph by a real number. Hence, they are referred to as topological descriptors. One major goal of defining such graph measures is to characterize the structure of a chemical graph (or of an arbitrary network) by means of a numerical topological index [6,15,169]. Topological indices are meant to reflect the size, shape, branching, and cyclicity patterns of a graph such as a molecular structure. Still, it is nearly impossible to capture all topological aspects of a graph by a single invariant and further identify every graph with a unique number. Topological indices are often viewed as a complexity measure of a graph, even though the term complexity is not very well defined [103].

Graph invariants have been defined using different structural properties of a graph and using concepts from other disciplines such as information theory. According to their underlying concepts, they can be divided into the following classes.

Distance-based measures are based on shortest paths between pairs of vertices, that is, on the distances between pairs of vertices. A prominent example thereof is the Wiener index [180]:

$$W(G) = \frac{1}{2} \sum_{v,w \in V(G)} d(v, w), \qquad (1.9)$$

where $d(v, w)$ is the distance between vertices v and w and the factor of 1/2 is used to avoid double counting of pairs. Further representatives of this family are the efficiency complexity [103] or the well-known Balaban-*J* index [6].

Degree-based measures are computed by using vertex degrees, often also respecting vertex adjacencies. Classical representatives are the Randíc index [136], also known as the *connectivity index*, given by

$$R(G) = \sum_{(v,w) \in E(G)} \frac{1}{\sqrt{\delta(v)\delta(w)}}. \tag{1.10}$$

$\delta(v), \delta(w)$ denote the degrees of vertices v and w. Further degree-based measures are the atom–bond connectivity (ABC) index [71] as well as the Zagreb indices [134] based on the previous two indices.

Information-theoretic approaches are often based on Shannon's entropy [53,151], defined by

$$E(\mathbf{p}) = \sum_{i=1}^{k} p_i \log p_i \tag{1.11}$$

for some finite probability vector $\mathbf{p} = (p_1, \ldots, p_k)$. Graph entropies can be divided into subclasses according to the definition of the probability vector $\mathbf{p}$. For partition-based entropies [20,127,139], the sizes of some topological set M, for example, vertices, partitioned according to some equivalence relation (e.g., automorphism groups) are used. Partition-independent entropies [38] depend on a vertex functional $f : V(G) \rightarrow \mathbb{R}$, where p_i is chosen proportional to $f(v_i)$ for every vertex $v_i \in V(G)$. Eigenvalue-based entropies [57] compute the entropy of the eigenvalues of some graph-theoretical matrix [98].

Eigenvalue-based measures are based eigenvalues inferred from graph-theoretical matrices [98] such as the adjacency or Laplacian matrix. Examples thereof are the graph energy [81] summing up the moduli of the eigenvalues, the Estrada index [70] using the exponential function, or the spectral radius [31].

Some topological indices do exist that cannot be assigned uniquely to one of the previously mentioned classes since they use different information, such as geometric–arithmetic indices [73,185] or the Hosoya index [94]. A vast amount of several hundreds of graph measures exists due to different motivations and problems. An overview can be found in [169].

Topological indices have been used extensively for the development of quantitative structure–activity relationships (QSARs) in which the biological activity or other properties of molecules are correlated with their chemical structure [61]. Also, they gained a lot of interest recently since a lot of disciplines besides chemistry such as physics, linguistics, and biology use network models to study complex systems. Still, the interpretation and relations or similarities among structural indices remain to be not very well understood, resulting in a lack of applicability to other problems and, hence, to the definition of even more measures. Several aspects have been studied aiming at a better understanding and interpretability of existing indices, some of which are discussed in the following.

1.3.2.1 Uniqueness of Graph Measures

A major aspect of topological indices is the quality of a measure, which can be defined from different viewpoints. One important topic in this sense is the discrimination power of a measure, also referred to as *uniqueness* or *degeneracy* [48]. This property indicates the ability to distinguish nonisomorphic graphs. A measure is called degenerate if it possesses the same value for more than one graph.

A major problem in quantitative graph theory is to study to which extent known topological indices are degenerate. In order to describe the structure of a graph quantitatively, a unique representation is often essential. Particularly to tackle the question of reconstructing the structure of a graph by its numerical invariants, degenerate measures cause problems. Also, for graph classification problems, degeneracy is not desirable. Up to today, no complete and efficient invariant, that is, a fully unique measure, has been found. Several studies have pointed out measures with high and low discrimination power on several graph classes [48].

First studies of degeneracy were performed on chemical graphs such as alkane trees or isomeric structures [6,112]. After defining the Balaban *J*-index in 1982 [6] as a branching index, Balaban proved that for alkane trees and other chemical graphs, this index is highly unique. For his statement, he quantified the discrimination power of a measure by considering the ratio between the number of graphs of size n, N_n, in a family and the number of distinct values T_I of the considered index I. Balaban showed in [6] and [7] that the *J* index outperforms other indices known at that time in terms of uniqueness on chemical graphs.

The following sensitivity index was defined by Konstantinova [112] to quantify the discrimination power, which is often considered instead of the ratio discussed earlier in recent studies. We define the sensitivity of measure I by

$$S(I) = \frac{|\mathcal{G}| - N_I}{|\mathcal{G}|}, \qquad (1.12)$$

where

$\mathcal{G}$ is the graph class

N_I is the number of graphs that cannot be distinguished by the index

This index, obviously, takes values between zero and one, where sensitivity one means a measure is completely unique. In [112], several distance-based indices have been studied on non-branched hexagonal systems. It has been proven that among others, the Balaban *J* index [6] is very unique on such graphs but that the information distance index introduced in [113] has a higher discrimination power.

Dehmer et al. [46,48] were the first who studied the discrimination power of several measures also on exhaustively generated, that is, nonisomorphic graphs without vertex and edge weights. Note that these sets have very large cardinalities, that is, $|N_9| = 261,080$, $|N_{10}| = 11,716,571$. It has been shown recently [48] that the Balaban *J* index, which was proven previously to be highly unique on chemical graphs, decreases its discrimination power tremendously when the sample size increases. Further, it has been shown that an information measure based on a vertex functional using degree–degree associations outperformed all other measures studied when considering graphs with no structural constraints [48]. Dehmer et al. [48] amplified their results and understanding of discrimination power on exhaustively generated graphs in [46,48], where they put the emphasis on degree-based and eigenvalue-based measures. They showed that several eigenvalue-based measures such as graph energy or Laplacian Estrada index have a better uniqueness on the sets N_8, N_9, N_{10} than the Balaban J index. Another class of measures that was of high interest are molecular ID numbers [45,121]. These descriptors were often claimed to be highly unique, but they were not evaluated on a large scale. To shed light on this problem, Dehmer et al. [45] found recently that these measures possess two major drawbacks. First, the computational effort is very high since all paths are often needed and the number of paths increases with $n!$. Second, their computation was quite demanding, and finally their discrimination power was often as good as (or even less than) other eigenvalue-based measures due to Dehmer et al. [45]. As a result, Dehmer et al. [45] found that the molecular ID numbers are not suitable to discriminate graphs on a large scale (e.g., if $n(G) > 9$).

Villas Boas et al. [175] used a different approach to study the degeneracy of certain classical measures such as shortest path length or clustering coefficients. They used graph perturbation operations such as edge deletions, additions, or rewiring to study the sensibility toward perturbations, that is, if and to what extend perturbed graphs can be distinguished from its original. Several random graph models such as Barabási graphs [8] or Watts–Strogatz graphs [179] have been studied for this purpose.

Note that the sensitivity index (1.12) and the ratio considered by Balaban do not provide information about the quantity of different values of the measure. To be precise, assume that all graphs of a family $\mathcal{G}$ have the same index $I(G) = a \, \forall \, G \in \mathcal{G}$, then $S(I) = 0$. If, on the other hand, for every graph $G \in \mathcal{G}$, there exists exactly one graph G' with the same value, $I(G) = I(G')$, that is, there are $|\mathcal{G}|/2$ different values of I, the sensitivity index is also zero, $S(I) = 0$. To avoid this loss of information and to evaluate the degree of the degeneracy precisely, Todeschini et al. [168] explored the equivalence relation where every class contains all graphs with the same value $I(G)$. Let $I_1, \ldots, I_g$ be these equivalence classes. We yield the information content of degeneracy by [168]:

$$U_H = 1 - \frac{\log n - H}{\log n}$$

$$H = \sum_{i=1}^{g} \frac{|I_i|}{n} \log \frac{|I_i|}{n}. \tag{1.13}$$

Several authors introduced new measures designed to have a high discrimination power. Hu and Xu [96] introduced EAID numbers as highly unique indices by using some families of chemical graphs. They proved that the EAID numbers have sensitivity index 1 for alkane trees, cyclic or polycyclic graphs, and other isomeric structures, proposing an index with the highest discrimination power known at that time. Yet a major drawback of these measures is that it is based on 3D information that is often difficult to infer.

The spectra of graphs were used as a starting point for novel descriptors by Dehmer et al. [55,57]. In [55], a new graph polynomial, the so-called information polynomial, has been defined, and its zeros have been used to define a new highly discriminating index. Dehmer and Sivakumar [57] used eigenvalues of several graph-related matrices directly to introduce families of measures. All measures defined have been proven to be highly unique on chemical graphs as well as on exhaustively generated graphs [45,57].

Since throughout the thorough study of the topic, no measure defined on a single kind of topological information was found to be completely unique, the idea of a super index arises. Such a measure combines several highly unique measures from different classes. A first super index has been defined by Bonchev et al. [19]. In [64], Diudea et al. defined such a super index based on shell-matrix operators [65] and polynomials, which were known to encode alkane structures well. They studied the degeneracy of this super index on atomic structures and isomers, where they proved that it is completely unique.

The aspect of degeneracy also stands in relation to comparative graph analysis described in Section 1.3.1, since it will be especially difficult for a measure to distinguish graphs that are very similar regarding the topological information used by the measure.

Note that Aigner and Triesch [3] discussed a different notion of uniqueness and asked the following question. Consider a finite sequence of structural invariants P and two graphs G and H. When does $P(G) = P(H)$ imply $G \cong H$, that is, G and H are isomorphic? They studied this problem for P being the degree sequence and neighborhood lists and proved that several such problems are *NP*-complete.

1.3.2.2   Information Inequalities

Another important problem in quantitative graph theory is the search for inequalities between graph measures, providing explicit bounds or relations between different measures. Information inequalities provide information about interactions and relations between information-based measures as well as information about their extremal behavior. The problem of information inequalities is rooted in information theory, where the problem has been addressed by Zhang and Yeung [181,184]. Dragomir and Goh [66] and several more have contributed to the field of communication theory by proving several information-theoretic inequalities such as Fisher's information inequalities.

When using information-theoretic measures for graphs (i.e., graph entropies), information inequalities (for graphs) can be divided into two classes [40]:

1. *Implicit information inequalities* describe relations between different information-theoretic graph measures. Formally, let I_1 and I_2 be two information-theoretic graph measures, $\mathcal{G}_1$ and $\mathcal{G}_2$ be two families of graphs, and f_1, f_2, g_1, g_2 be some algebraic functions. Then, under certain conditions,

$$f_1(I_1(G)) \leq f_2(I_2(G)) \ \forall \ G \in \mathcal{G}_1 \ \text{or}$$
$$g_1(I_1(G_1)) \leq g_2(I_1(G_2)) \ \forall \ G_1 \in \mathcal{G}_1, G_2 \in \mathcal{G}_2. \tag{1.14}$$

2. *Explicit information inequalities* give insight by providing explicit bounds and extremal properties of measures, such as

$$f(I(G)) \leq c \; \forall G \in \mathcal{G}, \tag{1.15}$$

where

$c \in \mathbb{R}$
f some algebraic function
I an information-theoretic graph measure
$\mathcal{G}$ a family of graphs.

These classification and idea can be generalized in a natural way to all topological graph measures. Explicit inequalities for graph measures have been studied for a large number of measures (see, e.g., [76,119,164,186]), while the question of implicit inequalities for noninformation-theoretic graph measures has been addressed by Kraus et al. [115].

Implicit information inequalities: Inspired by results from information theory, several studies have been performed aiming at similar results and using similar methods. First implicit inequalities were given by Dehmer et al. [40]. The authors proved several implicit inequalities for functional-based entropies based on the following Lemma [40]:

Lemma 1.3.1 *If the relation*

$$p_f(v_i) \overset{(>)}{<} \alpha p_{f*}(v_i)$$

holds for all $v_i \in V$, then we obtain

$$I_f(G) \overset{(<)}{>} \alpha I_{f*}(G) - \alpha \log \alpha.$$

Either α is a constant expression or it depends on the information functional.

Further implicit inequalities using parametric graph entropies have been proven in [52] and [56]. Dehmer and Mowshowitz [52] presented a method to establish interrelations between entropies together with some numerical results. Dehmer and Sivakumar proved inequalities analytically, where assumptions are made either on the information functional or on the graph. Further implicit inequalities have been proven by Sivakumar and Dehmer [152] who found interrelations between classic graph entropies based on Shannon's entropy and other measures defined by the Rényi entropy [140] of order α. Note that the Rényi entropy is a generalization of Shannon's entropy. As a result, Sivakumar and Dehmer [152] proved several theorems of similar shape as the following:

Theorem 1.3.3 *For $0 < \alpha < 1$,*

$$H_f(G) \leq H_{\alpha,f}(g) < H_f(g) + \frac{N(N-1)(1-\alpha)\rho^{\alpha-2}}{2\ln 2}. \tag{1.16}$$

When $\alpha > 1$,

$$H_f(G) - \frac{N(N-1)(\alpha-1)}{2\ln 2 \cdot \rho^{\alpha-2}} < H_{\alpha,f}(G) \leq H_f(G), \tag{1.17}$$

where

H_f denotes Shannon entropy

$H_{\alpha,f}$ denotes Rényi entropy of order α

$\rho = \max_{i,k} p(v_i)/p(v_k), \quad v_i, v_k \in V$.

One sees that inequalities for various partition-independent information measures have been obtained by using analytic methods, but no general relation has been found yet between entropies of different classes, for example, between partition-based and partition-independent entropies. Relations of this kind have been proven by Dehmer et al. [54] for special classes of graphs. When generalizing the idea of information inequalities to noninformation-theoretic measures, obviously, the question for inequalities between different classes of measures arises. Such interrelations would provide a deeper understanding of the mathematical apparatus of graph measures. Further, such knowledge will support the definition of super indices with high degeneracy, as described in Section 1.3.2.1, since measures can be chosen more systematically.

Explicit information inequalities: Tight upper bounds can be given by analytic estimations using techniques from various fields such as functional analysis and discrete mathematics. The search for extremal graphs, that is, graphs that take maximal and minimal value for given measures, is closely related to the topic of explicit information inequalities. In information theory, maximal and minimal entropies are well known; majorization provides a partial order allowing inequalities between certain probability vectors and their entropies. Dehmer and Kraus [49] generalized the concept to apply it to graph entropies and have been able to prove some specific extremal results. Later, Kraus et al. [116] defined a new class of graphs called *sphere-regular* graphs and proved that they are maximal for the entropy-based on the so-called sphere functional (see [38]). Immediately, this result provides a tight upper bound for the sphere entropy [116].

Upper and lower bounds, that is, explicit inequalities, also exist for several noninformation-theoretic measures. Zhou and Trinajstić found upper and lower bounds of the Balaban J index [186]. Tang and Deng characterized graphs with n vertices and n edges by using extremal values of a novel formula for calculating the Wiener index [164]. Also, there is a large body of literature that aims to determine extremal values of the Randić index [119] by using various graph classes. Chemical trees with extremal ABC have been found by Furtula et al. [76]. The concept of majorization was taken up by Bianchi et al. [34] to reprove and improve explicit bounds on the Randíc index and sum-connectivity and energy indices. Madaras and Mockovčiaková [122] proved bounds for the sum of an index of a bipartite graph G and its complement $\bar{G}$, for the Wiener index, Randíc index, and Zagreb indices.

Probabilistic inequalities for graph measures: Even though a lot of research has been going on in the field of inequalities for graph invariants, it turned out that the question of proving information inequalities analytically is very challenging (see [49,56,116]). Hence, a novel approach to information inequalities was presented by Kraus et al. [115], where statistical methods have been used to provide probabilistic inequalities. This approach for establishing probabilistic inequalities between structural graph measures is a promising step toward a better understanding of the underlying mathematical apparatus.

1.3.2.3   Correlation of Graph Measures

As already mentioned in Section 1.3.2.2, topological indices might stand in close relation to each other. This is evident to compute the correlation of different graph measures, given by the Pearson correlation coefficient. Let M_1 and M_2 be two measures and $\mathcal{G} = (G_1, \ldots, G_k)$ a family of graphs. Further, let $m_j = 1/k \sum_{i=1}^{k} M_j(G_i)$ be the arithmetic means of the values of M_j, for $j = 1, 2$. Then, the correlation has been expressed by [142]

$$C(M_1, M_2) = \frac{\sum_{i=1}^{k} \left(M_1(G_i) - m_1\right)\left(M_2(G_i) - m_2\right)}{\sqrt{\sum_{i=1}^{k} \left(M_1(G_i) - m_1\right)^2}\sqrt{\sum_{i=1}^{k} \left(M_2(G_i) - m_2\right)^2}}. \qquad (1.18)$$

A first attempt in this direction has been presented by Wagner [176], who computed the asymptotic covariances of the Merrifield–Simmons, the Hosoya, and the Wiener index as well as the number of subtrees known as the ρ-index, as the size n of the graphs tends to infinity. Wagner found that all correlations are of type

$$Cor(M, N) = \alpha \cdot \beta^n, \qquad (1.19)$$

for constants $\alpha \in [-1, 1]$ and $0 < \beta < 1$ (see [176]). Obviously, since these are asymptotic results, they hold only for very large values of n.

Hollas [88–91] studied the correlation and independence of several graph measures on chemical graphs analytically. Therefore, Hollas used random graphs as a concept to model molecules. For any random graph model, he shows in [88] that a class of distance-based topological indices may be strongly correlated. For degree-based measures, Hollas performed similar studies in [90]. In [89], he showed that for a random tree model, degree-based indices are uncorrelated under certain preliminaries. This result is of theoretical relevance only, since chemical graphs and other application-related networks contain cycles. In [91], he showed that indices may even be strongly correlated if they use independent vertex properties.

Computer-aided correlation studies have been performed by Basak et al. [9,10]. In [171], Trinajstić et al. studied the correlation of distance-based indices on chemical graphs. Horvat et al. [92] studied the pairwise correlation of 12 topological indices on benzenoid hydrocarbons with up to 6 rings, basically finding three independent subclasses of indices.

By finding that various measures are highly correlated, the question of their redundancy arises, since it causes doubt whether highly correlated indices describe different and meaningful properties of graphs. Note further that in the view of QSAR studies, high correlation is a problem.

1.3.3   Software for Quantitative Graph Analysis

1.3.3.1   R-Packages

So far, there is a lack of freely available tools to calculate quantitative network measures. A programming language that has been proven powerful for biological network analysis is R (see [1,129]). However, R-libraries such as `igraph` and `graph` contain only very few quantitative methods for characterizing graphs structurally.

Müller et al. developed the R-package `QuACN` [129,130] that contains over 150 structural network descriptors. This is a competitor of the commercial software `Dragon` [170] containing thousands of molecular descriptors [169]. From a graph-theoretical point of view, the given accuracy of the calculated descriptors by only three digits is `Dragon`'s considerable weak point. The insufficient accuracy makes this software unfeasible for characterizing graphs quantitatively. Thus, a consequence thereof is a high degeneracy of those measures (see Section 1.3.2.1).

To briefly sketch `QuACN`'s scope of application, we list the categories of quantitative network measures that have been implemented [129,130]:

- Distance-based measures and indices based on vertex degrees [68,153]

- Information-theoretic measures such as graph entropies [53,58,68]

- Spectrum-based measures [57]

- Subgraph-based and related measures [103]

A full description of the briefly mentioned quantitative network descriptors implemented in `QuACN` can be found in [131].

Further packages that provide quantitative analysis functions that may be useful are "NetworkAnalysis" and "statnet," both available from the CRAN repository. Finally, we would like to mention that "igraph" and "NetBioV" are packages that provide many different graph layout styles for visualizing networks and for their graphical exploration. Despite the lack of the latter two packages to provide, for example the implementations of network descriptors, the visualization of a network is usually a very beneficial step *before* a quantitative network analysis in order to generate hypotheses about the networks that can then be quantitatively investigated.

1.3.3.2   Conjecture Engines

Especially in view of inequalities described in Section 1.3.2.2, conjecture engines can be a very useful starting point for generating proofs. Given some topological indices and a set of graphs, a conjecture engine is supposed to find an implicit or explicit inequality and test its validity on the set of given graphs. A famous engine used by a large variety of graph theorists is `Graffiti` [59,60], but it is no longer maintained and available.

Graph invariant investigator, abbreviated as `GrInvIn` [135], is a freely available `Java` application with a graphical user interface, built to experiment with graphs and investigate their properties. `GrInvIn` has been developed at the university of Ghent and contains a small number of numerical graph invariants as well as a quite simple conjecture engine. `GrInvIn` can work with graph generation packages such as `geng`; code from other languages such as `R` can also be embedded. The engine computes selected graph indices on a sample of chosen graphs and proposes inequalities that are true on the tested graph sample. Unfortunately, the output consists of a single inequality that is "best" according to the program choice and cannot be replaced by a different one if desired.

1.4   Summary and Conclusion

In this chapter, we put the emphasis on quantitative graph theory. We pointed out that this intriguing field can be seen as a new branch of graph theory and should be treated as such. So far, most of the contributions in graph theory dealt with descriptive approaches for describing structural features of networks, see Section 1.1. By demonstrating the potential of quantitative approaches which have been widely spread over several disciplines, we believe that the discussed approaches will break new ground. An important finding of this chapter is that quantitative graph theory is per se strongly interdisciplinary. The mathematical apparatus belongs to applied mathematics but, as demonstrated, quantitative graph theory have been applied in various scientific areas.

References

1. R, software, a language and environment for statistical computing. www.r-project.org, 2011. R Development Core Team, Foundation for Statistical Computing, Vienna, Austria.

2. L. Adamic and B. Huberman. Power-law distribution of the world wide web. *Science*, 287:2115a, 2000.

3. M. Aigner and E. Triesch. Realizability and uniqueness in graphs. *Discrete Mathematics*, 136:3–20, 1994.

4. R. Albert, H. Jeong, and A. L. Barabási. Diameter of the world wide web. *Nature*, 401:130–131, 1999.

5. K. Anand and G. Bianconi. Entropy measures for networks: Toward an information theory of complex topologies. *Physical Review E, Statistical, Nonlinear, and Soft Matter Physics*, 80:045102(R), 2009.

6. A. T. Balaban. Highly discriminating distance-based topological index. *Chemical Physics Letters*, 89(5):399–404, 1982.

7. A. T. Balaban. Topological indices based on topological distances in molecular graphs. *Pure and Applied Chemistry*, 55(2):199–206, 1983.

8. A. L. Barabási and R. Albert. Emergence of scaling in random networks. *Science*, 286:509–512, 1999.

9. S. C. Basak, A. T. Balaban, G. D. Grunwald, and B. D. Gute. Topological indices: Their nature and mutual relatedness. *Journal of Chemical Information and Computer Sciences*, 40(4):891–898, 2000.

10. S. C. Basak, B. D Gute, and A. T Balaban. Interrelationship of major topological indices evidenced by clustering. *Croatica Chemica Acta*, 77:331–344, 2004.

11. A. Bavelas. Communication patterns in task-oriented groups. *Journal of the Acoustical Society of America*, 22:725–730, 1950.

12. M. Behzad, G. Chartrand, and L. Lesniak-Foster. *Graphs & Digraphs*. International Series. Prindle, Weber & Schmidt, Boston, MA, 1979.

13. V. Blondel, A. Gajardo, M. Heymans, P. Senellart, and P. Van Dooren. A measure of similarity between graph vertices: Applications to synonym extraction and web searching. *SIAM Review*, 46:647–666, 2004.

14. B. Bollabás. *Modern Graph Theory*. Graduate Texts in Mathematics. Springer, New York, 1998.

15. D. Bonchev. *Information Theoretic Indices for Characterization of Chemical Structures*. Research Studies Press, Chichester, U.K., 1983.

16. D. Bonchev. *Complexity in Chemistry. Introduction and Fundamentals*. Taylor & Francis, Boca Raton, FL, 2003.

17. D. Bonchev. Information theoretic measures of complexity. In R. Meyers, ed., *Encyclopedia of Complexity and System Science*, vol. 5, pp. 4820–4838. Springer, New York, 2009.

18. D. Bonchev and D. H. Rouvray. *Chemical Graph Theory. Introduction and Fundamentals*. Abacus Press, New York, 1991.

19. D. Bonchev, O. Mekenyan, and N. Trinajstić. Isomer discrimination by topological information approach. *Journal of Computational Chemistry*, 2:127–148, 1981.

20. D. Bonchev and N. Trinajstić. Information theory, distance matrix and molecular branching. *Journal of Chemical Physics*, 67:4517–4533, 1977.

21. M. Borgwardt. Graph kernels. PhD thesis, Ludwig-Maximilians-Universität München, Fakultät für Mathematik, Informatik und Statistik, 2007.

22. H. Bunke. What is the distance between graphs? *Bulletin of the EATCS*, 20:35–39, 1983.

23. H. Bunke. On a relation between graph edit distance and maximum common subgraph. *Pattern Recognition Letters*, 18(9):689–694, 1997.

24. H. Bunke. A graph distance metric based on the maximum common subgraph. *Pattern Recognition Letters*, 19(3):255–259, 1998.

25. H. Bunke. Error correcting graph matching: On the influence of the underlying cost function. *IEEE Transactions on Pattern Analysis and Machine Intelligence*, 21(9):917–921, 1999.

26. H. Bunke. Recent developments in graph matching. In *15th International Conference on Pattern Recognition*, Barcelona, Spain, vol. 2, pp. 117–124, 2000.

27. D. Buttler. A short survey of document structure similarity algorithms. In *International Conference on Internet Computing*, Las Vegas, NV, pp. 3–9, 2004.

28. S. Chakrabarti. Integrating the document object model with hyperlinks for enhanced topic distillation and information extraction. In *Proceedings of the 10th International World Wide Web Conference*, Hong Kong, China, pp. 211–220, May 1–5, 2001.

29. G. Chartrand, H. Gavlas, and H. Hevia. Rotation and jump distances between graphs. *Discussiones Mathematicae Graph Theory*, 17:285–300, 1997.

30. G. Chartrand, W. Goddard, M. A. Henning, L. Lesniak, H. C. Swart, and C. E. Wall. Which graphs are distance graphs? *Ars Combinatorica*, 29A:225–232, 1990.

31. L. Collatz and U. Sinogowitz. Spektren endlicher Grafen. *Abhandlungen aus dem Mathematischen Seminar der Universitat Hamburg*, 21:63–77, 1957.

32. D. Conte, F. Foggia, C. Sansone, and M. Vento. Thirty years of graph matching in pattern recognition. *International Journal of Pattern Recognition and Artificial Intelligence*, 18:265–298, 2004.

33. D. Cook and L. B. Holder. *Mining Graph Data*. Wiley-Interscience, Hoboken, NJ, 2007.

34. A. Cornaro, M. Bianchi, and A. Torriero. A majorization method for localizing graph topological indices. arXiv:1105.3631, 2011.

35. T. M. Cover and J. A. Thomas. *Elements of Information Theory*. Wiley Series in Telecommunications and Signal Processing. Wiley & Sons, Hoboken, NJ, 2006.

36. I. F. Cruz, S. Borisov, M. A. Marks, and T. R. Webb. Measuring structural similarity among web documents: Preliminary results. In *EP'98/RIDT'98: Proceedings of the Seventh International Conference on Electronic Publishing, Held Jointly with the Fourth International Conference on Raster Imaging and Digital Typography*, St. Malo, France, pp. 513–524. Springer-Verlag, London, U.K., 1998.

37. L. da, F. Costa, F. Rodrigues, and G. Travieso. Characterization of complex networks: A survey of measurements. *Advanced Physics*, 56:167–242, 2007.

38. M. Dehmer. Information processing in complex networks: Graph entropy and information functionals. *Applied Mathematics and Computation*, 201:82–94, 2008.

39. M. Dehmer, ed. *Structural Analysis of Complex Networks*. Birkhäuser Publishing, Boston, MA, 2010.

40. M. Dehmer, S. Borgert, and D. Bonchev. Information inequalities for graphs. *Symmetry: Culture and Science. Symmetry in Nanostructures (Special issue edited by M. Diudea)*, 19:269–284, 2008.

41. M. Dehmer, F. Emmert-Streib, and M. Grabner. A computational approach to construct a multivariate complete graph invariant. *Information Sciences*, 260(1):200–208, 2014.

42. M. Dehmer and F. Emmert-Streib. Comparing large graphs efficiently by margins of feature vectors. *Applied Mathematics and Computation*, 188(2): 1699–1710, 2007.

43. M. Dehmer and F. Emmert-Streib, eds. *Analysis of Complex Networks: From Biology to Linguistics*. Wiley-VCH Publishing, Weinheim, Germany, 2009.

44. M. Dehmer, F. Emmert-Streib, and M. Mehler, eds. *Towards an Information Theory of Complex Networks: Statistical Methods and Applications*. Birkhäuser Publishing, Basel, Switzerland, 2011.

45. M. Dehmer and M. Grabner. The discrimination power of molecular identification numbers revisited. *MATCH: Communications in Mathematical and in Computer Chemistry*, 69(3):785–794, 2013.

46. M. Dehmer, M. Grabner, and B. Furtula. Structural discrimination of networks by using distance, degree and eigenvalue-based measures. *PLoS ONE*, 7(7):e38564, 2012. doi:10.1371/journal.pone.0038564.

47. M. Dehmer, M. Grabner, A. Mowshowitz, and F. Emmert-Streib. An efficient heuristic approach to detecting graph isomorphism based on combinations of highly discriminating invariants. *Advances in Computational Mathematics*, 39(2):311–325, 2012.

48. M. Dehmer, M. Grabner, and K. Varmuza. Information indices with high discriminative power for graphs. *PLoS ONE*, 7(2):e31214, 2012. doi: 10.1371/journal.pone.0031214.

49. M. Dehmer and V. Kraus. On extremal properties of graph entropies. *MATCH: Communications in Mathematical and in Computer Chemistry*, 68:889–912, 2012.

50. M. Dehmer and A. Mehler. A new method of measuring similarity for a special class of directed graphs. *Tatra Mountains Mathematical Publications*, 36:39–59, 2007.

51. M. Dehmer, A. Mehler, and F. Emmert-Streib. Graph-theoretical characterizations of generalized trees. In *Proceedings of the International Conference on Machine Learning: Models, Technologies & Applications (MLMTA'07)*, Las Vegas, NV, pp. 113–117, 2007.

52. M. Dehmer and A. Mowshowitz. Inequalities for entropy-based measures of network information content. *Applied Mathematics and Computation*, 215:4263–4271, 2010.

53. M. Dehmer and A. Mowshowitz. A history of graph entropy measures. *Information Sciences*, 1:57–78, 2011.

54. M. Dehmer, A. Mowshowitz, and F. Emmert-Streib. Connections between classical and parametric network entropies. *PLoS ONE*, 6:e15733, 2011.

55. M. Dehmer, L.A.J. Mueller, and A. Graber. New polynomial-based molecular descriptors with low degeneracy. *PLoS ONE*, 5:e11393, 2010.

56. M. Dehmer and L. Sivakumar. Recent developments in quantitative graph theory: Information inequalities for networks. *PloS ONE*, 7(2):e31395, 2012.

57. M. Dehmer, L. Sivakumar, and K. Varmuza. Uniquely discriminating molecular structures using novel eigenvalue-based descriptors. *MATCH: Communications in Mathematical and in Computer Chemistry*, 67:147–172, 2012.

58. M. Dehmer, K. Varmuza, S. Borgert, and F. Emmert-Streib. On entropy-based molecular descriptors: Statistical analysis of real and synthetic chemical structures. *Journal of Chemical Information and Modeling*, 49:1655–1663, 2009.

59. E. DeLaVina. Graffiti.pc: A variant of graffiti. *DIMACS Series in Discrete Mathematics and Theoretical Computer Science: Graphs and Discovery*, 69:71–79, 2005.

60. E. DeLaVina. Some history of the development of graffiti. *DIMACS Series in Discrete Mathematics and Theoretical Computer Science: Graphs and Discovery*, 69:81–118, 2005.

61. J. Devillers and A. T. Balaban. *Topological Indices and Related Descriptors in QSAR and QSPR*. Gordon and Breach Science Publishers, Amsterdam, the Netherlands, 2000.

62. P. J. Dickinson, H. Bunke, A. Dadej, and M. Kraetzl. Matching graphs with unique node labels. *Pattern Analysis Applications*, 7:243–266, 2004.

63. M. V. Diudea, I. Gutman, and L. Jäntschi. *Molecular Topology*. Nova Publishing, New York, 2001.

64. M. V. Diudea, A. Ilić, K. Varmuza, and M. Dehmer. Network analysis using a novel highly discriminating topological index. *Complexity*, 16:32–39, 2011.

65. M. V. Diudea and O. Ursu. Layer matrices and distance property descriptors. *Indian Journal of Chemistry*, 42A:1283–1294, 2003.

66. S. S. Dragomir and C. J. Goh. Some bounds on entropy measures in information theory. *Applied Mathematics Letters*, 10:23–28, 1997.

67. F. Emmert-Streib. The chronic fatigue syndrome: A comparative pathway analysis. *Journal of Computational Biology*, 14(7):961–972, 2007.

68. F. Emmert-Streib and M. Dehmer. Networks for systems biology: Conceptual connection of data and function. *IET Systems Biology*, 5:185–207, 2011.

69. F. Emmert-Streib, M. Dehmer, and J. Kilian. Classification of large graphs by a local tree decomposition. In H. R. Arabnia et al., eds., *Proceedings of DMIN'05, International Conference on Data Mining*, Las Vegas, NV, pp. 200–207, 2006.

70. E. Estrada. Characterization of the folding degree of proteins. *Bioinformatics*, 18:697–704, 2002.

71. E. Estrada, L. Torres, L. Rodríguez, and I. Gutman. An atom-bond connectivity index: Modelling the enthalpy of formation of alkanes. *Indian Journal of Chemistry*, 37A:849–855, 1998.

72. L. Euler. Solutio problematis ad geometriam situs pertinentis. *Commentarii Academiae Petropolitanae*, 8:128–140, 1736.

73. G. Fath-Tabar, B. Furtula, and I. Gutman. A new geometric-arithmetic index. *Journal of Mathematical Chemistry*, 47:477–486, 2010.

74. J. Felsenstein. *Inferring Phylogenies*. Sinauer Associates, Sunderland, MA, 2003.

75. L. R. Foulds. *Graph Theory Applications*. Springer, New York, 1992.

76. B. Furtula, A. Graovac, and D. Vukičević. Atom-bond connectivity index of trees. *Discrete Applied Mathematics*, 157:2828–2835, 2009.

77. M. R. Garey and D. S. Johnson. *Computers and Intractability: A Guide to the Theory of NP-Completeness*. Series of Books in the Mathematical Sciences. W. H. Freeman, San Francisco, CA, 1979.

78. T. Gärtner. A survey of kernels for structured data. *SIGKDD Explorations*, 5(1):49–58, 2003.

79. J. Gasteiger and T. Engel. *Chemoinformatics—A Textbook*. Wiley-VCH, Weinheim, Germany, 2003.

80. J. L. Gross and T. W. Tucker. *Topological Graph Theory*. Wiley Interscience, New York, 1987.

81. I. Gutman, X. Li, and J. Zhang. Graph energy. In M. Dehmer and F. Emmert-Streib, eds., *Analysis of Complex Networks: From Biology to Linguistics*, pp. 145–174. Wiley-VCH, Weinheim, Germany, 2009.

82. I. Gutman and O. E. Polansky. *Mathematical Concepts in Organic Chemistry*. Springer, Berlin, Germany, 1986.

83. R. Halin. *Graphentheorie*. Akademie Verlag, Berlin, Germany, 1989.

84. F. Harary. Status and contrastatus. *Sociometry*, 22:23–43, 1959.

85. F. Harary. *Structural Models. An Introduction to the Theory of Directed Graphs*. Wiley, New York, 1965.

86. F. Harary. *Graph Theory*. Addison Wesley Publishing Company, Reading, MA, 1969.

87. M. Höchstmann, T. Töller, R. Giegerich, and S. Kurtz. Local similarity in RNA secondary structures. In *Proceedings of the IEEE Computational Systems Bioinformatics Conference (CSB'03)*, Stanford University, Stanford, CA, pp. 159–168, 2003.

88. B. Hollas. Correlations in distance-based descriptors. *MATCH: Communications in Mathematical and in Computer Chemistry*, (47):79–86, 2003.

89. B. Hollas. Asymptotically independent topological indices on random trees. *Journal of Mathematical Chemistry*, 38(3):379–387, 2005.

90. B. Hollas. The covariance of topological indices that depend on the degree of a vertex. *MATCH: Communications in Mathematical and in Computer Chemistry*, 54(1):177–187, 2005.

91. B. Hollas. An analysis of the redundancy of graph invariants used in chemoinformatics. *Discrete Applied Mathematics*, 154:2484–2498, 2006.

92. D. Horvat, A. Graovac, D. Plavšić, N. Trinajstić, and M. Strunje. On the intercorrelation of topological indices in benzenoid hydrocarbons. *International Journal of Quantum Chemistry*, 44(S26):401–408, 1992.

93. T. Horváth, T. Gärtner, and S. Wrobel. Cyclic pattern kernels for predictive graph mining. In *Proceedings of the 2004 ACM SIGKDD International Conference on Knowledge Discovery and Data Mining*, Seattle, WA, pp. 158–167, 2004.

94. H. Hosoya. A newly proposed quantity characterizing the topological nature of structural isomers of saturated hydrocarbons. *Bulletin of the Chemical Society of Japan*, 44:2332–2339, 1971.

95. S. M. Hsieh and C. C. Hsu. Graph-based representation for similarity retrieval of symbolic images. *Data & Knowledge Engineering*, 65(3):401–418, 2008.

96. C.-Y. Hu and L. Xu. On highly discriminating molecular topological index. *Journal of Chemical Information and Computer Sciences*, 36:82–90, 1996.

97. W. Imrich. On metric properties of tree-like spaces. In Sektion MAROEK der Technischen Hochschule Ilmenau, ed. *Beiträge zur Graphentheorie und deren Anwendungen*, Oberhof (DDR), Germany, pp. 129–156, 1977.

98. D. Janežić, A. Miležević, S. Nikolić, and N. Trinajstić. *Graph: Theoretical Matrices in Chemistry*. Mathematical Chemistry Monographs. University of Kragujevac and Faculty of Science Kragujevac, Kragujevac, Serbia, 2007.

99. T. Jiang, L. Wang, and K. Zhang. Alignment of trees — An alternative to tree edit. In *CPM'94: Proceedings of the Fifth Annual Symposium on Combinatorial Pattern Matching*, Asilomar, CA, pp. 75–86. Springer-Verlag, London, U.K., 1994.

100. S. Joshi, N. Agrawal, R. Krishnapuram, and S. Negi. A bag of paths model for measuring structural similarity in web documents. In *KDD'03: Proceedings of the Ninth ACM SIGKDD International Conference on Knowledge Discovery and Data Mining*, New York, pp. 577–582, 2003.

101. B. H. Junker and F. Schreiber. *Analysis of Biological Networks*. Wiley Series in Bioinformatics. Wiley-Interscience, Hoboken, NJ, 2008.

102. F. Kaden. Graphmetriken und Distanzgraphen. *ZKI-Informationen, Akad. Wiss. DDR*, 2(82):1–63, 1982.

103. J. Kim and T. Wilhelm. What is a complex graph? *Physica A*, 387:2637–2652, 2008.

104. G. Kirchhoff. Über die Auflösung von Gleichungen, auf welche man bei der Untersuchung der linearen Verteilungen galvanischer Ströme gef"uhrt wird. *Annalen der Physik und Chemie*, 72:497–508, 1847.

105. D. J. Klein. Graph geometry, graph metrics and wiener. *MATCH: Communications in Mathematical and in Computer Chemistry*, 35:7–27, 1997.

106. D. J. Klein. Resistance-distance sum rules. *Croatica Chemical Acta*, 75:633–649, 2002.

107. D. J. Klein and M. Randić. Resistance distance. *Journal of Mathematical Chemistry*, 12:81–95, 1993.

108. D. J. Klein and H.-Y. Zhu. Distances and volumina for graphs. *Journal of Mathematical Chemistry*, 23:179–195, 1998.

109. J. M. Kleinberg. Authoritative sources in a hyperlinked environment. *Journal of the ACM*, 46(5):604–632, 1999.

110. E. D. Kolaczyk. *Statistical Analysis of Network Data*. Springer Series in Statistics. Springer, New York, 2009.

111. D. König. *Theorie der endlichen und unendlichen Graphen*. Chelsea Publishing, New York, 1935.

112. E. Konstantinova. The discrimination ability of some topological and information distance indices for graphs of unbranched hexagonal systems. *Journal of Chemical Information and Computer Sciences*, 36:54–57, 1996.

113. E. Konstantinova and A.A. Paleev. Sensitivity of topological indices of polycyclic graphs. *Vychisl. Sistemy*, 136:38–48, 1990. in Russian.

114. E. V. Konstantinova, V. A. Skorobogatov, and M. V. Vidyuk. Applications of information theory in chemical graph theory. *Indian Journal of Chemistry*, 42:1227–1240, 2002.

115. V. Kraus, M. Dehmer, and F. Emmert-Streib. Probabilistic inequalities for evaluating structural network measures. submitted, 2013.

116. V. Kraus, M. Dehmer, and M. Schutte. On sphere-regular graphs and the extremality of information-theoretic network measures. *MATCH: Communications in Mathematical and in Computer Chemistry*, 70(3):885–900, 2013.

117. K. Kuratowski. Sur le problème des courbes gauches en topologie. *Fund. Math. Vol.*, 15:271–283, 1930.

118. V. I. Levenstein. Binary codes capable of correcting deletions, insertions, and reversals. *Soviet Physics—Doklady*, 10(8):707–710, February 1966.

119. X. Li and I. Gutman. *Mathematical Aspects of Randić-Type Molecular Structure Descriptors*. Mathematical Chemistry Monographs. University of Kragujevac and Faculty of Science Kragujevac, Kragujevac, Serbia, 2006.

120. X. Liu and D. J. Klein. The graph isomorphism problem. *Journal of Computational Chemistry*, 12(10):1243–1251, 1991.

121. M. Randić. On molecular identification numbers. *Journal of Chemical Information and Computer Sciences*, 24(3):164–175, 1984.

122. T. Madaras and M. Mockovčiaková. Notes on topological indices of graph and its complement. *Opuscula Mathematica*, 33(1):107–115, 2013.

123. G. M. Maggiora and V. Shanmugasundaram. Molecular similarity measures. In *Chemoinformatics: Concepts, Methods, and Tools for Drug Discovery*, Jürgen Bajorath (ed.), pp. 1–50. Humana Press, Totowa, NJ, 2004.

124. A. Mehler. Hierarchical orderings of textual units. In *Proceedings of the 19th International Conference on Computational Linguistics, COLING'02*, Taiwan, China, August 24–September 1, pp. 646–652. Morgan Kaufmann, San Francisco, CA, 2002.

125. A. Mehler, P. Weiß, and A. Lücking. A network model of interpersonal alignment. *Entropy*, 12(6):1440–1483, 2010.

126. S. Melnik, H. Garcia-Molina, and A. Rahm. Similarity flooding: A versatile graph matching algorithm and its application to schema matching. In *Proceedings of the 18th International Conference on Data Engineering*, San Jose, CA, 2002.

127. A. Mowshowitz. Entropy and the complexity of the graphs I: An index of the relative complexity of a graph. *Bulletin of Mathematical Biophysics*, 30:175–204, 1968.

128. A. Mowshowitz and M. Dehmer. A symmetry index for graphs. *Symmetry: Culture and Science*, 21(4):321–327, 2010.

129. L. A. J. Müller, M. Dehmer, and F. Emmert-Streib. Network-based methods for computational diagnostics by means of R. In Z. Trajanoski (ed.), *Computational Medicine*, pp. 185–197. Springer, Vienna, Austria, 2012.

130. L. A. J. Müller, K. G. Kugler, A. Dander, A. Graber, and M. Dehmer. QuACN—An R package for analyzing complex biological networks quantitatively. *Bioinformatics*, 27(1):140–141, 2011.

131. L. A. J. Müller, M. Schutte, K. G. Kugler, and M. Dehmer. *QuACN: Quantitative Analyze of Complex Networks*, 2012. R Package Version 1.6.

132. M. Neuhaus and H. Bunke. Automatic learning of cost functions for graph edit distance. *Information Sciences*, 177(1):239–247, 2007.

133. M. E. J. Newman, A. L. Barabási, and D. J. Watts. *The Structure and Dynamics of Networks*. Princeton Studies in Complexity. Princeton University Press, Princeton, NJ, 2006.

134. S. Nikolić, G. Kovačević, A. Miličević, and N. Trinajstić. The Zagreb index 30 years after. *Croatica Chemica Acta*, 76:113–124, 2003.

135. A. Peeters, K. Coolsaet, G. Brinkmann, N. Cleemput, and V. Fack. Grinvin in a nutshell. *Journal of Mathematical Chemistry*, 45:471–477, 2009.

136. M. Randić. On characterization of molecular branching. *Journal of the American Chemical Society*, 97:6609–6615, 1975.

137. M. Randić. Design of molecules with desired properties. molecular similarity approach to property optimization. In M. A. Johnson and G. Maggiora, eds., *Concepts and Applications of Molecular Similarity*, pp. 77–145. Wiley, New York, 1990.

138. M. Randić and C. L. Wilkins. Graph theoretical approach to recognition of structural similarity in molecules. *Journal of Chemical Information and Computer Sciences*, 19:31–37, 1979.

139. N. Rashevsky. Life, information theory and topology. *Bulletin of Mathematical Biology*, 17:229–235, 1955.

140. A. Rényi. On measures of information and entropy. *Proceedings of the Fourth Berkeley Symposium on Mathematics, Statistics and Probability*, Berkeley, CA, pp. 547–561, 1960.

141. A. Robles-Kelly and R. Hancock. Edit distance from graph spectra. In *Proceedings of the IEEE International Conference on Computer Vision*, Nice, France, pp. 234–241, 2003.

142. L. Rodgers and W. A. Nicewander. Thirteen ways to look at the correlation coefficient. *The American Statistician*, 42(1):59–66, 1988.

143. I. L. Ruiz, G. C. García, and M. A. Gómez-Nieto. Clustering chemical databases using adaptable projection cells and MCS similarity values. *Journal of Chemical Information and Modeling*, 45(5):1178–1194, 2005.

144. I. L. Ruiz and M. A. Gómez-Nieto. A java tool for the management of chemical databases and similarity analysis based on molecular graphs isomorphism. In M. Bubak, G. D. van Albada, J. Dongarra, and P. M. A. Sloot, eds., *Computational Science—ICCS 2008, Eighth International Conference*, Kraków, Poland, Lecture Notes in Computer Science, pp. 369–378, 2008.

145. I. L. Ruiz, M. Urbano-Cuadrado, and M. A. Gómez-Nieto. Advantages of the approximate similarity approach in the QSAR prediction of ligand activities for alzheimer disease detection. In *World Congress on Engineering*, London, U.K., pp. 165–170, 2007.

146. I. L. Ruiz, M. Urbano-Cuadrado, and M. A. Gómez-Nieto. Improving the development of QSAR prediction models with the use of approximate similarity approach. *Engineering Letters*, 16(1):36–43, 2008.

147. D. Sankoff, J. B. Kruskal, S. Mainville, and R. J. Cedergren. Fast algorithms to determine RNA secondary structures containing multiple loops. In D. Sankoff and J. Kruskal, eds., *Time Warps, String Edits and Macromolecules: The Theory and Practice of Sequence Comparison*, pp. 93–120. Addison-Wesley, Reading, MA, 1983.

148. B. Schölkopf, K. R. Müller, and A. J. Smola. Lernen mit Kernen: Support-Vektor-Methoden zur Analyse hochdimensionaler Daten. *Inform., Forsch. Entwickl.*, 14(3):154–163, 1999.

149. S. M. Selkow. The tree-to-tree editing problem. *Information Processing Letters*, 6(6):184–186, 1977.

150. L. B. Shams, M. J. Brady, and S. Schaal. Graph matching vs mutual information maximization for object detection. *Neural Networks*, 14(3):345–354, 2001.

151. C. E. Shannon and W. Weaver. *The Mathematical Theory of Communication*. University of Illinois Press, Urbana, IL, 1949.

152. L. Sivakumar and M. Dehmer. Towards information inequalities for generalized graph entropies. *PLoS ONE*, 7:e38159, 2012.

153. V. A. Skorobogatov and A. A. Dobrynin. Metrical analysis of graphs. *MATCH: Communications in Mathematical and in Computer Chemistry*, 23:105–155, 1988.

154. M. I. Skvortsova, I. I. Baskin, I. V. Stankevich, V. A. Palyulin, and N. S. Zefirov. Molecular similarity. 1. Analytical description of the set of graph similarity measures. *Journal of Chemical Information and Computer Sciences*, 38:785–790, 1998.

155. F. Sobik. Graphmetriken und Klassifikation strukturierter objekte. *ZKI-Informationen, Akad. Wiss. DDR*, 2(82):63–122, 1982.

156. F. Sobik. Graphmetriken und Charakterisierung von Graphenklassen. PhD thesis, Akademie der Wissenschaften der DDR, Berlin, Germany, 1985.

157. F. Sobik. Modellierung von Vergleichsprozessen auf der Grundlage von Ähnlichkeitsmaßen für Graphen. *ZKI-Informationen, Akad. Wiss. DDR*, 4:104–144, 1986.

158. O. Sokolsky, S. Kannan, and I. Lee. Simulation-based graph similarity. In *TACAS*, Vienna, Austria, pp. 426–440. Springer, Berlin, Germany, LNCS, 2006.

159. E. Sommerfeld. Modellierung kognitiver strukturtransformationen auf der grundlage von graphtransformationen. *ZKI-Informationen, Akad. Wiss. DDR*, 4:1–103, 1984.

160. E. Sommerfeld. Zur Stabilität der Korona Zweier Graphen. Graphs, hypergraphs and applications. *Proc. Conf. Graph Theory, Eyba/DDR, Teubner-Texte Math.*, 73:165–168, 1985.

161. E. Sommerfeld. Systematization and formalization of cognitive structure transformations on the basis of graph transformations. In T. Marek, ed., *Action and Performance: Models and Tests. Contributions to the Quantitative Psychology and Its Methodology*, pp. 105–120. Rodopi, Amsterdam, the Netherlands, 1990.

162. E. Sommerfeld. *Kognitive Strukturen. Mathematisch-psychologische Elementaranalysen der Wissensstrukturierung und Informationsverarbeitung.* Waxmann Publishing, Münster, Germany, 1994.

163. E. Sommerfeld and F. Sobik. Operations on cognitive structures—Their modeling on the basis of graph theory. In D. Albert, ed., *Knowledge Structures*, pp. 146–190. Springer, Berlin, Germany, 1994.

164. Z. Tang and H. Deng. The (n, n)-graphs with the first three extremal Wiener Indices. *Journal of Mathematical Chemistry*, 43(1):60–74, 2008.

165. Ch. Theoharatos, N. Laskaris, G. Economou, and S. Fotopoulos. A similarity measure for color image retrieval and indexing based on the multivariate two sample problem. In *Proceedings of EUSIPCO*, Vienna, Austria, 2004.

166. Ch. Theoharatos, V. K. Pothos, N. A. Laskaris, G. Economou, and S. Fotopoulos. Multivariate image similarity in the compressed domain using statistical graph matching. *Pattern Recognition*, 39(10):1892–1904, 2006.

167. S. Toda. Graph isomorphism: Its complexity and algorithms (abstract). In C. P. Rangan, V. Raman, and R. Ramanujam, eds., *FSTTCS, Foundations of Software Technology and Theoretical Computer Science, 19th Conference, Proceedings*, Chennai, India, December 13–15, vol. 1738 of Lecture Notes in Computer Science, pp. 341. Springer, Berlin, Germany, 1999.

168. R. Todeschini, V. Consonni, and A. Maiocchi. The *k* correlation index: Theory development and its application in chemometrics. *Chemometrics and Intelligent Laboratory Systems*, 46:13–29, 1999.

169. R. Todeschini, V. Consonni, and R. Mannhold. *Handbook of Molecular Descriptors*. Wiley-VCH, Weinheim, Germany, 2002.

170. R. Todeschini, V. Consonni, A. Mauri, and M. Pavan. Dragon, software for calculation of molecular descriptors. www.talete.mi.it, 2004. Talete srl, Milano, Italy.

171. N. Trinajstič, S. Nikolič, S. C. Basak, and I. Lukovits. Distance indices and their hyper-counterparts: Intercorrelation and use in the structure-property modeling. *SAR and QSAR in Environmental Research*, 12(1–2):31–54, 2001.

172. R. E. Ulanowicz. Quantitative methods for ecological network analysis. *Computational Biology and Chemistry*, 28:321–339, 2004.

173. J. R. Ullmann. An algorithm for subgraph isomorphism. *Journal of the ACM*, 23(1):31–42, 1976.

174. K. Varmuza and H. Scsibrany. Substructure isomorphism matrix. *Journal of Chemical Information and Computer Sciences*, 40:308–313, 2000.

175. P. R. Villas Boas, F. A. Rodrigues, G. Travieso, and L. da F. Costa. Sensitivity of complex networks measurements. *Journal of Statistical Mechanics*, 2010:P03009, 2010.

176. S. Wagner. Correlation of graph-theoretical indices. *SIAM Journal on Discrete Mathematics*, 21(1):33–46, 2007.

177. S. Wasserman and K. Faust. *Social Network Analysis: Methods and Applications*. Structural Analysis in the Social Sciences. Cambridge University Press, Cambridge, U.K., 1994.

178. D. J. Watts. *Small worlds: The Dynamics of Networks between Order and Randomness*. Princeton University Press, Princeton, NJ, 1999.

179. D. J. Watts and S. H. Strogatz. Collective dynamics of 'small-world' networks. *Nature*, 393:440–442, 1998.

180. H. Wiener. Structural determination of paraffin boiling points. *Journal of the American Chemical Society*, 69(17):17–20, 1947.

181. R. W. Yeung. A framework for linear information inequalities. *IEEE Transactions on Information Theory*, 43(6):1924–1934, 1997.

182. L. A. Zager and G. C. Verghese. Graph similarity scoring and matching. *Applied Mathematics Letters*, 21:86–94, 2008.

183. B. Zelinka. On a certain distance between isomorphism classes of graphs. *Časopis pro pěst. Mathematiky*, 100:371–373, 1975.

184. Z. Zhang and R. W. Yeung. On characterization of entropy functions via information inequalities. *IEEE Transactions on Information Theory*, 44(4):1440–1452, 1998.

185. B. Zhou, I. Gutman, B. Furtula, and Z. Du. On two types of geometric-arithmetic index. *Chemical Physics Letters*, 482:153–155, 2009.

186. B. Zhou and N. Trinajstić. Bounds on the Balaban index. *Croatica Chemica Acta*, 81(2):319–323, 2008.

Chapter 2

Localization of Graph Topological Indices via Majorization Technique

Monica Bianchi, Alessandra Cornaro, José Luis Palacios, and Anna Torriero

Contents

2.1 Introduction..36
2.2 Notations and Preliminaries...37
    2.2.1 Majorization...37
    2.2.2 Graph Theory..39
2.3 Extremal Elements with Respect to the Majorization Order.....................40
    2.3.1 Maximal Elements...41
    2.3.2 Minimal Elements...44
2.4 Topological Indices...47
    2.4.1 Degree-Based Indices over All Vertices...............................48
        2.4.1.1 First General Zagreb Index.....................................48
        2.4.1.2 First Multiplicative Zagreb Index..............................49
    2.4.2 Degree-Based Indices over All Edges.................................50
        2.4.2.1 General Randić Index..50
        2.4.2.2 Generalized Sum-Connectivity Index...........................52
    2.4.3 Eigenvalue-Based Indices...............................52
        2.4.3.1 Energy Index...53
        2.4.3.2 Laplacian and Normalized Laplacian Indices..................54
    2.4.4 Resistance-Based Indices...............................57
        2.4.4.1 Kirchhoff Index..57
        2.4.4.2 Multiplicative-Degree-Kirchhoff Index........................63
        2.4.4.3 Additive-Degree-Kirchhoff Index.............................64
2.5 Numerical Results..68
    2.5.1 General Randić Index...............................68
    2.5.2 Generalized Sum-Connectivity Index...................................70
    2.5.3 Kirchhoff Index...............................71
        2.5.3.1 Lower Bounds for d-Regular Graphs............................71
        2.5.3.2 Upper Bounds...72
    2.5.4 Additive Kirchhoff Index...............................72
        2.5.4.1 d-Regular Graph...72
        2.5.4.2 (a, b)-Semiregular Graph....................................73
        2.5.4.3 Full Binary Tree of Depth $d > 1$.............................73
References..74

2.1   Introduction

The notion of majorization ordering was introduced by Hardy et al. [34] and is closely connected with the economic theory of disparity indices [2]. It is worth pointing out that previous uses of the majorization partial order in chemistry and a general overview are given in Klein [36]. Indeed, the degree sequence partial order has previously been studied from a chemical perspective in [37,62].

Schur [63] first investigated functions that preserve the majorization order, the so-called Schur-convex functions, which can also be found in Karamata [35].

Using the order-preserving property and characterizing maximal and minimal vectors with respect to the majorization order under suitable constraints, Marshall et al. [52] derived many inequalities involving Schur-convex functions. In [9], the results given in [52] were extended to a variety of subsets of $\mathbb{R}^n$ determining their maximal and minimal elements.

Our purpose is to develop a quantitative method aimed to localize some relevant graph topological indices via majorization techniques. Through this method, a unified approach will be achieved. Many well-known bounds will be derived and new insights will be given for achieving new bounds of some relevant graph topological indices that can be expressed in terms of Schur-convex or Schur-concave functions.

This approach has been proven to be useful in solving certain nonlinear optimization problems aimed to localizing ordered sequences of real numbers as they occur in the problem of finding estimates of eigenvalues of a matrix [10,60,64,65]. The interest of methods based on majorization, compared with the classical one, lies in the fact that one can easily obtain a solution without extensive numerical computations performed through the application of Lagrange conditions. Significant applications of this methodology concern the area of mathematical chemistry, where the aim is to localize chemical molecular structure descriptors [32,33,66] and more generally in network analysis for studying assortativeness, that is, the preference for high-degree vertices to connect to other high-degree vertices [1,41,53]. The first use of majorization techniques for localizing topological indices can be found in [9,28] through the solution of suitable optimization problems under constraints. Further applications can be found in [4–8,24,46]. Significant results regarding the use of majorization theory have been provided in [21] where some graph entropy bounds have been obtained (see [19,20] for details). One major advantage of this approach is to provide a unified framework for recovering many well-known upper and lower bounds obtained with a variety of methods, as well as providing better ones. It is worth pointing out that the localization of topological indices is typically carried out by applying classical inequalities such as the Cauchy–Schwarz inequality or the arithmetic-geometric-harmonic mean inequalities.

After some preliminary definitions and notations, in Section 2.3, we perform a theoretical analysis aimed at determining maximal and minimal vectors with respect to the majorization order of suitable subsets of $\mathbb{R}^n$ extending the results obtained by Marshall et al. [52].

In Section 2.4, we illustrate how the general methodology discussed in the previous sections can be successfully applied to get upper and lower bounds for some relevant topological indices, namely, degree-based, eigenvalue-based, and resistance-based indices.

The review is not exhaustive though the technique described provides a more general tool suitable for application to many other graph topological indices in the literature. Finally, in Section 2.5, we present some numerical examples, comparing our results with those in the literature.

2.2 Notations and Preliminaries

In this section, we present the notations and some basic facts used in the sequel. For the sake of clarity, we have split the section in two subsections addressing the main topics of majorization and graphs.

2.2.1 Majorization

The main references about majorization order and Schur- convexity are the classical book [52] and the recent paper [9] for the notations and technique.

Let $\mathcal{D} = \{\mathbf{x} \in \mathbb{R}^n : x_1 \geq x_2 \geq \cdots \geq x_n\}$ and denote by $\left[x_1^{\alpha_1}, x_2^{\alpha_1}, \ldots, x_p^{\alpha_p}\right]$ a vector in $\mathbb{R}^n$ with α_i components equal to x_i, where $\sum_{i=1}^{p} \alpha_i = n$. If $\alpha_i = 1$, we use x_i instead of $x_i^{\alpha_i}$ for convenience, while x_i^0 means that the component x_i is not present.

Definition 2.2.1 *Given two vectors $\mathbf{y}, \mathbf{z} \in \mathcal{D}$, the majorization order $\mathbf{y} \trianglelefteq \mathbf{z}$ is defined as follows:*

$$\begin{cases} \langle \mathbf{y}, \mathbf{s^k} \rangle \leq \langle \mathbf{z}, \mathbf{s^k} \rangle, \ k = 1, \ldots, (n-1) \\ \langle \mathbf{y}, \mathbf{s^n} \rangle = \langle \mathbf{z}, \mathbf{s^n} \rangle \end{cases}$$

where $\langle \cdot, \ldots \rangle$ is the inner product in $\mathbb{R}^n$ and $\mathbf{s^j} = \left[1^j, 0^{n-j}\right], \quad j = 1, 2, \ldots, n$.

Fix a positive real number a and let

$$\Sigma_a = \mathcal{D} \cap \left\{ \mathbf{x} \in \mathbb{R}_+^n : \langle \mathbf{x}, \mathbf{s^n} \rangle = a \right\}.$$

Given a subset S of Σ_a, a vector $\mathbf{x}^*(S) \in S$ is said to be maximal for S with respect to the majorization order if $\mathbf{x} \trianglelefteq \mathbf{x}^*(S)$ for each $\mathbf{x} \in S$. Analogously, a vector $\mathbf{x}^*(S) \in S$ is said to be minimal for S with respect to the majorization order if $\mathbf{x}^*(S) \trianglelefteq \mathbf{x}$ for each $\mathbf{x} \in S$.

Remark 2.2.1 *Without loss of generality, in the sequel, we will consider only subsets of Σ_a. Indeed if S' is a subset of*

$$\Sigma' = \left\{ \mathbf{x} \in \mathbb{R}^n : x_1 \geq x_2 \geq \cdots \geq x_n \geq L, \langle \mathbf{x}, \mathbf{s^n} \rangle = a' > Ln \right\}$$

by considering the change of variable $y_i = x_i - L$, $1 \leq i \leq n$, it is easy to prove that $\mathbf{y}$ *belongs to a subset S of $\Sigma_{a'-Ln}$ with $(a' - Ln) > 0$. Then, easy computations show that*

$$\mathbf{x}^*(S') = \mathbf{x}^*(S) + L\mathbf{s}^n, \quad \mathbf{x}_*(S') = \mathbf{x}_*(S) + L\mathbf{s}^n,$$

that is, the maximal and minimal elements of S' can be easily deduced from the maximal and minimal elements of S adding to each component the constant L.

Our applications will deal with constrained optimization problems involving Schur-convex (or Schur-concave) objective functions and an admissible closed set $S \subseteq \Sigma_a$:

$$(P) \quad \begin{cases} \max \ (\min) \ \phi(\mathbf{x}) \\ \text{subject to } \mathbf{x} \in S \end{cases}$$

We are able to solve easily this kind of problems by means of majorization techniques. To this aim, let us recall the following definition.

Definition 2.2.2 *A symmetric function $\phi: A \to \mathbb{R}$, $A \subseteq \mathbb{R}^n$, is said to be Schur-convex on A if $\mathbf{x} \trianglelefteq \mathbf{y}$ implies $\phi(\mathbf{x}) \leq \phi(\mathbf{y})$. If in addition $\phi(\mathbf{x}) < \phi(\mathbf{y})$ for $\mathbf{x} \trianglelefteq \mathbf{y}$ but $\mathbf{x}$ is not a permutation of $\mathbf{y}$, ϕ is said to be strictly Schur-convex on A. A function ϕ is (strictly) Schur-concave on A if $-\phi$ is (strictly) Schur-convex on A.*

Thus, the set of S-convex functions preserves the ordering of majorization and the same do other classes of functions yielding S-convex functions:

Proposition 2.2.1 *Let $I \subset \mathbb{R}$ be an interval and let $\phi(\mathbf{x}) = \sum_{i=1}^n g(x_i)$, where $g : I \to \mathbb{R}$. If g is strictly convex on I, then ϕ is strictly Schur-convex on $I^n = \underbrace{I \times \cdots \times I}_{n-times}$.*

By the order-preserving property of Schur-convex (Schur-concave) functions, the solution of the constrained nonlinear optimization problem (P) can be obtained in a straightforward way. If the set S admits maximal vector $\mathbf{x}^*(S)$ and minimal vector $\mathbf{x}_*(S)$ with respect to the majorization order, the maximum and the minimum are attained at $\mathbf{x}^*(S)$ and $\mathbf{x}_*(S)$, respectively; the opposite holds if ϕ is a Schur-concave function. This allows us to solve problem (P) in a more direct way, avoiding the standard approach of Lagrange multipliers. Finally, it is worthwhile to consider the following result:

Proposition 2.2.2 *Let us consider two sets S'' and S', with $S'' \subseteq S'$, which admit maximal and minimal elements with respect to the majorization order. If ϕ is a strictly Schur-convex function, then*

$$\phi\left(\mathbf{x}^*\left(S''\right)\right) \leq \phi\left(\mathbf{x}^*\left(S'\right)\right)$$
$$\phi\left(\mathbf{x}_*\left(S'\right)\right) \leq \phi\left(\mathbf{x}_*\left(S''\right)\right)$$

and the equality holds if and only if $\mathbf{x}^*\left(S''\right) = \mathbf{x}^*\left(S'\right)$ and $\mathbf{x}_*\left(S''\right) = \mathbf{x}_*\left(S'\right)$.

2.2.2 Graph Theory

Let us now recall some basic concepts from graph theory (for more details, we refer the reader to [11,68]).

Let $G = (V, E)$ be a simple, connected, and undirected graph, where $V = \{v_1, \ldots, v_n\}$ is the set of vertices and $E \subseteq V \times V$ the set of edges. We consider graphs with fixed order $|V| = n$ and fixed size $|E| = m$.

Let $\pi = (d_1, d_2, \ldots, d_n)$ denote the degree sequence of G arranged in a nonincreasing order $d_1 \geq d_2 \geq \cdots \geq d_n$, where d_i is the degree of the vertex v_i. We recall that the sequences of integers that are degree sequences of a simple graph were characterized by Erdős and Gallai [25]. It is well known that $\sum_{i=1}^{n} d_i = 2m$. Also, if G is a tree, that is, a connected graph without cycles, then $m = n - 1$.

If π is a fixed degree sequence and $\mathbf{x} \in \mathbb{R}^m$ the vector whose components are $d_i^\alpha + d_j^\alpha$, $\alpha \neq 0$, $(v_i, v_j) \in E$, extending the result proved in [48], it is possible to show that

$$\sum_{i=1}^{m} x_i = \sum_{(v_i, v_j) \in E} \left(d_i^\alpha + d_j^\alpha \right) = \sum_{i=1}^{n} d_i^{\alpha+1} \tag{2.1}$$

and thus $\sum_{i=1}^{m} x_i$ is a constant.

Associated with a graph, there are certain types of matrices that have important properties related to their eigenvalues. Let $A(G)$ be the *adjacency matrix* of G and $D(G)$ be the diagonal matrix of vertex degrees. The matrix $L(G) = D(G) - A(G)$ is called the *Laplacian matrix* of G, while $\mathcal{L}(G) = D(G)^{-1/2}L(G)D(G)^{-1/2}$ is known as the *normalized Laplacian*.

Let $\lambda_1(L) \geq \lambda_2(L) \geq \cdots \geq \lambda_n(L)$ be the set of (real) eigenvalues of $L(G)$ and $\lambda_1(\mathcal{L}) \geq \lambda_2(\mathcal{L}) \geq \cdots \geq \lambda_n(\mathcal{L})$ be the (real) eigenvalues of $\mathcal{L}(G)$. The following properties of the spectra of $L(G)$ and $\mathcal{L}(G)$ hold:

$$\sum_{i=1}^{n} \lambda_i(L) = \text{tr}(L(G)) = 2m; \quad \lambda_1(L) \geq 1 + d_1 \geq \frac{2m}{n}; \quad \lambda_n(L) = 0, \lambda_{n-1}(L) > 0;$$

$$\sum_{i=1}^{n} \lambda_i(\mathcal{L}) = \text{tr}(\mathcal{L}(G)) = n; \quad \sum_{i=1}^{n} \lambda_i^2(\mathcal{L}) = \text{tr}(\mathcal{L}^2(G)) = n + 2 \sum_{(v_i, v_j) \in E} \frac{1}{d_i d_j};$$

$$\lambda_n(\mathcal{L}) = 0; \lambda_1(\mathcal{L}) \leq 2.$$

Note that the condition $\lambda_{n-1}(L) > 0$ characterizes the connected graphs.

Finally, we cite the *transition matrix* $P = D^{-1}A$, which arises in the simple random walk on G. This is the process that jumps from a vertex i to any adjacent vertex j with equal transition probabilities $1/d_i$. In other words, this process is the Markov chain with transition matrix P and its real eigenvalues are $1 = \lambda_1(P) > \lambda_2(P) \geq \cdots \geq \lambda_n(P) \geq -1$. For a bipartite graph, the spectrum of P is symmetric and, in particular, $\lambda_n(P) = -1$.

For any square matrix M of order n, let $\mu(M) = \text{tr}(M)/n$ and $\sigma^2(M) = (\text{tr}(M^2)/n) - (\text{tr}(M)/n)^2$. If M admits real eigenvalues $\lambda_1 \geq \lambda_2 \geq \cdots \geq \lambda_n$, the following inequalities hold [69]:

$$\mu(M) - \sigma(M)\sqrt{\frac{i-1}{n-i+1}} \leq \lambda_i \leq \mu(M) + \sigma(M)\sqrt{\frac{i-1}{n-i+1}}, \quad i = 1,\ldots,n. \quad (2.2)$$

And, in particular, more binding inequalities hold for the smallest and largest eigenvalues:

$$\lambda_1 \geq \mu(M) + \frac{\sigma(M)}{\sqrt{n-1}}, \quad \lambda_n \leq \mu(M) - \frac{\sigma(M)}{\sqrt{n-1}} \quad (2.3)$$

In the case of the normalized Laplacian, we get

$$\sigma^2(\mathcal{L}(G)) = \left(\frac{2}{n}\right) \sum_{(v_i,v_j)\in E} \frac{1}{d_i d_j},$$

and inequality (2.3) yields

$$\lambda_1(\mathcal{L}) \geq 1 + \sqrt{\frac{2}{n(n-1)} \sum_{(v_i,v_j)\in E} \frac{1}{d_i d_j}} \quad (2.4)$$

Notice that for every connected graph of order n, we have

$$1 > \sigma(\mathcal{L}(G)) \geq \frac{1}{\sqrt{n-1}},$$

and the right inequality is attained for the complete graph $G = K_n$.

2.3   Extremal Elements with Respect to the Majorization Order

Given a positive real number a, it is well known [52] that the maximal and the minimal elements of the set Σ_a with respect to the majorization order are

$$\mathbf{x}^*(\Sigma_a) = a\mathbf{e}^1 = \left[a, 0^{n-1}\right] \quad \text{and} \quad \mathbf{x}_*(\Sigma_a) = \frac{a}{n}\mathbf{s}^{\mathbf{n}} = \left[\left(\frac{a}{n}\right)^n\right].$$

In this section, we extend the previous result finding the maximal and the minimal elements, with respect to the majorization order, of the particular subset of Σ_a given by

$$S_a = \Sigma_a \cap \left\{\mathbf{x} \in \mathbb{R}^n : M_i \geq x_i \geq m_i, \ i = 1,\ldots,n\right\}, \quad (2.5)$$

where $\mathbf{m} = [m_1, m_2,\ldots,m_n]^T$ and $\mathbf{M} = [M_1, M_2,\ldots,M_n]^T$ are two assigned vectors arranged in a nonincreasing order with $0 \leq m_i \leq M_i$, for all $i = 1,\ldots,n$, and a is a positive real number such that $\langle \mathbf{m}, \mathbf{s}^n \rangle \le a \le \langle \mathbf{M}, \mathbf{s}^n \rangle$. Notice that the intervals $[m_i, M_i]$ are not necessarily disjointed unless the additional assumption $M_{i+1} < m_i$, $i = 1, \ldots, (n-1)$ is required.

The existence of maximal and minimal elements of S_a are ensured by the compactness of the set S_a and by the closure of the upper and level sets:

$$U(\mathbf{x}) = \{\mathbf{z} \in S_a : \mathbf{x} \trianglelefteq \mathbf{z}\}, \; L(\mathbf{x}) = \{\mathbf{z} \in S_a : \mathbf{z} \trianglelefteq \mathbf{x}\}$$

Let $\mathbf{v}^j = [0^j, 1^{n-j}]$, $j = 0, \ldots, n$. The Hadamard product of two vectors $\mathbf{x}, \mathbf{y} \in \mathbb{R}^n$ is defined as follows:

$$\mathbf{x} \circ \mathbf{y} = [x_1 y_1, x_2 y_2, \ldots, x_n y_n]^T .$$

It is easy to verify the following properties:

1. $\langle \mathbf{x} \circ \mathbf{y}, \mathbf{z} \rangle = \langle \mathbf{x}, \mathbf{y} \circ \mathbf{z} \rangle$

2. $\langle \mathbf{s}^h, \mathbf{v}^k \rangle = h - \min\{h, k\}$

3. $\mathbf{s}^k \circ \mathbf{s}^j = \mathbf{s}^h$, $h = \min\{k, j\}$

4. $\mathbf{v}^k \circ \mathbf{s}^j = \mathbf{s}^j - \mathbf{s}^h = \mathbf{v}^h - \mathbf{v}^j$, $h = \min\{k, j\}$

2.3.1   Maximal Elements

The following theorem states the main result concerning the structure of the maximal element of the set S_a. For the reader's convenience, we give also the proof.

Theorem 2.3.1 *[9] Let $k \ge 0$ be the smallest integer such that*

$$\left\langle \mathbf{M}, \mathbf{s}^k \right\rangle + \left\langle \mathbf{m}, \mathbf{v}^k \right\rangle \le a < \left\langle \mathbf{M}, \mathbf{s}^{k+1} \right\rangle + \left\langle \mathbf{m}, \mathbf{v}^{k+1} \right\rangle, \tag{2.6}$$

and $\theta = a - \left\langle \mathbf{M}, \mathbf{s}^k \right\rangle - \left\langle \mathbf{m}, \mathbf{v}^{k+1} \right\rangle$. Then

$$\mathbf{x}^*(S_a) = \mathbf{M} \circ \mathbf{s}^k + \theta \mathbf{e}^{k+1} + \mathbf{m} \circ \mathbf{v}^{k+1} = [M_1, M_2, \ldots, M_k, \theta, m_{k+2}, \ldots, m_n]. \tag{2.7}$$

Proof. First of all, we verify that $\mathbf{x}^*(S_a) \in S_a$. It easy to see that $\langle \mathbf{x}^*(S_a), \mathbf{s}^n \rangle = a$ and that $m_i \le \mathbf{x}_i^*(S_a) \le M_i$ for $i \ne k+1$. To prove that $m_{k+1} \le \mathbf{x}_{k+1}^*(S_a) \le M_{k+1}$, notice that from (2.6),

$$m_{k+1} = \left\langle \mathbf{m}, \mathbf{e}^{k+1} \right\rangle \le a - \left\langle \mathbf{M}, \mathbf{s}^k \right\rangle - \left\langle \mathbf{m}, \mathbf{v}^{k+1} \right\rangle = \theta < \left\langle \mathbf{M}, \mathbf{e}^{k+1} \right\rangle = M_{k+1}.$$

Now we show that $\mathbf{x} \trianglelefteq \mathbf{x}^*(S_a)$ for all $\mathbf{x} \in S_a$. By property (1) follows

$$\left\langle \mathbf{x}^*(S_a), \mathbf{s}^j \right\rangle = \left\langle \mathbf{M}, \mathbf{s}^k \circ \mathbf{s}^j \right\rangle + \theta \left\langle \mathbf{e}^{k+1}, \mathbf{s}^j \right\rangle + \left\langle \mathbf{m}, \mathbf{v}^{k+1} \circ \mathbf{s}^j \right\rangle, \; j = 1, \ldots, (n-1)$$

and by (3) and (4),

$$\left\langle \mathbf{x}^*(S_a), \mathbf{s^j} \right\rangle = \begin{cases} \left\langle \mathbf{M}, \mathbf{s^j} \right\rangle & 1 \le j \le k \\ \left\langle \mathbf{M}, \mathbf{s^k} \right\rangle + \theta + \left\langle \mathbf{m}, \mathbf{s^j} - \mathbf{s^{k+1}} \right\rangle & (k+1) \le j \le (n-1) \end{cases}.$$

Thus, given a vector $\mathbf{x} \in S_a$, for $1 \le j \le k$, we obtain

$$\left\langle \mathbf{x}, \mathbf{s^j} \right\rangle \le \left\langle \mathbf{M}, \mathbf{s^j} \right\rangle = \left\langle \mathbf{x}^*(S_a), \mathbf{s^j} \right\rangle,$$

while for $(k+1) \le j \le (n-1)$, by (3),

$$\left\langle \mathbf{x}, \mathbf{s^j} \right\rangle = \left\langle \mathbf{x}, \mathbf{s^n} \right\rangle - \left\langle \mathbf{x}, \mathbf{v^j} \right\rangle \le a - \left\langle \mathbf{m}, \mathbf{v^j} \right\rangle = \left\langle \mathbf{M}, \mathbf{s^k} \right\rangle + \theta + \left\langle \mathbf{m}, \mathbf{s^j} - \mathbf{s^{k+1}} \right\rangle = \left\langle \mathbf{x}^*(S_a), \mathbf{s^j} \right\rangle$$

and the result follows.   □

In the following corollaries, we characterize the maximal elements of particular subsets of S_a useful in the applications we will deal with in Section 2.4. We omit the proof of the results and we refer the reader to [9] for more details. We denote by $\lfloor x \rfloor$ the integer part of the real number x.

Corollary 2.3.1 *Given $1 \le h \le n$, let us consider the set*

$$S_a^{[h]} = \Sigma_a \cap \left\{ \mathbf{x} \in \mathbb{R}^n : M_1 \ge x_1 \ge \cdots \ge x_h \ge m_1, \quad M_2 \ge x_{h+1} \ge \cdots \ge x_n \ge m_2 \right\},$$
$$(2.8)$$

where $0 \le m_2 \le m_1$, $0 \le M_2 \le M_1$, $m_i < M_i, i = 1, 2,$ and

$$hm_1 + (n - h)m_2 \le a \le hM_1 + (n - h)M_2.$$

Let $a^* = hM_1 + (n - h)m_2$ and

$$k = \begin{cases} \left\lfloor \dfrac{a - h(m_1 - m_2) - nm_2}{M_1 - m_1} \right\rfloor & if \quad a < a^* \\ \left\lfloor \dfrac{a - h(M_1 - M_2) - nm_2}{M_2 - m_2} \right\rfloor & if \quad a \ge a^* \end{cases}$$

Then

$$\mathbf{x}^*\left(S_a^{[h]}\right) = \begin{cases} \left[M_1^k, \theta, m_1^{h-k-1}, m_2^{n-h} \right] & if \quad a < a^* \\ \left[M_1^h, M_2^{k-h}, \theta, m_2^{n-k-1} \right] & if \quad a \ge a^* \end{cases}$$

where θ is evaluated in order to entail $\mathbf{x}^*\left(S_a^{[h]}\right) \in \Sigma_a$, that is, $\theta = a - \left\langle \mathbf{M}, \mathbf{s^k} \right\rangle - \left\langle \mathbf{m}, \mathbf{v^{k+1}} \right\rangle$, $\mathbf{M} = [M_1^h, M_2^{n-h}]$, $\mathbf{m} = [m_1^h, m_2^{n-h}]$.

Remark 2.3.1 *The assumption $m_i < M_i$ in Corollary 2.3.1 can be relaxed to $m_i \leq M_i$. Indeed, if $m_i = M_i, i = 1, 2$, the set $S_a^{[h]}$ reduces to the singleton $\{m_1 s^h + m_2 v^h\}$, while if $m_1 = M_1, m_2 < M_2$, the first h components of any $x \in S_a^{[h]}$ are fixed and equal to m_1 and the maximal element of $S_a^{[h]}$ can be computed by the maximal element of $S_{a-hm_1} \in \mathbb{R}^{n-h}$ (see Corollary 2.3.2). The case $m_2 = M_2, m_1 < M_1$, is similar.*

The next proposition, proved in [52], immediately follows from Corollary 2.3.1 when $m_1 = m_2 = m$ and $M_1 = M_2 = M$.

Corollary 2.3.2 *Let $0 \leq m < M$ and $m \leq \dfrac{a}{n} \leq M$. Given the subset*

$$S_a^1 = \Sigma_a \cap \left\{ x \in \mathbb{R}^n : M \geq x_1 \geq x_2 \geq \cdots \geq x_n \geq m \right\}$$

we have

$$x^*\left(S_a^1\right) = M s^k + \theta e^{k+1} + m v^{k+1},$$

where
$$k = \left\lfloor \frac{a - nm}{M - m} \right\rfloor$$
$$\theta = a - Mk - m(n - k - 1).$$

In particular, when $m = 0$, we obtain

$$x^*\left(S_a^1\right) = M s^k + \theta e^{k+1},$$

where
$$k = \left\lfloor \frac{a}{M} \right\rfloor$$
$$\theta = a - Mk$$

Taking $m = m_n$ and $M = M_1$, it is worthwhile to notice that S_a is a subset of S_a^1. This clearly implies

$$x^*(S_a) \trianglelefteq x^*\left(S_a^1\right). \tag{2.9}$$

Finally, we recall the following result.

Corollary 2.3.3 *[10] Let $1 \leq h \leq n$ and $0 < \alpha \leq a/h$. Given the subset*

$$S_a^2 = \Sigma_a \cap \left\{ x \in \mathbb{R}^n : x_i \geq \alpha, \ i = 1, \ldots, h \right\},$$

we have $x^*(S_a^2) = (a - h\alpha) e^1 + \alpha s^h$.

2.3.2   Minimal Elements

The computation of the minimal element of the set S_a is more tangled. The minimal element of Σ_a is $\mathbf{x}^*(\Sigma_a) = [(a/n)^n]$. If it belongs to S_a, then it is its minimal element, too. Otherwise, we have to apply the following theorem.

Theorem 2.3.2 *[9] Let $k \geq 0$ and $d \geq 0$ be the smallest integers such that*

1. $k + d < n$.

2. $m_{k+1} \leq \rho \leq M_{n-d}$ *where* $\rho = \dfrac{a - \langle \mathbf{m}, \mathbf{s^k} \rangle - \langle \mathbf{M}, \mathbf{v^{n-d}} \rangle}{n - k - d}$.

Then

$$\mathbf{x}^*(S_a) = \mathbf{m} \circ \mathbf{s^k} + \rho \left(\mathbf{s^{n-d}} - \mathbf{s^k} \right) + \mathbf{M} \circ \mathbf{v^{n-d}}$$

$$= \left[m_1, \ldots, m_k, \rho^{n-d-k}, M_{n-d+1} \ldots, M_n \right].$$

Proof. The minimal element of the set Σ_a is $\mathbf{x}^*(\Sigma_a) = \dfrac{a}{n}\mathbf{s^n}$. If $m_1 \leq \mathbf{x}^*(\Sigma_a) \leq M_n$, then $\mathbf{x}^*(\Sigma_a) \in S_a$ and $\mathbf{x}^*(S_a) = \mathbf{x}^*(\Sigma_a)$ (notice that in this case, $k = d = 0$).

If $\mathbf{x}^*(\Sigma_a) \notin S_a$, let k and d be the smallest integers satisfying conditions (1) and (2) given earlier. It is easy to verify that $\mathbf{x}^*(S_a) \in S_a$. In order to prove that it is the minimal element, we must show that for all $\mathbf{x} \in S_a$,

$$\left\langle \mathbf{x}^*(S_a), \mathbf{s^h} \right\rangle \leq \left\langle \mathbf{x}, \mathbf{s^h} \right\rangle, \quad h = 1, \ldots, (n-1). \tag{2.10}$$

We distinguish three cases:

1. $1 \leq \mathbf{h} \leq \mathbf{k}$. Since $\left\langle \mathbf{x}^*(S_a), \mathbf{s^h} \right\rangle = \left\langle \mathbf{m}, \mathbf{s^h} \right\rangle$, the inequality (2.10) is straightforward.

2. $\mathbf{k + 1} \leq \mathbf{h} \leq \mathbf{n - d}$. We prove the inequality (2.10) for $h = k + 1$. By induction, similar arguments can be applied to prove the inequality for $h = k + 2, \ldots,$ $(n - d)$.

    By contradiction, let us assume that there exists $\mathbf{x} \in S_a$ such that

    $$\left\langle \mathbf{x}_*(S_a), \mathbf{s^{k+1}} \right\rangle = \left\langle \mathbf{m}, \mathbf{s^k} \right\rangle + \rho > \left\langle \mathbf{x}, \mathbf{s^k} \right\rangle + x_{k+1}.$$

    Then $x_j \leq x_{k+1} < \langle \mathbf{m}, \mathbf{s^k} \rangle + \rho - \langle \mathbf{x}, \mathbf{s^k} \rangle$, for $j = k + 2, \ldots n$ and thus

    $$\left\langle \mathbf{x}, \mathbf{s^{n-d}} \right\rangle = \left\langle \mathbf{x}, \mathbf{s^k} \right\rangle + \left\langle \mathbf{x}, \mathbf{s^{n-d}} - \mathbf{s^k} \right\rangle$$

    $$< \left\langle \mathbf{x}, \mathbf{s^k} \right\rangle + (n - d - k) \left(\left\langle \mathbf{m}, \mathbf{s^k} \right\rangle + \rho - \left\langle \mathbf{x}, \mathbf{s^k} \right\rangle \right).$$

    Taking into account that

    $$\left\langle \mathbf{x}, \mathbf{s^{n-d}} \right\rangle = a - \left\langle \mathbf{x}, \mathbf{v^{n-d}} \right\rangle \geq a - \left\langle \mathbf{M}, \mathbf{v^{n-d}} \right\rangle,$$

we get

$$a - \left\langle \mathbf{M}, \mathbf{v}^{n-d} \right\rangle < (1 - n + d + k) \left\langle \mathbf{x}, \mathbf{s}^k \right\rangle + (n - d - k) \left(\left\langle \mathbf{m}, \mathbf{s}^k \right\rangle + \rho \right).$$

Using the expression of ρ, we obtain

$$0 < (1 - n + d + k) \left(\left\langle \mathbf{x}, \mathbf{s}^k \right\rangle - \left\langle \mathbf{m}, \mathbf{s}^k \right\rangle \right).$$

Since $(1 - n + d + k) \leq 0$ and $\left\langle \mathbf{x}, \mathbf{s}^k \right\rangle \geq \left\langle \mathbf{m}, \mathbf{s}^k \right\rangle$, the inequality given earlier is false, and we have reached the contradiction.

3. $\mathbf{n} - \mathbf{d} + \mathbf{1} \leq \mathbf{h} < \mathbf{n}$. For any $\mathbf{x} \in S_a$, we have

$$
\begin{aligned}
\left\langle \mathbf{x}^*(S_a), \mathbf{s}^h \right\rangle &= \left\langle \mathbf{x}^*(S_a), \mathbf{s}^{n-d} \right\rangle + \left\langle \mathbf{x}^*(S_a), \mathbf{s}^h - \mathbf{s}^{n-d} \right\rangle \\
&= \left\langle \mathbf{m}, \mathbf{s}^k \right\rangle + (n - d - k)\rho + \left\langle \mathbf{M}, \mathbf{s}^h - \mathbf{s}^{n-d} \right\rangle \\
&= a - \left\langle \mathbf{M}, \mathbf{v}^{n-d} \right\rangle + \left\langle \mathbf{M}, \mathbf{s}^h - \mathbf{s}^{n-d} \right\rangle \\
&= a - \left\langle \mathbf{M}, \mathbf{s}^n - \mathbf{s}^h \right\rangle \\
&= \left\langle \mathbf{x}, \mathbf{s}^h \right\rangle + \left\langle \mathbf{x}, \mathbf{s}^n - \mathbf{s}^h \right\rangle - \left\langle \mathbf{M}, \mathbf{s}^n - \mathbf{s}^h \right\rangle \\
&\leq \left\langle \mathbf{x}, \mathbf{s}^h \right\rangle.
\end{aligned}
$$

$\square$

In the following corollaries, we characterize the minimal element of particular subsets of S_a.

Corollary 2.3.4 *[9] Given $1 \leq h \leq n$, let us consider the set*

$$S_a^{[h]} = \frac{\Sigma_a \cap \{\mathbf{x} \in \mathbb{R}^n : M_1 \geq x_1 \geq \cdots \geq x_h \geq m_1,}{M_2 \geq x_{h+1} \geq \cdots \geq x_n \geq m_2\}}.$$

where $0 \leq m_2 \leq m_1$, $0 \leq M_2 \leq M_1$, $m_i < M_i, i = 1, 2$, and

$$hm_1 + (n - h)m_2 \leq a \leq hM_1 + (n - h)M_2.$$

If $m_1 \leq \dfrac{a}{n} \leq M_2$, we have $\mathbf{x}^* \left(S_a^{[h]} \right) = \frac{a}{n} \mathbf{s}^n$. Otherwise, let $\tilde{a} = hm_1 + (n - h)M_2$.

If $\begin{cases} a < m_1 n \\ a \leq \tilde{a} \end{cases}$, given $\rho = \dfrac{a - hm_1}{n - h}$, we have

$$\mathbf{x}^* \left(S_a^{[h]} \right) = m_1 \mathbf{s}^h + \rho \mathbf{v}^h = \left[m_1^h, \rho^{n-h} \right].$$

If $\begin{cases} a > M_2 n \\ a \geq \tilde{a} \end{cases}$, given $\rho = \dfrac{a - M_2(n - h)}{h}$, we have

$$\mathbf{x}^* \left(S_a^{[h]} \right) = \rho \mathbf{s}^h + M_2 \mathbf{v}^h = \left[\rho^h, M_2^{n-h} \right].$$

Corollary 2.3.4 distinguishes the minimal element of $S_a^{[h]}$ whether

$$\begin{cases} a < m_1 n \\ a \leq \tilde{a} \end{cases} \quad \text{or} \quad \begin{cases} a > M_2 n \\ a \geq \tilde{a} \end{cases}.$$

We note that if $m_1 \leq M_2$, the first inequality in the aforementioned systems is always stronger than the second one, while if $M_2 < m_1$, the second one is stronger than the first. Thus, we can summarize the minimal element of $S_a^{[h]}$ in a more accessible way according to the following scheme:

1. If $m_1 \leq M_2$, then

$$\mathbf{x}_* \left(S_a^{[h]} \right) = \begin{cases} \left[\left(\frac{a}{n} \right)^n \right] & \text{if } m_1 \leq \frac{a}{n} \leq M_2 \\[2ex] \left[m_1^h, \left(\frac{a - hm_1}{n-h} \right)^{n-h} \right] & \text{if } \frac{a}{n} < m_1 \\[2ex] \left[\left(\frac{a - M_2(n-h)}{h} \right)^h, M_2^{n-h} \right] & \text{if } \frac{a}{n} > M_2. \end{cases} \tag{2.11}$$

2. If $M_2 < m_1$, then

$$\mathbf{x}^* \left(S_a^{[h]} \right) = \begin{cases} \left[m_1^h, \left(\dfrac{a - hm_1}{n-h} \right)^{n-h} \right] & \text{if } a < \tilde{a} \\[2ex] \left[\left(\dfrac{a - M_2(n-h)}{h} \right)^h, M_2^{n-h} \right] & \text{if } a \geq \tilde{a}. \end{cases} \tag{2.12}$$

Remark 2.3.2 *We note that the minimal element of the set $S_a^{[h]}$ does not necessarily have integer components, while this is not the case for the maximal element. For our purposes, it is crucial to find the minimal vector in S_a with integer components that can be constructed by the following procedure (see Remark 12 in [9]). Let us consider, for instance, the vector $\mathbf{x}^* \left(S_a^{[h]} \right) = (a/n)^n$ that corresponds to the case $m_1 \leq \dfrac{a}{n} \leq M_2$. If a/n is not an integer, let us find the index k, $1 \leq k \leq n$, such that*

$$\left(\left\lfloor \frac{a}{n} \right\rfloor + 1 \right) k + \left\lfloor \frac{a}{n} \right\rfloor (n - k) = a$$

that is, $k = a - \left\lfloor \dfrac{a}{n} \right\rfloor n$. The vector

$$\mathbf{x}_*^1 = \left[\left(\left\lfloor \frac{a}{n} \right\rfloor + 1 \right)^k, \left(\left\lfloor \frac{a}{n} \right\rfloor \right)^{n-k} \right]$$

is the minimal element of $S_a^{[h]}$ with integer components. With slight modifications, the same procedure can be applied also in the other cases discussed in (2.11), (2.12), or in Theorem 2.3.2.

To complete our analysis, from Corollary 2.3.4, particular cases can be deduced. Assuming $m_1 = m_2$, $M_1 = M_2$, or $h = n$ we get

Corollary 2.3.5 *[52] Let $0 \leq m < M$ and $m \leq \dfrac{a}{n} \leq M$. Given the subset*

$$S_a^1 = \Sigma_a \cap \left\{ \mathbf{x} \in \mathbb{R}^n : M \geq x_1 \geq \cdots \geq x_{n-1} \geq x_n \geq m \right\}$$

we have $\mathbf{x}^* \left(S_a^1 \right) = \dfrac{a}{n} \mathbf{s^n} = \left[\left(\dfrac{a}{n} \right)^n \right]$.

By similar arguments applied to the maximal element in (2.9), the following inequality holds:

$$\mathbf{x}^* \left(S_a^1 \right) \trianglelefteq \mathbf{x}^*(S_a). \tag{2.13}$$

Assuming $m_1 = \alpha$, $m_2 = 0$, $M_1 = M_2 = a$ or $m_1 = m_2 = 0$, $M_2 = \alpha$, $M_1 = a$, we easily obtain the following two corollaries.

Corollary 2.3.6 *[10] Let $1 \leq h \leq n$ and $0 < \alpha \leq a/h$. Given the subset*

$$S_a^2 = \Sigma_a \cap \left\{ \mathbf{x} \in \mathbb{R}^n : x_i \geq \alpha, \ i = 1, \dots, h \right\},$$

we have

$$\mathbf{x}_* \left(S_a^2 \right) = \begin{cases} \left[\left(\dfrac{a}{n} \right)^n \right] & \text{if} \quad \alpha \leq \dfrac{a}{n} \\ \left(\alpha^h, \rho^{n-h} \right) \text{ with } \rho = \dfrac{a - \alpha h}{n - h} & \text{if} \quad \alpha > \dfrac{a}{n} \end{cases}$$

Corollary 2.3.7 *[10] Let $1 \leq h \leq (n-1)$ and $0 < \alpha < a$. Given the subset*

$$S_a^3 = \Sigma_a \cap \left\{ \mathbf{x} \in \mathbb{R}^n : x_i \leq \alpha, \ i = h+1, \dots, n \right\},$$

we have

$$\mathbf{x}^*(S_a^3) = \begin{cases} \left[\left(\dfrac{a}{n} \right)^n \right] & \text{if} \quad \alpha \geq \dfrac{a}{n} \\ \left(\rho^h, \alpha^{n-h} \right) \text{ with } \rho = \dfrac{a - (n-h)\alpha}{h} & \text{if} \quad \alpha < \dfrac{a}{n}. \end{cases}$$

2.4 Topological Indices

Structural properties of graphs can be characterized in terms of graph invariants, that is, descriptors representing properties that are preserved by a graph isomorphism. Among them, we will focus on topological indices, which are numerical parameters describing the graph topology and mainly used in mathematical chemistry, computational biology, and, more generally, in network analysis.

It is possible to classify topological indices according to the mathematical object they are based on. We will restrict our attention to those indices that can be formulated as Schur-convex (Schur-concave) functions of the degree sequence π as well as of the eigenvalues of some matrices associated to the graph G such as $L(G)$, $\mathcal{L}(G)$, and $P(G)$. The first category of descriptors can be in turn grouped in two subsets: the degree-based indices over all vertices (e.g., the first general Zagreb index, the first multiplicative Zagreb index, the inverse degree) and the degree-based indices over all edges (e.g., the general Randić index, the generalized sum-connectivity index). On the other hand, to the eigenvalues-based indices belong, for example, the energy index, the (normalized) Laplacian index. Let us notice that the latter category also includes some resistance-based indices that can be reformulated in terms of eigenvalues of suitable matrices associated to G. We refer, for example, to the Kirchhoff index as well as to some of its generalizations like the additive-/multiplicative-degree-Kirchhoff indices.

In the next subsections, we compute the upper and lower bounds for graph topological indices via the general methodology we have discussed in Section 2.2. For any category of indices, we introduce the variables $x_1, x_2, \ldots, x_k$ on which they are based and the corresponding subsets $S \subset \mathbb{R}^k$ where these variables are valued. Notice that the majorization technique requires that $\sum_{i=1}^{k} x_i = a$.

Let $F(x_1, x_2, \ldots, x_k)$ be any topological index that is a Schur-convex function of its arguments, defined on the subset $S \subset \mathbb{R}^k$. Then, by the order-preserving property of the Schur-convex functions, we get

$$F(\mathbf{x}^*(S)) \leq F(x_1, x_2, \ldots, x_k) \leq F(\mathbf{x}^*(S)).$$

Analogously, if F is a Schur-concave function, then

$$F(\mathbf{x}^*(S)) \leq F(x_1, x_2, \ldots, x_k) \leq F(\mathbf{x}_*(S)).$$

By characterizing both the variables $x_1, x_2, \ldots, x_k$ and the function $F(x_1, x_2, \ldots, x_k)$, we provide in the following explicit upper and lower bounds for some relevant topological indices. A minor caveat: our list is not exhaustive; our interest lies on showing the novelty of the methodology that can be successfully applied to other cases amenable to this framework.

2.4.1   Degree-Based Indices over All Vertices

A class of topological indices of particular interest found in the literature and depending on the degree sequence of a graph over all vertices is represented by the *first general Zagreb indices* and the *first multiplicative Zagreb index*.

2.4.1.1   First General Zagreb Index

The first general Zagreb index is defined as

$$M_1^\alpha = \sum_{i=1}^{n} d_i^\alpha \tag{2.14}$$

TABLE 2.1:   Bounds for M_1^α ($\alpha < 0 \vee \alpha > 1$)

c	Lower Bounds	Upper Bounds
1	$(n-1)^\alpha + 2^{\alpha+1} + (n-3)$	$n\,(2^\alpha)$
2	$(n-1)^\alpha + 3^\alpha + 2^{\alpha+1} + (n-4)$	$2\,(3^\alpha) + (2^\alpha)\,(n-2)$
3	$4\,(3^\alpha) + (2^\alpha)\,(n-4)$	$(n-1)^\alpha + 4^\alpha + 3\,(2^\alpha) + (n-5)$
		$(n-1)^\alpha + 3^{\alpha+1} + (n-4)$
4	$2\,(3^{\alpha+1}) + (2^\alpha)\,(n-6)$	$(n-1)^\alpha + 5^\alpha + 2^{\alpha+2} + (n-6)$
		$(n-1)^\alpha + 2^\alpha(2^{\alpha+1}) + 2\,(3^\alpha) + (n-5)$
5	$8\,(3^\alpha) + (n-8)\,2^\alpha$	$(n-1)^\alpha + 6^\alpha + 5\,(2^\alpha) + (n-7)$
		$(n-1)^\alpha + 5^\alpha + 2\,(3^\alpha) + 2\,(2^\alpha) + (n-6)$
		$(n-1)^\alpha + 2^{2\alpha+1} + 2\,(3^\alpha) + (n-8)$
6	$10\,(3^\alpha) + (n-10)\,2^\alpha$	$(n-1)^\alpha + 7 + 6\,(2^\alpha)\,(n-8)$
		$(n-1)^\alpha + 6 + 2\,(3^\alpha) + 3\,(2^\alpha) + (n-7)$
		$(n-1)^\alpha + 5 + 4 + 2\,(3^\alpha) + 2 + (n-6)$
		$(n-1)^\alpha + 2^{2\alpha+2} + (n-5)$

where α is an arbitrary real number except 0 and 1. It was firstly introduced by Li and Zheng [44]. For $\alpha = 2$, we get the *first Zagreb index*, while for $\alpha = -1$, the *inverse degree*. Notice that M_1^α is a Schur-convex function of the degree sequence either for $\alpha < 0$ or $\alpha > 1$, while it is a Schur-concave function of the degree sequence for $0 < \alpha < 1$.

In [6], the authors studied upper and lower bounds of M_1^α for the classes of c-cyclic graphs with $0 \le c \le 6$. More specifically, after characterizing c-cyclic graphs as those whose degree sequences belongs to particular subsets of $\mathbb{R}^n$, the maximal and minimal elements of these subsets with respect to the majorization order have been identified and the upper and lower bounds have been evaluated (see Table 2.1). For convenience, the analysis has been restricted to the first general Zagreb index with either $\alpha < 0$ or $\alpha > 1$ and $n \ge c + 2$. In the case $0 < \alpha < 1$, the upper and lower bounds are turned over.

In the case of bicyclic graphs, we recover the same results as in [73], Theorems 1, 5, 7, and 8.

Notice that when more maximal elements are identified, the best choice depends on α. In [6], we discussed in detail the case $\alpha = -1$ [16,17,43].

2.4.1.2   First Multiplicative Zagreb Index

The *first multiplicative Zagreb index*, defined as

$$\ln M_1 = 2 \sum_{i=1}^{n} \ln(d_i) \qquad (2.15)$$

and introduced by Gutman [29], is a Schur-concave function of the degree sequence.

For c-cyclic graphs, $0 \le c \le 6$, bounds for the first multiplicative Zagreb index can be obtained following similar steps to those of the first general Zagreb index.

2.4.2   Degree-Based Indices over All Edges

A class of topological indices depending on the degrees of nodes linked by an edge is represented by the *general Randić index* and the *generalized sum-connectivity index*.

2.4.2.1   General Randić Index

With respect to the degree sequence, one of the most popular index is the general Randić index:

$$R_\alpha(G) = \sum_{(v_i,v_j)\in E} \left(d_i d_j\right)^\alpha,$$

where α is a nonzero real number [12]. For a specific value of α, some very well-known indices can be obtained: $\alpha = 1$, for example, corresponds to the Zagreb index $M_2(G)$ [54], while $\alpha = -(1/2)$ and $\alpha = -1$ to the branching indices [61]. Easy computations show that the generalized Randić index can be equivalently expressed as

$$R_\alpha(G) = \frac{1}{2} \left(\sum_{(v_i,v_j)\in E} \left(d_i^\alpha + d_j^\alpha\right)^2 - \sum_{i=1}^n d_i^{2\alpha+1} \right).$$

Let $\pi = (d_1, d_2, \ldots, d_n)$ be a fixed degree sequence and $\mathbf{x} \in \mathbb{R}^m$ be the vector whose components are $d_i^\alpha + d_j^\alpha$, with $(v_i, v_j) \in E$. Notice that by (2.1), $\sum_{i=1}^m x_i$ is a constant. Since $\sum_{i=1}^n d_i^{2\alpha+1}$ is also a constant, $R_\alpha(G)$ is a Schur-convex function of $\mathbf{x}$ and it is minimal (maximal) if and only $f(\mathbf{x}) = \sum_{i=1}^m x_i^2 = \|\mathbf{x}\|_2^2$ is minimal (maximal).

Hence, considering a closed subset S of $\Sigma_a \subseteq \mathbb{R}^m$, where $a = \sum_{i=1}^n d_i^{\alpha+1}$, which admits $\mathbf{x}_*(S)$ and $\mathbf{x}^*(S)$ as extremal vectors with respect to the majorization order, the function f attains its minimum and maximum on S at $f(\mathbf{x}_*(S))$ and $f(\mathbf{x}^*(S))$, respectively. The general Randić index can be consequently bounded as follows:

$$\frac{\|\mathbf{x}_*(S)\|_2^2 - \sum_{i=1}^n d_i^{2\alpha+1}}{2} \leq R_\alpha(G) \leq \frac{\|\mathbf{x}^*(S)\|_2^2 - \sum_{i=1}^n d_i^{2\alpha+1}}{2}. \tag{2.16}$$

Using the information available on the degree sequence of G and characterizing suitably the set S, different numerical bounds can be derived.

For the particular case $\alpha = 1$, which corresponds to the Zagreb index $M_2(G)$, this methodology was applied in [28] and, more recently, in [9] where the authors get sharper bounds for the index M_2 in the case of a particular class of graphs having exactly h pendant vertices, that is, vertices with degree one.

We briefly recall the results in [9]. Let C_π be the class of graphs $G = (V, E)$ with h pendant vertices and degree sequence

$$\pi = (d_1, d_2, \ldots, \underbrace{d_{n-h-1}, d_{n-h}}_{n-h}, \underbrace{1, \ldots, 1}_{h}), \quad n \geq 4, n - h \geq 2, h \geq 1$$

and let us consider graphs $G \in C_\pi$ with maximum vertex degree upper bounded by $d_{n-h} + d_{n-h-1}$, that is, $d_1 < d_{n-h} + d_{n-h-1}$, or equivalently

$$1 + d_1 \le d_{n-h} + d_{n-h-1}. \tag{2.17}$$

For $G \in C_\pi$, we note that this constraint is always satisfied, for example, if the maximum vertex degree is at most three, as for some graphs of chemical interest where the maximum degree is four.

We observe that for $i, j = 1, \ldots, n - h$ and $(v_i, v_j) \in E$,

$$d_{n-h} + d_{n-h-1} \le d_i + d_j \le d_1 + d_2,$$

while for $i = n - h + 1, \ldots, n, \; j = 1, \ldots, n - h$ and $(v_i, v_j) \in E$,

$$1 + d_{n-h} \le d_i + d_j \le 1 + d_1.$$

Furthermore, inequality (2.17) assures that the aforementioned intervals are concatenated so that the vector $\mathbf{x} \in \mathbb{R}^m$ can be arranged in a nonincreasing order with the h pendant vertices in the last h positions.

Setting $m_1 = d_{n-h} + d_{n-h-1}, \; m_2 = 1 + d_{n-h}, \; M_1 = d_1 + d_2, \; M_2 = 1 + d_1$, we face the set

$$S_a^{[m-h]} = \Sigma_a \cap \left\{ \mathbf{x} \in \mathbb{R}^n : M_1 \ge x_1 \ge \cdots x_{m-h} \ge m_1, M_2 \ge x_{m-h+1} \ge \cdots x_m \ge m_2 \right\}.$$

Applying Corollaries 2.3.1 and 2.3.4, we can compute maximal and minimal elements of $S_a^{[m-h]}$ with respect to the majorization order, and from (2.16), we obtain

$$\frac{\left\| \mathbf{x}^* \left(S_a^{[m-h]} \right) \right\|_2^2 - \sum_{i=1}^n d_i^3}{2} \le M_2(G) \le \frac{\left\| \mathbf{x}^* \left(S_a^{[m-h]} \right) \right\|_2^2 - \sum_{i=1}^n d_i^3}{2}. \tag{2.18}$$

In spite of inequalities (2.9) and (2.13), these bounds can't be worse than those in [28], and they are often sharper. It is noteworthy that both equalities in (2.18) are attained if and only if the set $S_a^{[m-h]}$ reduces to a singleton, that is, by Remark 2.3.1, $m_i = M_i, i = 1, 2$.

The condition $m_2 = 1 + d_{n-h} = M_2 = 1 + d_1$ implies that in $G = (V, E)$, all nonpendant vertices have the same degree. Some examples of this kind of graphs are as follows:

1. Trees T_π with degree sequence

$$\pi = \left(\underbrace{k, \ldots, k}_{r}, \underbrace{1, \ldots, 1}_{rk-2r+2} \right), \tag{2.19}$$

including, as particular case, for $k = 2$, the path.

2. Graphs $G_{\pi'}$ obtained by adding the same number s of pendant vertices to each vertex of a k−regular graph on r vertices, being kr even, $2 \le k \le r - 1$, that is, graphs with degree sequence

$$\pi' = \left(\underbrace{k+s, \ldots, k+s}_{r}, \underbrace{1, \ldots, 1}_{sr} \right).$$

Computing the Zagreb index, from Remark 2.3.1 and (2.18), we get the exact values $M_2(T_\pi) = k\,(2kr - 2r - k + 2)$ and $M_2(G_{\pi'}) = (1/2)r\left(2s + ks + k^2\right)$ $(k + s)$.

In Section 2.5, we give some significant examples, computing bounds for graphs belonging to C_π and satisfying (2.17).

2.4.2.2   Generalized Sum-Connectivity Index

In the recent contribution [23], the generalized sum-connectivity index

$$\chi_\alpha(G) = \sum_{(v_i, v_j) \in E} \left(d_i + d_j\right)^\alpha$$

has been proposed. For $\alpha = 1$, $\chi_1(G)$ reduces to the first Zagreb index

$$M_1(G) = \sum_{(v_i, v_j) \in E} \left(d_i + d_j\right) = \sum_{i=1}^{n} d_i^2,$$

while for $\alpha = -(1/2)$, we recover the sum-connectivity index defined in [78].

Let π be a fixed degree sequence and $\mathbf{x} \in \mathbb{R}^m$ be the vector whose components are $(d_i + d_j)$, $(v_i, v_j) \in E$. The function $f(\mathbf{x}) = \sum_{i=1}^{m} x_i^\alpha$ is strictly Schur-convex for $\alpha > 1$ or $\alpha < 0$, while it is strictly Schur-concave for $0 < \alpha < 1$. Thus, taking into account that $\sum_{i=1}^{m} x_i = \sum_{i=1}^{n} d_i^2$ is a constant and considering a closed subset S of Σ_a, where $a = \sum_{i=1}^{n} d_i^2$, for $\alpha > 1$ or $\alpha < 0$, we get

$$\left\| \mathbf{x}^*(S) \right\|_\alpha^\alpha \le \chi_\alpha(G) \le \left\| \mathbf{x}^*(S) \right\|_\alpha^\alpha, \tag{2.20}$$

where $\|\cdot\|_\alpha$ stands for the l_α-norm. For $0 < \alpha < 1$, the bounds are exchanged. As for the previous index, different bounds, depending on the choice of the set S, can be obtained.

2.4.3   Eigenvalue-Based Indices

In this subsection, we deal with classes of topological indices that can be formulated as Schur-convex (Schur-concave) functions of eigenvalues of some particular matrices associated to the graph G, like adjacency, Laplacian, and normalized Laplacian matrices. The topological indices based on the eigenvalue of the transition matrix P will be investigated in the next section dedicated to resistance indices.

2.4.3.1   Energy Index

The energy index [30] is given by

$$E(G) = \sum_{i=1}^{n} |\lambda_i(A)| = \sum_{i=1}^{n} \sqrt{\lambda_i^2(A)}$$

where $\lambda_i(A)$ are the eigenvalues of the adjacency matrix. This index is a Schur-concave function of the variables $\lambda_i^2(A), i = 1,\dots,n$. It is well known that $\sum_{i=1}^{n} \lambda_i^2(A) = 2m$ and $\lambda_1(A) \geq 2m/n$ [39,40].

If a sharper lower bound for $\lambda_1(A)$ is available, that is, $\lambda_1(A) \geq k(\geq 2m/n)$, introducing the new variables x_i as the square of the eigenvalue $\lambda_i(A)$ arranged in nondecreasing order, we get $x_1 = \lambda_1^2(A)$ and

$$x_1 \geq k^2 \geq \left(\frac{2m}{n}\right)^2 \geq \frac{2m}{n}.$$

Let us consider the set

$$S_E = \Sigma_{2m} \cap \{\mathbf{x} \in \mathbb{R}^n : x_1 \geq k^2\}. \tag{2.21}$$

Applying Corollary 2.3.6 with $a = 2m$, $h = 1$, $\alpha = k^2$, since we are in the case $\alpha \geq a/n$, we get

$$\mathbf{x}^*(S_E) = \left[k^2, \underbrace{\frac{2m-k^2}{n-1}, \dots, \frac{2m-k^2}{n-1}}_{n-1} \right].$$

Thus, the following upper bound for the energy index holds:

$$E(G) \leq k + \sqrt{(n-1)\left(2m - k^2\right)}.$$

For a bipartite graphs, by the equality $\lambda_1(A) = -\lambda_n(A)$, we get $x_1 = x_2$. In this case, we face the set

$$S_E^1 = \Sigma_{2m} \cap \{\mathbf{x} \in \mathbb{R}^n : x_i \geq k^2, i = 1, 2\}.$$

Applying again Corollary 2.3.6, the minimal element becomes

$$\mathbf{x}^*(S_E^1) = \left[k^2, k^2, \underbrace{\frac{2m-2k^2}{n-2}, \dots, \frac{2m-2k^2}{n-2}}_{n-2} \right].$$

and the following upper bound

$$E(G) \leq 2k + \sqrt{(n-2)\left(2m - 2k^2\right)}$$

holds. Notice that for $k = 2m/n$, we get the bounds in [39,40], while for $k = \left(\sum_{i=1}^{n} d_i^2\right)/n$, the bounds in [75]. Different choices of k allow us to obtain all the other bounds in [45]. New bounds for $E(G)$ can be derived as soon as a sharper lower bound of $\lambda_1(A)$ is available.

2.4.3.2   Laplacian and Normalized Laplacian Indices

The Laplacian index is given by the sum of the αth power of the nonzero Laplacian eigenvalues [47,76]

$$s_\alpha(G) = \sum_{i=1}^{n-1} \lambda_i(L)^\alpha, \quad \alpha \neq 0, 1.$$

Consider the set

$$S_L = \Sigma'_{2m} \cap \{\lambda \in \mathbb{R}^{n-1} : \lambda_1 \geq 1 + d_1\} \tag{2.22}$$

where $\Sigma'_n = \{\lambda \in \mathbb{R}^{n-1} : \lambda_1 \geq \lambda_2 \geq \cdots \geq \lambda_{n-1} \geq 0, \ \sum_{i=1}^{n-1} \lambda_i = n\}$.

Applying Corollary 2.3.6 with $a = 2m$, $\alpha = 1 + d_1$, $h = 1$ and taking into account that $2m/(n-1) \leq (1 + d_1)$, the minimal element of S_L is given by

$$\mathbf{x}^*(S_L) = \left[1 + d_1, \underbrace{\frac{2m - 1 - d_1}{n - 2}, \ldots, \frac{2m - 1 - d_1}{n - 2}}_{n-2} \right].$$

By the Schur-convexity or Schur-concavity of the functions $s_\alpha(G)$, the bounds in [76], Theorem 3, can be easily recovered.

The same approach can be used to find the bounds for bipartite graphs given in [76], Theorem 5, observing that in this case, $\lambda_1(L) \geq 2\sqrt{\left(\sum_{i=1}^{n} d_i^2\right)}/n$.

Taking into account further information on the localization of the eigenvalues, the aforementioned bounds can be improved. For instance, since $\lambda(L)_2 \geq d_2$ (see [14]), we can consider the set

$$S_L^1 = \Sigma_{2m} \cap \{\lambda \in \mathbb{R}^{n-1} : \lambda_1 \geq 1 + d_1, \ \lambda_2 \geq d_2\}. \tag{2.23}$$

The minimal element of this set can be computed applying the general result of Theorem 2.3.2. Under the assumption $2m \leq 1 + d_1 + (n-2)d_2$, we get

$$\mathbf{x}^*\left(S_L^1\right) = \left[1 + d_1, d_2, \underbrace{\frac{2m - 1 - d_1 - d_2}{n - 3}, \ldots, \frac{2m - 1 - d_1 - d_2}{n - 3}}_{n-3} \right].$$

Thus, the following bounds on $s_\alpha(G)$ hold:

Theorem 2.4.1 *Let G be a simple connected graph such that* $2m \leq 1 + d_1 + (n-2)d_2$,

1. *If* $\alpha < 0$ *or* $\alpha > 1$, *then*

$$s_\alpha(G) \geq (1 + d_1)^\alpha + d_2^\alpha + \frac{(2m - 1 - d_1 - d_2)^\alpha}{(n-3)^{\alpha-1}}.$$

2. *If* $0 < \alpha < 1$, *then*

$$s_\alpha(G) \leq (1 + d_1)^\alpha + d_2^\alpha + \frac{(2m - 1 - d_1 - d_2)^\alpha}{(n-3)^{\alpha-1}}.$$

In the same manner, we obtain and improve the bounds of the normalized Laplacian index

$$s_\alpha^*(G) = \sum_{i=1}^{n-1} \lambda_i^\alpha(\mathcal{L}), \quad \alpha \neq 0, 1$$

given by the sum of the αth powers of the nonzero normalized Laplacian eigenvalues, first introduced by Bozkurt and Bozkurt [13] and studied by the authors in [4].

To this aim, let

$$S_\mathcal{L} = \Sigma_n' \cap \{\lambda \in \mathbb{R}^{n-1} : \lambda_1 \geq Q\},$$

where $Q \geq n/(n-1)$. By Corollary 2.3.6, it follows that, for $n \geq 3$, the minimal element of $S_\mathcal{L}$ is

$$\mathbf{x}^*(S_\mathcal{L}) = \left[Q, \underbrace{\frac{n-Q}{n-2}, \ldots, \frac{n-Q}{n-2}}_{n-2} \right].$$

By the Schur-concavity or Schur-convexity of the function $s_\alpha^*(G)$, the following result holds.

Theorem 2.4.2 *Let G be a simple connected graph with* $n \geq 3$ *vertices.*

1. *If* $\alpha < 0$ *or* $\alpha > 1$, *then*

$$s_\alpha^*(G) \geq Q^\alpha + \frac{(n-Q)^\alpha}{(n-2)^{\alpha-1}}.$$

2. *If* $0 < \alpha < 1$, *then*

$$s_\alpha^*(G) \leq Q^\alpha + \frac{(n-Q)^\alpha}{(n-2)^{\alpha-1}}.$$

In particular, we can consider

$$Q = 1 + \sqrt{\frac{2}{n(n-1)} \sum_{(v_i,v_j)\in E} \frac{1}{d_i d_j}} \,. \tag{2.24}$$

In virtue of (2.4), we get $\lambda_1(\mathcal{L}) \geq Q$.

Let us point out that in Theorem 2.4.2 making use of (2.24), we recover the same bounds as in Theorems 3.3 and 3.7 in [13].

Notice that for the complete graph K_n, the spectra of $\mathcal{L}$ is given by

$$spec(\mathcal{L}) = \left\{ \frac{n}{n-1}, \ldots, \frac{n}{n-1}, 0 \right\}$$

which implies

$$s_\alpha^*(K_n) = \frac{n^\alpha}{(n-1)^\alpha} \,.$$

Being $Q = n/(n-1)$, by easy computation, we get that bounds in Theorem 2.4.2 are attained for $G = K_n$.

Now we show how the aforementioned bounds can be improved taking into account additional information on the localization of the eigenvalues.

In the following, we consider a noncomplete graph. If we know that

$$\lambda_2(\mathcal{L}) \geq \beta$$

with $\beta \leq Q$, Theorem 2.3.2 allows us to compute the minimal element of the set

$$S_{\mathcal{L}}^1 = \Sigma'_n \cap \{\lambda \in \mathbb{R}^{n-1} : \lambda_1 \geq Q, \ \lambda_2 \geq \beta\}.$$

To this aim, since the condition $(n-1)Q > n$ is always satisfied, if

$$Q + \beta(n-2) > n,$$

the minimal element of $S_{\mathcal{L}}^1$ is given by

$$\mathbf{x}^*\left(S_{\mathcal{L}}^1\right) = \left[Q, \beta, \underbrace{\frac{n-Q-\beta}{n-3}, \ldots, \frac{n-Q-\beta}{n-3}}_{n-3} \right].$$

By the Schur-concavity or Schur-convexity of the function $s_\alpha^*(G)$, we get the following bounds.

Theorem 2.4.3 *Let G be a simple connected graph with* $n \geq 4$ *vertices that is not complete and* $\lambda_2(\mathcal{L}) \geq \beta$ *with* $Q + \beta(n-2) > n$.

1. *If* $\alpha > 1$ *or* $\alpha < 0$, *then* $s_\alpha^*(G) \geq Q^\alpha + \beta^\alpha + \dfrac{(n - Q - \beta)^\alpha}{(n-3)^{\alpha-1}}$.

2. *If* $0 < \alpha < 1$, *then* $s_\alpha^*(G) \leq Q^\alpha + \beta^\alpha + \dfrac{(n - Q - \beta)^\alpha}{(n-3)^{\alpha-1}}$.

Notice that, due to Proposition 2.2.2, the bounds in the previous theorem perform equal or better than (16) and (17) in [13].

Finally, in case of bipartite graphs, bounds in Theorems 2.4.2 and 2.4.3 can be improved. We refer to [4] for a detailed discussion.

2.4.4   Resistance-Based Indices

Among the various indices in mathematical chemistry, those indices based on the effective resistance R_{ij} between the node i and j of a connected undirected graph $G = (V, E)$ have received a lot of attention in the literature. The resistance indices, namely, the Kirchhoff index and its generalizations, have undergone intense scrutiny in recent years because they have proven to be useful in discriminating among chemical molecules according to their cyclicity. A variety of techniques have been used, including graph theory, algebra (the study of the Laplacian and of the normalized Laplacian), electric networks, probabilistic arguments involving hitting times of random walks, and discrete potential theory (equilibrium measures and Wiener capacities). The references that follow are a sample, by no means exhaustive, of these diverse techniques, whose end results usually follow either of these two paths: on the one hand, exact values for the index are obtained for graphs endowed with some form of symmetry or special property [3,27,55,71]. On the other hand, general bounds for the resistance indices, not containing effective resistances but a few invariants such as $|V|$, $|E|$, the degrees d_i, etc., and sometimes extremal graphs, are found for specific families of graphs, as in [56,58,59,67,72,77,79].

In what follows, we adopt this latter approach, finding upper and lower bounds by using majorization techniques. The application of the majorization relies on the fact that these indices can be written as Schur-convex functions whose variables are eigenvalues of particular matrices as well as vertices degrees of the graphs. This subsection deals with the *Kirchhoff index* and some of its modifications, namely, the *multiplicative- degree-Kirchhoff index* and the *additive-degree-Kirchhoff index*.

2.4.4.1   Kirchhoff Index

The Kirchhoff index $R(G)$ of a simple connected graph $G = (V, E)$ was defined by Klein and Randić in [38] as

$$R(G) = \sum_{i<j} R_{ij}, \qquad (2.25)$$

where R_{ij} is the effective resistance between vertices i and j, which can be computed using Ohm's law. In addition to its original definition, it was shown in [31,80] that

$$R(G) = n \sum_{i=1}^{n-1} \frac{1}{\lambda_i(L)}, \tag{2.26}$$

where $\lambda_i(L)$ are the nonzero eigenvalues of the Laplacian matrix L.

If G is d-regular, then $L = dI - A$, $P = D^{-1}A = I - (1/d)L$, and

$$\lambda_{n-i+1}(P) = 1 - \frac{\lambda_i(L)}{d} \quad i = 1, \dots, n. \tag{2.27}$$

In this case, from (2.26), the alternative expression

$$R(G) = \frac{n}{d} \sum_{i=2}^{n} \frac{1}{1 - \lambda_i(P)}$$

in terms of the eigenvalues of the transition matrix P holds [57].

In case G is arbitrary, we do not have such a compact expression, but still, we have the bounds given in [59], Corollary 2:

$$\left(\frac{n}{d_1}\right) \sum_{i=2}^{n} \left(\frac{1}{1 - \lambda_i(P)}\right) \leq R(G) \leq \left(\frac{n}{d_n}\right) \sum_{i=2}^{n} \left(\frac{1}{1 - \lambda_i(P)}\right). \tag{2.28}$$

All these expressions of $R(G)$ in terms of sums of inverses of eigenvalues can be used to find upper and lower bounds, as was done in [57,77,79].

In order to get bounds for $R(G)$, we want to apply the majorization technique to the summations in (2.28), and we must deal with vectors arranged in nonincreasing order. To this aim, let us make a change of variable setting

$$\nu_i = 1 - \lambda_{n-i+1}(P), \quad i = 1, \dots, (n-1).$$

For the vector $\mathbf{v} = [\nu_1, \nu_2, \dots, \nu_{n-1}]$, we have

$$0 < \nu_{n-1} \leq \nu_{n-2} \leq \cdots \leq \nu_1 \leq 2$$

and $\sum_{i=1}^{n-1} \nu_i = n$ since

$$\text{tr}(P) = \sum_{i=1}^{n} \lambda_i(P) = 0 \Rightarrow \sum_{i=2}^{n} \lambda_i(P) = -1.$$

In order to tackle the inequalities in (2.28), we evaluate the extremal values of the Schur-convex function

$$f(\nu_1, \nu_2, \dots, \nu_{n-1}) = \sum_{i=1}^{n-1} \frac{1}{\nu_i}. \tag{2.29}$$

Let us consider the sets

$$S = \left\{ \mathbf{v} \in \mathbb{R}^{n-1} : \sum_{i=1}^{n-1} v_i = n, \ 0 < v_{n-1} \le v_{n-2} \le \cdots \le v_1 \le 2 \right\}$$

and

$$S_0 = \left\{ \mathbf{v} \in \mathbb{R}^{n-1} : \sum_{i=1}^{n-1} v_i = n, \ 0 \le v_{n-1} \le v_{n-2} \le \cdots \le v_1 \le 2 \right\}$$

By Corollary 2.3.5, we know that the minimal element of S_0 with respect to the majorization order is given by

$$\left(\underbrace{\frac{n}{n-1}, \frac{n}{n-1}, \dots, \frac{n}{n-1}}_{n-1} \right).$$

The function f attains the minimum value $(n-1)^2/n$ at this point. Since the minimum point belongs also to S, we have $\min_S f = \min_{S_0} f = (n-1)^2/n$. By (2.28), we get

$$R(G) \ge \frac{(n-1)^2}{d_1} \qquad (2.30)$$

which is the bound given in [59], Corollary 4. Notice that the lower bound is attained if and only if $G = K_n$.

Analogously, we can obtain the lower bound for bipartite graphs. In this case, $\lambda_n = -1$, which implies $v_1 = 2$ and so we face the set

$$S_0^b = \left\{ \mathbf{v} \in \mathbb{R}^{n-2} : \sum_{i=2}^{n-1} v_i = n - 2, \ 0 \le v_{n-1} \le v_{n-2} \le \cdots \le v_2 \le 2 \right\} \qquad (2.31)$$

and the Schur-convex function

$$f^b(v_1, \dots, v_{n-1}) = \frac{1}{2} + \sum_{i=2}^{n-1} \frac{1}{v_i}. \qquad (2.32)$$

Since the minimal element with respect to the majorization order of S_0^b is given by

$$\left(\underbrace{1, 1, \dots, 1}_{n-2} \right),$$

the function f attains the minimum value $(2n-3)/2$ at this point. Again by (2.28), we get the bound in [59], Corollary 3:

$$R(G) \ge \frac{n(2n-3)}{2d_1}. \qquad (2.33)$$

To obtain better bounds by means of majorization techniques, some subsets of S_0 should be considered. Indeed, if we have more information on the localization of the eigenvalues $\lambda_i(P)$ of the transition matrix P, we can improve both the lower bound, by using Corollaries 2.3.6 and 2.3.7, and the upper bound by Corollary 2.3.2. For the reader's convenience, in the sequel, we distinguish the lower and the upper bounds. Moreover, we analyze only nonbipartite graphs and we refer to [5] for the discussion of bipartite graphs.

Lower Bounds

Case 1: Assume we have the additional eigenvalue bound:

$$\lambda_n(P) \leq -\beta < 0.$$

We get

$$\nu_1 = 1 - \lambda_n(P) \geq 1 + \beta = \alpha \geq \frac{n}{n-1}.$$

In case of $\alpha > n/(n-1)$, it is possible to get sharper bounds for the Kirchhoff index by applying Corollary 2.3.6. We consider the subset of S_0 given by

$$S_0^1 = \{\mathbf{v} \in S_0 : \nu_1 \geq \alpha\}$$

whose minimal element is

$$\left(\alpha, \underbrace{\frac{n-\alpha}{n-2}, \frac{n-\alpha}{n-2}, \ldots, \frac{n-\alpha}{n-2}}_{n-2} \right).$$

Thus, the Schur-convex function f in (2.29) has its minimum value in S_0^1 given by $\frac{1}{\alpha} + \frac{(n-2)^2}{n-\alpha}$ and this is also the minimum value of f on

$$S^1 = \{\mathbf{v} \in S : \nu_1 \geq \alpha\}.$$

We can thus infer

$$R(G) \geq \frac{n}{d_1} \left[\frac{1}{\alpha} + \frac{(n-2)^2}{n-\alpha} \right] \tag{2.34}$$

or, equivalently

$$R(G) \geq \frac{n}{d_1} \left[\frac{1}{1+\beta} + \frac{(n-2)^2}{n-1-\beta} \right]. \tag{2.35}$$

Case 2: Assume now that

$$\lambda_2(P) \geq \beta > 0. \tag{2.36}$$

We get

$$\nu_{n-1} = 1 - \lambda_2\,(P) \leq 1 - \beta = \alpha < \frac{n}{n-1},$$

and we face the set

$$T_0^1 = \{\mathbf{v} \in S_0 : \nu_{n-1} \leq \alpha\}.$$

By Corollary 2.3.6, the minimal vector of T_0^1 is given by

$$\left(\underbrace{\frac{n-\alpha}{n-2}, \frac{n-\alpha}{n-2}, \ldots, \frac{n-\alpha}{n-2}}_{n-2}, \alpha \right),$$

and, by the same arguments as before, we get the bound (2.34), which, in terms of β, is now given by

$$R(G) \geq \frac{n}{d_1} \left[\frac{1}{1-\beta} + \frac{(n-2)^2}{n-1+\beta} \right]. \tag{2.37}$$

Now, we exploit Case 1 given earlier in order to get a general lower bound. In Section 2.2, we recall that for every matrix M with real eigenvalues $\lambda_1(M) \geq \lambda_2(M) \geq \cdots \geq \lambda_n(M)$, the following inequality is well known

$$\lambda_n(M) \leq \mu - \frac{\sigma}{\sqrt{n-1}} \tag{2.38}$$

where

$$\mu = \frac{\text{tr}(M)}{n}$$
$$\sigma^2 = \frac{\text{tr}(M^2)}{n} - \left(\frac{\text{tr}(P)}{n} \right)^2$$

If M is a transition matrix P of a connected graph G, we observe that $\text{tr}(P) = 0$ and $\text{tr}(P^2) = 2 \sum_{(v_i,v_j) \in E} \frac{1}{d_i d_j}$. Then $\mu = 0$ and

$$\sigma^2 = \frac{2}{n} \sum_{(v_i,v_j) \in E} \frac{1}{d_i d_j} = \left(\frac{2}{n} \right) R_{-1}(G),$$

where $R_{-1}(G)$ is the general Randić index for $\alpha = -1$ [43,61]. Moreover, by the equality

$$\sigma^2 = \frac{\text{tr}(P^2)}{n} = \frac{1 + \sum_{i=2}^{n} \lambda_i^2}{n}$$

and the conditions on the eigenvalues of P, it easily follows that P has at least one eigenvalue whose absolute value is less than one. This gives $\sigma^2 < 1$. Notice that the upper bound $\sigma = 1$ is attained by any unconnected graph with an even number n of vertices and $n/2$ connected components, each of which is of order two. In this case,

the spectrum of P is $\left\{\underbrace{-1, -1, \ldots, -1}_{n/2}, \underbrace{1, 1, \ldots, 1}_{n/2}\right\}$ and consequently $\sigma = 1$.

It is also worth noting that $1/(n-1)$ is the minimal value attainable by σ^2 among all connected graph of order n. This follows by applying the majorization technique to the set

$$S = \left\{\lambda_2(P) \geq \lambda_3(P) \geq \cdots \lambda_n(P) \geq -1 : \sum_{i=2}^{n} \lambda_i(P) = -1\right\}.$$

Indeed, taking into account Remark 2.2.1, the minimal element of the set S is

$\left(\underbrace{\frac{1}{n-1}, \ldots, -\frac{1}{n-1}}_{(n-1)\,times}\right)$, which yields $\sigma = 1/(\sqrt{n-1})$. Notice that this value cor-

responds to the variance of the spectrum of the transition matrix P associated to the complete graph K_n.

Applying now (2.35) with $\beta = \sigma/(\sqrt{n-1})$, we get the following:

Proposition 2.4.1 *[5] For any simple connected graph G*

$$R(G) \geq \frac{n}{d_1}\left[\frac{1}{1 + \frac{\sigma}{\sqrt{n-1}}} + \frac{(n-2)^2}{n-1 - \frac{\sigma}{\sqrt{n-1}}}\right]. \tag{2.39}$$

The next proposition contributes to show that the new bound (2.39) always performs better than (2.30) except in the case where $G = K_n$ for which the two bounds coincide.

Proposition 2.4.2 *[5] Let G be a simple connected graph on n vertices, with $n \geq 3$. The lower bound of $R(G)$ in (2.39) is an increasing function of σ for $1/(\sqrt{n-1}) \leq \sigma < 1$, where the equality in the left side holds if and only if $G = K_n$.*

Notice that, due to (2.13), the new bound (2.39) always performs better than (2.30). Indeed, for the complete graph K_n, the bounds are equal, while for all other graphs, since $\sigma > 1/(\sqrt{n-1})$ by Proposition 2.4.2, the minimal element of the set S^1 majorizes the minimal element of the set S_0 and the bound improves.

Upper Bounds

Taking into account the domain of the function f in (2.29), to get an upper bound, we must obtain a maximal element with non null components. To this aim, let us consider the set

$$S_\beta = \left\{\mathbf{v} \in \mathbb{R}^{n-1} : \sum_{i=1}^{n-1} v_i = n, \; 0 < \beta \leq v_{n-1} \leq v_{n-2} \leq \cdots \leq v_1 \leq 2\right\}.$$

For a $d-$regular graph, Palacios and Renom [57] found the following upper bound where, for simplicity, we write $\lambda_2(P) = \lambda_2$:

$$R(G) \leq \frac{n(n-1)}{d(1-\lambda_2)}. \tag{2.40}$$

The quantity $(1 - \lambda_2)$ is known as *spectral gap*. It is noteworthy to underline that the bound (2.40) holds in general, for $d = d_n$, as can be seen from the r.h.s. (2.28).

By applying our procedure, it is possible to get an upper bound in terms of the spectral gap. Taking $1 - \lambda_2 = v_{n-1}$, let us consider the set

$$S_{\lambda_2} = \left\{ \mathbf{v} \in \mathbb{R}^{n-2} : \sum_{i=1}^{n-2} v_i = (n-1+\lambda_2), \ 0 < 1 - \lambda_2 \leq v_{n-2} \leq \cdots \leq v_1 \leq 2 \right\}.$$

From Corollary 2.3.2 the maximal element of S_{λ_2} is given by

$$\left(\underbrace{2, 2, \ldots, 2}_{k}, \theta, \underbrace{1 - \lambda_2, 1 - \lambda_2, \ldots, 1 - \lambda_2}_{n-k-3} \right)$$

where

$$k = \left\lfloor \frac{\lambda_2(n-1)+1}{\lambda_2+1} \right\rfloor \quad \text{and} \quad \theta = \lambda_2(n-k-2) - k + 2.$$

Using now (2.28), we get the following:

Proposition 2.4.3 *[5] For any simple connected graph G, we have*

$$R(G) \leq \frac{n}{d_n} \left(\frac{n-k-2}{1-\lambda_2} + \frac{k}{2} + \frac{1}{\theta} \right). \tag{2.41}$$

In Section 2.5, we will consider some examples of particular graphs whose spectral gap is well known.

The Kirchhoff index spawned a family of resistance indices such as the *multiplicative-degree-Kirchhoff index and the additive- degree-Kirchhoff index.* Here is another brief illustration of the majorization method applied to these indices.

2.4.4.2 Multiplicative-Degree-Kirchhoff Index

The multiplicative-degree-Kirchhoff index, proposed by Chen and Zhang [15], is defined as

$$R^*(G) = \sum_{i<j} d_i d_j R_{ij}.$$

This index was looked at in [59], where the following expression in terms of the eigenvalues of the transition matrix P was given:

$$R^*(G) = 2|E| \sum_{j=2}^{n} \frac{1}{1 - \lambda_j(P)}. \tag{2.42}$$

Furthermore, it was shown that

$$R^*(G) \geq \frac{2|E|(n-1)^2}{n}, \tag{2.43}$$

which is basically bound (2.30), after replacing n/d_1 with $2|E|$. An upper bound of order n^5 for this index that is attained (up to the constant of the leading term) by the barbell graph was provided in [59] also. With electrical network techniques, the lower bound was improved in [58] to

$$R^*(G) \geq 2|E| \left(n - 2 + \frac{1}{d_1 + 1} \right). \tag{2.44}$$

It is clear, by looking at the expression (2.42), that we can obtain new upper and lower bounds for $R^*(G)$ by using the bounds on $R(G)$. Among these, we mention explicitly the following results related to (2.39).

Proposition 2.4.4 *[5] For any simple connected graph G, we have*

$$R^*(G) \geq 2|E| \left[\frac{1}{1 + \frac{\sigma}{\sqrt{n-1}}} + \frac{(n-2)^2}{n - 1 - \frac{\sigma}{\sqrt{n-1}}} \right]. \tag{2.45}$$

This bound improves (2.44) if G has at least one vertex with degree $n - 1$.

The only thing left to show is that (2.45) improves (2.44) under the given condition, which is clear because in that case, (2.44) becomes (2.43), which is always less than (2.45), by the arguments after Proposition 2.4.2.

2.4.4.3   Additive-Degree-Kirchhoff Index

Gutman et al. defined in [26] the *additive-degree-Kirchhoff index* as

$$R^+(G) = \sum_{i<j} (d_i + d_j) R_{ij}, \tag{2.46}$$

and worked on the identification of graphs with lowest such degree among unicyclic graphs.

The additive-degree-Kirchhoff index is motivated by the degree distance of a graph as defined by Dobrynin and Kochatova [22]. It coincides with the additive-degree-Kirchhoff index whenever the graph is a tree, and the former is always greater than or equal to the latter, because $d(i,j) \geq R_{ij}$ holds for any i,j in any graph G (see [22] for other details).

Recently, Palacios showed in [56], using Markov chain theory, that for any graph G,

$$R^+(G) \geq 2(N-1)^2, \tag{2.47}$$

and the lower bound is attained by the complete graph. Also, in [56], it was shown that for any G

$$R^+(G) \leq \frac{1}{3}(N^4 - N^3 - N^2 + N),$$

and it was conjectured that the maximum of $R^+(G)$ over all graphs is attained by the $(1/3, 1/3, 1/3)$ barbell graph, which consists of two complete graphs on $N/3$ vertices united by a path of length $N/3$, and for which $R^+(G) \sim (2/27)N^4$.

We show next how majorization can be applied to bound the Additive-degree-Kirchhoff index. This approach can be pursued if we can identify a set of variables with constant sum and a Schur-convex function f to be optimized on the set S of these variables. In this case, we know that the global minimum (maximum) of f is attained at the minimum (maximum) element of the set S with respect to the majorization order. For the reader's convenience, we insert also the proof of the result.

Theorem 2.4.4 *[7] For any graph G with degree sequence $d_1 \leq d_2 \leq \cdots \leq d_N$, let*

$$\sum_{i<j} \frac{d_j}{d_i} = H. \tag{2.48}$$

Then

$$R^+(G) \geq N(N-3) + H + \left[\frac{N(N-1)}{2}\right]^2 \frac{1}{H}. \tag{2.49}$$

Proof. Let us consider the $(N(N-1))/2$ variables $x_{ij} = d_j/d_i$, with $i < j$. The function

$$f(x_{12}, x_{13}, \ldots, x_{(N-1)N}) = \sum_{i<j} \left(\frac{d_i}{d_j} + \frac{d_j}{d_i}\right) = \sum_{i=1}^{N-1} \sum_{j=i+1}^{N} \left(x_{ij} + \frac{1}{x_{ij}}\right)$$

is Schur-convex in the variables x_{ij}. The minimal element of the set

$$\Sigma_H = \left\{\mathbf{w} \in \mathbb{R}^{N(N-1)/2} : w_1 \geq w_2 \geq \cdots w_{N(N-1)/2} \geq 0, \sum_i w_i = H\right\}$$

with respect to the majorization order is $\frac{2H}{N(N-1)}\mathbf{s}^{N(N-1)/2}$, where $\mathbf{s}$ is the unit vector of length $N(N-1)/2$. Thus, we get the lower bound:

$$\sum_{i=1}^{N-1} \sum_{j=i+1}^{N} \left(x_{ij} + \frac{1}{x_{ij}}\right) \geq H + \left[\frac{N(N-1)}{2}\right]^2 \frac{1}{H}. \tag{2.50}$$

By effective resistance inequalities, it has been proved in [7] that

$$R^+(G) \geq N(N-3) + \sum_{i<j} \left(\frac{d_i}{d_j} + \frac{d_j}{d_i} \right)$$

and by (2.50), we obtain the expected bound.   □

Remark 2.4.1 *The majorization technique would work just as fine if we considered the $N(N-1)/2$ variables $\dfrac{1}{x_{ij}} = \dfrac{d_i}{d_j}$ and the invariant quantity*

$$\sum_{i<j} \frac{d_i}{d_j} = H^*. \tag{2.51}$$

Following the same steps as given earlier, we have another lower bound

$$\sum_{i=1}^{N-1} \sum_{j=i+1}^{N} \left(x_{ij} + \frac{1}{x_{ij}} \right) \geq H^* + \left[\frac{N(N-1)}{2} \right]^2 \frac{1}{H^*}. \tag{2.52}$$

Except for d-regular graphs for which the bounds (2.50) and (2.52) coincide, because $H = H^*$, in all other cases, the first bound is always better than the second one. In fact, by means of the harmonic mean–arithmetic mean inequality, we get

$$H^* \cdot H \geq \left[\frac{N(N-1)}{2} \right]^2$$

If G is not a d-regular graph, the inequality $H > \frac{N(N-1)}{2} > H^*$ holds. Multiplying both sides of the last inequality by $(H - H^*)$ yields

$$(H - H^*) \cdot H \cdot H^* \geq (H - H^*) \left[\frac{N(N-1)}{2} \right]^2.$$

It follows that

$$(H - H^*) \geq \left(\frac{H - H^*}{H \cdot H^*} \right) \left[\frac{N(N-1)}{2} \right]^2 = \left(\frac{1}{H^*} - \frac{1}{H} \right) \left[\frac{N(N-1)}{2} \right]^2.$$

Rearranging this inequality, we conclude that the lower bound in (2.50) is always better than the one in (2.52).

The usefulness of (2.49) is limited by the computation of the graph invariant H. We will list in Section 2.5 some examples of graphs for which this computation can be easily handled and compare (2.49) with the other bounds.

Majorization is also the main argument in yet another possible approach for obtaining lower bounds. In reference [56], it was shown that the following relationship between the additive- and multiplicative-degree-Kirchhoff indices holds:

$$R^+(G) = \frac{N}{2|E|} R^*(G) + \sum_{j=1}^{N} \sum_{i \neq j} \pi_i E_i T_j, \tag{2.53}$$

where $E_i T_j$ is the expected value of the number of steps T_j that the random walk on G, started from vertex i, takes to reach vertex j. We recall that this random walk moves from a vertex v to any neighboring vertex w with uniform probabilities $p(v, w) = 1/d_v$ and that the $N \times N$ matrix $P = [p(v, w)]$ of transition probabilities has a unique probabilistic left eigenvector $\pi = (\pi_i)$ (the stationary distribution), which is present in the summation in (2.53), and a spectrum $1 = \lambda_1 > \lambda_2 \geq \lambda_3 \geq \cdots \geq \lambda_N \geq -1$ in terms of which $R^*(G)$ can be expressed as in (2.42) [59]. With the preceding remarks and notation in [7], the authors proved the following result:

Theorem 2.4.5 *For any graph G,*

$$R^+(G) \geq N \left[\frac{1}{1 + \frac{\sigma}{\sqrt{N-1}}} + \frac{(N-2)^2}{N - 1 - \frac{\sigma}{\sqrt{N-1}}} \right] + (N-1)^2. \qquad (2.54)$$

We remark that we recover the universal bound (2.47) for the complete graph, for which $\sigma = 1/\sqrt{N-1}$, and for all other graphs, the bound is better than the universal one (2.47) (for details, see [5]).

Finally, we turn to the analysis of some significant upper bounds that can be obtained by combining ideas from Markov chains and majorization. These bounds can be suitably expressed in terms of the spectral gap.

Recall that from (2.53) and subsequent comments, we have

$$R^+(G) = \frac{N}{2|E|} R^*(G) + \sum_{j=1}^{N} \sum_{i=1}^{N} \pi_i E_i T_j = N \sum_{i=2}^{N} \frac{1}{1 - \lambda_i} + \sum_{j=1}^{N} \sum_{i=1}^{N} \pi_i E_i T_j \qquad (2.55)$$

We want to find an upper bound for the summation with the hitting times in (2.55), for which we use some Markov chain theory found in reference [49], specifically

$$\sum_i \pi_i E_i T_j = \frac{1}{\pi_j} \sum_{k=2}^{N} \frac{1}{1 - \lambda_k} v_{kj}^2, \qquad (2.56)$$

where v_{kj} is the jth component of the eigenvector v_k associated to the eigenvalue λ_k (the vectors v_k can be chosen to be orthonormal), and

$$\sum_{k=2}^{N} v_{kj}^2 = 1 - \pi_j.$$

It is clear that (2.56) can be bounded as follows:

$$\frac{1}{\pi_j} \sum_{k=2}^{N} \frac{1}{1 - \lambda_k} v_{kj}^2 \leq \frac{1}{(1 - \lambda_2)\pi_j} \sum_{k=2}^{N} v_{kj}^2 = \frac{1}{1 - \lambda_2} \frac{1 - \pi_j}{\pi_j}.$$

And so the sum of expected hitting times can be bounded as

$$\sum_{j=1}^{N} \sum_{i=1}^{N} \pi_i E_i T_j \leq \frac{1}{1 - \lambda_2} \sum_j \frac{1 - \pi_j}{\pi_j} = \frac{1}{1 - \lambda_2} \left(2|E| \sum_j \frac{1}{d_j} - N \right).$$

Now use in (2.55) the upper bounds in [5], Section 3.2 for $\sum_{i=2}^{N} 1/1 - \lambda_i$, to obtain the following corollaries:

Corollary 2.4.1 *[7] For any G, we have*

$$R^+(G) \leq N \left(\frac{N-k-2}{1-\lambda_2} + \frac{k}{2} + \frac{1}{\theta} \right) + \frac{1}{1-\lambda_2} \left(2|E| \sum_j \frac{1}{d_j} - N \right), \qquad (2.57)$$

where

$$k = \left\lfloor \frac{\lambda_2(N-1)+1}{\lambda_2+1} \right\rfloor$$
$$\theta = \lambda_2(N-k-2) - k + 2$$

Corollary 2.4.2 *[7] For any bipartite G, we have*

$$R^+(G) \leq N \left(\frac{1}{2} + \frac{N-k-3}{1-\lambda_2} + \frac{k}{2} + \frac{1}{\theta} \right) + \frac{1}{1-\lambda_2} \left(2|E| \sum_j \frac{1}{d_j} - N \right), \qquad (2.58)$$

where k and θ are defined earlier.

For the N-star graph, we have that $\lambda_2 = 0$, $k = 1$, and $\theta = 1$ and therefore the bound (2.58) becomes $3N^2 - 7N + 4$ and the actual value is attained. This can be extended to the complete bipartite graph $K_{r,s}$, for arbitrary r, s, for which the bound (2.58) becomes

$$3r^2 + 3s^2 + 2rs - 3r - 3s, \qquad (2.59)$$

whose order is always N^2 and improves the bound $2|E|(N-1)D = 4rs(r+s-1)$. The smallest value of (2.59) occurs for $r = s = N/2$, where it takes the value $N(2N-3)$, which is equal to the actual value $N(2N-3)$ of $R^+(G)$.

2.5   Numerical Results

In this section, we present some numerical examples that illustrate the majorization technique developed throughout the chapter. The numerical results obtained by our new method have been compared to those of the literature. We restrict our attention to computing bounds of some particular graphs. For more details, we refer the reader to [4–9].

2.5.1   General Randić Index

Example 2.1 Consider a tree T with the following degree sequence:

$$\pi = (5, 3, 3, 3, 3, 3, 1, 1, 1, 1, 1, 1, 1, 1, 1, 1).$$

TABLE 2.2: Lower and Upper Bounds for $R_\alpha(T)$

α	Ref./Formula	Lower Bound	Upper Bound
$-\frac{1}{2}$	[42]	5.89	7.49
	(2.18)	6.77	7.06
-1	[42]	1	4.89
	(2.18)	3.2	3.67

Table 2.2 shows that our bounds (2.18) are sharper than bounds given by Li and Shi [42], Theorems 2.6 and 2.24 for the case $\alpha = -(1/2)$ and Theorem 3.7 for the case $\alpha = -1$.

Example 2.2 Let us consider a unicyclic graph G, that is, a graph with $n = m$, with the degree sequence

$$\pi = (3, 3, 3, 3, 2, 2, 2, 2, 2, 1, 1, 1, 1).$$

Being (2.17) satisfied, by Remark 2.3.2, (2.18) gives

$$64 \leq M_2(G) \leq 74.$$

The comparison (see Tables 2.3 and 2.4) with bounds in [12,18,28,50,70] shows that our bounds always perform better. Indeed, we obtain the following:

Example 2.3 Consider the graphs G and H with degree sequences $\pi_1 = (3, 2, 2, 1)$ and $\pi_2 = (3, 3, 3, 3, 2, 1, 1)$, respectively, as in Examples 2.2 and 2.3 in [28]. Besides

TABLE 2.3: Lower and Upper Bounds for $M_2(G)$ (Example 2.2)

Ref./Formula	Lower Bound	Upper Bound
(2.18)	64	74
[12]	—	277.9
[18]	—	182
[28]	61.462	77
[50]	28	76
[70]	64	92

TABLE 2.4: Lower and Upper Bounds for $M_2(G)$ (Example 2.3)

Ref./Formula	Lower Bound	Upper Bound
(2.18)	19	20
[12]	—	22.5
[18]	—	20
[28]	18.5	20
[50]	18	22
[70]	19	19

TABLE 2.5:   Lower and Upper Bounds for $M_2(H)$

Ref./Formula	Lower Bound	Upper Bound
(2.18)	54	58
[12]	—	99.75
[18]	—	80
[28]	51.25	58
[50]	40	59
[74]	50	68

the bounds discussed in [28], we add the comparison with those in [18,70,74]. Observing that G is a unicyclic graph ($m = n$) and H is a bicyclic graph ($m = n + 1$), both with pendant vertices, bounds in [70,74] can also be, respectively, properly applied. Computing bounds for $M_2(G)$, we have the following:

Our bounds are sharper than [12,28,50]. The best one is provided by [70] as it has been specifically constructed for this class of graph.

Computing bounds for $M_2(H)$, we have the following:

Note that our bounds perform better than all the others and in particular better than [74], which is properly designed for bicyclic graphs as H is (Table 2.5).

2.5.2   Generalized Sum-Connectivity Index

The next examples show how the majorization technique can provide better bounds.

Example 2.4 Let us consider a graph G with degree sequence $\pi = (3, 2, 2, 1)$.

We compare in Table 2.6 our results obtained by using formula (21) and Corollary 7 with those provided by Du et al. [23], Theorem 1, in case when n is even and $\alpha = -5$.

Example 2.5 Let us consider a tree T with the following degree sequence:

$$\pi = (4, 4, 4, 3, 1, 1, 1, 1, 1, 1, 1, 1, 1).$$

The comparison between our results and [23] is shown in Table 2.7 when n is odd and $\alpha = -5$.

TABLE 2.6:   Upper Bounds for $\chi_\alpha(G)$ with n Even

Ref./Formula	Upper Bound
[23]	9.21×10^{-3}
(2.20)	5.08×10^{-3}

TABLE 2.7:   Upper Bounds for $\chi_\alpha(T)$ with n Odd

Ref./Formula	Upper Bound
[23]	2.49×10^{-2}
(2.20)	4.94×10^{-3}

2.5.3 Kirchhoff Index

2.5.3.1 Lower Bounds for d-Regular Graphs

For a d-regular graph K_d, we have further information about the eigenvalues of the transition matrix P. Indeed, by the fact that $\lambda_1(L) \geq 1 + d$ and (2.27), we get $\lambda_n(P) \leq -\dfrac{1}{d}$, which is tighter than the bound $\lambda_n(P) \leq -\dfrac{\sigma}{\sqrt{n-1}} = -\dfrac{1}{\sqrt{d(n-1)}}$.

Applying (2.35), we have

$$R(G) \geq \frac{n}{d} \left[\frac{1}{1 + \dfrac{1}{d}} + \frac{(n-2)^2}{n - 1 - \dfrac{1}{d}} \right] = \frac{n}{1+d} + \frac{n(n-2)^2}{nd - 1 - d}. \tag{2.60}$$

Notice that (2.60) is equal to (2) in [57], which corresponds to bound (1) in [78] for the particular case of d-regular graphs.

Bound (2.60) can be strengthened if some tighter bounds on λ_1 are available. In [51], Corollary 9, it is shown that for a d-regular graph of diameter D,

$$\lambda_1(L) \geq d + \frac{2D}{D+1}.$$

For $D > 1$, this bound is tighter than $\lambda_1(L) \geq d+1$, while the case $D = 1$ corresponds to the complete graph. For all graphs with a known diameter $D > 1$, we can thus improve bound (2.60) with the following:

$$R(G) \geq \frac{n}{d} \left[\frac{1}{1 + \frac{2D}{d(D+1)}} + \frac{(n-2)^2}{n - 1 - \frac{2D}{d(D+1)}} \right] \tag{2.61}$$

The n-cycle and the d-cube are particular type of d-regular graphs. The first is a 2-regular graph, with $D = n/2$ for n even and $D = (n-1)/2$ for n odd, while the second one is a d-regular graph with $n = 2^d$ and $D = d$.

For these two classes of d-regular graphs in the following tables, we summarize the lower bounds given by (2.60) and (2.61) (Tables 2.8 and 2.9):

TABLE 2.8:   Lower Bounds for the n-Cycle

n	Formula (2.60)	Formula (2.61)
3	2	2
4	4.53	4.63
5	8.09	8.25
6	12.67	13.01
7	18.24	18.67
8	24.82	25.45
50	1204.30	1211.78
75	2743.88	2755.51

TABLE 2.9:   Lower Bounds for the d-Cube

d	Formula (2.60)	Formula (2.61)
3	16.4	16.54
4	56.35	56.56
5	192.34	192.62

TABLE 2.10:   Upper Bounds for the n-Cycle

n	Formula (2.40)	Formula (2.41)
3	2	2
4	6	5
5	14.47	10.03
6	30	18
7	55.77	33.26
8	95.60	50.53

TABLE 2.11:   Upper Bound for the d-Cube

d	Formula (2.40)	Formula (2.41)
2	6	5
3	28	20.66
4	120	84.66
5	496	334.5

2.5.3.2   Upper Bounds

In what follows, we consider some examples of particular graphs whose spectral gap is well known. It is worth noting that, due to Proposition 2.5, our bounds always perform equal or better than bound (2.40) (Tables 2.10 and 2.11).

1. **The n-cycle**
   The n-cycle graph is a 2-regular graph whose second largest eigenvalue is $\lambda_2 = \cos\left(\dfrac{2\pi}{n}\right)$.

2. **The d-cube**
   The second largest eigenvalue of the d-cube is $\lambda_2 = \dfrac{d-2}{d}$.

2.5.4   Additive Kirchhoff Index

We report some examples when H can be easily computed.

2.5.4.1 d-Regular Graph

For d-regular graphs, we have $H = N(N-1)/2$. The lower bound (2.49) becomes $N(N-1)$, which is worse than bound (2.47).

TABLE 2.12:   Lower Bound for $R^+(G)$
(Example 2.6)

Formula	Lower Bound
(2.47)	338
(2.54)	338.03
(2.49)	359.64

2.5.4.2 (a,b)-Semiregular Graph

Let us consider a semiregular graph G that has N_1 vertices with degree a and N_2 vertices with degree b, $a < b$, $N = N_1 + N_2$. Then $H = \frac{N(N-1)}{2} + \left(\frac{b}{a} - 1\right) N_1 N_2$.

We deal with two examples: (1) a semiregular bipartite graph and (2) a semiregular not bipartite graph.

Example 2.6 Let us consider a semiregular bipartite graph with $N_1=10$ vertices with degree $a = 4$ and $N_2 = 4$ vertices with degree $b = 10$. For this graph, we have $H = 151$, $\sigma = 0.47$, which imply showing that the bound (2.49) performs better than the others (Table 2.12).

Example 2.7 Let us take a semiregular graph G on N vertices (N even ≥ 8) that is the union of a complete $K_{N/2}$ and a $N/2$-cycle such that vertex i of the cycle is linked to vertex i of the complete graph with a single edge, for $1 \leq i \leq N/2$.

This graph has $N_1 = N_2 = N/2$, $a = 3$, $b = N/2$, thus, $H = \frac{1}{24}N\left(6N + N^2 - 12\right)$. By (2.49), we get

$$R^+(G) \geq \frac{N\left(228N^2 - 1152N + 36N^3 + N^4 + 1152\right)}{24\left(6N + N^2 - 12\right)} \qquad (2.62)$$

By Calculus, it is easy to show that , for $N > 8$, the bound (2.62) is better than (2.47).

Table 2.13 give a comparison between all lower bounds applicable to this example, for $N = 20$:

The bound (2.62) performs always better than (2.54), which in turn improves (2.47).

2.5.4.3 Full Binary Tree of Depth $d > 1$

We consider a full binary tree T of depth $d > 1$, which has $N_1 = 2^d$ vertices of degree 1, one vertex (the root) of degree 2, and $N_2 = 2^d - 2$ vertices of degree 3.

TABLE 2.13:   Lower Bound for $R^+(G)$
(Example 2.7)

Formula	Lower Bound
(2.47)	722
(2.54)	722.001
(2.62)	848.61

TABLE 2.14:   Lower Bounds for $R^+(T)$

Formula	Lower
(2.47)	392
(2.54)	392.14
(2.49)	406

Then $H = \dfrac{N(N-1)}{2} + 2N_1 + \frac{3}{2}N_2 + 2N_1N_2$. Taking $d = 3$, we obtain the results summarized in the following table, which shows that our new bounds are better than the universal one (2.47) (Table 2.14):

References

1. D.L. Alderson and L. Li. Diversity of graphs with highly variable connectivity. *Physical Review E*, 75:1–11, 2007.

2. A.B. Atkinson. On the measurement of inequality. *Journal of Economic Theory*, 2:224–263, 1970.

3. E. Bendito, A. Carmona, A.M. Encinas, and J.M. Gesto. A formula for the Kirchhoff index. *International Journal of Quantum Chemistry*, 108:1200–1206, 2008.

4. M. Bianchi, A. Cornaro, J.L. Palacios, and A. Torriero. Bounding the sum of powers of normalized Laplacian eigenvalues of graphs through majorization methods. *MATCH: Communications in Mathematical and in Computer Chemistry*, 70(2):707–716, 2013.

5. M. Bianchi, A. Cornaro, J.L. Palacios, and A. Torriero. Bounds for the Kirchhoff index via majorization techniques. *Journal of Mathematical Chemistry*, 51(2):569–587, 2013.

6. M. Bianchi, A. Cornaro, J.L. Palacios, and A. Torriero. New bounds of degree-based topological indices for some classes of c-cyclic graphs. *arXiv:1311.5691 [math.CO]*, 2013.

7. M. Bianchi, A. Cornaro, J.L. Palacios, and A. Torriero. New upper and lower bounds for the additive degree-Kirchhoff index. *Croatica Chemica Acta*, 86(4):363–370, 2013.

8. M. Bianchi, A. Cornaro, and A. Torriero. A majorization method for localizing graph topological indices. *Discrete Applied Mathematics*, 161:2731–2739, 2013.

9. M. Bianchi, A. Cornaro, and A. Torriero. Majorization under constraints and bounds of the second Zagreb index. *Mathematical Inequalities and Applications*, 16(2):329–347, 2013.

10. M. Bianchi and A. Torriero. Some localization theorems using a majorization technique. *Journal of Inequalities and Applications*, 5:443–446, 2000.

11. B. Bollobás. *Graph Theory: An Introductory Course.* Springer Verlag, New York, 1990.

12. B. Bollobás and P. Erdös. Graphs of extremal weights. *Ars Combinatoria*, 50:225–233, 1998.

13. S.B. Bozkurt and D. Bozkurt. On the sum of powers of normalized Laplacian eigenvalues of graphs. *MATCH: Communications in Mathematical and in Computer Chemistry*, 68:817–930, 2012.

14. A.E. Brouwer and W.H. Haemers. A lower bound for the Laplacian eigenvalue of a graph—Proof of a conjecture by Guo. *Linear Algebra and Its Applications*, 429:2131–2135, 2008.

15. H. Chen and F. Zhang. Resistance distance and the normalized Laplacian spectrum. *Discrete Applied Mathematics*, 155:654–661, 2007.

16. P. Dankelmann, A. Hellwig, and L. Volkmann. Inverse degree and edge-connectivity. *Discrete Mathematics*, 309:2943–2947, 2009.

17. P. Dankelmann, H.C. Swart, and P. Van den Berg. Diameter and inverse degree. *Discrete Mathematics*, 308:670–673, 2008.

18. K.Ch. Das, I. Gutman, and B. Zhou. New upper bounds on Zagreb indices. *Journal of Mathematical Chemistry*, 46:514–521, 2009.

19. M. Dehmer, S. Borgert, and F. Emmert-Streib. Entropy bounds for molecular hierarchical networks. *PLoS ONE*, 3(8):e3079, 2008.

20. M. Dehmer, F. Emmert-Streib, and J. Kilian. A similarity measure for graphs with low computational complexity. *Applied Mathematics and Computation*, 182(1):447–459, 2006.

21. M. Dehmer and V. Kraus. On extremal properties of graph entropies. *MATCH: Communications in Mathematical and in Computer Chemistry*, 68(3):889–912, 2012.

22. A.A. Dobrynin and A.A. Kochatova. Degree distance of a graph: A degree analogue of the Wiener index. *Journal of Chemical Information and Computer Sciences*, 34:1082–1086, 1994.

23. Z. Du, B. Zhou, and N. Trinajstic. On the general sum-connectivity index of trees. *Applied Mathematics Letters*, 24:402–405, 2011.

24. M. Eliasi. A simple approach to order the multiplicative Zagreb indices of connected graphs. *Transactions on Combinatorics*, 1(4):17–24, 2012.

25. P. Erdős and T. Gallai. Graphs with prescribed degrees of nodes. *Hungarian Matematikai Lapok*, 11:64–274, 1960.

26. L. Feng, I. Gutman, and L. Yu. Degree resistance distance of unicyclic graphs. *Transactions on Combinatorics*, 1:27–40, 2010.

27. X. Gao, Y. Luo, and W. Liu. Resistance distances and the Kirchhoff index in Cayley graphs. *Discrete Applied Mathematics*, 159:2050–2057, 2011.

28. R. Grassi, S. Stefani, and A. Torriero. Extremal properties of graphs and eigen-centrality in trees with a given degree sequence. *The Journal of Mathematical Sociology*, 34(2):115–135, 2010.

29. I. Gutman. Multiplicative Zagreb indices of trees. *Bulletin of International Mathematical Virtual Institute*, 1:13–19, 2011.

30. I. Gutman. The energy of a graph: Old and new results. In: A. Betten, A. Kohnert, R. Laue, and A. Wasserman (eds.). *Algebraic Combinatorics and Applications*, pp. 196–211. Springer, Berlin, Germany, 2011.

31. I. Gutman and B. Mohar. The quasi-Wiener and the Kirchhoff indices coincide. *Journal of Chemical Information and Computer Sciences*, 36:982–985, 1996.

32. I. Gutman, B. Ruščić, N. Trinajstić, and C. F. Wilcox. Graph theory and molecular orbitals. XII. Acyclic polyenes. *Journal of Chemical Physics*, 62:3399–3405, 1975.

33. I. Gutman and N. Trinajstić. Graph theory and molecular orbits. Total π-electron energy of alternant hydrocarbons. *Chemical Physics Letters*, 17:535–538, 1972.

34. G.H. Hardy, E. Littlewood, and G. Polya. Some simple inequalities satisfied by convex functions. *Messanger of Mathematics*, 58:145–152, 1929.

35. J. Karamata. Sur une inégalité rélative aux fonctions convexes (in French). *Publ. Math. Univ. Belgrade*, 1:145–148, 1932.

36. D.J. Klein. Prolegomenon on partial orderings in chemistry. *MATCH: Communications in Mathematical and in Computer Chemistry*, 42:7–21, 2000.

37. D.J. Klein and D. Babić. Partial orderings in chemistry. *Journal of Chemical Information and Computer Sciences*, 37:656–671, 1997.

38. D.J. Klein and M. Randić. Resistance distance. *Journal of Mathematical Chemistry*, 12:81, 1993.

39. J.H. Koolen and V. Moulton. Maximal energy graphs. *Advances in Applied Mathematics*, 26:47–52, 2001.

40. J.H. Koolen and V. Moulton. Maximal energy bipartite graphs. *Graphs and Combinatorics*, 19:131–135, 2003.

41. L. Li, D. Alderson, J.C. Doyle, and W. Willinger. Supplemental material: The $S(G)$ metric and assortativity. *Internet Mathematics*, 2(4):1–6, 2005.

42. X. Li and Y.T. Shi. A Survey on the Randić index. *MATCH: Communications in Mathematical and in Computer Chemistry*, 59:127–156, 2008.

43. X. Li and Y.T. Shi. On the diameter and inverse degree. *Ars Combinatoria*, 101:481–487, 2011.

44. X. Li and J. Zheng. A unified approach to the extremal trees for different indices. *MATCH: Communications in Mathematical and in Computer Chemistry*, 54:195–208, 2004.

45. H. Liu, M. Lu, and F. Tian. Some upper bounds for the energy of graphs. *Journal of Mathematical Chemistry*, 41(1):45–57, 2007.

46. M. Liu. A simple approach to order the first Zagreb indices of connected graphs. *MATCH: Communications in Mathematical and in Computer Chemistry*, 63:425–432, 2010.

47. M. Liu and B. Liu. A note on sum of powers of the Laplacian eigenvalues of a graphs. *Applied Mathematics Letters*, 24:249–252, 2011.

48. L. Lovász. *Combinatorial Problems and Exercises*. North-Holland, Amsterdam, the Netherlands, 1993.

49. L. Lovász. Random walks on graphs: A survey. *Bolyai Society Mathematical Studies, Combinatorics, Paul Erdős is Eighty*, 2:1–46, 1993.

50. M. Lu, H. Liu, and F. Tian. The connectivity index. *MATCH: Communications in Mathematical and in Computer Chemistry*, 51:149–154, 2004.

51. M. Lu, F. Tian, and L. Zhang. Lower bounds of the Laplacian spectrum of graphs based on diameters. *Linear Algebra and Its Applications*, 420:400–406, 2007.

52. A.W. Marshall, I. Olkin, and B. Arnold. *Inequalities: Theory of Majorization and Its Applications*. Springer, New York, 2011.

53. M.E.J. Newman. Assortative mixing in networks. *Physical Review Letters*, 89(20):1–4, 2002.

54. S. Nikolić, G. Kovačević, A. Miličević, and N. Trinajstić. The Zagreb indices 30 years after. *Croatica Chemica Acta*, 76:113–124, 2003.

55. J.L. Palacios. Closed form formulas for Kirchhoff index. *International Journal of Quantum Chemistry*, 81:135–140, 2001.

56. J.L. Palacios. Upper and lower bounds for the additive degree-Kirchhoff index. *MATCH: Communications in Mathematical and in Computer Chemistry*, 70(2):651–655, 2013.

57. J.L. Palacios and J.M. Renom. Bounds for the Kirchhoff index of regular graphs via the spectra of their random walks. *International Journal of Quantum Chemistry*, 110:1637–1641, 2010.

58. J.L. Palacios and J.M. Renom. Another look at the degree-Kirchhoff index. *International Journal of Quantum Chemistry*, 111:3453–3455, 2011.

59. J.L. Palacios and J.M. Renom. Broder and Karlin's formula for hitting times and the Kirchhoff index. *International Journal of Quantum Chemistry*, 111:35–39, 2011.

60. C.T. Pan. A vector majorization method for solving a nonlinear programming problem. *Linear Algebra and Its Applications*, 119:129–139, 1989.

61. M. Randić. On characterization of molecular branching. *Journal of the American Chemical Society*, 97:6609–6615, 1975.

62. E. Ruch and I. Gutman. The branching extent of graphs. *Journal Combinatorics, Informatiion & System Sciences*, 4:285–295, 1979.

63. I. Schur. Über ein Klass von Mittelbindungen mit Anwendungen auf Determinantentheorie. *Sitzer. Berl. Math. Ges.*, 22:9–20, 1923.

64. P. Tarazaga. Eigenvalue estimate for symmetric matrices. *Linear Algebra and Its Applications*, 135:171–179, 1990.

65. P. Tarazaga. More estimate for eigenvalues and singular values. *Linear Algebra and Its Applications*, 149:97–110, 1991.

66. R. Todeschini and V. Consonni. *Handbook of Molecular Descriptor*. Wiley-VHC, Weinheim, Germany, 2000.

67. H. Wang, H. Hua, and D. Wang. Cacti with minimum, second-minimum, and third-minimum Kirchhoff indices. *Mathematical Communications*, 15:347–358, 2010.

68. R.J. Wilson. *Introduction to Graph Theory*. Addison-Wesley, Essex, U.K., 1996.

69. H. Wolkowicz and G.P.H. Styan. Bounds for eigenvalues using trace. *Linear Algebra and Its Applications*, 29:471–506, 1980.

70. H. Yan, Z. Liu, and H. Liu. Sharp bound for the second Zagreb index of unicyclic graphs. *Journal of Mathematical Chemistry*, 42(3):565–574, 2007.

71. Y. Yang and H. Zhang. Kirchhoff index of linear hexagonal chains. *International Journal of Quantum Chemistry*, 108:503–512, 2008.

72. Y. Yang, H. Zhang, and D. Klein. New Nordhaus-Gaddum-type results for the Kirchhoff index. *Journal of Mathematical Chemistry*, 49:1–12, 2011.

73. S. Zhang, W. Wang, and T.C.E. Cheng. Byciclic graphs with the first three smallest and largest values of the first general Zagreb index. *MATCH: Communications in Mathematical and in Computer Chemistry*, 56:579–592, 2006.

74. Q. Zhao and S. Li. Sharp bounds for the Zagreb indices of bicyclic graphs with k-pendant nodes. *Discrete Applied Mathematics*, 158(17):1953–1962, 2010.

75. B. Zhou. Energy of a graph. *MATCH: Communications in Mathematical and in Computer Chemistry*, 51:111–118, 2004.

76. B. Zhou. On a sum of powers of the Laplacian eigenvalues of a graphs. *Linear Algebra and Its Applications*, 429:2239–2246, 2008.

77. B. Zhou and N. Trinajstić. A note on Kirchhoff index. *Chemical Physics Letters*, 455:120–123, 2008.

78. B. Zhou and N. Trinajstić. On a novel connectivity index. *Journal of Mathematical Chemistry*, 64:1252–1270, 2009.

79. B. Zhou and N. Trinajstić. On resistance-distance and Kirchhoff index. *Journal of Mathematical Chemistry*, 46:283–289, 2009.

80. H.Y. Zhu, D.J. Klein, and I. Lukovits. Extensions of the Wiener number. *Journal of Chemical Information and Computer Sciences*, 36:420–428, 1996.

Chapter 3

Wiener Index of Hexagonal Chains with Segments of Equal Length

Andrey A. Dobrynin

Contents

3.1   Introduction...81
3.2   Hexagonal Chains..83
3.3   Representation of Hexagonal Chains.......................................85
3.4   Linked Hexagonal Chains...86
     3.4.1   Complete Families...87
     3.4.2   Linked Families...87
     3.4.3   Symmetrically Linked Families......................................89
3.5   Hexagonal Chains with Extremal Wiener Index.............................89
3.6   Wiener Index of Linked Chains...90
3.7   Wiener Index of Complete Families...92
3.8   Wiener Index of Linked Families...94
3.9   Wiener Index of Symmetrically Linked Families...........................95
3.10 Expansion of the Wiener Index..98
3.11 Degeneracy of the Wiener Index...102
Acknowledgment..106
References..106

3.1   Introduction

Graphs and their generalizations are successfully used to model structural objects, relations, and process dynamics in chemical, physical, biological, social, and information systems. Methods of graph theory are necessary tools for the analysis of structure of complex networks arising in these fields [7,8]. One of the approaches in studying graph structure is quantifying structural information by various quantitative measures. Usually, a quantitative graph measure is a graph invariant that maps a set of graphs to a set of numbers such that invariant values coincide for isomorphic graphs. In this chapter, we deal with quantitative methods for analyzing structures of molecular graphs that are the standard representation of structure of chemical compounds in organic chemistry. Many invariants belong to the molecular structure descriptors, called topological indices, that are nowadays extensively used in theoretical chemistry for the characterization of molecular complexity and for the design of quantitative structure–property relations and quantitative structure–activity relations including pharmacologic and biological activities [2–5,14,28,44–46,49].

As a graph measure, we will use the Wiener index, which is a well-known distance-based topological index introduced as structural descriptor for acyclic organic molecules by the American physicochemist Harold Wiener in 1947 [50]. It is defined as the sum of distances between all unordered pairs of vertices of an undirected connected graph G with vertex set $V(G)$:

$$W(G) = \sum_{\{u,v\} \subseteq V(G)} d(u, v),$$

where distance $d(u, v)$ is the number of edges in the shortest path connecting vertices u and v in G. The average distance of a graph G with n vertices, $W(G)/\binom{n}{2}$, is used in analysis of large complex networks. The Wiener index and its modifications are extensively used in theoretical chemistry. The bibliography on the Wiener index and its applications can be found in books [3,14,28,29,33,48,49] and reviews [15,21,23, 24,35,36,38,41–45].

In the present paper, we consider the Wiener index for graphs of hexagonal systems that include molecular graphs of benzenoid hydrocarbons. Graphs of hexagonal systems consist of mutually fused hexagons. Since this class of chemical compounds is attracting the great attention of theoretical chemists, the theory of the Wiener index of the respective molecular graphs has been intensively developed in the last four decades. Benzenoid hydrocarbons are important raw materials of the chemical industry (used, for instance, for the production of dyes and plastics) but are also dangerous pollutants [26,27,51]. Progress in physics and chemistry inspires great interest to new classes of molecules consisting of hexagonal rings, which can be regarded as hexagonal networks. As an example, we mention fullerenes, carbon nanotubes, and graphene that have promising properties for applications.

Consider a hexagonal network in the plain shown in the top part of Figure 3.1. Suppose that this network is destroyed into a family of hexagonal chains $\mathcal{G}$ as depicted in the bottom part of Figure 3.1. To characterize quantitatively the destroyed network, we shall use the Wiener index. Since the family of chains $\mathcal{G}$ forms a disconnected graph, it is naturally defined as the Wiener index of $\mathcal{G}$ by the expression

$$W(\mathcal{G}) = \sum_{G \in \mathcal{G}} W(G).$$

If hexagonal chains of $\mathcal{G}$ have arbitrary structures, the quantity $W(\mathcal{G})$ can be obtained by computer calculations only. We are interested in the following question: are there existing families of hexagonal chains for which $W(\mathcal{G})$ can be calculated by simple methods (without computer)? This means that the corresponding calculating formulas for $W(\mathcal{G})$ should not include values of $W(G)$ or similar information for individual chains $G \in \mathcal{G}$. In this case, quantity $W(\mathcal{G})$ does not reflect the property of Wiener index of any particular hexagonal chain of $\mathcal{G}$ but a collective property of sets of such graphs.

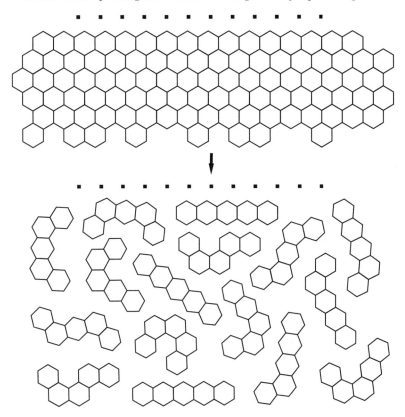

FIGURE 3.1:   Destruction of a hexagonal network.

3.2 Hexagonal Chains

A hexagonal system is a connected plane graph in which every inner face is bounded by hexagon. An inner face with its hexagonal bound is called a *hexagonal ring* (or simply *ring*). Two hexagonal rings are either disjoint or have exactly one common edge (adjacent rings), and no three rings share a common edge. A vertex of a hexagonal system belongs to at most three hexagonal rings. A hexagonal system is called *catacondensed* if it does not possess three hexagonal rings sharing a common vertex. A ring having exactly one adjacent ring is called *terminal*. A catacondensed hexagonal system having exactly two terminal rings is called a *hexagonal chain*. A ring adjacent to exactly two other rings has two vertices of degree 2. If these two vertices are adjacent, then the ring is angularly annelated; if these two vertices are not adjacent, then it is linearly annelated. An example of hexagonal chain G is shown in Figure 3.2. The *linear chain* L_h with h hexagons is the catacondensed system without angularly annelated rings. The *helicene* H_h does not contain linearly annelated rings.

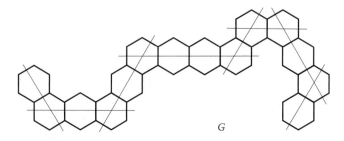

FIGURE 3.2:   A hexagonal chain and its segments.

A *segment* is a maximal subchain in which all rings are linearly annelated. A segment including a terminal hexagon is a *terminal segment*. The number of hexagons in a segment S is called its *length* and is denoted by $\ell(S)$, $2 \leq \ell(S) \leq h$. We say that G consists of the set of segments $S_1, S_2, \ldots, S_n$ with lengths $\ell(S_i) = \ell_i$ for some $n \geq 3$. Since two neighboring segments have always one hexagon in common, the number of hexagons of G is $h = \ell_1 + \ell_2 + \cdots + \ell_n - n + 1$. The hexagonal chain G in Figure 3.2 has eight segments of length 2, 3, 3, 4, 2, 2, 3, 2.

Denote by $\mathcal{G}_{n,\ell}$ the set of all hexagonal chains having n segments of equal length, $\ell_1 = \ell_2 = \cdots = \ell_n = \ell$. Then $h = n(\ell - 1) + 1$, and a condition for the existence of such systems is that the number of segments $n = (h - 1)/(\ell - 1)$ must be an integer. As an illustration, all hexagonal chains having five segments of length 3 are shown in Figure 3.3.

Denote by $\mathcal{A}_{n,\ell}$ and $\mathcal{S}_{n,\ell}$ the sets of all asymmetrical and all symmetrical hexagonal chains from $\mathcal{G}_{n,\ell}$, respectively. Hexagonal chains of $\mathcal{G}_{n,2}$ are known as *fibonacenes* [1]. The name of these chains comes from the fact that the number of perfect matchings of any fibonacene relates with the Fibonacci numbers. This special class of hexagonal chains will be also denoted by $\mathcal{F}_n$. Detailed information about properties of fibonacenes can be found in [1,19,20,26,31]. Repeatedly increasing segments'

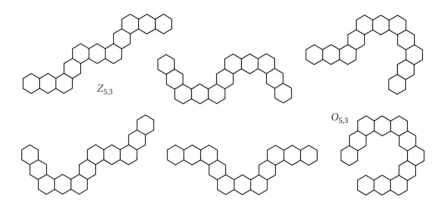

FIGURE 3.3:   All hexagonal chains of $\mathcal{G}_{5,3}$ with five segments of length 3.

length of all graphs of $\mathcal{F}_n$, we get a sequence of bijections between sets of graphs with equal numbers of segments:

$$\mathcal{F}_n = \mathcal{G}_{n,2} \rightarrow \mathcal{G}_{n,3} \rightarrow \mathcal{G}_{n,4} \rightarrow \cdots \mathcal{G}_{n,\ell-1} \rightarrow \mathcal{G}_{n,\ell} \rightarrow \cdots$$

Since the number of fibonacenes is known [1], we have

$$|\mathcal{G}_{n,\ell}| = \begin{cases} 2^{n-3} + 2^{\frac{n-4}{2}}, & \text{if } n \text{ is even} \\ 2^{n-3} + 2^{\frac{n-3}{2}}, & \text{if } n \text{ is odd} \end{cases}$$

$$|\mathcal{A}_{n,\ell}| = \begin{cases} 2^{n-3} - 2^{\frac{n-4}{2}}, & \text{if } n \text{ is even} \\ 2^{n-3} - 2^{\frac{n-3}{2}}, & \text{if } n \text{ is odd} \end{cases}$$

$$|\mathcal{S}_{n,\ell}| = \begin{cases} 2^{\frac{n-2}{2}}, & \text{if } n \text{ is even} \\ 2^{\frac{n-1}{2}}, & \text{if } n \text{ is odd.} \end{cases}$$

Throughout this chapter, n, ℓ, and h always denote the number of segments, length of segments, and the number of hexagonal rings of chains, respectively.

3.3 Representation of Hexagonal Chains

A way of segment attachment defines the structure of hexagonal chains. We distinguish between two kinds of segments. Consider a nonterminal segment S with two neighboring segments embedded into the regular hexagonal lattice in the plane and draw a line through the centers of the hexagons of S as shown in Figure 3.4. If the neighboring segments of S lie on different sides of the line, then S is called a *zigzag* segment. If these segments lie on the same side, then S is said to be a *nonzigzag* segment. Nonzigzag segments correspond to kinks in a chain. It is convenient to consider that the terminal segments are zigzag segments. For a hexagonal chain G, denote by $U(G)$ the set of its nonzigzag segments, $u = |U(G)|$. Since $0 \le u \le n - 2$, there are two extremal chains with respect to the number of segments. The *zigzag* hexagonal

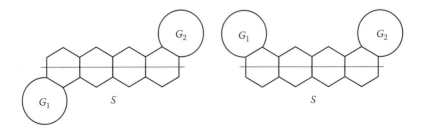

FIGURE 3.4:   Two types of segments in a hexagonal chain.

chain $Z_{n,\ell} \in \mathcal{G}_{n,\ell}$ contains only zigzag segments. All segments of the *spiral* hexagonal chain $O_{n,\ell} \in \mathcal{G}_{n,\ell}$ are nonzigzag with the exception of the terminal segments. The zigzag and spiral hexagonal chains of $\mathcal{G}_{5,3}$ are shown in Figure 3.3.

Based on two types of segments, hexagonal chains of $\mathcal{G}_{n,\ell}$ can be naturally represented by binary codes. We assume that all segments of a chain G are sequentially numbered by $0, 1, \ldots, n-1$ beginning from a terminal segment. Since a way of attachment of nonterminal segments of G completely defines its structure, codes of all hexagonal chains of $\mathcal{G}_{n,\ell}$ have length $n-2$, $n \geq 3$. Assume that every nonterminal nonzigzag segment of $U(G)$ corresponds to 1 in the code $\mathbf{r}(G)$ while every nonterminal zigzag segment corresponds to 0. A chain with nontrivial symmetry has symmetrical code. For instance, the central hexagonal chains in Figure 3.3 have the following codes: (101) and (010). The zigzag and the spiral chains have codes (00..0) and (11..1), respectively.

By family of hexagonal chains, we mean a set or multiset, that is, a family may contain isomorphic graphs. A hexagonal chain induced by a binary vector $\mathbf{r}$ will be denoted by $G(\mathbf{r})$. Let V_n be the set of all binary vectors of length $n-2$. If $X \subseteq V_n$, then $W(X) = \sum_{\mathbf{r} \in X} W(G(\mathbf{r}))$. The family $\mathcal{B}_{n,\ell} = \{G(\mathbf{r}) \mid \mathbf{r} \in V_n\}$ contains all possible hexagonal chains with segment length ℓ induced by binary vectors of V_n, $|\mathcal{B}_{n,\ell}| = 2^{n-2}$.

Components of the *reverse code* $\mathbf{r}^*$ of a binary code $\mathbf{r}$ are defined as $\mathbf{r}_i^* = \mathbf{r}_{n-1-i}$, $1 \leq i \leq n-2$. It is clear that hexagonal chains $G(\mathbf{r})$ and $G(\mathbf{r}^*)$ are always isomorphic. We will assume that $\mathbf{r}(G)$ corresponds to one of two possible codes of G (it is not important how to choose $\mathbf{r}(G)$).

Codes of hexagonal chains of a family $\mathcal{G} = \{G(\mathbf{r}_1), G(\mathbf{r}_2), \ldots, G(\mathbf{r}_k)\}$ form a binary $k \times (n-2)$ matrix $\mathbf{M}(\mathcal{G})$ with rows $\mathbf{r}_1, \mathbf{r}_2, \ldots, \mathbf{r}_k$. The complement of a family $\mathcal{G}$ is defined as $\overline{\mathcal{G}} = \{G(\overline{\mathbf{r}}_1), G(\overline{\mathbf{r}}_2), \ldots, G(\overline{\mathbf{r}}_k)\}$, where $\overline{\mathbf{r}}$ denotes the bitwise negation of $\mathbf{r}$.

3.4   Linked Hexagonal Chains

In this section, we define three families of hexagonal chains for which their Wiener indices can be easily calculated.

Two hexagonal chains $G(\mathbf{r})$ and $G'(\mathbf{r}')$ are *linked* if $\mathbf{r}' = \overline{\mathbf{r}}$. Denote by $\overline{G}$ the linked chain for a hexagonal chain G. It is clear that $\overline{\overline{G}} = G$. A hexagonal chain G is *self-linked* if $G(\mathbf{r}) \cong G(\overline{\mathbf{r}})$. Examples of linked hexagonal chains and their codes are shown in Figure 3.5. A self-linked chain G exists if its number of segments n is even and $\mathbf{r}(G) = (\mathbf{r}_1 \overline{\mathbf{r}}_1^*)$ where $\mathbf{r}_1$ is the first half of $\mathbf{r}(G)$. The set of all self-linked chains of $\mathcal{G}_{n,\ell}$ will be denoted by $\mathcal{L}_{n,\ell}$. It is clear that a self-linked chain cannot be symmetrical, that is, $\mathcal{S}_{n,\ell} \cap \mathcal{L}_{n,\ell} = \emptyset$ and, therefore, $\mathcal{L}_{n,\ell} \subset \mathcal{A}_{n,\ell}$. Since the first half of a self-linked chain defines the second one, the number of nonisomorphic self-linked chains with n segments is $|\mathcal{L}_{n,\ell}| = 2^{(n-2)/2}/2 = 2^{(n-4)/2}$.

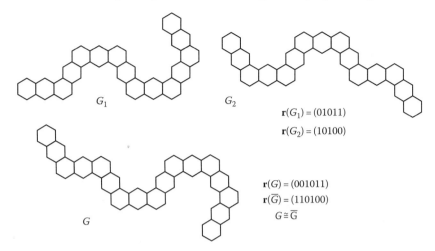

$\mathbf{r}(G_1) = (01011)$
$\mathbf{r}(G_2) = (10100)$

$\mathbf{r}(G) = (001011)$
$\mathbf{r}(\overline{G}) = (110100)$
$G \cong \overline{G}$

FIGURE 3.5: Linked and self-linked hexagonal chains.

3.4.1 Complete Families

A set of hexagonal chains $\mathcal{G}$ is called *complete* if for every chain $G \in \mathcal{G}$, the set always contains its linked graph $\overline{G}$. A complete family $\mathcal{G}$ can be decomposed into two subfamilies, $\mathcal{G} = \mathcal{G}_1 \cup \mathcal{G}_2$, where $\mathcal{G}_2$ contains all self-linked hexagonal chains of $\mathcal{G}$ and $\mathcal{G}_1 = \mathcal{G} \backslash \mathcal{G}_2 = \cup \{G, \overline{G}\}$.

It is clear that if $\mathcal{G}_1$ and $\mathcal{G}_2$ are arbitrary complete families, then $\mathcal{G}_1 \cup \mathcal{G}_2$, $\mathcal{G}_1 \cap \mathcal{G}_2$, and $\mathcal{G}_1 \backslash \mathcal{G}_2$ are also complete families.

Examples

1. Since a symmetrical hexagonal chain cannot be self-linked, $\mathcal{S}_{n,\ell} = \cup \{G, \overline{G}\}$, that is, the set of all symmetrical chains $\mathcal{S}_{n,\ell}$ is complete.

2. Any set of self-linked hexagonal chains is complete. In particular, the set $\mathcal{L}_{n,\ell}$ is complete.

3. The set of all asymmetrical hexagonal chains $\mathcal{A}_{n,\ell}$ is complete, $\mathcal{A}_{n,\ell} = \cup \{G, \overline{G}\} \cup \mathcal{L}_{n,\ell}$.

4. The set of all hexagonal chains $\mathcal{G}_{n,\ell}$ is complete, $\mathcal{G}_{n,\ell} = \mathcal{S}_{n,\ell} \cup \mathcal{A}_{n,\ell}$.

3.4.2 Linked Families

A family of chains $\mathcal{G} = \{G_1, G_2, \ldots, G_k\}$ is called *a-linked* if the sum of units in every columns of $\mathbf{M}(\mathcal{G}) = [m_{i,j}]$ is equal to a, that is, $a = \sum_{i=1}^{k} m_{i,j}$ for every $j = 1, 2, \ldots, n - 2$.

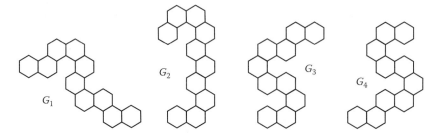

FIGURE 3.6:   2-Linked family of fibonacenes.

A 2-linked family of fibonacenes $\mathcal{F} = \{G_1, G_2, G_3, G_4\}$ with nine segments shown in Figure 3.6 has the following matrix:

$$\mathbf{M}(\mathcal{F}) = \begin{bmatrix} \mathbf{r}(G_1) \\ \mathbf{r}(G_2) \\ \mathbf{r}(G_3) \\ \mathbf{r}(G_4) \end{bmatrix} = \begin{bmatrix} 0\ 1\ 1\ 0\ 1\ 0\ 0 \\ 1\ 0\ 0\ 0\ 0\ 1\ 1 \\ 1\ 1\ 0\ 1\ 1\ 0\ 0 \\ 0\ 0\ 1\ 1\ 0\ 1\ 1 \end{bmatrix}.$$

Let $\mathcal{G}$ be arbitrary a-linked family of hexagonal chains, $|\mathcal{G}| = k$. If $a = 0$ and $k \geq 1$, then $\mathcal{G}$ contains k zigzag chains $Z_{n,\ell}$. If $a = k$, then $\mathcal{G}$ contains k spiral chains $O_{n,\ell}$. If $Z_{n,\ell} \notin \mathcal{G}$ then $a \leq |\mathcal{G}| \leq a(n-2)$.

Let $\mathcal{G}, \mathcal{G}'$ be arbitrary a- and a'-linked families of hexagonal chains, respectively. It is clear that $\mathcal{G} \cup \mathcal{G}'$ is always $(a + a')$-linked family.

Examples

1. A hexagonal chain G and its linked chain $\overline{G}$ form a 1-linked set.

2. Any complete family of hexagonal chains $\mathcal{G}$ with odd number of segments is $|\mathcal{G}|/2$-linked, $\mathcal{G} = \cup\{G, \overline{G}\}$.

3. If $\mathcal{G}$ is an a-linked family of hexagonal chains, then the complement $\overline{\mathcal{G}}$ is a $(|\mathcal{G}| - a)$-linked family.

4. The set of all symmetrical hexagonal chains $\mathcal{S}_{n,\ell}$ is $|\mathcal{S}_{n,\ell}|/2$-linked.

5. The set of all hexagonal chains $|\mathcal{G}_{n,\ell}|$ is $|\mathcal{G}_{n,\ell}|/2$-linked for odd n, $\mathcal{G}_{n,\ell} = \cup\{G, \overline{G}\}$. If n is even, then $\mathcal{G}'_{n,\ell} = \mathcal{G}_{n,\ell} \backslash \mathcal{L}_{n,\ell} = \cup\{G, \overline{G}\}$ is a $|\mathcal{G}'_{n,\ell}|/2$-linked set.

6. The family of hexagonal chains induced by all binary vectors of length $n - 2$ can be represented as $\mathcal{B}_{n,\ell} = \cup\{G, \overline{G}\}$ (graphs of $\mathcal{B}_{n,\ell}$ may be isomorphic). Then $\mathcal{B}_{n,\ell}$ is a $|\mathcal{B}_{n,\ell}|/2$-linked family.

7. Suppose that a code $\mathbf{r}(G)$ of a hexagonal chain G with n segments has a units. Then the family of chains generated by all cyclic shifts of bits in $\mathbf{r}(G)$ is a-linked.

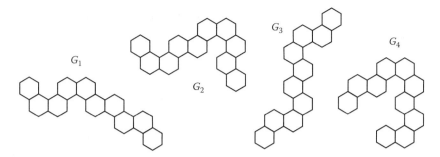

FIGURE 3.7: Symmetrically 3-linked family of fibonacenes.

If a complete set of chains does not contain self-linked graphs, then it is *a*-linked for some *a*. To construct a complete nonlinked family, it is sufficient to add a self-linked graph to such *a*-linked family.

3.4.3 Symmetrically Linked Families

Two columns j_1 and j_2 of a code matrix **M** are called *symmetrical* if $j_2 = n - j_1 - 1$ for $1 \leq j_1 \leq \lfloor \frac{n-2}{2} \rfloor$. A family of hexagonal chains $\mathcal{G} = \{G_1, G_2, \ldots, G_k\}$ is called *symmetrically a-linked* if the sum of units in every pairs of symmetrical columns of $\mathbf{M}(\mathcal{G}) = [m_{i,j}]$ is equal to a (if the number of segments n is odd, then the central column of $\mathbf{M}(\mathcal{G})$ does not take into account), that is, $a = \sum_{i=1}^{k} m_{i,j} + \sum_{i=1}^{k} m_{i,n-j-1}$ for every $j = 1, 2, \ldots, \lfloor (n-2)/2 \rfloor$ and $a \leq 2k$.

A symmetrically 3-linked family of chains $\mathcal{F} = \{G_1, G_2, G_3, G_4\}$ with nine segments shown in Figure 3.7 has the following matrix:

$$\mathbf{M}(\mathcal{F}) = \begin{bmatrix} \mathbf{r}(G_1) \\ \mathbf{r}(G_2) \\ \mathbf{r}(G_3) \\ \mathbf{r}(G_4) \end{bmatrix} = \begin{bmatrix} 1\,0\,1\,0\,0\,0\,0 \\ 1\,0\,0\,0\,1\,1\,0 \\ 0\,1\,0\,0\,0\,1\,0 \\ 0\,0\,1\,1\,0\,0\,1 \end{bmatrix} .$$

It is clear that an arbitrary *a*-linked family of hexagonal chains is always symmetrically 2*a*-linked.

3.5 Hexagonal Chains with Extremal Wiener Index

Among all hexagonal chains with a fixed number of rings h, two are extremal with regard to their Wiener indices: the helicene H_h and the linear chain L_h. Their Wiener indices have the minimal and the maximal values, respectively [25]:

$$W(H_h) = \frac{1}{3}\left(8h^3 + 72h^2 - 26h + 27\right),$$

$$W(L_h) = \frac{1}{3}\left(16h^3 + 24h^2 + 62h - 21\right).$$

Similarly, the spiral chain $O_{n,\ell}$ and the zigzag chain $Z_{n,\ell}$ are extremal graphs among all hexagonal chains of $\mathcal{G}_{n,\ell}$ [23]. Their Wiener indices have the minimal $W(O_{n,\ell}) = W_{min}$ and the maximal $W(Z_{n,\ell}) = W_{max}$ values: $W(O_{n,\ell}) < W(G) < W(Z_{n,\ell})$ for all $G \in \mathcal{G}_{n,\ell} \setminus \{O_{n,\ell}, Z_{n,\ell}\}$ where

$$
\begin{aligned}
W_{min} &= \frac{1}{3(\ell - 1)}\Big(8h^3(2\ell - 3) + 24h^2(2\ell - 1) - 2h(2\ell^2 - 5\ell + 15) \\
&\quad + 4\ell^2 + 7\ell - 3\Big) \\
&= \frac{1}{3}\Big(8n^3(\ell - 1)^2(2\ell - 3) + 96n^2(\ell - 1)^2 - 2n(\ell - 1)(2\ell - 75) + 81\Big), \\
W_{max} &= \frac{1}{3}\Big(16h^3 + 24h^2 + 2h(6\ell + 19) - 12\ell + 3\Big) \\
&= \frac{1}{3}\Big(16n^3(\ell - 1)^3 + 72n^2(\ell - 1)^2 + n(\ell - 1)(12\ell + 134) + 81\Big).
\end{aligned}
$$

Denote by W_s the average value of W_{min} and W_{max}, that is,

$$
\begin{aligned}
W_s &= \frac{1}{2}(W_{min} + W_{max}) \\
&= \frac{1}{3(\ell - 1)}\Big(4h^3(4\ell - 5) + 12h^2(3\ell - 2) + 2h(2\ell^2 + 9\ell - 17) \\
&\quad - 4\ell^2 + 11\ell - 3\Big) \\
&= \frac{1}{3}\Big(4n^3(\ell - 1)^2(4\ell - 5) + 84n^2(\ell - 1)^2 + 2n(\ell - 1)(2\ell + 71) + 81\Big).
\end{aligned}
$$

The difference between extremal values is

$$
W_{max} - W_{min} = \frac{8}{3(\ell - 1)}(h - 1)(h - \ell)(h - 2\ell + 1) = 16(\ell - 1)^2\binom{n}{3}.
$$

Further, these W-values will be considered only for hexagonal chains with n segments of length ℓ. To distinguish these quantities for chains with distinct number of segments or distinct lengths of segments, we will use additional subscripts: $W_{max,n}$ or $W_{min,\ell}$.

3.6   Wiener Index of Linked Chains

To calculate the Wiener index, it is convenient to apply a formula based on segments of hexagonal chains [18]. Denote by l_S and r_S the numbers of hexagonal rings of two chains G_1 and G_2 obtained by deleting a segment S from G (see Figure 3.4).

Proposition 3.6.1 *For the Wiener index of an arbitrary hexagonal chain G with h rings and n segments of lengths $\ell_1, \ell_2, \ldots, \ell_n$,*

$$W(G) = W(L_h) - 16 \sum_{S \in U(G)} l_S r_S - 4 \left(h^2 + n - 1 - \sum_{i=1}^{n} \ell_i^2 \right). \quad (3.1)$$

If a hexagonal chain has segments of equal length, then formula (3.1) has a more simple form.

Corollary 3.6.1 *For a hexagonal chain $G \in \mathcal{G}_{n,\ell}$,*

$$W(G) = W_{max} - 16 \sum_{S \in U(G)} l_S r_S. \quad (3.2)$$

Proof. Let $G \in \mathcal{G}_{n,\ell}$. We have $W(L_h) = W_{max} + 4(h-1)(h-\ell)$ and $n = (h-1)/(\ell-1)$. Then $4(h-1)(h-\ell) - 4(h^2 + n - 1 - n\ell^2) = 4(h-1)(h-\ell) - 4[(h-1)(h+1) - (h-1)(\ell+1)] = 0$, and corollary follows from formula (3.1). □

Corollary 3.6.2 *For a hexagonal chain $G \in \mathcal{G}_{n,\ell}$,*

$$W(G) = W_{max} - 16(\ell - 1)^2 \sum_{S_i \in U(G)} (i-1)(n-i). \quad (3.3)$$

Proof. For *i*th segment of a hexagonal chain $G \in \mathcal{G}_{n,\ell}$, we can write $l_S r_S = (i-1)(\ell-1) \cdot (n-i)(\ell-1) = (\ell-1)^2 (i-1)(n-i)$. □

The main property of linked graphs of $\mathcal{G}_{n,\ell}$ is that the sum of their Wiener indices depends on the number of segments and their length.

Proposition 3.6.2 *For a hexagonal chain $G \in \mathcal{G}_{n,\ell}$,*

$$W(G) + W(\overline{G}) = W_{min} + W_{max}.$$

Proof. By formulas (3.2) and (3.3), we have

$$W(G) + W(\overline{G}) = 2W_{max} - 16 \sum_{S \in U(G)} l_S r_S - 16 \sum_{S \in U(\overline{G})} l_S r_S$$

$$= 2W_{max} - 16 \sum_{S \in U(O_{n,\ell})} l_S r_S = 2W_{max} - 16(\ell-1)^2 \sum_{i=1}^{n-2} (i-1)(n-i)$$

$$= 2W_{max} - 16(\ell-1)^2 n(n-1)(n-2)/6$$

$$= 2W_{max} - \frac{8}{3(l-1)}(h-1)(h-\ell)(h-2\ell+1) = W_{max} + W_{min}.$$

The last equality follows from the expression for the index difference $W_{max} - W_{min}$ (see Section 3.5).   □

The aforementioned property of linked chains completely defines the Wiener indices of all self-linked hexagonal chains.

Corollary 3.6.3 *For a self-linked hexagonal chain* $G \in \mathcal{G}_{n,\ell}$,

$$W(G) = W_s.$$

This is the main property of hexagonal chains from $\mathcal{L}_{n,\ell}$: their Wiener indices are the same and coincide with the arithmetic mean of W_{min} and W_{max}.

3.7   Wiener Index of Complete Families

For the first time, collective properties of the Wiener index of hexagonal chains have been studied for fibonacenes [19,20]. Similar properties are also valid for hexagonal chains with segments of equal length.

Proposition 3.7.1 *For a complete family of hexagonal chains* $\mathcal{G}$,

$$W(\mathcal{G}) = \sum_{G \in \mathcal{G}} W(G) = |\mathcal{G}| \, W_s. \tag{3.4}$$

Proof. Let $\mathcal{G} = \mathcal{G}_1 \cup \mathcal{G}_2$ be decomposition of a complete family $\mathcal{G}$ into disjoint subfamilies where $\mathcal{G}_2$ contains all self-linked hexagonal chains of $\mathcal{G}$. Since $\sum_{G \in \mathcal{G}_1} W(G) = \sum_{G \in \mathcal{G}_1} W(\overline{G})$, we can write

$$\sum_{G \in \mathcal{G}} W(G) = \frac{1}{2} \left(\sum_{G \in \mathcal{G}_1} W(G) + \sum_{G \in \mathcal{G}_1} W(\overline{G}) \right) + \sum_{G \in \mathcal{G}_2} W(G)$$

$$= \sum_{G \in \mathcal{G}_1} \left(W(G) + W(\overline{G}) \right) / 2 + W_s \, |\mathcal{G}_2|$$

$$= W_s \, |\mathcal{G}_1| + W_s \, |\mathcal{G}_2| = W_s \, |\mathcal{G}|. \qquad \qquad \square$$

For the average value W_{avr} of the Wiener index of hexagonal chains of the complete set $\mathcal{G}_{n,\ell}$,

$$W_{avr} = \frac{W(\mathcal{G}_{n,\ell})}{|\mathcal{G}_{n,\ell}|} = W_s.$$

This fact has been established for fibonacenes in [22]. The equality $W_{avr} = W_s$ shows that every self-linked hexagonal chain is an *average* graph in complete families with respect to the Wiener index. The quantity W_s is also the expected value of the Wiener index of random fibonacenes [30].

Example By distance $d_H(G_1, G_2)$ between hexagonal chains G_1 and G_2, we mean the Hamming distance $d_H(\mathbf{r}(G_1), \mathbf{r}(G_2))$ between their binary codes $\mathbf{r}(G_1)$ and $\mathbf{r}(G_2)$, that is, $d_H(G_1, G_2)$ is equal to the number of codes' components for which $r_i(G_1) \neq r_i(G_2)$, $i = 1, 2, \ldots, n - 2$. Let $Sp(G_0)$ be the sphere of radius 1 with the center in a hexagonal chain G_0: $Sp(G_0) = \{G \in \mathcal{G}_{n,\ell} \mid d_H(G_0, G) = 1\}$, $|Sp(G_0)| = n - 2$. It is clear that $Sp(\overline{G}) = \overline{Sp(G)}$. Therefore, $Sp(G) \cup Sp(\overline{G})$ is a complete family of hexagonal chains, and therefore, $W(Sp(G) \cup Sp(\overline{G})) = 2(n - 2)W_s$.

Corollary 3.7.1 *Let $\mathcal{G}_1, \mathcal{G}_2$ be complete families of hexagonal chains. Then $W(\mathcal{G}_1) = W(\mathcal{G}_2)$ if and only if $|\mathcal{G}_1| = |\mathcal{G}_2|$.*

Proof. Corollary immediately follows from equality (3.4). □

Based on this result, it is possible to get relations between Wiener indices of some families of hexagonal chains:

$$W(\mathcal{B}_{n,\ell}) = 2^{\frac{n}{2}} \left(2^{\frac{n-2}{2}} + 1 \right)^{-1} W(\mathcal{G}_{n,\ell}) \quad \text{for even } n,$$

$$W(\mathcal{B}_{n,\ell}) = 2^{\frac{n-1}{2}} \left(2^{\frac{n-3}{2}} + 1 \right)^{-1} W(\mathcal{G}_{n,\ell}) \quad \text{for odd } n,$$

$$W(\mathcal{G}_{n,\ell}) = \frac{1}{2} \left(2^{\frac{n-2}{2}} + 1 \right) W(\mathcal{S}_{n,\ell}) \quad \text{for even } n,$$

$$W(\mathcal{G}_{n,\ell}) = \frac{1}{2} \left(2^{\frac{n-3}{2}} + 1 \right) W(\mathcal{S}_{n,\ell}) \quad \text{for odd } n,$$

$$W(\mathcal{A}_{n,\ell}) = \frac{1}{2} \left(2^{\frac{n-2}{2}} - 1 \right) W(\mathcal{S}_{n,\ell}) \quad \text{for even } n,$$

$$W(\mathcal{A}_{n,\ell}) = \frac{1}{2} \left(2^{\frac{n-3}{2}} - 1 \right) W(\mathcal{S}_{n,\ell}) \quad \text{for odd } n \geq 5,$$

$$W(\mathcal{B}_{n,\ell}) = 2^{\frac{n-2}{2}} W(\mathcal{S}_{n,\ell}) \quad \text{for even } n,$$

$$W(\mathcal{B}_{n,\ell}) = 2^{\frac{n-3}{2}} W(\mathcal{S}_{n,\ell}) \quad \text{for odd } n,$$

$$W(\mathcal{S}_{n,\ell}) = 2W(\mathcal{L}_{n,\ell}) \quad \text{for even } n.$$

In particular, the Wiener index of the family $\mathcal{G}_{n,\ell}$ can be represented as a relation of the Wiener indices of symmetrical and asymmetrical parts of $\mathcal{G}_{n,\ell}$:

$$W(\mathcal{G}_{n,\ell}) = \frac{1}{4} \left(2^{n-2} - 1 \right) \frac{W^2(\mathcal{S}_{n,\ell})}{W(\mathcal{A}_{n,\ell})} \quad \text{for even } n,$$

$$W(\mathcal{G}_{n,\ell}) = \frac{1}{4} \left(2^{n-3} - 1 \right) \frac{W^2(\mathcal{S}_{n,\ell})}{W(\mathcal{A}_{n,\ell})} \quad \text{for odd } n \geq 5.$$

3.8   Wiener Index of Linked Families

To calculate the Wiener index of linked families, it is sufficient to know their characteristics.

Proposition 3.8.1 *For an a-linked family of chains* $\mathcal{G} = \{G_1, G_2, \ldots, G_k\}$ *with n segments of length* ℓ,

$$W(\mathcal{G}) = |\mathcal{G}|W_{max} - a(W_{max} - W_{min})$$

$$= |\mathcal{G}|W_{max} - 16a(\ell - 1)^2 \binom{n}{3}.$$

Proof. Applying formula (3.2) to every hexagonal chain of $\mathcal{G}$, we have

$$W(\mathcal{G}) = kW_{max} - 16 \left(\sum_{S \in U(G_1)} l_S r_S + \cdots + \sum_{S \in U(G_k)} l_S r_S \right)$$

$$= (k - a)W_{max} + a \left(W_{max} - 16 \sum_{S \in U(O_{n,\ell})} l_S r_S \right)$$

$$= (k - a)W_{max} + aW(O_{n,\ell})$$

$$= kW_{max} - a(W_{max} - W_{min}).$$

Proposition follows after simplification of the expression $W_{max} - W_{min}$.   □

Consider the family of 2-linked fibonacenes $\mathcal{F}$ shown in Figure 3.6. By computer calculations, $W(\mathcal{F}) = 5661 + 5917 + 5533 + 5533 = 22644$. Using Proposition 3.8.1, we get $W(\mathcal{F}) = 4 \cdot 6,333 - 16 \cdot 2(2 - 1)^2 \binom{9}{3} = 25,332 - 2,688 = 22,644$.

Is there exists a linked family of chains for which the calculation of its Wiener index would be so simple as for complete sets? The next result answers this question in the affirmative.

Corollary 3.8.1 *For a* $|\mathcal{G}|/2$-*linked family of hexagonal chains* $\mathcal{G}$,

$$W(\mathcal{G}) = |\mathcal{G}| W_s.$$

Proof. By Proposition 3.8.1, $W(\mathcal{G}) = |\mathcal{G}|W_{max} - \frac{|\mathcal{G}|}{2}(W_{max} - W_{min}) = |\mathcal{G}|(W_{max} + W_{min})/2 = |\mathcal{G}|W_s.$ □

A family of hexagonal chains $\mathcal{G}$ is called *combined* if it can be represented as disjoint union $\mathcal{G} = \mathcal{G}_1 \cup \mathcal{G}_2$ where $\mathcal{G}_1$ is a $|\mathcal{G}_1|/2$-linked family of chains and $\mathcal{G}_2$ consists of self-linked chains. By Corollaries 3.6.3 and 3.8.1, $W(\mathcal{G}) = |\mathcal{G}| W_s$ for an arbitrary combined family $\mathcal{G}$.

Corollary 3.8.2 *Let $\mathcal{G}_1$ and $\mathcal{G}_2$ be a-linked families of hexagonal chains of $\mathcal{G}_{n,\ell}$. Then $W(\mathcal{G}_1) = W(\mathcal{G}_2)$ if and only if $|\mathcal{G}_1| = |\mathcal{G}_2|$.*

Proof. By Proposition 3.8.1, $W(\mathcal{G}_1) - W(\mathcal{G}_2) = W_{max}(|\mathcal{G}_1| - |\mathcal{G}_2|)$. $\qquad\qquad\square$

This property is also valid for a combined family of chains.

Example Let $\mathcal{G}$ be a family of hexagonal chains with even number of segments n induced by all binary vectors of length $n - 2$ having exactly $(n - 2)/2$ units. The cardinality of $\mathcal{G}$ is equal to $\binom{n-2}{\frac{1}{2}(n-2)}$. The number of hexagonal chains $G \in \mathcal{G}$ with $\mathbf{r}_i(G) = 1$ is equal to $\binom{n-3}{\frac{1}{2}(n-4)}$ for every $i = 1, 2, \ldots, n - 2$. Therefore, $\mathcal{G}$ is a $\binom{n-3}{\frac{1}{2}(n-4)}$–linked family. Since $\binom{n-3}{\frac{1}{2}(n-4)} = \frac{1}{2}\binom{n-2}{\frac{1}{2}(n-2)}$, the family $\mathcal{G}$ is $|\mathcal{G}|/2$-linked, and we can apply Corollary 3.8.1. Therefore, $W(\mathcal{G}) = |\mathcal{G}|W_s = \binom{n-2}{\frac{1}{2}(n-2)}W_s$.

3.9 Wiener Index of Symmetrically Linked Families

Since a hexagonal chain has two terminal segments, jth and $n - j - 1$ nonzigzag segments make equal contributions to formula (3.2) for every $j = 1, 2, \ldots, n-2$. This fact can be applied for constructing more general families of hexagonal chains that include linked families.

Proposition 3.9.1 *For a symmetrically a-linked family of hexagonal chains $\mathcal{G}$ with n segments of length ℓ (with b units in the central column of $\mathbf{M}(\mathcal{G})$ for odd n),*

$$W(\mathcal{G}) = |\mathcal{G}|W_{max} - \frac{1}{2}a\,(W_{max} - W_{min}) - \phi(a, b, n, \ell)$$

$$= |\mathcal{G}|W_{max} - 8a(\ell - 1)^2\binom{n}{3} - \phi(a, b, n, \ell)$$

where
$\phi(a, b, n, \ell) = 2(n - 1)^2(\ell - 1)^2(2b - a)$ *for odd n*
$\phi(a, b, n, \ell) = 0$ *for even n.*

Proof. To calculate $W(\mathcal{G})$, we apply formula (3.2) for graphs of $\mathcal{G}$. Then

$$W(\mathcal{G}) = |\mathcal{G}|W_{max} - 16 \sum_{G \in \mathcal{G}} \sum_{S \in U(G)} l_S r_S.$$

Let c_i be the sum of units in ith column of $\mathbf{M}(\mathcal{G})$. For ith segment S_i, we shall write $l_{S_i} = l_i$ and $r_{S_i} = r_i$. Let n be odd.

Since $l_i r_i = l_{n-i-1} r_{n-i-1}$ for all i and $l_{\frac{n-1}{2}} = r_{\frac{n-1}{2}} = \frac{(n-1)(\ell-1)}{2}$,

$$
\sum_{G \in \mathcal{G}} \sum_{S \in U(G)} l_S r_S = c_1 l_1 r_1 + c_2 l_2 r_2 + \cdots + c_{n-2} l_{n-2} r_{n-2}
$$

$$
= l_1 r_1 [c_1 + c_{n-2}] + l_2 r_2 [c_2 + c_{n-3}] + \cdots
$$

$$
+ l_{\frac{n-3}{2}} r_{\frac{n-3}{2}} \left[c_{\frac{n-3}{2}} + c_{\frac{n+1}{2}} \right] + c_{\frac{n-1}{2}} l_{\frac{n-1}{2}} r_{\frac{n-1}{2}}
$$

$$
= a \left[l_1 r_1 + l_2 r_2 + \cdots + l_{\frac{n-3}{2}} r_{\frac{n-3}{2}} \right] + b l_{\frac{n-1}{2}} r_{\frac{n-1}{2}}
$$

$$
= \frac{1}{2} a \sum_{S \in U(O_{n,\ell})} l_S r_S - \frac{1}{2} a l_{\frac{n-1}{2}} r_{\frac{n-1}{2}} + b l_{\frac{n-1}{2}} r_{\frac{n-1}{2}}
$$

$$
= \frac{1}{2} a \sum_{S \in U(O_{n,\ell})} l_S r_S
$$

$$
- \frac{1}{8} a(n-1)^2 (\ell-1)^2 + \frac{1}{4} b(n-1)^2 (\ell-1)^2.
$$

Then

$$
W(\mathcal{G}) = |\mathcal{G}| W_{max} - \frac{a}{2} W_{max} + \frac{a}{2} \left(W_{max} - 16 \sum_{S \in U(O_{n,\ell})} l_S r_S \right)
$$

$$
- 2(n-1)^2 (\ell-1)^2 (a-2b)
$$

$$
= |\mathcal{G}| W_{max} - \frac{a}{2} (W_{max} - W_{min}) + 2(n-1)^2 (\ell-1)^2 (2b-a).
$$

For even n, all calculations are almost the same. $\square$

Consider the family of symmetrically 3-linked fibonacenes $\mathcal{F}$ shown in Figure 3.7. By computer calculations, $W(\mathcal{F}) = 5,981 + 5,789 + 5,949 + 5,725 = 23,444$. Applying Proposition 3.9.1 for chains of $\mathcal{F}$, we have $W(\mathcal{F}) = 4 \cdot 6,333 - 8 \cdot 3$ $(2-1)^2 \binom{9}{3} - 2(9-1)^2 (2-1)^2 (2 \cdot 1 - 3) = 25,332 - 2,016 + 128 = 23,444$.

For some families of hexagonal chains $\mathcal{G}$, the Wiener index can be easily calculated.

Corollary 3.9.1 *For a symmetrically $|\mathcal{G}|$-linked family of hexagonal chains (with $b = |\mathcal{G}|/2$ for odd n),*

$$
W(\mathcal{G}) = |\mathcal{G}| W_s.
$$

Now we use Proposition 3.9.1 for finding the Wiener index of families of growing hexagonal chains. Let a chain G be obtained by fusing terminal parts (ring, segment, or two segments) of two hexagonal chains G_1 and G_2 such that all segments of G have also equal length. An example of such graph transformation t_1, t_2, and t_3 is shown in Figure 3.8. Fused parts in the resulting graph are depicted by bold lines.

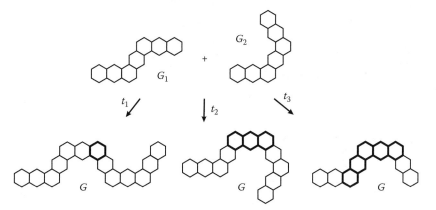

FIGURE 3.8: Types of graph transformations.

The resulting graph $G = t_i(G_1, G_2)$, $i = 1, 2, 3$, has the following numbers of hexagonal rings and segments:

t_1: $h(G) = h(G_1) + h(G_2) - 1$ and $n(G) = n(G_1) + n(G_2)$;
t_2: $h(G) = h(G_1) + h(G_2) - \ell$ and $n(G) = n(G_1) + n(G_2) - 1$;
t_3: $h(G) = h(G_1) + h(G_2) - 2\ell$ and $n(G) = n(G_1) + n(G_2) - 2$.

A code of G can be represented through codes $\mathbf{r}_1$ and $\mathbf{r}_2$ of chains G_1 and G_2. Namely, $\mathbf{r}(G) = (\mathbf{r}_1 b_1 b_2 \mathbf{r}_2)$, $\mathbf{r}(G) = (\mathbf{r}_1 b_1 \mathbf{r}_2)$, and $\mathbf{r}(G) = (\mathbf{r}_1 \mathbf{r}_2)$ for transformations t_1, t_2, and t_3, respectively. A way of chains fusing defines values of new bits b_1 and b_2 of code $\mathbf{r}(G)$.

Let $\mathcal{G}_{n,\ell}$, $\mathcal{G}_{n_1,\ell}$, and $\mathcal{G}_{n_2,\ell}$ be symmetrically a-, a_1- and a_2-linked families, respectively. Let every chain of $\mathcal{G}_{n,\ell}$ be obtained from a chain of $\mathcal{G}_{n_1,\ell}$ and a chain of $\mathcal{G}_{n_2,\ell}$ by the aforementioned graph transformations. We are interested in representing the value of $W(\mathcal{G}_{n,\ell})$ through the quantities $W(\mathcal{G}_{n_1,\ell})$ and $W(\mathcal{G}_{n_2,\ell})$.

Corollary 3.9.2 *For the Wiener index of a symmetrically a-linked family of hexagonal chains $\mathcal{G}_{n,\ell}$ (n, n_1, and n_2 are even for simplicity),*

$$W(\mathcal{G}_{n,\ell}) = W(\mathcal{G}_{n_1,\ell}) + W(\mathcal{G}_{n_2,\ell})$$
$$+ |\mathcal{G}_{n,\ell}|W_{max,n} - |\mathcal{G}_{n_1,\ell}|W_{max,n_1} - |\mathcal{G}_{n_2,\ell}|W_{max,n_2}$$
$$- 8(\ell - 1)^2 \left[a\binom{n}{3} - a_1\binom{n_1}{3} - a_2\binom{n_2}{3} \right].$$

Proof. Corollary follows from Proposition 3.9.1. □

As an illustration, consider transformation t_1 when $|\mathcal{G}_{n_1,\ell}| = |\mathcal{G}_{n_2,\ell}| = |\mathcal{G}_{n,\ell}|$ and $n_1 = n_2$, $a = a_1 = a_2 (b_1 = b_2 = a_1/2$ if n_1, n_2 are odd). Then

$$W(\mathcal{G}_{n,\ell}) = W(\mathcal{G}_{n_1,\ell}) + W(\mathcal{G}_{n_2,\ell}) + |\mathcal{G}_{n_1,\ell}|(W_{max,2n_1} - 2W_{max,n_1})$$
$$- 8(\ell - 1)^2 a_1 n_1^2 (n_1 - 1). \tag{3.5}$$

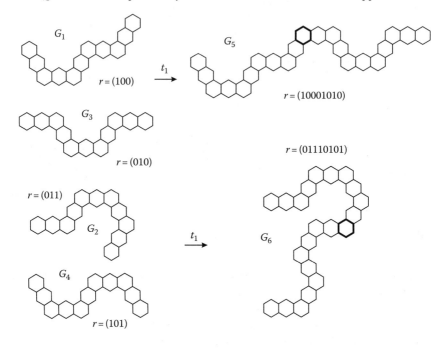

FIGURE 3.9:   Growing hexagonal chains.

Growing hexagonal chains are shown in Figure 3.9. Here, $\mathcal{G}_{n_1,3} = \{G_1, G_2\}$, $\mathcal{G}_{n_2,3} = \{G_3, G_4\}$ are symmetrically 2-linked families of chains, $n_1 = n_2 = 5$. The resulting symmetrically 2-linked family $\mathcal{G}_{10,3} = \{G_5, G_6\}$ consists of graphs with 10 segments: $G_5 = t_1(G_1, G_3)$ and $G_6 = t_1(G_2, G_4)$. By direct computing, $W(\mathcal{G}_{n_1,3}) = 8,135 + 7,879 = 16,014$; $W(\mathcal{G}_{n_2,3}) = 8,071 + 7,943 = 16,014$; $W(\mathcal{G}_{10,3}) = 50,739 + 48,435 = 99,174$; and $W_{max,5} = 8,327$, $W_{max,10} = 53,427$. Using formula (3.5), we have

$$W(\mathcal{G}_{10,3}) = 16,014 + 16,014 + 2(53,427 - 2 \cdot 8,327) - 8(3 - 1)^2 2 \cdot 5^2(5 - 1)$$
$$= 32,028 + 2 \cdot 36,773 - 6,400 = 105,574 - 6,400 = 99,174.$$

3.10   Expansion of the Wiener Index

Codes of hexagonal chains can be considered as vectors in a suitable vector space. Let $\mathbf{Z}_2$ be a field with two elements. For two binary vectors, $\mathbf{x} = (x_1, x_1, \ldots, x_n)$ and $\mathbf{y} = (y_1, y_1, \ldots, y_n)$ are defined two operations: $\mathbf{x} + \mathbf{y} = (x_1 + y_1, x_2 + y_2, \ldots, x_n + y_n)$, where $0 + 0 = 1 + 1 = 0$ and $0 + 1 = 1 + 0 = 1$, and $\lambda \cdot \mathbf{x} = (\lambda \cdot x_1, \lambda \cdot x_2, \ldots, \lambda \cdot x_n)$ for $\lambda \in \mathbf{Z}_2$, where $0 \cdot 0 = 0 \cdot 1 = 1 \cdot 0 = 0$ and $1 \cdot 1 = 1$. Denote by $\mathbf{e}_i$ the binary vector $\mathbf{e}_i = (0, 0, \ldots, 0, \overset{i}{1}, 00 \ldots, 0)$ of length $n - 2$ for $n \geq 3$. These vectors form the standard basis for the vector space $\mathbf{B}$ of dimension $n - 2$ over $\mathbf{Z}_2$. Then a code $\mathbf{r}(G)$ of a chain $G \in \mathcal{G}_{n,\ell}$ can be represented as a linear combination of the basis vectors $\mathbf{e}_i$, $i = 1, 2, \ldots, n - 2$:

$$\mathbf{r}(G) = \lambda_1 \mathbf{e}_1 + \lambda_2 \mathbf{e}_2 + \cdots + \lambda_{n-2} \mathbf{e}_{n-2},$$

where $\lambda_1, \lambda_2, \ldots, \lambda_{n-2} \in \mathbf{Z}_2$. For example, chain's code $\mathbf{r}(G) = (10011)$ has the following expansion:

$$\mathbf{r}(G) = 1 \cdot \mathbf{e}_1 + 0 \cdot \mathbf{e}_2 + 0 \cdot \mathbf{e}_3 + 1 \cdot \mathbf{e}_4 + 1 \cdot \mathbf{e}_5.$$

In order to get expansion for codes of the spiral chain $O_{n,\ell}$ or for the zigzag chain $Z_{n,\ell}$, one can assume $\lambda_i = 1$ or $\lambda_i = 0$ for all $i = 1, 2, \ldots, n - 2$, respectively.

Denote by C_i the *basis hexagonal chains* of $\mathcal{G}_{n,\ell}$ corresponding to the basis vectors $\mathbf{e}_i$. Since positions of units in vectors $\mathbf{e}_i$ and $\mathbf{e}_{n-i-1}$ are symmetrical, $C_i \cong C_{n-i-1}$ for every $i = 1, 2, \ldots, n - 2$. Let $W_i = W(C_i)$.

Proposition 3.10.1 *For an arbitrary hexagonal chain $G \in \mathcal{G}_{n,\ell}$ with code $\mathbf{r}(G) = \lambda_1 \mathbf{e}_1 + \lambda_2 \mathbf{e}_2 + \cdots + \lambda_{n-2} \mathbf{e}_{n-2}$,*

$$W(G) = \lambda_1 W_1 + \lambda_2 W_2 + \cdots + \lambda_{n-2} W_{n-2} - (u - 1) W_{max},$$

where $u = \lambda_1 + \lambda_2 + \cdots + \lambda_{n-2} = |U(G)|$.

Proof. Applying formula (3.2) to every basis hexagonal chain C_i, we have

$$W(C_1) = W_{max} - 16 l_1 r_1$$
$$W(C_2) = W_{max} - 16 l_2 r_2$$

$$\vdots$$

$$W(C_{n-2}) = W_{max} - 16 l_{n-2} r_{n-2}.$$

Multiplying ith equation by λ_i for every $i = 1, 2, \ldots, n - 2$ and then summing the obtained equations, we get

$$\sum_{i=1}^{n-2} \lambda_i W_i = (u - 1) W_{max} + W_{max} - 16 \sum_{S \in U(G)} l_S r_S.$$

The two last terms in the right-hand part of the aforementioned equation give the Wiener index of G according to formula (3.2). $\quad\square$

Assuming $\lambda_i = 1$ for all $i = 1, 2, \ldots, n - 2$, one can calculate the sum of the Wiener indices for all basis graphs C_i.

Proposition 3.10.2 *For the sum of Wiener indices of basis hexagonal chains $C_1, C_2, \ldots, C_{n-2} \in \mathcal{G}_{n,\ell}$,*

$$W_1 + W_2 + \cdots + W_{n-2} = W_{min} + (n - 3) W_{max}$$

$$= \frac{1}{3(\ell - 1)} (h - 2\ell + 1) \left(16h^3 + 16h^2 + 2h(10\ell + 23) - 20\ell + 3 \right).$$

Proof. Since equalities $\lambda_1 = \lambda_2 = \cdots = \lambda_{n-2} = 1$ are valid for the spiral chain $O_{n,\ell}$, one can apply Proposition 3.10.1 with $u(O_{n,\ell}) = n - 2$.   □

Proposition 3.10.1 can be used for calculating the Wiener indices of families of chains.

Examples

1. Let $\mathcal{K}_{n,\ell,2k} \subseteq \mathcal{S}_{n,\ell}$ be the set of all symmetrical hexagonal chains G with even number of segments n and $u(G) = 2k$, where $1 \le k \le (n-2)/2$. Codes of these chains contain $2k$ units. Since the first half of chain's code completely defines the second half, $N = |\mathcal{K}_{n,\ell,2k}| = \binom{\frac{1}{2}(n-2)}{k}$.

Proposition 3.10.3 *For hexagonal chains of $\mathcal{K}_{n,\ell,2k}$ with even number of segments n and $u = 2k$,*

$$W(\mathcal{K}_{n,\ell,2k}) = \binom{\frac{1}{2}(n-4)}{k-1}(W_1 + W_2 + \cdots + W_{n-2})$$
$$- \binom{\frac{1}{2}(n-2)}{k}(2k-1)W_{max}.$$

Proof. Applying formula of Proposition 3.10.1 to every hexagonal chains $G_1, G_2, \ldots, G_N \in \mathcal{K}_{n,\ell,2k}$, we get

$$W(G_1) = \lambda_1^1 W_1 + \lambda_2^1 W_2 + \cdots + \lambda_{n-3}^1 W_{n-2} - (2k-1)W_{max}$$
$$W(G_2) = \lambda_1^2 W_1 + \lambda_2^2 W_2 + \cdots + \lambda_{n-3}^2 W_{n-2} - (2k-1)W_{max}$$

$$\vdots$$

$$W(G_N) = \lambda_1^N W_1 + \lambda_2^N W_2 + \cdots + \lambda_{n-3}^N W_{n-2} - (2k-1)W_{max}.$$

It is easy to see that 1 occurs $\binom{\frac{1}{2}(n-2)-1}{k-1}$ times in ith column $\lambda_i^1, \lambda_i^2, \ldots, \lambda_i^N$ for every $i = 1, 2, \ldots, n-2$. The proof is completed by summing these equalities.   □

As an illustration, consider symmetrical hexagonal chains of set $\mathcal{K}_{10,3,4}$. This set contains six graphs with codes (11000011), (10100101), (10011001), (01100110), (01011010), and (00111100). By computer calculations, $W(\mathcal{K}_{10,3,4}) = 50,611 + 50,099 + 49,843 + 49,331 + 49,075 + 48,563 = 297,522$. Using Proposition 3.10.3, we get $W(\mathcal{K}_{10,3,4}) = 3 \cdot 419,736 - 6 \cdot 3 \cdot 53,427 = 1,259,208 - 961,686 = 297,522$. Since

$$\binom{\frac{1}{2}(n-4)}{k-1} = \frac{1}{2^{k-1}}\frac{1}{(k-1)!}(n-4)(n-6)\cdots(n-2k),$$

$$\binom{\frac{1}{2}(n-2)}{k} = \frac{1}{2^k}\frac{1}{k!}(n-2)(n-4)\cdots(n-2k),$$

$$(n-3)W_{max} + W_{min} = (n-2)W_{max} - 16(\ell - 1)^2\binom{n}{3},$$

the expression for $W(\mathcal{K}_{n,\ell,2k})$ can be rewritten in the form

$$W(\mathcal{K}_{n,\ell,2k}) = \frac{1}{3\,k!\,2^k}\,(n-2)(n-4)(n-6)\cdots(n-2k)\,\phi(n,\ell,k), \qquad (3.6)$$

where $\phi(n,\ell,k) = 3W_{max} - 16k(\ell-1)^2 n(n-1)$. For small k and ℓ, we have the following equalities:

$W(\mathcal{K}_{n,2,2}) = (n-2)(16n^3 + 56n^2 + 174n + 81)/6,$
$W(\mathcal{K}_{n,2,4}) = (n-2)(n-4)(16n^3 + 40n^2 + 190n + 81)/24,$
$W(\mathcal{K}_{n,2,6}) = (n-2)(n-4)(n-6)(16n^3 + 24n^2 + 206n + 81)/144,$
$W(\mathcal{K}_{n,3,2}) = (n-2)(128n^3 + 224n^2 + 404h + 81)/6,$
$W(\mathcal{K}_{n,3,4}) = (n-2)(n-4)(128n^3 + 160n^2 + 468n + 81)/24,$
$W(\mathcal{K}_{n,3,6}) = (n-2)(n-4)(n-6)(128n^3 + 96n^2 + 532n + 81)/144,$
$W(\mathcal{K}_{n,4,2}) = (n-2)(144n^3 + 168n^2 + 230h + 27)/2,$
$W(\mathcal{K}_{n,4,4}) = (n-2)(n-4)(144n^3 + 120n^2 + 278n + 27)/8,$
$W(\mathcal{K}_{n,4,6}) = (n-2)(n-4)(n-6)(144n^3 + 72n^2 + 326n + 27)/48.$

2. Let $\mathbf{X}_{n,\ell,k}$ be the set of vectors of length $n-2$ with k units, $N = |\mathbf{X}_{n,\ell,k}| = \binom{n-2}{k}$. For example, $\mathbf{X}_{n,\ell,1}$ induces all basis hexagonal chains, and the spiral chain is induced by $\mathbf{X}_{n,\ell,n-2}$.

Proposition 3.10.4 *For the Wiener index of hexagonal chains induced by vectors of* $\mathbf{X}_{n,\ell,k}$,

$$W(\mathbf{X}_{n,\ell,k}) = \binom{n-3}{k-1}(W_1 + W_2 + \cdots + W_{n-2}) - \binom{n-2}{k}(k-1)W_{max}.$$

Proof. Applying Proposition 3.10.1 to all graphs $\mathbf{r}(G_1), \mathbf{r}(G_2), \ldots, \mathbf{r}(G_N)$,

$$W(\mathbf{r}(G_1)) = \lambda_1^1 W_1 + \lambda_2^1 W_2 + \cdots + \lambda_{n-3}^1 W_{n-2} - (k-1)W_{max}$$
$$W(\mathbf{r}(G_2)) = \lambda_1^2 W_1 + \lambda_2^2 W_2 + \cdots + \lambda_{n-3}^2 W_{n-2} - (k-1)W_{max}$$
$$\vdots$$
$$W(\mathbf{r}(G_N)) = \lambda_1^N W_1 + \lambda_2^N W_2 + \cdots + \lambda_{n-3}^N W_{n-2} - (k-1)W_{max}.$$

The value 1 occurs $\binom{n-3}{k-1}$ times in ith column $\lambda_i^1, \lambda_i^2, \ldots, \lambda_i^N$ of the aforementioned equations for every $i = 1, 2, \ldots, n-2$. $\qquad \square$

Consider the set $X_{6,3,2} = \{(1100), (1010), (1001), (0110), (0101), (0011)\}$. By computing, $W(X_{6,3,2}) = 12,739 + 12,739 + 12,867 + 12,611 + 12,739 + 12,739 = 76,434$. By the aforementioned formula, we have $W(X_{6,3,2}) = 3 \cdot 52236 - 6 \cdot 1 \cdot 13,379 = 156,708 - 80,274 = 76,434$.

The equality of Proposition 3.10.4 can be rewritten in the form

$$W(X_{n,\ell,k}) = \frac{1}{3k!}\,(n-2)(n-3)(n-4)\ldots(n-k-1)\,\phi(n,\ell,k), \qquad (3.7)$$

where $\phi(n, \ell, k) = 3W_{max} - 8k(\ell - 1)^2 n(n - 1)$. For small k and ℓ, we have the following equalities:

$$W(X_{n,2,1}) = (n - 2)(16n^3 + 64n^2 + 166n + 81)/3,$$
$$W(X_{n,2,2}) = (n - 2)(n - 3)(16n^3 + 56n^2 + 174n + 81)/6,$$
$$W(X_{n,2,3}) = (n - 2)(n - 3)(n - 4)(16n^3 + 48n^2 + 182n + 81)/18,$$
$$W(X_{n,3,1}) = (n - 2)(128n^3 + 256n^2 + 372n + 81)/3,$$
$$W(X_{n,3,2}) = (n - 2)(n - 3)(128n^3 + 224n^2 + 404n + 81)/6,$$
$$W(X_{n,3,3}) = (n - 2)(n - 3)(n - 4)(128n^3 + 192n^2 + 436n + 81)/18,$$
$$W(X_{n,4,1}) = (n - 2)(144n^3 + 192n^2 + 206n + 27),$$
$$W(X_{n,4,2}) = (n - 2)(n - 3)(144n^3 + 168n^2 + 230n + 27)/2,$$
$$W(X_{n,4,3}) = (n - 2)(n - 3)(n - 4)(144n^3 + 144n^2 + 254n + 27)/6.$$

From formulas (3.6) and (3.7), one can obtain a relation between the Wiener indices of hexagonal chains of $\mathcal{K}_{n,\ell,2k} \subseteq \mathcal{S}_{n,\ell}$ and hexagonal chains induced by the vector set $X_{n,\ell,2k}$:

$$(k + 1)(k + 2)\ldots\cdot 2k\, W(X_{n,\ell,2k}) = 2^k (n - 3)(n - 5)\ldots(n - 2k - 1)W(\mathcal{K}_{n,\ell,2k}).$$

For small k, we have the following expressions:

$$W(X_{n,\ell,2}) = (n - 3)W(\mathcal{K}_{n,\ell,2})$$
$$3W(X_{n,\ell,4}) = (n - 3)(n - 5)W(\mathcal{K}_{n,\ell,4})$$
$$15W(X_{n,\ell,6}) = (n - 3)(n - 5)(n - 7)W(\mathcal{K}_{n,\ell,6}).$$

3.11   Degeneracy of the Wiener Index

A problem of the uniqueness representation of graph's structure by their invariants attracts the great attention of researches in graph theory and its applications [4,5,37,48]. Many efforts have been devoted to find invariants with good ability to distinguish between nonisomorphic graphs. An information-theoretic approach to design topological indices has been developed [3,6,11]. New information-theoretic measures for graphs based on information functionals have been recently proposed [9,10,12,13]. The derived graph invariants have high discriminating ability even in the case of general graphs. A graph invariant is called degenerate if it possesses the same value for more than one graph. A set of graphs with the same value of a given invariant forms a degeneracy class. Since a graph invariant can be regarded as a measure of structural similarity of molecular graphs, the finding of information on degeneracy classes can be useful for applications.

In this section, we will describe some properties of degeneracy classes for the Wiener index of hexagonal chains with segments of equal length. The discriminating ability of the Wiener index for hexagonal chains was studied in [16,17,32,39,40,47]. By a *degeneracy class*, we will mean a subset of $\mathcal{G}_{n,\ell}$ consisting of all hexagonal chains with the same Wiener index. A *trivial* degeneracy class contains the unique hexagonal chain. For example, the zigzag and the spiral chains form trivial degeneracy classes. The cardinality of $\mathcal{G}_{n,\ell}$ grows as 2^n, while the number of possible values of the Wiener index for chains from $\mathcal{G}_{n,\ell}$ grows only as $(nl)^3$ when n increases. Therefore, for each value of W, the average cardinality of the corresponding degeneracy class has exponential growth.

Proposition 3.11.1 *For hexagonal chains $G_1 \in \mathcal{G}_{n,\ell_1}$ and $G_2 \in \mathcal{G}_{n,\ell_2}$ with $\mathbf{r}(G_1) = \mathbf{r}(G_2)$,*

$$W(G_1) = W_{max,\ell_1} - \left(\frac{\ell_1 - 1}{\ell_2 - 1}\right)^2 \left(W_{max,\ell_2} - W(G_2)\right). \tag{3.8}$$

Proof. By Corollary 3.6.2, we can write

$$W(G_1) = W_{max,\ell_1} - 16(\ell_1 - 1)^2 \sum_{S_i \in U(G_1)} (i - 1)(n - i),$$

$$W(G_2) = W_{max,\ell_2} - 16(\ell_2 - 1)^2 \sum_{S_i \in U(G_2)} (i - 1)(n - i).$$

Since codes of hexagonal chains G_1 and G_2 coincide, the sums of these equations are the same. $\square$

The Wiener index of an arbitrary hexagonal chain can be calculated from the Wiener index of the corresponding fibonacene.

Corollary 3.11.1 *For a fibonacene $F \in \mathcal{F}_n$ and for a hexagonal chain $G \in \mathcal{G}_{n,\ell}$ with $\mathbf{r}(F) = \mathbf{r}(G)$,*

$$W(G) = (\ell - 1)^2 W(F) + W_{max,\ell} - (\ell - 1)^2 W_{max,2}.$$

This result shows that characteristics of degeneracy classes of $\mathcal{G}_{n,\ell}$ are the same for hexagonal chains with the fixed number of segments n and arbitrary segment length ℓ. Table 3.1 contains data for sets of hexagonal chains $\mathcal{G}_{n,\ell}$. Here,

n is the number of segments in hexagonal chains;

$h_2, h_3, \ldots, h_\ell$ are the numbers of rings in hexagonal chains with segments of length $2, 3, \ldots, \ell$;

N_{self} is the number of self-linked hexagonal chains;

N_{max} is the maximal cardinality of degeneracy classes;

N_{all} is the number of all degeneracy classes;

N_{tr} is the number of all nontrivial degeneracy classes.

It is well-known that $W(G_1) \equiv W(G_2) \pmod 8$ for graphs of arbitrary catacondensed hexagonal systems G_1, G_2 with the same number of rings, and W-values of

TABLE 3.1: Degeneracy Classes of Hexagonal Chains with n Segments of Length ℓ

n	h_2	h_3	$\ldots$	h_ℓ	$|\mathcal{G}_{n,\ell}|$	N_{self}	N_{Ws}	N_{max}	N_{all}	N_{tr}
3	4	7	$\ldots$	$3\ell - 2$	2	0	0	1	2	0
4	5	9	$\ldots$	$4\ell - 3$	3	1	0	1	3	0
5	6	11	$\ldots$	$5\ell - 4$	6	0	0	1	6	0
6	7	13	$\ldots$	$6\ell - 5$	10	2	0	2	9	1
7	8	15	$\ldots$	$7\ell - 6$	20	0	0	2	18	2
8	9	17	$\ldots$	$8\ell - 7$	36	4	0	4	21	11
9	10	19	$\ldots$	$9\ell - 8$	72	0	2	4	47	19
10	11	21	$\ldots$	$10\ell - 9$	136	8	0	8	49	35
11	12	23	$\ldots$	$11\ell - 10$	272	0	0	10	102	68
12	13	25	$\ldots$	$12\ell - 11$	528	16	4	20	91	73
13	14	27	$\ldots$	$13\ell - 12$	1,056	0	4	20	209	165
14	15	29	$\ldots$	$14\ell - 13$	2,080	32	16	48	155	133
15	16	31	$\ldots$	$15\ell - 14$	4,160	0	0	47	350	314
16	17	33	$\ldots$	$16\ell - 15$	8,256	64	54	118	241	221
17	18	35	$\ldots$	$17\ell - 16$	16,512	0	80	125	547	501
18	19	37	$\ldots$	$18\ell - 17$	32,896	128	184	312	359	339
19	20	39	$\ldots$	$19\ell - 18$	65,792	0	0	312	800	758
20	21	41	$\ldots$	$20\ell - 19$	131,328	256	626	882	505	483
21	22	43	$\ldots$	$21\ell - 19$	262,656	0	430	888	1133	1083
22	23	45	$\ldots$	$22\ell - 20$	524,800	512	1928	2440	699	667
23	24	47	$\ldots$	$23\ell - 22$	1,049,600	0	0	2670	1542	1472
24	25	49	$\ldots$	$24\ell - 23$	2,098,176	1024	6720	7744	923	889
25	26	51	$\ldots$	$25\ell - 24$	4,196,352	0	7788	7813	2029	1969
26	27	53	$\ldots$	$26\ell - 25$	8,390,656	2048	23344	25392	1193	1157
27	28	55	$\ldots$	$27\ell - 26$	16,781,312	0	0	24781	2614	2554
28	29	57	$\ldots$	$28\ell - 27$	33,558,528	4096	74682	78778	1519	1487
29	30	59	$\ldots$	$29\ell - 28$	67,117,056	0	57586	80778	3297	3229

graphs of this class are always odd [25,34]. The following congruences restrict the possible values of the Wiener index for hexagonal chains having segments with equal length.

Proposition 3.11.2 *For arbitrary hexagonal chains $G_1, G_2 \in \mathcal{G}_{n,\ell}$,*

1. $W(G_1) \equiv W(G_2) \pmod{16(\ell - 1)^2}$.

2. *If n is even, then* $W(G_1) \equiv W(G_2) \pmod{32(\ell - 1)^2}$.

3. *If n is odd, then*

    3.1. *If all segments of $U(G_1)$ and $U(G_2)$ are in odd positions, then $W(G_1) \equiv W(G_2) \pmod{64(\ell - 1)^2}$.*

    3.2. *If the numbers $|U(G_1)|$ and $|U(G_2)|$ have the same parity, then $W(G_1) \equiv W(G_2) \pmod{32(\ell - 1)^2}$.*

Proof. Properties (1) and (2) directly follow from formula (3.3): if n is even, then $(i - 1)(n - i)$ is always even. Let n be odd. Then $(i - 1)(n - i)$ is even for odd i and is odd for even i. Therefore, the equal parity of $|U(G_1)|$ and $|U(G_2)|$ implies that $\sum_{S_i \in U(G_1)} (i - 1)(n - i) - \sum_{S_j \in U(G_2)} (j - 1)(n - j)$ is even. □

Two hexagonal chains G_1 and G_2 of $\mathcal{G}_{n,\ell}$ are called W_s-linked if $W(G_1) = W(G_2) = W_s$ and G_1, G_2 are linked but non-self-linked chains, that is, $\mathbf{r}(G_1) = \bar{\mathbf{r}}(G_2)$ and $G_1 \not\cong G_2$. The numbers of W_s-linked chains for small numbers of segments are shown in Table 3.1 (column *Ws*).

Based on properties of fibonacenes [19], we formulate the corresponding statements for hexagonal chains of $\mathcal{G}_{n,\ell}$.

Proposition 3.11.3 *For hexagonal chains with h rings and equal segment length, the following apply:*

1. *W_s-linked hexagonal chains exist for every odd $h \geq 13$.*

2. *W_s-linked hexagonal chains do not exist for $h \equiv 0 \pmod 4$.*

3. *The existence of W_s-linked hexagonal chains for $h \equiv 2 \pmod 4$ is a conjecture for large numbers of rings.*

The numbers of W_s-linked chains are presented in Table 3.1 for small number of segments.

In practice, it is quite difficult to get information about degeneracy classes for graphs with arbitrary number of vertices or rings without computer calculations. Because of Corollary 3.11.1, general properties of the degeneracy classes for hexagonal chains of $\mathcal{G}_{n,\ell}$, $\ell \geq 3$, are the same as for fibonacenes of $\mathcal{F}_n$ [19].

Proposition 3.11.4 *For the set of hexagonal chains $\mathcal{G}_{n,\ell}$,*

1. *If $n \geq 4$ is even, then self-linked and W_s-linked hexagonal chains form one degeneracy class. Its cardinality is more than $2^{(n-4)/2}$ (equality holds for $n \leq 10$).*

2. *If $n \geq 9$ is odd, then W_s-linked hexagonal chains form one degeneracy class when ℓ is even and $\ell - n \equiv 1 \pmod 4$. For the number of rings, a relation $h \equiv 2 \pmod 4$ holds.*

   If ℓ is even and $\ell - n \equiv 3 \pmod 4$, then W_s is a nonrealizable value of the Wiener index. For this case, $h \equiv 0 \pmod 4$.

3. *If hexagonal chains (except for graphs of points 1 and 2) form a degeneracy class, then their linked graphs also form a degeneracy class with the same cardinality.*

A graphical illustration of this proposition is presented in Figure 3.10. Black circles represent hexagonal chains that are grouped in degeneracy classes. Table 3.1 contains data of degeneracy classes for hexagonal chains with up to 29 segments.

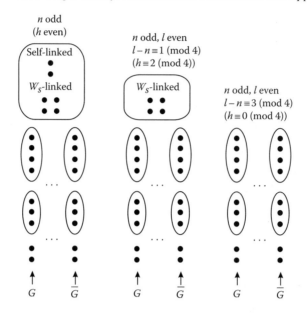

FIGURE 3.10: Degeneracy classes of hexagonal chains of $\mathcal{G}_{n,\ell}$.

Acknowledgment

This work was supported by the Russian Foundation for Basic Research (project number 12-01-00631) and grant NSh-1939.2014.1 of President of the Russian Federation for Leading Scientific Schools.

References

1. A.T. Balaban. Chemical graphs. L. Symmetry and enumeration of fibonacenes (unbranched catacondensed benzenoids isoarithmic with helicenes and zigzag catafusenes). *MATCH Communications in Mathematical and in Computer Chemistry*, 24:9–38, 1989.

2. A.T. Balaban, I. Motoc, D. Bonchev, and O. Mekenyan. Topological indices for structure-activity correlations. *Topics in Current Chemistry*, 114:21–55, 1983.

3. D. Bonchev. *Information Theoretic Indices for Characterization of Chemical Structure*. Research Studies Press, Chichester, U.K., 1983.

4. D. Bonchev and D.H. Rouvray, eds. *Chemical Graph Theory—Introduction and Fundamentals*. Gordon & Breach, New York, 1991.

5. D. Bonchev and D.H. Rouvray, eds. *Complexity in Chemistry, Biology, and Ecology. Mathematical and Computational Chemistry.* Springer, New York, 2005.

6. D. Bonchev and N. Trinajstić. Information theory, distance matrix and molecular branching. *Journal of Chemical Physics*, 67:4517–4533, 1977.

7. M. Dehmer, ed. *Structural Analysis of Complex Networks*. Birkhäuser Publishing, Basel, Switzerland, 2011.

8. M. Dehmer and F. Emmert-Streib, eds. *Analysis of Complex Networks: From Biology to Linguistics*. Wiley-VCH Publishing, Weinheim, Germany, 2009.

9. M. Dehmer and M. Grabner. The discrimination power of molecular identification numbers revisited. *MATCH Communications in Mathematical and in Computer Chemistry*, 69(3):785–794, 2013.

10. M. Dehmer, M. Grabner, A. Mowshowitz, and F. Emmert-Streib. An efficient heuristic approach to detecting graph isomorphism based on combinations of highly discriminating invariants. *Advances in Computational Mathematics*, 39:311–325, 2012. DOI: 10.1007/s10444-012-9281-0.

11. M. Dehmer and A. Mowshowitz. A history of graph entropy measures. *Information Sciences*, 1:57–78, 2011.

12. M. Dehmer, A. Mowshowitz, and K. Varmuza. Information indices with high discrimination power for arbitrary graphs. *PLoS ONE*, 7:e31214, 2012.

13. M. Dehmer, L. Sivakumar, and K. Varmuza. On distance-based entropy measures, pp. 123–138. In: I. Gutman and B. Furtula, eds., *Distance in Molecular Graphs— Theory*. Mathematical Chemistry Monographs, 12. University of Kragujevac and Faculty of Science Kragujevac, Kragujevac, Serbia, 2012.

14. J. Devillers and A.T. Balaban, eds. *Topological Indices and Related Descriptors in QSAR and QSPR*. Gordon & Breach, Reading, MA, 1999.

15. M.V. Diudea and I. Gutman. Wiener-type topological indices. *Croatica Chemica Acta*, 71(1):21–51, 1998.

16. A.A. Dobrynin. Distribution of the distance numbers of unbranched hexagonal system graphs. *Vychisl. Sistemy*, 136:61–141, 1990 (in Russian).

17. A.A. Dobrynin. Graph distance numbers of nonbranched hexagonal systems. *Siberian Advances in Mathematics*, 2:121–134, 1990.

18. A.A. Dobrynin. A simple formula for the calculation of the Wiener index of hexagonal chains. *Computers & Chemistry*, 23(1):43–48, 1999.

19. A.A. Dobrynin. On the Wiener index of fibonacenes. *MATCH Communications in Mathematical and in Computer Chemistry*, 64(3):707–726, 2010.

20. A.A. Dobrynin. On the Wiener index of certain families of fibonacenes. *MATCH Communications in Mathematical and in Computer Chemistry*, 70(2):565–574, 2013.

21. A.A. Dobrynin, R. Entringer, and I. Gutman. Wiener index of trees: Theory and applications. *Acta Applicandae Mathematicae*, 66(3):211–249, 2001.

22. A.A. Dobrynin and I. Gutman. The average Wiener index of hexagonal chains. *Computers & Chemistry*, 23(6):571–576, 1999.

23. A.A. Dobrynin, I. Gutman, S. Klavžar, and P. Žigert. Wiener index of hexagonal systems. *Acta Applicandae Mathematicae*, 72(3):247–294, 2002.

24. R. Entringer. Distance in graphs: Trees. *Journal of Combinatorial Mathematics and Combinatorial Computing*, 24:65–84, 1997.

25. I. Gutman. Wiener numbers of benzenoid hydrocarbons: Two theorems. *Chemical Physics Letters*, 136:134–136, 1987.

26. I. Gutman. Topological properties of benzenoid systems. *Topics in Current Chemistry*, 162:21–28, 1992.

27. I. Gutman and S.J. Cyvin. *Introduction to the Theory of Benzenoid Hydrocarbons*. Springer-Verlag, Berlin, Germany, 1989.

28. I. Gutman and B. Furtula, eds. *Distance in Molecular Graphs—Applications*. Mathematical Chemistry Monographs, 13. University of Kragujevac and Faculty of Science Kragujevac, Kragujevac, Serbia, 2012.

29. I. Gutman and B. Furtula, eds. *Distance in Molecular Graphs—Theory*. Mathematical Chemistry Monographs, 12. University of Kragujevac and Faculty of Science Kragujevac, Kragujevac, Serbia, 2012.

30. I. Gutman, J.W. Kennedy, and L.V. Quintas. Wiener numbers of random benzenoid chains. *Chemical Physics Letters*, 173:403–408, 1990.

31. I. Gutman and S. Klavžar. Chemical graph theory of fibonacenes. *MATCH Communications in Mathematical and in Computer Chemistry*, 55:39–54, 2006.

32. I. Gutman, S. Marković, D. Luković, V. Radivojević, and S. Rančić. On Wiener numbers of benzenoid hydrocarbons. *Collection of Scientific Papers of the Faculty of Science Kragujevac*, 8:15–34, 1987.

33. I. Gutman and O.E. Polansky. *Mathematical Concepts in Organic Chemistry*. Springer-Verlag, Berlin, Germany, 1986.

34. I. Gutman and O.E. Polansky. Wiener numbers of polyacenes and related benzenoid molecules. *MATCH Communications in Mathematical and in Computer Chemistry*, 20:115–123, 1986.

35. I. Gutman and J.H. Potgieter. Wiener index and intermolecular forces. *Journal of the Serbian Chemical Society*, 62:185–192, 1997.

36. I. Gutman, Y.N. Yeh, S.L. Lee, and Y.L. Luo. Some recent results in the theory of the Wiener number. *Indian Journal of Chemistry*, 32A:651–661, 1993.

37. F. Harary. *Graph Theory*. Addison-Wesley, Reading, MA, 1969.

38. S. Klavžar and I. Gutman. Wiener number of vertex-weighted graphs and a chemical application. *Discrete Applied Mathematics*, 80:73–81, 1997.

39. E.V. Konstantinova. The discrimination ability of some topological and information distance indices for graphs of unbranched hexagonal systems. *Journal of Chemical Information and Computer Sciences*, 36:54–57, 1996.

40. E.V. Konstantinova and A.A. Paleev. On sensitivity of topological indices for polycyclic graphs. *Vychisl. Sistemy*, 136:38–48, 1990 (in Russian).

41. S. Nikolić, N. Trinajstić, and Z. Mihalić. The Wiener index: Developments and applications. *Croatica Chemica Acta*, 68:105–129, 1995.

42. O.E. Polansky and D. Bonchev. The Wiener number of graphs. I. General theory and changes due to some graph operations. *MATCH Communications in Mathematical and in Computer Chemistry*, 21:133–186, 1986.

43. O.E. Polansky and D. Bonchev. Theory of the Wiener number of graphs. II. Transfer graphs and some of their metric properties. *MATCH Communications in Mathematical and in Computer Chemistry*, 25:3–39, 1990.

44. D.H. Rouvray. Should we have designs on topological indices?, pp. 159–177. In: R.B. King, ed., *Chemical Application of Topology and Graph Theory*. Studies in Physical and Theoretical Chemistry, 28. Elsevier, Amsterdam, the Netherlands, 1983.

45. D.H. Rouvray. The modelling of chemical phenomena using topological indices. *Journal of Computational Chemistry*, 8:470–480, 1987.

46. A. Sabljić and N. Trinajstić. Quantitative structure–activity relationships: The role of topological indices. *Acta Pharmaceutica Jugoslavica*, 31:189–214, 1981.

47. V.A. Skorobogatov, E.V. Mzhelskaya, and N.M. Meyrmanova. A study of metric characteristics of cata-condensed polybenzenoids. *Vychisl. Sistemy*, 127:40–91, 1988 (in Russian).

48. R. Todeschini and V. Consonni. *Handbook of Molecular Descriptors*. Wiley-VCH, Weinheim, Germany, 2000.

49. N. Trinajstić. *Chemical Graph Theory*. CRC Press, Boca Raton, FL, 1983; 2nd edn., 1992.

50. H. Wiener. Structural determination of paraffin boiling points. *Journal of the American Chemical Society*, 69:17–20, 1947.

51. M. Zander. *Polycyclische Aromaten*. Teubner, Stuttgart, Germany, 1995.

Chapter 4

Metric-Extremal Graphs

Ivan Gutman and Boris Furtula

Contents

4.1 Introduction: Quantifying a Graph..111
    4.1.1 Mathematical Preliminaries...112
    4.1.2 Metric in Graph..112
    4.1.3 Matrix Representations of Graphs....................................113
    4.1.4 Metric Invariants of Graphs...114
4.2 Definition and Main Properties of Metric Invariants W, WW, and H............115
    4.2.1 Wiener Index...115
    4.2.2 Hyper-Wiener Index...117
    4.2.3 Harary Index...118
    4.2.4 Connections to Hosoya Polynomial...................................118
    4.2.5 More Metric Invariants: Szeged and Terminal Wiener Indices..........119
4.3 Metric-Extremal Graphs...120
    4.3.1 Metric-Extremal General Graphs.....................................120
    4.3.2 Metric-Extremal Trees..122
    4.3.3 Metric-Extremal Unicyclic and Bicyclic Graphs......................126
4.4 Concluding Remarks...128
References..129

4.1 Introduction: Quantifying a Graph

A graph is a mathematical object defined as an ordered pair (V, E) of sets V and E, where V is a (finite or infinite) set of some unspecified elements, called vertices, and E is a set of some (ordered or unordered) pairs of elements of V, called edges. Thus, a graph is an abstract set-theoretical concept. As such, it cannot be viewed as something *quantitative*.* Yet, for most applications of graphs, pertinent numerical indicators are needed, which requires that their structural aspects be *quantified* (see, for instance, [5,11,16,29,30,33,37,81,91,93,122]). This can be done in many different ways, depending on the nature of the intended application, or on the mathematical apparatus preferred.

A direct and straightforward quantification of graphs is by associating with each vertex $v \in V$ and with each edge $e \in E$ real (or, in rare cases, complex) numbers

---

* Under *quantitative*, we understand something related with, or consisting of, numbers.

$w(v)$ and $w(e)$, referred to as the *weights* of the respective vertex or edge. The so constructed *weighted graphs* have found countless applications in various areas of natural, technical, social, and computer sciences, communication, transportation, business, military, medicine, etc.

In this chapter, we follow a different, metric-based, direction of graph quantification. It is conceived so as to reflect the actual structure of the underlying graph and appears to be suitable for *pure* and *applied* mathematics-based studies, especially for those aiming at establishing structure–property relations in chemistry. In other words, in this chapter, we provide a survey of some relations between special metric-based indices of chemical interest.

4.1.1   Mathematical Preliminaries

Throughout this chapter, we restrict our considerations to graphs with finite vertex sets and with edge sets consisting of unordered vertex pairs. In addition, we require that a vertex pair (v, v) is never contained in the edge set. Graphs of this kind are referred to as *simple* or *schlicht*.

Let $G = (V, E)$ be a simple graph. Then its vertex and edge sets are $V = V(G) = \{v_1, \ldots, v_n\}$ and $E = E(G) = \{e_1, \ldots, e_m\}$, respectively. Thus, the number of vertices and edges of G is denoted by n and m, respectively.

A *path* in a graph G is a sequence of its mutually distinct vertices $v_1, v_2, \ldots, v_{k+1}$, such that $(v_i, v_{i+1}) \in E(G)$ for all $i = 1, 2, \ldots, k$. The *length* of this path is k, and it connects the vertices v_1 and v_{k+1}.

A graph is said to be *connected* if any two of its vertices are connected by at least one path. Throughout this chapter, if not stated otherwise, the graphs considered are assumed to be connected.

In what follows, we denote by C_n, P_n, S_n, and K_n the cycle graph, the path graph, the star graph, and the complete graph of order n, respectively. Other undefined notation and terminology can be found in [11,81]. For concepts related to applications of graphs in chemistry, especially for the concept of *molecular graph*, the reader is directed to the books [76,138].

4.1.2   Metric in Graph

At this point, we recall the mathematical definition of a metric.

Let S be a set. A *metric* on this set is a function f that maps a real number to any pair of elements of S, such that for any $x, y, z \in S$,

$$f(x, y) \geq 0 \tag{4.1}$$

$$f(x, y) = 0 \text{ if and only if } x = y \tag{4.2}$$

$$f(x, y) = f(y, x) \tag{4.3}$$

$$f(x, z) \leq f(x, y) + f(y, z). \tag{4.4}$$

The function f with properties (4.1) through (4.4) is said to be a *distance* on the set S. For more details of the distance concept, see [37].

A metric on a graph can be defined in many different ways. The intuitively most obvious and the most frequently used is a metric on the vertex set, referred to as the ordinary *distance between vertices* [14,131].

Although in what follows we shall be concerned with this kind of metric, it should be noted that other kinds of graph metrics exist [37]. For instance, one can consider distances or similarities between graphs [32,157].

The metric employed throughout this chapter is defined in the following standard manner [14,131]. This is a metric of the vertex set.

Let G be a connected simple graph. The *distance* $d(u, v|G)$ (or simply $d(u, v)$ when no misunderstanding could occur) between the vertices u and v of G is equal to the length of (=number of edges in) a shortest path connecting u and v. It is easy to check that the distance so defined satisfies the conditions (4.1) through (4.4). Details on distance in graph theory can be found in the books [14,25,60].

Metric in graphs may be defined also on the edge set. Here, however, the situation is not as straightforward as in the case of vertices [63,88]. There are several possible ways to define the distance between two edges in a graph [88]. The most usual [23–25] is the following:

If G is a simple graph with edge set $E(G)$, then its *line graph* $L(G)$ is the graph whose vertex set is $E(G)$ [11,81]. Two vertices of $L(G)$ are adjacent if the corresponding edges of G have a common end vertex.

Let e and e' be two edges of the graph G, that is, $e, e' \in E(G)$. Let these correspond to the vertices v and v' of the line graph $L(G)$. Then the distance between the edges e and e' in the graph G is equal to the distance between the vertices v and v' in the line graph $L(G)$. In other words, $d(e, e'|G) = d(v, v'|L(G))$.

4.1.3 Matrix Representations of Graphs

A pair of vertices of G is either contained in $E(G)$ or not. In view of this, we can associate the number 1 to any pair contained in $E(G)$ and the number 0 to any pair not contained in $E(G)$. If the elements of the vertex set are labeled by $v_1, v_2, \ldots, v_n$, then the numbers associated with the edges form a $(0, 1)$ matrix, referred to as the *adjacency matrix* and denoted by $A = A(G) = \|a_{ij}\|$. If the graph G is simple, then $A(G)$ is a symmetric square matrix of order n with zero diagonal.

Evidently, the graph G is fully determined by its adjacency matrix, or—as mathematicians use to say—the adjacency matrix $A(G)$ determines the graph G up to isomorphism.

By means of the adjacency matrix, the powerful apparatus of the matrix theory and linear algebra are connected with graphs and can be used in their study. In fact, any property of a graph G can be, directly or less directly, expressed as a function of the elements of the adjacency matrix $A(G)$. This, in particular, applies to real-number graph invariants, reflecting certain structural features of G. Many of these invariants, especially those discussed in the latter parts of this chapter, are easier to express by means of the following, more advanced type of graph quantification.

Countless other matrices can be associated with graphs [91]. We specify here only the *distance matrix* $D = D(G) = \|d_{ij}\|$.

If G is a connected graph with vertex set $V(G) = \{v_1, v_2, \ldots, v_n\}$, then its distance matrix $D(G)$ is the square matrix of order n whose (i, j) entry is equal to $d(v_i, v_j | G)$.

It is not difficult to see that also the distance matrix determines the graph up to isomorphism. Indeed, whenever $d_{ij} = 1$, then the respective vertices v_i and v_j are adjacent, that is, $a_{ij} = 1$; whenever $d_{ij} \neq 1$, then the vertices v_i and v_j are not adjacent, that is, $a_{ij} = 0$.

4.1.4   Metric Invariants of Graphs

When the structural aspects of a graph are quantified, then one can calculate its appropriate real-number invariants. These invariants reflect certain structural details of the underlying graph. Although such invariants are capable of carrying only a partial information on the graph structure, in practical applications, they may prove to be of great value. In particular, in chemical applications, pertinently chosen graph invariants are referred to as *topological indices* or *molecular structure descriptors* [135,136].

A large variety of metric-based graph invariants (often referred to as distance-based topological indices) has been considered in the literature. Of these, in this chapter, we focus our attention to the following:

- Wiener index W

- Hyper-Wiener index WW

- Harary index H

We will be mainly concerned with the extremal (minimal and maximal) values that these metric invariants assume in various classes of graphs. For a more detailed survey on this matter, see [151]. Note that, for example, in [31], extremal properties of graph entropies based on the so-called information functionals have been studied. Proving extremal properties of such graph measures remains intricate, especially for finding analytical results when characterizing graphs whose entropy is minimal [31].

By determining the metric-extremal members of a given class of graphs, we can establish several important features of the underlying graph invariant.

For instance, by knowing that among n-vertex trees, the star S_n and the path P_n have minimal and maximal Wiener indices and that $W(S_n) = (n - 1)^2$ and $W(P_n) = n(n^2 - 1)/6$, then we see that there may be at most $(n^6 - 6n^2 + 11n)/6$ distinct Wiener indices. Since this is far smaller than the number of distinct n-vertex trees, we conclude that there must exist large families of trees having equal W value [74,78].

Although until the present moment a generally accepted measure of branching does not exist, any graph invariant (GI) acceptable as a measure of branching must satisfy the inequalities [7,9,10,124]

$$GI(S_n) < GI(T) < GI(P_n) \quad \text{or} \quad GI(P_n) < GI(T) < GI(S_n)$$

for any tree T of order $n \geq 5$ different from S_n and P_n. In what follows, we will see that the distance-based indices W, WW, and H satisfy this basic requirement and thus may serve as measures of branching of graphs or molecules.

4.2 Definition and Main Properties of Metric Invariants W, WW, and H

4.2.1 Wiener Index

The history of distance-based graph invariants begins in 1947, when the physical chemist (and future psychiatrist) Harold Wiener [143] proposed the following formula to calculate the boiling point t_B of alkanes:

$$t_B = a\,W(G) + b\,W_P(G) + c.$$

In this formula, a, b, c are constants for a given isomeric group and $W(G)$ is (in modern terminology, different from what Wiener originally used) equal to the sum of distances of all unordered vertex pairs in the molecular graph G, whereas $W_P(G)$ is the number of unordered vertex pairs at distance 3 in G, that is, $d(G, 3)$. Initially, Wiener's ideas did not attract the attention of the chemical community, but as time passes, the quantity W became one of the most popular molecular structure descriptors. It found numerous applications for designing quantitative structure–property relationships (QSPR) [36,93]. Besides, it also was applied in crystallography, communication theory, facility location, cryptology, etc. [8,79,119]. Eventually, this graph invariant became known under the name *Wiener index* or *Wiener number*; for details, see [125,126].

Mathematicians started with the study of $W(G)$ almost three decades after chemists [49], initially without any knowledge of Wiener's earlier works. Anyway, also in contemporary mathematical literature, $W(G)$ is usually referred to as the Wiener index [39,42,48,59,97,129,140].

In the previously specified notation, the Wiener index is defined as

$$W(G) = \sum_{\{u,v\} \subseteq V(G)} d(u, v | G). \tag{4.5}$$

Note that Equation 4.5, explicitly using the graph–distance concept, was first given by Hosoya [84]. Wiener himself spoke about the number of carbon–carbon bonds separating two carbon atoms [143].

Evidently,

$$\overline{W}(G) = \binom{n}{2}^{-1} W(G) = \binom{n}{2}^{-1} \sum_{\{u,v\} \subseteq V(G)} d(u, v | G)$$

is the average values of the distances between vertices of the graph G. In the mathematical literature, this quantity was also much studied, under the name of *mean distance* [13,46,82,98,115,144] or *average distance* [1,3,19–22,26,27,50,101,130].

At this point also, the edge-Wiener index should be mentioned. It is defined in analogy to Equation 4.5 as [88]

$$W_e(G) = \sum_{\{e,e'\} \subseteq E(G)} d(e, e'|G). \tag{4.6}$$

As explained earlier, the distance between two edges in a graph is equal to the distance between the respective two vertices in the line graph. Therefore [23,24], the edge-Wiener index of a graph coincides with the ordinary Wiener index of the line graph:

$$W_e(G) = W(L(G)).$$

Yet some independent studies of the edge-Wiener index were recently communicated [4,23,45,153].

After the Wiener index was invented, a large number of other metric-based graph invariants have been proposed and considered in the chemical and mathematico-chemical literature; for more information and additional references, see [6,38,67, 68,70,71,102,106,109–112,135,136,151]. All distance-based graph invariants can be derived from the distance matrix or some closely related distance-based matrices; for more information on this matter, see [18,89–91,112,114,123].

It is worth noting that the Wiener index is contained in a concealed manner in many other metric-based graph invariants. First of all, Buckley showed [13] that the Wiener index of a tree T of order n and its line graph $L(T)$ are related as

$$W(T) = W(L(T)) - \binom{n}{2}.$$

This automatically implies the simple relation

$$W_e(T) = W(T) + \binom{n}{2}$$

between the Wiener and edge-Wiener index of a tree.

For cycle-containing graphs, the situation is somewhat more complicated. For unicyclic graphs [62], $W_e(G) \leq W(G)$, with equality if and only if $G \cong C_n$. For bicyclic graphs [73,75], all the three cases $W_e < W$, $W_e = W$, and $W_e > W$ may happen. If the graph has no pendent vertices, then [17,145] $W_e \geq W$ holds.

A weighted version of the Wiener index is the *degree distance* defined as [43]

$$DD(G) = \sum_{\{u,v\} \subseteq V(G)} [\deg(u) + \deg(v)]d(u, v|G)$$

where $\deg(u)$ is the degree (number of first neighbors) of the vertex u.

The same quantity $DD(G)$ was examined in the paper [61] under the name *Schultz index*. Namely, somewhat earlier, Schultz [127] proposed a structure descriptor named *molecular topological index*, defined as

$$MTI(G) = \sum_{i=1}^{n} \mathbf{X}[A(G) + D(G)]_i$$

where $\mathbf{X}$ is the vector of vertex degrees. It can be easily shown that

$$MTI(G) = \sum_{u \in V(G)} \deg(u)^2 + \sum_{\{u,v\} \subseteq V(G)} [\deg(u) + \deg(v)] d(u, v|G)$$

which means that

$$MTI(G) = M_1(G) + DD(G)$$

with $M_1(G)$ standing for the well-known first Zagreb index (cf. [135,136]). Klein et al. [96] discovered the simple relation between degree distance and Wiener index:

$$DD(G) = 4W(G) - n(n - 1) \tag{4.7}$$

which holds for trees with n vertices.

In [61], a novel proof of the relation (4.7) was put forward. Then it was noticed that a fully analogous relation can be obtained for the multiplicative variant of the degree distance, namely, for

$$Sch(G) = \sum_{\{u,v\} \subseteq V(G)} \deg(u)\, \deg(v)\, d(u, v|G). \tag{4.8}$$

This relation reads [61]

$$Sch(G) = 4W(G) - (2n - 1)(n - 1)$$

and also holds for trees of order n. The author of [61] called $Sch(G)$ the *Schultz index of the second kind* and by no means proposed it for a novel molecular structure descriptor. Unfortunately, not carefully reading the paper [61], the authors of the seminal handbooks [135,136] included $S(G)$ among the topological indices and, even worse, named it *Gutman index*; see [2,53,116].

4.2.2 Hyper-Wiener Index

In 1993, Milan Randić [121] introduced a distance-based quantity that he named hyper-Wiener index and denoted by WW. His definition could be applied only to trees and was impossible to use for cycle-containing graphs. In 1995, Klein, Lukovits, and one of the present authors [95] showed that Randić's WW (for trees) satisfies the identity

$$WW(G) = \frac{1}{2} \sum_{\{u,v\} \subseteq V(G)} \left[d(u, v|G) + d(u, v|G)^2 \right] \tag{4.9}$$

which could be applied to all connected graphs. Since then, Equation 4.9 is used as the definition of hyper-Wiener index.

4.2.3   Harary Index

In 1993, Plavšić et al. [120] and Ivanciuc et al. [89] independently introduced the Harary index, named in honor of Frank Harary on the occasion of his 70th birthday. Actually, the Harary index was first proposed in 1992 by Mihalić and Trinajstić [113] as

$$H_{old}(G) = \sum_{\{u,v\} \subseteq V(G)} \frac{1}{d(u,v|G)^2}.$$

In spite of this, the Harary index is nowadays defined as [89,120]

$$H(G) = \sum_{\{u,v\} \subseteq V(G)} \frac{1}{d(u,v|G)}.$$

4.2.4   Connections to Hosoya Polynomial

The number of vertex pairs of the graph G, whose distance is k, will be denoted by $d(G,k)$.

If the graph G is not connected, and the vertices u and v belong to its different components, then their distance $d(u,v|G)$ is undefined. On the other hand, even if the graph G is disconnected, the quantities $d(G,k)$ are well defined for all $k \geq 0$. In most applications of the vertex-distance concept of disconnected graphs, this difficulty can be overcome by employing $d(G,k)$ instead of $d(u,v|G)$. All the following expressions (4.10) through (4.19) hold for both connected and disconnected graphs.

Concluding this section, we mention the *Hosoya polynomial*, defined as

$$Hos(G,\lambda) = \sum_{k \geq 1} d(G,k)\,\lambda^k . \tag{4.10}$$

Details on this distance-based polynomial can be found in the survey [80], in the references quoted therein, and in the recent paper [12].

Recall that the topological indices W, WW, and H can be expressed via the numbers $d(G,k)$ as

$$W(G) = \sum_{k \geq 1} k\,d(G,k) \tag{4.11}$$

$$WW(G) = \sum_{k \geq 1} \frac{k + k^2}{2}\,d(G,k) \tag{4.12}$$

$$H(G) = \sum_{k \geq 1} \frac{1}{k}\,d(G,k) . \tag{4.13}$$

A similar distance-based invariant was introduced in 1990 [137] and is usually referred to as the Tratch–Stankevich–Zefirov index *TSZ*. It is defined as

$$TSZ(G) = \sum_{k \geq 1} \left(\frac{1}{6} k^3 + \frac{1}{2} k^2 + \frac{1}{3} k \right) d(G,k). \tag{4.14}$$

These indices satisfy the following peculiar identities:

$$W(G) = \left. \frac{\partial}{\partial \lambda} Hos(G,\lambda) \right|_{\lambda=1} \tag{4.15}$$

$$WW(G) = \frac{1}{2} \frac{\partial^2}{\partial \lambda^2} \lambda \, Hos(G,\lambda) \Big|_{\lambda=1} \tag{4.16}$$

$$H(G) = \int_0^1 \frac{Hos(G,\lambda)}{\lambda} \, d\lambda \tag{4.17}$$

$$TSZ(G) = \frac{1}{3!} \frac{\partial^3}{\partial \lambda^3} \lambda^2 \, Hos(G,\lambda) \Big|_{\lambda=1}. \tag{4.18}$$

Formula (4.15) was discovered by Hosoya [85]. It seems that the formulas (4.16) through (4.18) were first reported in [12]. The earlier expressions suggest that also for $k > 3$, the invariants (4.19) deserve some attention in mathematical chemistry:

$$\frac{1}{k!} \frac{\partial^k}{\partial \lambda^k} \lambda^{k-1} H(G,\lambda) \Big|_{\lambda=1}. \tag{4.19}$$

4.2.5 More Metric Invariants: Szeged and Terminal Wiener Indices

In Wiener's first paper [143], a remarkable result can be found (although in a concealed form). We present it in modern mathematical language.

Let G be a graph and e its edge, connecting the vertices u and v. Define the sets $\mathcal{N}_1(e|G) = \{x \in V(G)|d(x,u) < d(x,v)\}$ and $\mathcal{N}_2(e|G) = \{x \in V(G)|d(x,u) > d(x,v)\}$. Denote by $n_1(e|G)$ and $n_2(e|G)$ the cardinalities of the sets $\mathcal{N}_1(e|T)$ and $\mathcal{N}_2(e|T)$, respectively. In words, $n_1(e|G)$ is the number of vertices of the graph G, lying closer to one end of the edge e than to its other end. The meaning of $n_2(e|G)$ is analogous. Then Wiener's result reads as follows:

Let T be a tree. Then its Wiener index satisfies the identity

$$W(T) = \sum_{e \in E(T)} n_1(e|T) \, n_1(e|T). \tag{4.20}$$

The right-hand side of formula (4.20) is well defined for all graphs, but in the general case, it differs from the Wiener index. Bearing this in mind, the extension of formula (4.20) to all graphs was put forward in [40] and eventually named *Szeged index*. Thus, the definition of the Szeged index is

$$Sz(G) = \sum_{e \in E(G)} n_1(e|G) \, n_1(e|G). \tag{4.21}$$

Its main property is that in the case of trees, it coincides with the Wiener index. There, however, is a more interesting property: The Szeged index and the Wiener index of a graph G are equal if and only if all blocks of G are complete graphs [40,41].

More details on the theory of the Szeged index can be found in the survey [65] and in the recent papers [94,117,118]. Also the edge version of the Szeged index was considered [99].

The concept of *terminal Wiener index* was put forward motivated by Zaretskii's intriguing result [156] that a tree is completely characterized by the distances between its vertices of degree one (the so-called pendent vertices or leaves); see also [132]. The respective definition, Equation 4.22, was proposed by Petrović and the present authors [72]. Somewhat later, but independently, Székely, Wang, and Wu arrived at the same idea [133].

Let $V_1(G) \subset V(G)$ be the set of vertices of the graph G whose degree is equal to one (the so-called pendent vertices or leaves). Then TW is defined in full analogy with the Wiener index, Equation 4.5, as

$$TW(G) = \sum_{\{u,v\} \subseteq V_1(G)} d(u, v|G). \qquad (4.22)$$

Thus, the terminal Wiener index consists of the sum of distances between pendent vertices. If the graph G has no pendent vertices, or just one such vertex, then $TW(G) = 0$. The application of this molecular structure descriptor is purposeful mainly for graphs with many pendent vertices, especially trees [35,66,69].

4.3   Metric-Extremal Graphs

In this section, we report some extremal results with respect to the distance-based invariants W, WW, and H in different classes of graphs. For a more detailed survey, see [151].

4.3.1   Metric-Extremal General Graphs

We start with a few elementary and well-known results:

Theorem 4.3.1 *Let G be a connected graph and $e \notin E(G)$. Then, $W(G) > W(G+e)$, $WW(G) > WW(G+e)$, and $H(G) < H(G+e)$.*

Theorem 4.3.2 *Let G be a connected graph of order n, different from the complete graph K_n and the path P_n. Then*

$$W(K_n) < W(G) < W(P_n)$$
$$WW(K_n) < WW(G) < WW(P_n)$$
$$H(P_n) < H(G) < H(K_n).$$

A connected graph G is called a *cactus* if each block of G is either an edge or a cycle. Denote by $Cact(n, t)$ the set of connected cacti possessing n vertices and t cycles. Let $C^0(n, t)$ be the cactus obtained from the star S_n by adding to it t independent edges.

Theorem 4.3.3 *Among graphs in $Cact(n, t)$, $C^0(n, t)$ is the unique graph having minimal Wiener index* [103], *minimal hyper-Wiener index* [57], *and maximal Harary index* [141].

Let $1 \leq k < n$ and K_n^k be the graph obtained by attaching k pendent vertices to a vertex of the complete graph K_{n-k}.

Theorem 4.3.4 *Among connected graphs with n vertices and k cut edges, K_n^k is the unique graph having minimal Wiener index* [86], *minimal hyper-Wiener index* [152], *and maximal Harary index* [152].

The *kite graph* $K_{n,k}$ is obtained by identifying one vertex of K_k with one pendent vertex of P_{n-k+1}. The *Turán graph* $T_n(k)$ is the complete k-partite graph of order n in which any two partition sets differ in size by at most one.

Theorem 4.3.5 *Among connected graphs with n vertices and clique number k*

- $T_n(k)$ *is the unique graph with minimal Wiener index* [56], *minimal hyper-Wiener index* [56], *and maximal Harary index* [148].

- $K_{n,k}$ *is the unique graph with minimal Harary index* [148], *maximal Wiener index* [56], *and maximal hyper-Wiener index* [56].

Theorem 4.3.6 *Among connected graphs with n vertices and chromatic number k*

- $T_n(k)$ *is the unique graph with minimal Wiener index* [56], *minimal hyper-Wiener index* [56], *and maximal Harary index* [148].

- $K_{n,k}$ *is the unique graph with maximal Wiener index* [56], *maximal hyper-Wiener index* [56], *and minimal Harary index* [148].

Denote by M_1 the sum of squares of the vertex degrees of the graph G. Recall that M_1 is the well-known first Zagreb index (see [64,77]).

The *Moore graph* is an r-regular graph with diameter k whose order attains the upper bound

$$1 + r \sum_{i=0}^{k-1} (r - 1)^i.$$

Hoffman and Singleton [83] proved that every r-regular Moore graph G with diameter 2 must have $r \in \{2, 3, 7, 57\}$. They pointed out that $G \cong C_5$ if $r = 2$, G is the Petersen graph for $r = 3$, whereas for $r = 7$, G is known under the name Hoffman–Singleton graph. For $r = 57$, it is not known whether such a graph does exist or not.

Theorem 4.3.7 *Let G be a connected triangle- and quadrangle-free graph with n vertices and m edges. Then*

- $\frac{3n(n-1)}{2} - \frac{1}{2}M_1 - m \leq W(G)$ *with equality holding if and only if the diameter of G is d $\leq$ 3* [159].

- $3n(n-1) - \frac{3}{2}M_1 - 2m \leq WW(G)$ *with equality holding if and only if the diameter of G is d $\leq$ 3* [158].

- $H(G) \leq \frac{n(n-1)}{4} + \frac{m}{2}$ *with equality holding if and only if G is a star or a Moore graph of diameter 2* [158].

4.3.2   Metric-Extremal Trees

Again, we start with a few elementary and well-known results:

Theorem 4.3.8 *Let T be a tree of order n, different from the path P_n and the star S_n. Then*

$$W(S_n) < W(T) < W(P_n)$$
$$WW(S_n) < WW(T) < WW(P_n)$$
$$H(P_n) < H(T) < H(S_n).$$

Let $T_n(n_1, n_2, \ldots, n_m)$ be the tree of order n obtained by inserting, respectively, $n_1 - 1, n_2 - 1, \ldots, n_m - 1$ vertices into the m edges of the star S_{m+1}, where $n_1 + n_2 + \cdots + n_m = n - 1$. For convenience, we denote $T_{16}(2, 2, 3, 3, 5)$ by $T_{16}(2^2, 3^2, 5)$.

Let $T_1, T_2, \ldots, T_7$ be the trees of order $n \geq 14$ and T_D depicted in Figure 4.1.

Theorem 4.3.9 (1) *Let T be a tree of order $n \geq 28$, different from those encountered in the following set of inequalities. Then* [34,107],

$$
\begin{aligned}
W(P_n) &> W(T_n(n-3, 1^2)) > W(T_n(n-4, 2, 1)) > W(T_n(1^2; 1^2)) \\
&> W(T_n(n-5, 3, 1)) > W(T_n(n-4, 1^3)) > W(T_n(1^2; 2, 1)) \\
&> W(T_n(n-6, 4, 1)) > W(T_n(n-5, 2^2)) > W(T_n(1^2; n-5, 1)) \\
&> W(T_n(1^2; 3, 1)) > W(T_n(2, 1; 2, 1)) > W(T_n(1^2; 1^3)) \\
&> W(T_n(n-7, 5, 1)) > W(T_n(1^2; n-6, 1)) > W(T_n(1^2; 4, 1)) \\
&> W(T_n(n-5, 2, 1^2)) > W(T_n(1^2; 2^2)) > W(T_n(2, 1; 3, 1)) \\
&> W(T_D) > W(T).
\end{aligned}
$$

(2) *Let T be a tree of order $n \geq 20$, different from those encountered in the below set of inequalities. Then* [105],

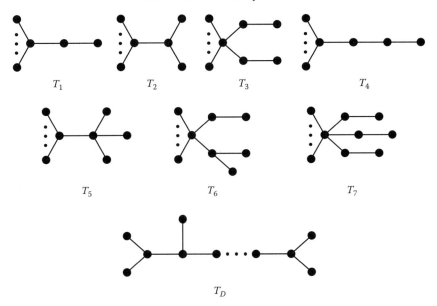

FIGURE 4.1: The trees $T_1, T_2, \ldots, T_7$ and T_D encountered in Theorems 4.3.9 and 4.3.10.

$$WW(P_n) > WW(T_n(n-3, 1^2)) > WW(T_n(n-4, 2, 1)) > WW(T_n(1^2; 1^2))$$
$$> WW(T_n(n-5, 3, 1)) > WW(T_n(n-4, 1^3)) > WW(T_n(1^2; 2, 1))$$
$$> WW(T_n(n-6, 4, 1)) > WW(T_n(n-5, 2^2)) > WW(T_n(1^2; n-5, 1))$$
$$> WW(T_n(1^2; 3, 1)) > WW(T_n(2, 1; 2, 1)) > WW(T_n(1^2; 1^3))$$
$$> WW(T_n(n-7, 5, 1)) > WW(T_n(1^2; n-6, 1))) > WW(T).$$

(3) *Let T be a tree of order $n \geq 16$, different from those encountered in the following set of inequalities. Then* [147],

$$H(P_n) < H(T_n(n-3, 1^2)) < H(T_n(n-4, 2, 1)) < H(T_n(1^2; 1^2))$$
$$< H(T_n(n-5, 3, 1)) < H(T_n(1^2; 2, 1)) < H(T_n(n-4, 1^3)) < H(T).$$

Theorem 4.3.10 (1) *Let T be a tree of order $n \geq 24$, different from those encountered in the following set of inequalities. Then* [44,108],

$$W(T) > W(T_4) > W(T_7) = W(T_6) > W(T_5) > W(T_3)$$
$$> W(T_2) > W(T_1) > W(S_n).$$

(2) *Let T be a tree of order $n \geq 18$, different from those encountered in the following set of inequalities. Then* [105],

$$WW(T) > WW(T_4) > WW(T_7) > WW(T_6) > WW(T_5) > WW(T_3)$$
$$> WW(T_2) > WW(T_1) > WW(S_n).$$

(3) *Let T be a tree of order $n \geq 14$, different from those encountered in the following set of inequalities. Then* [147],

$$H(T) < H(T_7) < H(T_6) < H(T_5) < H(T_4) < H(T_3)$$
$$< H(T_2) < H(T_1) < H(S_n).$$

The *dumbbell* $D(n, p, q)$ is a tree obtained from the path P_{n-p-q} by adding to its two terminal vertices p and q pendent vertices.

In the subsequent two theorems, we assume that $n - 1 = kq + r$ with $0 \leq r < k$, that is, $q = \lfloor n/k \rfloor$.

Theorem 4.3.11 [15,128] *Let T be a tree with n vertices and k pendent vertices, where $2 \leq k \leq n - 2$. Then,*

$$W\left(T_n\left(\left\lceil\frac{n}{k}\right\rceil^r, \left\lfloor\frac{n}{k}\right\rfloor^{k-r}\right)\right) \leq W(T) \leq W(D(n, \lfloor k/2 \rfloor, \lceil k/2 \rceil))$$

with the left equality holding if and only if $T \cong T_n(\lceil n/k \rceil^r, \lfloor n/k \rfloor^{k-r})$ and the right equality holding if and only if $T \cong D(n, \lfloor k/2 \rfloor, \lceil k/2 \rceil)$.

Theorem 4.3.12 [155] *Let T be a tree with n vertices and k pendent vertices, where $2 \leq k \leq n - 2$.*

- $WW(T_n(\lceil n/k \rceil^r, \lfloor n/k \rfloor^{k-r})) \leq WW(T)$, *with equality holding if and only if* $T \cong T_n(\lceil n/k \rceil^r, \lfloor n/k \rfloor^{k-r})$.

- $H(T) \leq H(T_n(\lceil n/k \rceil^r, \lfloor n/k \rfloor^{k-r}))$, *with equality holding if and only if* $T \cong T_n(\lceil n/k \rceil^r, \lfloor n/k \rfloor^{k-r})$.

Let $A_{n,\beta}$ be the tree obtained by attaching a pendent vertex to $\beta - 1$ noncentral vertices of the star $S_{n-\beta+1}$. Clearly, the matching number of $A_{n,\beta}$ is β, and the star S_n is a unique tree with n vertices and matching number $\beta = 1$.

Theorem 4.3.13 *Let T be a tree with n vertices and matching number $2 \leq \beta \leq \lfloor n/2 \rfloor$.*

- $W(A_{n,\beta}) \leq W(T)$, *with equality if and only if* $T \cong A_{n,\beta}$ [47].

- $WW(A_{n,\beta}) \leq WW(T)$, *with equality if and only if* $T \cong A_{n,\beta}$ [155].

- $H(T) \leq H(A_{n,\beta})$, *with equality if and only if* $T \cong A_{n,\beta}$ [28,87].

For a bipartite graph G of order n with matching number β and independence number α, it is well known [11,81] that $\alpha + \beta = n$. Therefore, Theorem 4.3.13 implies [28,87,155]:

Theorem 4.3.14 *Let T be a tree with n vertices and independence number α. Then,*

- $W(A_{n,n-\alpha}) \leq W(T)$, *with equality if and only if* $T \cong A_{n,n-\alpha}$.

- $WW(A_{n,n-\alpha}) \leq WW(T)$, *with equality if and only if* $T \cong A_{n,n-\alpha}$,

- $H(T) \leq H(A_{n,n-\alpha})$, *with equality if and only if* $T \cong A_{n,n-\alpha}$.

For $2 \leq \Delta \leq n - 1$, the *Volkmann tree* $V_{n,\Delta}$ is defined as follows [58,59,100]: If $n = \Delta + 1$, then $V_{n,\Delta}$ is just a star of order n.
For $n > \Delta + 1$, define n_i as

$$n_i = 1 + \sum_{j=1}^{i} \Delta(\Delta - 1)^j$$

for $i = 1, 2, \ldots,$ and choose k such that $n_{k-1} < n \leq n_k$.
Then, calculate the parameters m and h so that

$$m = \frac{n - n_{k-1}}{\Delta - 1} \quad \text{and} \quad h = n - n_{k-1} - (\Delta - 1)m.$$

The vertices of $V_{n,\Delta}$ are arranged into $k + 1$ levels. In level 0, there is only one vertex labeled as $v_{0,1}$. In level i for $i = 1, 2, \ldots, k - 1$, there are $\Delta(\Delta - 1)^i$ vertices labeled as $v_{i,1}, v_{i,2}, \ldots, v_{i,\Delta(\Delta-1)^i}$. These are connected (in that order) to the vertices in level i, so that $\Delta - 1$ vertices from level i are adjacent to each vertex from level $i - 1$. At level k, there are $n - n_{k-1}$ vertices, labeled as $v_{k,1}, v_{k,2}, \ldots, v_{k,n-n_{k-1}}$. They are connected (in that order) to the vertices in level $k - 1$, so that $\Delta - 1$ vertices from level k are adjacent to vertices $v_{k-1,1}, v_{k-1,1}, \ldots, v_{k-1,m}$. The remaining h vertices at level k (if any) are connected to the vertex $v_{k-1,m+1}$ in level $k - 1$.
In Figure 4.2, we illustrate the structure of the Volkmann trees $V_{n,\Delta}$ by the example with $n = 22$ and $\Delta = 4$.

Theorem 4.3.15 [59,92,100] *Let T be a tree of order n, with maximum vertex degree $\Delta \geq 3$. Then*

$$W(V_{n,\Delta}) \leq W(T)$$

with equality holding if and only if $T \cong V_{n,\Delta}$.

The *broom* $B_{n,\Delta}$ is the tree obtained by attaching $\Delta - 1$ pendent vertices to one pendent vertex of the path $P_{n-\Delta+1}$.

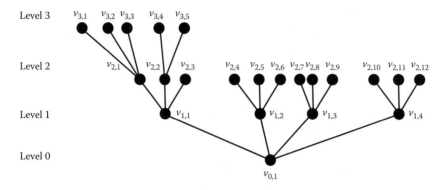

FIGURE 4.2: The Volkmann tree $V_{22,4}$ with its vertices labeled.

Theorem 4.3.16 *Let T be a tree with n vertices and maximum degree $\Delta \geq 3$. Then*

- $W(T) \leq W(B_{n,\Delta})$, *with equality if and only if $T \cong B_{n,\Delta}$* [39].

- $WW(T) \leq WW(B_{n,\Delta})$, *with equality if and only if $T \cong B_{n,\Delta}$* [155].

- $H(B_{n,\Delta}) \leq H(T) \leq H(V_{n,\Delta})$, *with the left equality holding if and only if $T \cong B_{n,\Delta}$ and the right equality holding if and only if $T \cong V_{n,\Delta}$* [87,139].

Let $C_{n,d}$ be the tree obtained from the path $P_{d+1} = v_0v_1 \ldots v_d$, by attaching $n - d - 1$ pendent vertices to the vertex $v_{\lfloor d/2 \rfloor}$.

Theorem 4.3.17 *Let T be a tree with n vertices and diameter d, where $3 \leq d \leq n - 2$. Then*

- $W(C_{n,d}) \leq W(T)$, *with equality if and only if $T \cong C_{n,d}$* [104,142].

- $WW(C_{n,d}) \leq WW(T)$, *with equality if and only if $T \cong C_{n,d}$* [54,155].

- $H(T) \leq H(C_{n,d})$, *with equality if and only if $T \cong C_{n,d}$* [87].

4.3.3 Metric-Extremal Unicyclic and Bicyclic Graphs

Recall that a unicyclic graph is a connected graph with n vertices and n edges and a bicyclic graph is a connected graph with n vertices and $n + 1$ edges.

In order to characterize the unicyclic graphs extremal w.r.t. the distance-based invariants, we first introduce some necessary notations.

Denote by $C_k\left(n_1^{\ell_1}, n_2^{\ell_2}, \ldots, n_m^{\ell_m}\right)$ the unicyclic graph obtained by attaching ℓ_1 paths of length n_1; ℓ_2 paths of length $n_2, \ldots$; *and* ℓ_m paths of length n_k, respectively, to one vertex of C_k, where $n_1 > n_2 > \cdots > n_m$. For example, the graph $C_7(5^1, 4^2, 2^4)$ is shown in Figure 4.3.

There are exactly two unicyclic graphs of order 4, namely, C_4 and $C_3(1^1)$. We have $W(C_4) = W(C_3(1^1))$, $WW(C_4) = WW(C_3(1^1))$, and $H(C_4) = H(C_3(1^1))$.

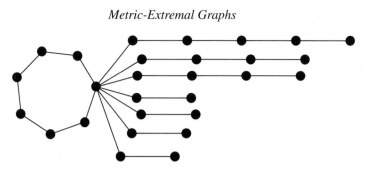

FIGURE 4.3: The graph $C_7(5^1, 4^2, 2^4)$.

Theorem 4.3.18 *Let G be a unicyclic graph of order $n \geq 5$.*

- $W(C_3(1^{n-3})) \leq W(G) \leq W(C_3((n-3)^1))$, *where the left equality holds if and only if $G \cong C_3(1^{n-3})$ for $n \geq 6$ and $G \cong C_3(1^{n-3})$ or $G \cong C_5$ for $n = 5$ and the right equality holds if and only if $G \cong C_3((n-3)^1)$* [134,154].

- $WW(C_3(1^{n-3})) \leq WW(G) \leq WW(C_3((n-3)^1))$, *where the left equality holds if and only if $G \cong C_3(1^{n-3})$ for $n \geq 6$ and $G \cong C_3(1^{n-3})$ or $G \cong C_5$ for $n = 5$ and the right equality holds if and only if $G \cong C_3((n-3)^1)$* [52,146].

- $H(C_3((n-3)^1)) \leq H(G) \leq H(C_3(1^{n-3}))$, *where the left equality holds if and only if $G \cong C_3((n-3)^1)$ and the right equality holds if and only if $G \cong C_3(1^{n-3})$ for $n \geq 6$ and $G \cong C_3(1^{n-3})$ or $G \cong C_5$ for $n = 5$* [149].

For $2 \leq \beta \leq \lfloor n/2 \rfloor$, we denote by $U_{n,m}$ the unicyclic graph obtained by attaching $n - 2m + 1$ pendent edges and $m - 2$ pendent paths of length 2 to one vertex of C_3.

Theorem 4.3.19 *Let G be a unicyclic graph with $n \geq 9$ vertices and matching number β.*

- *If $\beta \geq 2$, then $W(U_{n,\beta}) \leq W(G)$, with equality if and only if $G \cong U_{n,\beta}$* [47].

- *If $\beta \geq 2$, then $WW(U_{n,\beta}) \leq WW(G)$, with equality if and only if $G \cong U_{n,\beta}$* [51].

- *If $\beta \geq 3$, then $H(G) \leq H(U_{n,\beta})$, with equality if and only if $G \cong U_{n,\beta}$* [150].

In order to state the next group of results, let $\mathcal{B}(n)$ denote the set of bicyclic graphs of order n. In this class, we first define the graph $\Theta_{k,\ell,m}$ obtained by identifying the endpoints of the path graphs P_{k+1}, $P_{\ell+1}$, and P_{m+1}. Thus, $\Theta_{k,\ell,m}$ has $k + \ell + m - 1$ vertices. Note that $\Theta_{2,2,2}$ is just the complete bipartite graph $K_{2,3}$.

For $n \geq 5$, let $B_n^{(0)} \in \mathcal{B}(n)$ be the graph obtained by attaching a path of length $n - 4$ to one vertex of degree 2 of $\Theta_{1,2,2}$.

For $n \geq 5$, let $B_n^{(1)} \in \mathcal{B}(n)$ be the graph obtained by attaching $n-4$ pendent vertices to a vertex of degree 3 of $\Theta_{1,2,2}$.

For $n \geq 6$, let $B_n^{(2)} \in \mathcal{B}(n)$ be the graph obtained by adding two independent edges to the star S_n.

Note that $B_n^{(1)}$ and $B_n^{(2)}$ are graphs of diameter 2. Therefore, for $i \in \{1, 2\}$,

$$d\left(B_n^{(i)}, 1\right) = n + 1$$

$$d\left(B_n^{(i)}, 2\right) = \binom{n}{2} - (n + 1)$$

$$d\left(B_n^{(i)}, j\right) = 0 \quad \text{for } k \geq 3.$$

From Equations 4.11 through 4.13, it then follows that

$$W\left(B_n^{(1)}\right) = W\left(B_n^{(2)}\right), \qquad WW\left(B_n^{(1)}\right) = WW\left(B_n^{(2)}\right), \qquad H\left(B_n^{(1)}\right) = H\left(B_n^{(2)}\right).$$

Theorem 4.3.20 *Let G be a bicyclic graph of order $n \geq 5$ and $i \in \{1, 2\}$.*

- $W(B_n^{(i)}) \leq W(G)$, *with equality if and only if* $G \cong B_n^{(i)}$ *for* $n \geq 7$ *and* $G \cong B_n^{(i)}$ *or* $G \cong \Theta_{2,2,3}$ *for* $n = 6$ *and* $G \cong B_n^{(i)}$ *or* $G \cong \Theta_{2,1,3}$ *or* $G \cong K_{2,3}$ *for* $n = 5$ [55].

- $WW(B_n^{(i)}) \leq WW(G)$, *with equality if and only if* $G \cong B_n^{(i)}$ *for* $n \geq 7$ *and* $G \cong B_n^{(i)}$ *or* $G \cong \Theta_{2,2,3}$ *for* $n = 6$ *and* $G \cong B_n^{(i)}$ *or* $G \cong \Theta_{2,1,3}$ *or* $G \cong K_{2,3}$ *for* $n = 5$ [55].

- $H(G) \leq H(B_n^{(i)})$, *with equality if and only if* $G \cong B_n^{(i)}$ *for* $n \geq 7$ *and* $G \cong B_n^{(i)}$ *or* $G \cong \Theta_{2,2,3}$ *for* $n = 6$ *and* $G \cong B_n^{(i)}$ *or* $G \cong \Theta_{2,1,3}$ *or* $G \cong K_{2,3}$ *for* $n = 5$ [149].

Theorem 4.3.21 *Let G be a bicyclic graph of order $n \geq 5$.*

- $W(G) \leq W(B_n^{(0)})$, *with equality if and only if* $G \cong B_n^{(0)}$ [55].

- $WW(G) \leq WW(B_n^{(0)})$, *with equality if and only if* $G \cong B_n^{(0)}$ [55].

- $H(B_n^{(0)}) \leq H(G)$, *with equality if and only if* $G \cong B_n^{(0)}$ [149].

4.4   Concluding Remarks

In view of the limited available space, the theorems stated in the preceding sections have been selected so as to show the analogy between the three considered metric invariants. There, however, exist many cases when their behavior fails to be parallel. Even more numerous are the results established only for some, and not for all, of these invariants. In particular, the Wiener index is much more extensively studied than the hyper-Wiener and Harary indices, not to mention the plethora of other distance-based invariants. For more details, the interested readers are directed to the bibliography that follows, which, although far from complete, may well serve its purpose.

References

1. I. Althöfer. Average distance in undirected graphs and the removal of vertices. *Journal of Combinational Theory*, B48:140–142, 1990.

2. V. Andova, D. Dimitrov, J. Fink, and R. Škrekovski. Bounds on Gutman index. *MATCH Communications in Mathematical and in Computer Chemistry*, 67:515–524, 2012.

3. M. Aouchiche and P. Hansen. Automated results and conjectures on average distance in graphs. In A. Bondy, J. Fonlupt, J. L. Fouquet, J. C. Fournier, and J. L. Ramírez Alfonsín, eds., *Graph Theory in Paris*, pp. 21–36. Birkhauser, Basel, Switzerland, 2007.

4. M. Azari and A. Iranmanesh. Computation of the edge Wiener indices of the sum of graphs. *Ars Combinatoria*, 100:113–128, 2011.

5. A. T. Balaban, ed. *Chemical Applications of Graph Theory*. Academic Press, London, U.K., 1976.

6. A. T. Balaban. Topological indices based on topological distances in molecular graphs. *Pure and Applied Chemistry*, 55:199–206, 1983.

7. S. H. Bertz. Branching in graphs and molecules. *Discrete Applied Mathematics*, 19:65–83, 1988.

8. D. Bonchev. The Wiener number some applications and new developments. In D. H. Rouvray and R. B. King, eds., *Topology in Chemistry—Discrete Mathematics of Molecules*, pp. 58–88. Horwood, Chichester, U.K., 2002.

9. D. Bonchev, J. V. Knop, and N. Trinajstić. Mathematical models of branching. *MATCH Communications in Mathematical and in Computer Chemistry*, 6:21–47, 1979.

10. D. Bonchev and N. Trinajstić. Information theory, distance matrix, and molecular branching. *Journal of Chemical Physics*, 67:4517–4533, 1977.

11. J. A. Bondy and U. S. R. Murty. *Graph Theory with Applications*. Macmillan, New York, 1976.

12. F. M. Brückler, T. Došlić, A. Graovac, and I. Gutman. On a class of distance–based molecular structure descriptors. *Chemical Physics Letters*, 503:336–338, 2011.

13. F. Buckley. Mean distance in line graphs. *Congressus Numerantium*, 32:153–162, 1981.

14. F. Buckley and F. Harary. *Distance in Graphs*. Addison-Wesley, Redwood, CA, 1990.

15. K. Burns and R. C. Entringer. A graph-theoretic view of the United States postal service. In A. J. Schwenk, ed., *Graph Theory, Combinatorics, and Algorithms*, pp. 323–334. Wiley, New York, 1995.

16. S. P. Chan. *Introductory Topological Analysis of Electrical Networks*. Holt, New York, 1969.

17. N. Cohen, D. Dimitrov, R. Krakovski, R. Škrekovski, and V. Vukašinović. On Wiener index of graphs and their line graphs. *MATCH Communications in Mathematical and in Computer Chemistry*, 64:683–698, 2010.

18. Z. Cui and B. Liu. On Harary matrix, Harary index and Harary energy. *MATCH Communications in Mathematical and in Computer Chemistry*, 68:815–823, 2012.

19. M. Cygan, M. Pilipczuk, and R. Škrekovski. Relation between Randić index and average distance of trees. *MATCH Communications in Mathematical and in Computer Chemistry*, 66:605–612, 2011.

20. P. Dankelmann. Average distance and independence number. *Discrete Applied Mathematics*, 51:75–83, 1994.

21. P. Dankelmann. Average distance and domination number. *Discrete Applied Mathematics*, 80:21–35, 1997.

22. P. Dankelmann and R. C. Entringer. Average distance, minimum degree and spanning trees. *Journal of Graph Theory*, 33:1–13, 2000.

23. P. Dankelmann, I. Gutman, S. Mukwembi, and H. C. Swart. The edge-Wiener index of a graph. *Discrete Mathematics*, 309:3452–3457, 2009.

24. P. Dankelmann, I. Gutman, S. Mukwembi, and H. C. Swart. On the degree distance of a graph. *Discrete Applied Mathematics*, 157:2773–2777, 2009.

25. P. Dankelmann and S. Mukwembi. The distance concept and distance in graphs. In I. Gutman and B. Furtula, eds., *Distance in Molecular Graphs—Theory*, pp. 3–48. University of Kragujevac, Kragujevac, Serbia, 2012.

26. P. Dankelmann, S. Mukwembi, and H. C. Swart. Average distance and edge-connectivity II. *SIAM: Journal on Discrete Mathematics*, 21:1035–1052, 2007.

27. P. Dankelmann, S. Mukwembi, and H. C. Swart. Average distance and edge-connectivity I. *SIAM: Journal on Discrete Mathematics*, 22:92–101, 2008.

28. K. C. Das, K. Xu, and I. Gutman. On Zagreb and Harary indices. *MATCH Communications in Mathematical and in Computer Chemistry*, 70:301–314, 2013.

29. M. Dehmer and F. Emmert-Streib, eds. *Analysis of Complex Networks—From Biology to Linguistics*. Wiley, Weinheim, Germany, 2009.

30. M. Dehmer, F. Emmert-Streib, and A. Mehler, eds. *Towards an Information Theory of Complex Networks*. Birkhäuser, New York, 2011.

31. M. Dehmer and V. Kraus. On extremal properties of graph entropies. *MATCH Communications in Mathematical and in Computer Chemistry*, 68:889–912, 2012.

32. M. Dehmer and A. Mehler. A new method of measuring similarity for a special class of directed graphs. *Tatra Mountains Mathematical Publications*, 36: 39–59, 2007.

33. M. Dehmer, K. Varmuza, and D. Bonchev, eds. *Statistical Modelling of Molecular Descriptors in QSAR/QSPR*. Wiley, Weinheim, Germany, 2012.

34. H. Deng. The trees on $n > 9$ vertices with the first to seventeenth greatest Wiener indices are chemical trees. *MATCH Communications in Mathematical and in Computer Chemistry*, 57:393–402, 2007.

35. X. Deng and J. Zhang. Equiseparability on terminal Wiener index. *Applied Mathematics Letters*, 25:580–585, 2012.

36. J. Devillers and A. T. Balaban, eds. *Topological Indices and Related Descriptors in QSAR and QSPR*. Gordon & Breach, Amsterdam, the Netherlands, 1999.

37. M. M. Deza and E. Deza. *Dictionary of Distances*. Elsevier, Amsterdam, the Netherlands, 2006.

38. M. V. Diudea and I. Gutman. Wiener-type topological indices. *Croatica Chemica Acta*, 71:21–51, 1998.

39. A. A. Dobrynin, R. C. Entringer, and I. Gutman. Wiener index of trees: Theory and applications. *Acta Applicandae Mathematicae*, 66:211–249, 2001.

40. A. A. Dobrynin and I. Gutman. On a graph invariant related to the sum of all distances in a graph. *Publications de l'Institut Mathématique (Beograd)*, 56:18–22, 1994.

41. A. A. Dobrynin and I. Gutman. Solving a problem connected with distances in graphs. *Graph Theory Notes New York*, 28:21–23, 1995.

42. A. A. Dobrynin, I. Gutman, S. Klavžar, and P. Žigert. Wiener index of hexagonal systems. *Acta Applicandae Mathematicae*, 72:247–294, 2002.

43. A. A. Dobrynin and A. A. Kochetova. Degree distance of a graph: A degree analogue of the Wiener index. *Journal of Chemical Information and Computer Sciences*, 34:1082–1086, 1994.

44. H. Dong and X. Guo. Ordering trees by their Wiener indices. *MATCH Communications in Mathematical and in Computer Chemistry*, 56:527–540, 2006.

45. Y. Dou, H. Bian, H. Gao, and H. Yu. The polyphenyl chains with extremal edge–Wiener indices. *MATCH Communications in Mathematical and in Computer Chemistry*, 64:757–766, 2010.

46. J. K. Doyle and J. E. Graver. Mean distance in a graph. *Discrete Applied Mathematics*, 17:147–154, 1977.

47. Z. Du and B. Zhou. Minimum Wiener indices of trees and unicyclic graphs of given matching number. *MATCH Communications in Mathematical and in Computer Chemistry*, 63:101–112, 2010.

48. M. Eliasi, G. Raeisi, and B. Taeri. Wiener index of some graph operations. *Discrete Applied Mathematics*, 160:1333–1344, 2012.

49. R. C. Entringer, D. E. Jackson, and D. A. Snyder. Distance in graphs. *Czechoslovak Mathematical Journal*, 26:283–296, 1976.

50. R. C. Entringer, D. J. Kleitman, and L. A. Székely. A note on spanning trees with minimum average distance. *Bulletin of the Institute of Combinatorics and Its Applications*, 17:71–78, 1996.

51. L. Feng. The hyper-Wiener index of unicyclic graphs with given matching number. *Ars Combinatoria*, 100:9–17, 2011.

52. L. Feng, A. Ilić, and G. Yu. The hyper-Wiener index of unicyclic graphs. *Utilitas Mathematica*, 82:215–226, 2010.

53. L. Feng and B. Liu. The maximal Gutman index of bicyclic graphs. *MATCH Communications in Mathematical and in Computer Chemistry*, 66:699–708, 2011.

54. L. Feng and W. Liu. The hyper-Wiener index of graphs with given diameter. *Utilitas Mathematica*, 88:3–12, 2012.

55. L. Feng, W. Liu, and K. Xu. The hyper-Wiener index of bicyclic graphs. *Utilitas Mathematica*, 84:97–104, 2011.

56. L. Feng, G. Yu, and W. Liu. The hyper-Wiener index of graphs with a given chromatic (clique) number. *Utilitas Mathematica*, 88:399–407, 2012.

57. L. Feng and G. Yu. On the hyper-Wiener index of cacti. *Utilitas Mathematica*, 93:57–64, 2014.

58. M. Fischermann, I. Gutman, A. Hoffmann, D. Rautenbach, D. Vidović, and L. Volkmann. Extremal chemical trees. *Zeitschrift für Naturforschung*, 57a:49–52, 2002.

59. M. Fischermann, A. Hoffmann, D. Rautenbach, L. A. Székely, and L. Volkmann. Wiener index versus maximum degree in trees. *Discrete Applied Mathematics*, 122:127–137, 2002.

60. W. Goddard and O. R. Oellermann. Distance in graphs. In M. Dehmer, ed., *Structural Analysis of Complex Networks*, pp. 49–72. Birkhäuser, Dordrecht, the Netherlands, 2011.

61. I. Gutman. Selected properties of the Schultz molecular topological index. *Journal of Chemical Information and Computer Sciences*, 34:1087–1089, 1994.

62. I. Gutman. Distance of line graphs. *Graph Theory Notes of New York*, 31:49–52, 1996.

63. I. Gutman. Edge versions of topological indices. In I. Gutman and B. Furtula, eds., *Novel Molecular Structure Descriptors—Theory and Applications II*, pp. 3–20. University of Kragujevac, Kragujevac, Serbia, 2010.

64. I. Gutman and K. C. Das. The first Zagreb index 30 years after. *MATCH Communications in Mathematical and in Computer Chemistry*, 50:83–92, 2004.

65. I. Gutman and A. A. Dobrynin. The Szeged index—A success story. *Graph Theory Notes of New York*, 34:37–44, 1998.

66. I. Gutman, M. Essalih, M. El Marraki, and B. Furtula. Why plerograms are not used in chemical graph theory? The case of terminal-Wiener index. *Chemical Physics Letters*, 568:195–197, 2013.

67. I. Gutman and B. Furtula, eds. *Novel Molecular Structure Descriptors—Theory and Applications I*. University of Kragujevac, Kragujevac, Serbia, 2010.

68. I. Gutman and B. Furtula, eds. *Novel Molecular Structure Descriptors—Theory and Applications II*. University of Kragujevac, Kragujevac, Serbia, 2010.

69. I. Gutman and B. Furtula. A survey on terminal Wiener index. In I. Gutman and B. Furtula, eds., *Novel Molecular Structure Descriptors—Theory and Applications I*, pp. 173–190. University of Kragujevac, Kragujevac, Serbia, 2010.

70. I. Gutman and B. Furtula, eds. *Distance in Molecular Graphs—Applications*. University of Kragujevac, Kragujevac, Serbia, 2012.

71. I. Gutman and B. Furtula, eds. *Distance in Molecular Graphs—Theory*. University of Kragujevac, Kragujevac, Serbia, 2012.

72. I. Gutman, B. Furtula, and M. Petrović. Terminal Wiener index. *Journal of Mathematical Chemistry*, 46:522–531, 2009.

73. I. Gutman, V. Jovašević, and A. A. Dobrynin. Smallest graphs for which the distance of the graph is equal to the distance of its line graph. *Graph Theory Notes New York*, 33:19–19, 1997.

74. I. Gutman, Y. L. Luo, and S. L. Lee. The mean isomer degeneracy of the Wiener index. *Journal of the Chinese Chemical Society*, 40:195–198, 1993.

75. I. Gutman and L. Pavlović. More on distance of line graphs. *Graph Theory Notes New York*, 33:14–18, 1997.

76. I. Gutman and O. E. Polansky. *Mathematical Concepts in Organic Chemistry*. Springer, Berlin, Germany, 1986.

77. I. Gutman and N. Trinajstić. Graph theory and molecular orbitals. III. Total π-electron energy of alternant hydrocarbons. *Chemical Physics Letters*, 17:535–538, 1972.

78. I. Gutman and L. Šoltés. The range of the Wiener index and its mean isomer degeneracy. *Zeitschrift für Naturforschung*, 46A:865–868, 1991.

79. I. Gutman, Y. N. Yeh, S. L. Lee, and Y. L. Luo. Some recent results in the theory of the Wiener number. *Indian Journal of Chemistry*, 32A:651–661, 1993.

80. I. Gutman, Y. Zhang, M. Dehmer, and A. Ilić. Altenburg, Wiener, and Hosoya polynomials. In I. Gutman and B. Furtula, eds., *Distance in Molecular Graphs— Theory*, pp. 49–70. University of Kragujevac, Kragujevac, Serbia, 2012.

81. F. Harary. *Graph Theory*. Addison-Wesley, Reading, MA, 1969.

82. G. R. T. Hendry. On mean distance in certain classes of graphs. *Networks*, 19:451–457, 1989.

83. A. J. Hoffman and R. R. Singleton. On Moore graphs with diameters 2 and 3. *IBM Journal of Research and Development*, 4:497–504, 1960.

84. H. Hosoya. Topological index. A newly proposed quantity characterizing the topological nature of structural isomers of saturated hydrocarbons. *Bulletin of the Chemical Society of Japan*, 44:2332–2339, 1971.

85. H. Hosoya. On some counting polynomials in chemistry. *Discrete Applied Mathematics*, 19:239–257, 1988.

86. H. Hua. Wiener and Schultz molecular topological indices of graphs with specified cut edges. *MATCH Communications in Mathematical and in Computer Chemistry*, 61:643–651, 2009.

87. A. Ilić, G. Yu, and L. Feng. The Harary index of trees. *Utilitas Mathematica*, 87:21–32, 2012.

88. A. Iranmanesh, I. Gutman, O. Khormali, and A. Mahmiani. The edge versions of the Wiener index. *MATCH Communications in Mathematical and in Computer Chemistry*, 61:663–672, 2009.

89. O. Ivanciuc, T. S. Balaban, and A. T. Balaban. Reciprocal distance matrix, related local vertex invariants and topological indices. *Journal of Mathematical Chemistry*, 12:309–318, 1993.

90. O. Ivanciuc, T. Ivanciuc, and A. T. Balaban. The complementary distance matrix, a new molecular graph metric. *ACH Models in Chemistry*, 137:57–82, 2000.

91. D. Janežič, A. Miličević, S. Nikolić, and N. Trinajstić. *Graph Theoretical Matrices in Chemistry*. University of Kragujevac, Kragujevac, Serbia, 2007.

92. F. Jelen and E. Triesch. Superdominance order and distance of trees with bounded maximum degree. *Discrete Applied Mathematics*, 125:225–233, 2003.

93. M. Karelson. *Molecular Descriptors in QSAR/QSPR*. Wiley-Interscience, New York, 2000.

94. S. Klavžar and M. J. Nadjafi-Arani. Wiener index versus Szeged index in networks. *Discrete Applied Mathematics*, 161:1150–1153, 2013.

95. D. J. Klein, I. Lukovits, and I. Gutman. On the definition of the hyper-Wiener index for cycle-containing structures. *Journal of Chemical Information and Computer Sciences*, 35:50–52, 1995.

96. D. J. Klein, Z. Mihalić, D. Plavšić, and N. Trinajstić. Molecular topological index: A relation with the Wiener index. *Journal of Chemical Information and Computer Sciences*, 32:304–305, 1992.

97. M. Knor, P. Potočnik, and R. Škrekovski. The Wiener index in iterated line graphs. *Discrete Applied Mathematics*, 160:2234–2245, 2012.

98. M. Kouider and P. Winkler. Mean distance and minimum degree. *Journal of Graph Theory*, 25:95–99, 1997.

99. J. Li. A relation between the edge Szeged index and the ordinary Szeged index. *MATCH Communications in Mathematical and in Computer Chemistry*, 70:621–625, 2013.

100. W. Li, X. Li, and I. Gutman. Volkmann trees and their molecular structure descriptors. In I. Gutman and B. Furtula, eds., *Novel Molecular Structure Descriptors—Theory and Applications II*, pp. 231–246. University of Kragujevac, Kragujevac, Serbia, 2010.

101. X. Li and Y. Shi. Randić index, diameter and average distance. *MATCH Communications in Mathematical and in Computer Chemistry*, 64:425–431, 2010.

102. B. Liu and M. Liu. Some recent results on the variable Wiener index of graphs. In I. Gutman and B. Furtula, eds., *Distance in Molecular Graphs—Theory*, pp. 231–282. University of Kragujevac, Kragujevac, Serbia, 2012.

103. H. Liu and M. Lu. A unified approach to cacti for different indices. *MATCH Communications in Mathematical and in Computer Chemistry*, 58:193–204, 2007.

104. H. Liu and X. Pan. On the Wiener index of trees with fixed diameter. *MATCH Communications in Mathematical and in Computer Chemistry*, 60:85–94, 2008.

105. M. Liu and B. Liu. Trees with the seven smallest and fifteen greatest hyper-Wiener indices. *MATCH Communications in Mathematical and in Computer Chemistry*, 63:151–170, 2010.

106. M. Liu and B. Liu. A survey on recent results of variable Wiener index. *MATCH Communications in Mathematical and in Computer Chemistry*, 69:491–520, 2013.

107. M. Liu, B. Liu, and Q. Li. Erratum to 'the trees on $n > 9$ vertices with the first to seventeenth greatest Wiener indices are chemical trees'. *MATCH Communications in Mathematical and in Computer Chemistry*, 64:743–756, 2010.

108. S. Liu, L. Tong, and Y. Yeh. Trees with the minimum Wiener number. *International Journal of Quantum Chemistry*, 78:331–340, 2000.

109. B. Lučić, I. Lukovits, S. Nikolić, and N. Trinajstić. Distance–related indexes in the quantitative structure–property relationship modeling. *Journal of Chemical Information and Computer Sciences*, 41:527–535, 2001.

110. B. Lučić, S. Nikolić, and N. Trinajstić. Distance–related molecular descriptors. *Internet Electronic Journal of Molecular Design*, 7:195–206, 2008.

111. A. K. Madan and H. Dureja. Eccentricity based descriptors for QSAR/QSPR. In I. Gutman and B. Furtula, eds., *Novel Molecular Structure Descriptors—Theory and Applications II*, pp. 91–138. University of Kragujevac, Kragujevac, Serbia, 2010.

112. Z. Mihalić, S. Nikolić, and N. Trinajstić. Comparative study of molecular descriptors derived from the distance matrix. *Journal of Chemical Information and Computer Sciences*, 32:28–36, 1992.

113. Z. Mihalić and N. Trinajstić. A graph-theoretical approach to structure–property relationships. *Journal of Chemical Education*, 69:701–712, 1992.

114. Z. Mihalić, D. Veljan, D. Amić, S. Nikolić, D. Plavšić, and N. Trinajstić. The distance matrix in chemistry. *Journal of Mathematical Chemistry*, 11:223–258, 1992.

115. B. Mohar. Eigenvalues, diameter, and mean distance in graphs. *Graphs and Combinatorics*, 7:53–64, 1991.

116. S. Mukwembi. On the upper bound of Gutman index of graphs. *MATCH Communications in Mathematical and in Computer Chemistry*, 68:343–348, 2012.

117. M. J. Nadjafi-Arani, H. Khodashenas, and A. R. Ashrafi. On the differences between Szeged and Wiener indices of graphs. *Discrete Mathematics*, 311:2233–2237, 2011.

118. M. J. Nadjafi-Arani, H. Khodashenas, and A. R. Ashrafi. Graphs whose Szeged and Wiener numbers differ by 4 and 5. *Mathematical and Computer Modelling*, 55:1644–1648, 2012.

119. S. Nikolić, N. Trinajstić, and Z. Mihalić. The Wiener index: Development and applications. *Croatica Chemica Acta*, 68:105–129, 1995.

120. D. Plavšić, S. Nikolić, N. Trinajstić, and Z. Mihalić. On the Harary index for the characterization of chemical graphs. *Journal of Mathematical Chemistry*, 12:235–250, 1993.

121. M. Randić. Novel molecular descriptor for structure–property studies. *Chemical Physics Letters*, 211:478–483, 1993.

122. K. H. Rosen. *Discrete Mathematics and Its Applications*. McGraw-Hill, New York, 2003.

123. D. H. Rouvray. The role of the topological distance matrix in chemistry. In N. Trinajstić, ed., *Mathematics and Computational Concepts in Chemistry*, pp. 295–306. Horwood, Chichester, U.K., 1986.

124. D. H. Rouvray. The challenge of characterizing branching in molecular species. *Discrete Applied Mathematics*, 19:317–338, 1988.

125. D. H. Rouvray. Harry in the limelight: The life and times of Harry Wiener. In D. H. Rouvray and R. B. King, eds., *Topology in Chemistry—Discrete Mathematics of Molecules*, pp. 1–15. Horwood, Chichester, U.K., 2002.

126. D. H. Rouvray. The rich legacy of half century of the Wiener index. In D. H. Rouvray and R. B. King, eds., *Topology in Chemistry—Discrete Mathematics of Molecules*, pp. 16–37. Horwood, Chichester, U.K., 2002.

127. H. P. Schultz. Topological organic chemistry. 1. Graph theory and topological indices of alkanes. *Journal of Chemical Information and Computer Sciences*, 29:227–228, 1989.

128. R. Shi. The average distance of trees. *Journal of Systems Science and Mathematical Sciences*, 6:18–24, 1993.

129. A. V. Sills and H. Wang. On the maximal Wiener index and related questions. *Discrete Applied Mathematics*, 160:1615–1623, 2012.

130. S. Sivasubramanian. Average distance in graphs and eigenvalues. *Discrete Mathematics*, 309:3458–3468, 2009.

131. V. A. Skorobogatov and A. A. Dobrynin. Metrical analysis of graphs. *MATCH Communications in Mathematical and in Computer Chemistry*, 23:105–155, 1988.

132. E. A. Smolenskii, E. V. Shuvalova, L. K. Maslova, I. V. Chuvaeva, and M. S. Molchanova. Reduced matrix of topological distances with a minimum number of independent parameters: Distance vectors and molecular codes. *Journal of Mathematical Chemistry*, 45:1004–1020, 2009.

133. L. A. Székely, H. Wang, and T. Wu. The sum of distances between the leaves of a tree and the 'semi-regular' property. *Discrete Mathematics*, 311:1197–1203, 2011.

134. Z. Tang and H. Deng. The (n, n)-graphs with the first three extremal Wiener indices. *Journal of Mathematical Chemistry*, 43:60–74, 2008.

135. R. Todeschini and V. Consonni. *Handbook of Molecular Descriptors*. Wiley-VCH, Weinheim, Germany, 2000.

136. R. Todeschini and V. Consonni. *Molecular Descriptors for Chemoinformatics*. Wiley-VCH, Weinheim, Germany, 2009.

137. S. S. Tratch, M. I. Stankevich, and N. S. Zefirov. Combinatorial models and algorithms in chemistry. The expanded Wiener number—A novel topological index. *Journal of Computational Chemistry*, 11:899–908, 1990.

138. N. Trinajstić. *Chemical Graph Theory*. CRC Press, Boca Raton, FL, 1983.

139. S. Wagner, H. Wang, and X. Zhang. Distance–based graph invariants of trees and the Harary index. *Filomat*, 27:39–48, 2013.

140. H. Wang. The extremal values of the Wiener index of a tree with given degree sequence. *Discrete Applied Mathematics*, 156:2647–2654, 2009.

141. H. Wang and L. Kang. More on the Harary index of cacti. *Journal of Applied Mathematics and Computing*, 43:369–386, 2013.

142. S. Wang and X. Guo. Trees with extremal Wiener indices. *MATCH Communications in Mathematical and in Computer Chemistry*, 60:609–622, 2008.

143. H. Wiener. Structural determination of paraffin boiling points. *Journal of the American Chemical Society*, 69:17–20, 1947.

144. P. Winkler. Mean distance in a tree. *Discrete Applied Mathematics*, 27:179–185, 1990.

145. B. Wu. Wiener index of line graphs. *MATCH Communications in Mathematical and in Computer Chemistry*, 64:699–706, 2010.

146. R. Xing, B. Zhou, and X. Qi. Hyper-Wiener index of unicyclic graphs. *MATCH Communications in Mathematical and in Computer Chemistry*, 66:315–328, 2011.

147. K. Xu. Trees with the seven smallest and eight greatest Harary indices. *Discrete Applied Mathematics*, 160:321–331, 2012.

148. K. Xu and K. C. Das. On Harary index of graphs. *Discrete Applied Mathematics*, 159:1631–1640, 2011.

149. K. Xu and K. C. Das. Extremal unicyclic and bicyclic graphs with respect to Harary index. *Bulletin of the Malaysian Mathematical Sciences Society*, 36:373–383, 2013.

150. K. Xu, K. C. Das, H. Hua, and M. V. Diudea. Maximal Harary index of unicyclic graphs with given matching number. *Studia Universitatis Babeş-Bolyai Chemia*, 58:71–86, 2013.

151. K. Xu, M. Liu, K. C. Das, I. Gutman, and B. Furtula. A survey on graphs extremal with respect to distance-based topological indices. *MATCH Communications in Mathematical and in Computer Chemistry*, 71:461–508, 2014.

152. K. Xu and N. Trinajstić. Hyper-Wiener and Harary indices of graphs with cut edges. *Utilitas Mathematica*, 84:153–163, 2011.

153. H. Yousefi-Azari, M. H. Khalifeh, and A. R. Ashrafi. Calculating the edge Wiener and edge Szeged indices of graphs. *Journal of Computational and Applied Mathematics*, 235:4866–4870, 2011.

154. G. Yu and L. Feng. On the Wiener index of unicyclic graphs with given girth. *Ars Combinatoria*, 94:361–369, 2010.

155. G. Yu, L. Feng, and A. Ilić. The hyper-Wiener index of trees with given parameters. *Ars Combinatoria*, 96:395–404, 2010.

156. K. A. Zaretskii. Construction of trees from distances between pendent vertices. *Uspekhi Math. Nauk.*, 20(6):90–92 (in Russian), 1965.

157. B. Zelinka. On a certain distance between isomorphism classes of graphs. *Časopis Pěst. Mat.*, 100:371–373, 1975.

158. B. Zhou, X. Cai, and N. Trinajstić. On Harary index. *Journal of Mathematical Chemistry*, 44:611–618, 2008.

159. B. Zhou and I. Gutman. Relations between Wiener, hyper-Wiener and Zagreb indices. *Chemical Physics Letters*, 394:93–95, 2004.

Chapter 5

Quantitative Methods for Nowhere-Zero Flows and Edge Colorings

Martin Kochol

Contents

5.1 Introduction..141
5.2 Nowhere-Zero Flows...143
    5.2.1 Graphs, Networks, and Flows...143
    5.2.2 Flows with Integral Values...146
    5.2.3 Dual Graphs...148
5.3 General Counting Principles...150
    5.3.1 Counting for Simple Networks..150
    5.3.2 First Reduction Principle...152
    5.3.3 Planar Case...154
    5.3.4 Numbers of Partitions..155
    5.3.5 Flows on Wheels...156
5.4 Nowhere-Zero 5-Flows...158
    5.4.1 Restrictions of Cyclical Edge Connectivity............................159
    5.4.2 Forbidden Networks...162
    5.4.3 Restrictions of Girths...163
    5.4.4 Reductions by Dihedral Groups.......................................165
    5.4.5 Discussion about Computations.......................................166
    5.4.6 Open Problems..169
5.5 Second Reduction Principle..170
    5.5.1 Reductions by Superproper Permutations.............................170
    5.5.2 Application..172
5.6 Edge Colorings of Cubic Graphs...174
Acknowledgments...178
References...178

5.1 Introduction

Nowhere-zero flows on graphs were introduced by Tutte [30–32] as a dual concept to graph coloring problems. A graph has a nowhere-zero k-flow if its edges can be oriented and assigned numbers $\pm 1, \ldots, \pm(k-1)$, so that for each vertex, the sum value on the incoming edges equals the sum on the outgoing ones. By Tutte, a planar

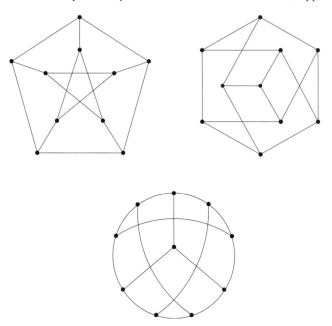

FIGURE 5.1:   Three different drawings of the Petersen graph.

graph has a k-coloring if and only if its dual has a nowhere-zero k-flow. Thus, the dual form of the four-color theorem (every planar graph is four colorable; see [2]) is that every bridgeless planar graph has a nowhere-zero 4-flow.

There are three celebrated unsolved conjectures dealing with nowhere-zero flows in bridgeless graphs, all due to Tutte. The first is the 5-flow conjecture of [30] that every such graph admits a nowhere-zero 5-flow. The 4-flow conjecture of [32] suggests that if the graph does not contain a subgraph contractible to the Petersen graph (see Figure 5.1), then it has a nowhere-zero 4-flow. This in fact generalizes the dual form of the four-color theorem, because the Petersen graph is not planar. Finally, the 3-flow conjecture is that if the graph does not contain a 3-edge cut, then it has a nowhere-zero 3-flow. Jaeger [8] and Kilpatrick [11] proved that every bridgeless graph admits a nowhere-zero 8-flow. The value 8 was improved to 6 by Seymour [27]. Thomassen [29] proved that every 8-edge-connected graph admits a nowhere-zero 3-flow.

The aim of this chapter is to present quantitative methods for nowhere-zero flows in graphs. We study the relation among numbers of nowhere-zero flows with prescribed values on an edge cut in a graph and evaluate these numbers as a dot product of an integral vector (determined by the graph) and a zero–one vector (corresponding to the prescribed value). Applying this principle, we show that the smallest counterexample to the 5-flow conjecture must be cyclically 6-edge connected and has girth at least 11. Furthermore, we give an alternative proof of the dual form to Birkhoff's theorem [3] that a smallest counterexample to the four-color theorem must be an internally 6-connected graph.

5.2   Nowhere-Zero Flows

5.2.1   Graphs, Networks, and Flows

We deal with finite unoriented graphs with multiple edges and loops. A graph $G = (V, E)$ consists of a set of vertices $V(G) = V$ and a set of edges $E(G) = E$. V and E are disjoint. We say that a graph H is a *subgraph* of G if $V(H) \subseteq V(G)$ and $E(H) \subseteq E(G)$. Furthermore, if $V(H) = V(G)$, then H is called a *spanning* subgraph of G. If e is an edge of a graph G, then $G - e$ and G/e denote the graphs obtained from G after deleting and contracting the edge e, respectively. Note that $G/e = G - e$ if e is a loop.

Every edge e of the graph G determines two opposite arcs arising from it after endowing e with two distinct orientations. All arcs obtained in this way are called *arcs* of G, and the set of them is called the *arc set* of G and denoted by $D(G)$. Clearly, $|D(G)| = 2|E(G)|$. If x is an arc of G, then denote by x^{-1} the second arc arising from the same edge. Clearly $(x^{-1})^{-1} = x$ and $x \neq x^{-1}$ for every arc x of G. If $X \subseteq D(G)$, then denote by $X^{-1} = \{x \in D(G); x^{-1} \in X\}$. By an *orientation* of G, we mean every $X \subseteq D(G)$ such that $X \cup X^{-1} = D(G)$ and $X \cap X^{-1} = \emptyset$. In Figure 5.2, we indicate a graph G, its arc set $D(G)$, and its orientation X.

If $W \subseteq V(G)$, then the set of the arcs of $D(G)$ directed from W to $V(G) \backslash W$ is denoted by $\omega_G^+(W)$. Furthermore, define $\omega_G^-(W) = (\omega_G^+(W))^{-1}$. We write $\omega_G^+(v)$, $\omega_G^-(v)$ instead of $\omega_G^+(\{v\})$, $\omega_G^-(\{v\})$, respectively.

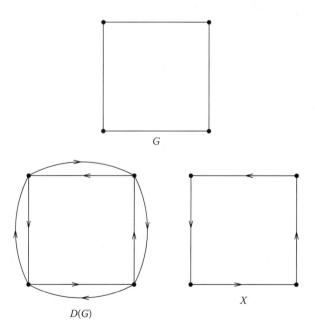

FIGURE 5.2:   Graph G, its arc set $D(G)$, and its orientation X.

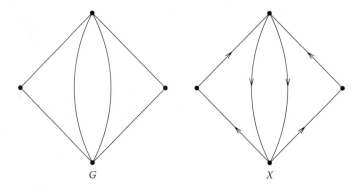

FIGURE 5.3:   A cycle G and a directed cycle X.

A *circuit* is a connected graph with all vertices of valency two. A *cycle* is a graph with all vertices of even valency. *A directed circuit* (*directed cycle*) is an orientation of a circuit (cycle) such that for every vertex v, the number of the arcs entering v equals the number of the arcs leaving v. See Figure 5.3.

By a *multiterminal network*, briefly a *network*, we mean a pair (G, U) where G is a graph and $U = (u_1, \ldots, u_n)$ is an ordered set of pairwise different vertices of G. If no confusions can occur, we denote by U also the set $\{u_1, \ldots, u_n\}$. (We apply this convention writing formulas $u \in U$, $U \cap W$, or $W \backslash U$ for $W \subseteq V(G)$.) The vertices from U and $V(G) \backslash U$ are called the *outer* and *inner* vertices of the network (G, U). We allow $n = 0$, that is, $U = \emptyset$. In this case, $(G, \emptyset)$ is usually identified with the graph G.

We consider only additive Abelian groups. If G is a graph and A is an Abelian group, then an *A-chain* in G is a mapping $\varphi : D(G) \to A$ such that $\varphi(x^{-1}) = -\varphi(x)$ for every $x \in D(G)$. In order to determine an *A*-chain φ in G, it suffices to describe all values $\varphi(x)$ where x belongs to an orientation X of G. For example, in Figure 5.4 is indicated a graph G with a $\mathbb{Z}$-chain φ as a mapping on $D(G)$ and a mapping on an orientation X of G.

If φ is an *A*-chain in G, then the mapping $\partial \varphi : V(G) \to A$ such that

$$\partial \varphi(v) = \sum_{x \in \omega_G^+(v)} \varphi(x) \quad (v \in V(G)) \tag{5.1}$$

is called the *boundary* of φ. *Support* of φ, denoted by $\sigma(\varphi)$, is the set of the edges associated with the arcs of G having nonzero values in φ. An *A*-chain φ is called *nowhere-zero* if $\sigma(\varphi) = E(G)$.

Lemma 5.2.1 *Let φ be an A-chain in a graph G and $W \subseteq V(G)$. Then*

$$\sum_{x \in \omega_G^+(W)} \varphi(x) = \sum_{v \in W} \partial \varphi(v).$$

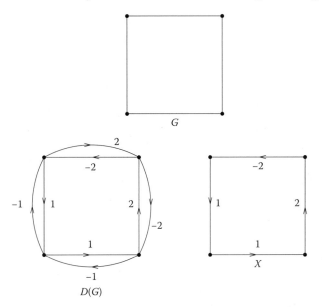

FIGURE 5.4: An Z-chain on $D(G)$ and an orientation X of G.

Proof. Since $\varphi(x^{-1}) = -\varphi(x)$ for every $x \in D(G)$, we have

$$\sum_{x \in \omega_G^+(W)} \varphi(x) = \sum_{v \in W} \sum_{x \in \omega_G^+(v)} \varphi(x) = \sum_{v \in W} \partial\varphi(v).$$

□

If (G, U) is a network, then a (nowhere-zero) A-chain φ in G is called a *(nowhere-zero) A-flow* in (G, U) if $\partial\varphi(v) = 0$ for every inner vertex v of G.

Lemma 5.2.2 *Let (G, U) be a network, $W \subseteq V(G)$, and φ be an A-flow in (G, U). Then*

$$\sum_{x \in \omega_G^+(W)} \varphi(x) = \sum_{v \in U \cap W} \partial\varphi(v).$$

Proof. Follows from Lemma 5.2.1 and the fact that $\partial\varphi(v) = 0$ for every $v \in W \backslash U$.

□

Lemma 5.2.3 *Let (G, U) be a network and φ be an A-flow in (G, U). Then*

$$\sum_{u \in U} \partial\varphi(u) = 0.$$

Proof. Follows directly from Lemma 5.2.2 after setting $W = V(G)$.

□

By a (nowhere-zero) A-flow in a graph G, we mean a (nowhere-zero) A-flow in the network $(G, \emptyset)$. Let $X \subseteq D(G)$ be an orientation of a graph G. Then a mapping

$\psi : X \to A$ can be extended into an A-flow φ in (G, U) if and only if the following equality is satisfied for every inner vertex v of (G, U):

$$\sum_{x \in \omega_G^+(v) \cap X} \psi(x) = \sum_{x \in \omega_G^-(v) \cap X} \psi(x). \qquad (5.2)$$

The extension is defined by $\varphi(x^{-1}) = -\varphi(x)$ for each $x \in X$. Thus, our concept of nowhere-zero flows in graphs coincides with the usual definition of nowhere-zero flows as presented in Diestel [5], Jaeger [8,9], Seymour [27], Younger [35], and Zhang [36]. The only difference is that instead of a fixed (but arbitrary) orientation of a graph G, we use the set $D(G)$ as a domain for a flow.

5.2.2   Flows with Integral Values

If k is an integer ≥ 2, then by a (*nowhere-zero*) *k-flow* φ in a network (G, U), we mean a (nowhere-zero) $\mathbb{Z}$-flow in (G, U) such that $|\varphi(x)| < k$ for every $x \in D(G)$. Note that this notion coincides with the usual definition of (nowhere-zero) k-flows in graphs (see, e.g., Jaeger [8,9] or Zhang [36]).

Now we show that a classical equivalence theorem of Tutte [32] known for graphs can be generalized for multiterminal networks.

Theorem 5.2.1 *Let (G, U) be a network. Then the following statements are pairwise equivalent:*

(1) *(G, U) has a nowhere-zero $\mathbb{Z}_k$-flow.*

(2) *(G, U) has a nowhere-zero A-flow for any Abelian group A of order k.*

(3) *(G, U) has a nowhere-zero k-flow.*

(4) *(G, U) has a nowhere-zero k-flow φ such that $|\partial \varphi(u)| < k$ for each outer vertex u of (G, U).*

Proof. We prove that (1) and (2) are equivalent. Let A be an Abelian group of order k. Suppose that $F_{G,U}(A)$ denotes the number of nowhere-zero A-flows in (G, U). We assume that $F_{G,U}(A) = 1$ when $E(G)$ is an empty set. If e is a loop, then $F_{G,U}(A) = (k - 1)F_{G-e,U}(A)$. Otherwise, $F_{G,U}(A) = F_{G/e,U}(A) - F_{G-e,U}(A)$. Thus, applying induction by $|E(G)|$, we get that $F_{G,U}(A) = F_{G,U}(\mathbb{Z}_k)$. (Notice that in this way, we can prove that $F_{G,U}(\mathbb{Z}_k)$ is a polynomial function of k; see cf. [15].)

Clearly, (4) implies (3).

If (G, U) has a nowhere-zero k-flow, then considering its values modulo k, we get a nowhere-zero $\mathbb{Z}_k$-flow in (G, U). Thus, (3) implies (1).

We prove that (1) implies (4). Let φ be a nowhere-zero $\mathbb{Z}_k$-flow in (G, U). For any orientation X of G, define by φ_X the $\mathbb{Z}$-chain in G so that for each $x \in X$, we have $\varphi_X(x) = \varphi(x) = -\varphi_X(x^{-1})$, supposing that $\varphi(x) \in \mathbb{Z}_k$ is considered as an integer $0 \leq \varphi(x) < k$. Then $\partial \varphi_X(v) \equiv \partial \varphi(v) \mod k$ for every vertex v of G and $0 < |\varphi(x)| < k$ for every arc x of G. Choose X so that $\sum_{v \in V(G)} |\partial \varphi_X(v)|$ is minimal.

We claim that $|\partial \varphi_X(v)| < k$ for each vertex v of G. If not, then there exists a vertex w of (G, U) such that $|\partial \varphi_X(w)| \geq k$. Moreover, we can assume that $\partial \varphi_X(w) > 0$ (otherwise, we can replace X by X^{-1} because $\varphi_X(x) = -\varphi_{X^{-1}}(x)$ for each arc x of G and $\partial \varphi_X(v) = -\partial \varphi_{X^{-1}}(v)$ for each vertex v of G).

Let W be the set of the vertices v of G for which there exists a path in X oriented from w to v (note that w belongs to W). Then $\omega_G^+(W) \subseteq X^{-1}$, and by Lemma 5.2.2,

$$\sum_{v \in W} \partial \varphi_X(v) = \sum_{x \in \omega_G^+(W)} \varphi_X(x) \leq 0.$$

Thus, there exists $w' \in W$ satisfying $\partial \varphi_X(w') < 0$. Let P be a path in X oriented from w to w'. Changing the orientation of the arcs from P, we get X' from X. Moreover, $\partial \varphi_{X'}(w) = \partial \varphi_X(w) - k$, $\partial \varphi_{X'}(w') = \partial \varphi_X(w') + k$, and $\partial \varphi_{X'}(v) = \partial \varphi_X(v)$ for every vertex v of G, $v \neq w, w'$. Since $\partial \varphi_X(w) \geq k$ and $\partial \varphi_X(w') < 0$, then $|\partial \varphi_{X'}(w)| = |\partial \varphi_X(w)| - k$ and $|\partial \varphi_{X'}(w')| = |\partial \varphi_X(w') + k| < |\partial \varphi_X(w')| + k$. Therefore,

$$\sum_{v \in V(G)} |\partial \varphi_{X'}(v)| < \sum_{v \in V(G)} |\partial \varphi_X(v)|,$$

a contradiction with the choice of X. Thus, $|\partial \varphi_X(v)| < k$ for each vertex v of G. Furthermore, for each inner vertex u of (G, U), $\partial \varphi_X(u) = 0$ because it must be an integral multiple of k and 0 is the unique multiple of k with absolute value smaller than k. Hence, φ_X is a nowhere-zero k-flow satisfying (4). This implies the statement. □

Notice that in the proof, we have followed the ideas presented in Jaeger [9], Welsh [34], and Younger [35]. This approach differs from the original proof of Tutte [30–32].

Theorem 5.2.2 *If a network (G, U) has a nowhere-zero k-flow, $k \geq 2$, then it has a nowhere-zero $(k + 1)$-flow.*

Proof. Follows from the fact that every nowhere-zero k-flow is also a nowhere-zero $(k + 1)$-flow. □

An edge cut of a graph is called *cyclic* if after deleting its edges we get at least two components having cycles. A graph is called *cyclically k-edge connected* if it does not have a cyclic edge cut of cardinality smaller than k. A *Snark* is a nontrivial cubic (3-regular) graph without 3-edge coloring. By nontrivial, we mean cyclically 4-edge connected and with girth (length of the shortest circuit) at least 5. The term snark was introduced by Gardner [7] borrowing this name from Lewis Carroll's ballad "The Hunting of the Snarks." Snarks are studied very intensively and several methods have been developed for their constructions (see cf. Kochol [12,14,20,21]). Let us note that the smallest known snark is the Petersen graph (see Figure 5.1).

Suppose that $\overline{G}$ is a smallest counterexample to the 5-flow conjecture. If $\overline{G}$ has a vertex of degree at least 4, we can split this into two vertices, one of them having degree 2, and then suppressing all vertices of degree 2. Thus, the graph must be cubic.

If $\overline{G}$ is 3-edge colorable, we can color the edges by nonzero elements of the group $\mathbb{Z}_2 \times \mathbb{Z}_2$. This coloring corresponds to a nowhere-zero $\mathbb{Z}_2 \times \mathbb{Z}_2$-flow φ after setting $\varphi(x) = \varphi(x^{-1})$ to be the color corresponding to the edge accompanied with x. From the arithmetic in the group $\mathbb{Z}_2 \times \mathbb{Z}_2$, it follows that φ is a $\mathbb{Z}_2 \times \mathbb{Z}_2$-flow. Then by Theorem 5.2.1, $\overline{G}$ has a nowhere-zero 4-flow, which is also a nowhere-zero 5-flow, which is not possible. Thus, $\overline{G}$ is cubic graph without a 3-edge coloring. Similarly, we can check that it has girth at least 5; thus, $\overline{G}$ is a snark. In fact, the main reason for interest in snarks is that among them must be the smallest counterexamples to many important conjectures about graphs, in particular the 5-flow conjecture and the cycle double cover conjecture (every bridgeless graph has a family of circuits that together cover each edge twice; see [14,36,37]).

5.2.3   Dual Graphs

We identify a planar graph G with its embedding in the plane and denote by G^* its dual. In other words, G^* is the planar graph that we obtain if we replace the faces of G by the vertices of G^* and for every edge e of G we join the two vertices associated with the faces neighboring e by an edge e^* of G^* so that e and e^* intersect exactly once. In Figure 5.5 is indicated a planar graph and by dotted lines its dual.

If G is connected, then $(G^*)^*$ can be considered to be identical with G. If x is an arc of G arising from the edge e of G and directed from left to right, then denote by x^* the arc of G^* arising from e^* and directed downward. Let us note that $(x^*)^* = x^{-1}$. If $X \subseteq D(G)$, then denote by $X^* = \{x^*; x \in X\} \subseteq D(G^*)$. In Figure 5.6 is indicated an orientation X of a planar graph G and by dotted lines indicated X^*, an orientation of G^*. In Figure 5.7 is indicated an orientation X^* from Figure 5.6 and by dotted lines indicated $(X^*)^*$, an orientation of G. Notice that $(X^*)^* = X^{-1}$.

Now we can introduce the second equivalence theorem of Tutte [31].

Theorem 5.2.3 *A planar graph is k-colorable if and only if its dual admits a nowhere-zero k-flow.*

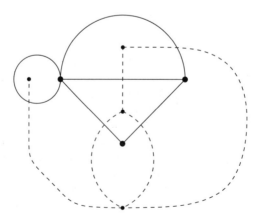

FIGURE 5.5:   A planar graph with its dual.

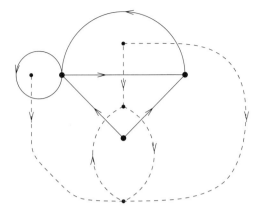

FIGURE 5.6:   An orientation X of a graph and X^*.

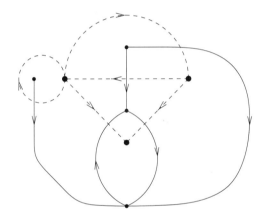

FIGURE 5.7:   Orientations X^* and $(X^*)^*$.

Proof. Clearly, a graph is k-colorable (admits a nowhere k-flow) if and only if each of its components is k-colorable (admits a nowhere k-flow). Thus, without abuse of generality, we can assume that G is connected. Let A be an Abelian group of order k.

Let G is k-colorable, that is, there exists a mapping $\lambda : V(G) \rightarrow A$ such that $\lambda(v_1) \neq \lambda(v_2)$ for any two adjacent vertices v_1, v_2 of G. Notice that G cannot have a loop. Define $\varphi : D(G^*) \rightarrow A$ so that if x is the arc of G directed from v_1 to v_2, then $\varphi(x^*) = \lambda(v_2) - \lambda(v_1)$. Notice that there can be several arcs directed from v_1 to v_2 (corresponding to various multiple edges), but the value $\varphi(x^*)$ is the same for all of them. Clearly, φ is a nowhere-zero A-chain in G^*.

Consider a vertex of w of G^*. It corresponds to a face f_w of G, and let v_1, ..., v_n be the vertices from the boundary of f_w in the anticlockwise order. Then $\partial \varphi(w) = \lambda(v_1) - \lambda(v_2) + \lambda(v_2) - \lambda(v_3) + \cdots + \lambda(v_n) - \lambda(v_1) = 0$. Thus, φ is a nowhere-zero A-flow in G^*.

On the other hand, suppose that G^* has a nowhere-zero A-flow φ. We construct a mapping $\lambda : V(G) \rightarrow A$ corresponding to a k-coloring of G. G is connected and thus contains a spanning tree T (a connected subgraph containing no circuit and covering all vertices of G). We pick up a vertex v_0 of G. For each vertex v of G, there exists a unique path $v_0 e_1 v_1 \ldots v_{n-1} e_n v_n$ $(v_n = v)$ in T. For $i = 1, \ldots, n$, let x_i be the arc corresponding to e_i and directed from v_{i-i} to v_i. Define by (v_0, v)-*path* the sequence $(x_1, \ldots, x_n)$. Define by $\lambda(v) = \sum_{i=1}^{n} \varphi(x_i^*)$. Notice that (v_0, v_0)-path is $\emptyset$ and $\lambda(v_0) = 0$.

Suppose that v and v' are the two ends of an edge $\bar{e}$ of G and $\bar{x}$ is the arc of G directed from v to v'. We claim that $\lambda(v') - \lambda(v) = \varphi(\bar{x}^*)$. If $\bar{e}$ is from T, the claim holds true by definition of λ. If $\bar{e}$ is not from T, there is a (v_0, v)-path $(x_1, \ldots, x_r, x_{r+1}, \ldots, x_n)$ and a (v_0, v')-path $(x_1, \ldots, x_r, x'_{r+1}, \ldots, x'_{n'})$. Let $e_{r+1}, \ldots, e_n$, $e'_{r+1}, \ldots, e'_{n'}$ be the edges of G corresponding to the arcs $x_{r+1}, \ldots, x_n, x'_{r+1}, \ldots, x'_{n'}$, respectively. Then edges $e_{r+1}, \ldots, e_n, \bar{e}, e'_{n'}, \ldots, e'_{r+1}$ induce a circuit C in G. Let W be the set of vertices of G^* corresponding to the faces of G contained inside the region bounded by C. Then the set of arcs $\{(x_{r+1}^*)^{-1}, \ldots, (x_n^*)^{-1}, \bar{x}^{-1}, (x'_{n'})^*, \ldots, (x'_{r+1})^*\}$ equals either $\omega_W^+(G^*)$ or $\omega_W^-(G^*)$, whence by Lemma 5.2.2,

$$\sum_{i=1}^{n'} \varphi(x_{n'}^{'*}) - \sum_{i=1}^{n} \varphi(x_i^*) - \varphi(\bar{x}) - = 0.$$

From this equality and the definition of λ, we have that $\lambda(v') - \lambda(v) = \varphi(\bar{x}^*)$, and the claim holds true in general.

Therefore, $\lambda(v) \neq \lambda(v')$ for any two neighboring vertices v and v' of G, that is, λ is a k-coloring of G by elements of the group A.   □

Refer to [5,8–10,13,15,16,26,35–37] for more details about nowhere-zero flows on graphs.

5.3   General Counting Principles

5.3.1   Counting for Simple Networks

A network (G, U), $U = (u_1, \ldots, u_n)$, is simple if the vertices $u_1, \ldots, u_n$ have degree 1. If φ is a nowhere-zero A-flow in (G, U), then denote by

$$\partial \varphi(U) = (\partial \varphi(u_1), \ldots, \partial \varphi(u_n)). \tag{5.3}$$

By Lemma 5.2.3, $\sum_{i=1}^{n} \partial \varphi(u_i) = 0$. Furthermore, $\partial \varphi(u_i) \neq 0$ because u_i has degree 1 $(i = 1, \ldots, n)$. Thus, $\partial \varphi(U)$ belongs to the set

$$A_n = \{(s_1, \ldots, s_n); \; s_1, \ldots, s_n \in A \backslash \{0\},$$
$$s_1 + \cdots + s_n = 0\}. \tag{5.4}$$

Note that $A_1 = \emptyset$. For every $s \in A_n$, denote by $\Phi_{G,U}(s)$ the set of nowhere-zero A-flows φ in (G, U) satisfying $\partial \varphi(U) = s$. Furthermore, denote $F_{G,U}(s) = |\Phi_{G,U}(s)|$.

Let $P = \{Q_1, \ldots, Q_r\}$ be a partition of the set $\{1, \ldots, n\}$. If one of $Q_1, \ldots, Q_r$ is a singleton, P is called *trivial*; otherwise, it is called *nontrivial*. Let $\mathcal{P}_n$ denote the set of nontrivial partitions of $\{1, \ldots, n\}$.

If $s = (s_1, \ldots, s_n) \in A_n$, $P = \{Q_1, \ldots, Q_r\} \in \mathcal{P}_n$, and $\sum_{i \in Q_j} s_i = 0$ for $j = 1, \ldots, r$, then we say that P and s are *compatible* (e.g., setting $A = \mathbb{Z}_5$, then $\{\{1, 2\}, \{3, 4, 5\}\} \in \mathcal{P}_5$ is compatible with $(1, 4, 1, 2, 2) \in A_5$ but not with $(1, 1, 2, 4, 2) \in A_5$).

If $|\mathcal{P}_n| = p_n$, then $\mathcal{P}_n$ can be considered as a p_n-tuple $(P_{n,1}, \ldots, P_{n,p_n})$. For any $s \in A_n$, denote by $\chi_n(s)$ the integral vector $(\chi_{s,1}, \ldots, \chi_{s,p_n})$ such that $\chi_{s,i} = 1$ ($\chi_{s,i} = 0$) if $P_{n,i}$ is (is not) compatible with s, $i = 1, \ldots, p_n$. The following statement was proved in Kochol [17].

Theorem 5.3.1 *Let (G, U), $U = (u_1, \ldots, u_n)$, be a simple network and $|\mathcal{P}_n| = p_n$. Then there exists an integral vector $\mathbf{x}_{G,U} = (x_1, \ldots, x_{p_n})$ such that for every $s \in A_n$,*

$$F_{G,U}(s) = \chi_n(s) \cdot \mathbf{x}_{G,U} = \sum_{i=1}^{p_n} \chi_{s,i} x_i.$$

Proof. We can assume that no two outer vertices of (G, U) are joined by an edge (otherwise, subdivide each such edge by a new vertex of valency 2) and G has no isolated vertex. For such networks, we apply induction on $|E(G)| \geq n$.

If $|E(G)| = n$, then G is bipartite with partition sets $\{u_1, \ldots, u_n\}$ and $\{v_1, \ldots, v_r\}$. Thus, there exists a partition $P = \{Q_1, \ldots, Q_r\}$ of $\{1, \ldots, n\}$ such that $i \in Q_j$ if and only if u_i is adjacent to v_j ($i \in \{1, \ldots, n\}, j \in \{1, \ldots, r\}$). If P is trivial (i.e., (G, U) has an inner vertex of valency 1), then $F_{G,U}(s) = 0$ for every $s \in A_n$, and we can choose $\mathbf{x}_{G,U}$ to be the zero vector of $\mathbb{Q}^{p_n}$. If $P \in \mathcal{P}_n$, then there exists $j \in \{1, \ldots, p_n\}$ such that $P = P_{n,j}$. Now $F_{G,U}(s) = \chi_{s,j}$ for every $s \in A_n$. Thus $\mathbf{x}_{G,U}$ equals the jth standard basis vector of $\mathbb{Q}^{p_n}$ satisfies the assumptions of theorem.

If $|E(G)| > n$, then there exists an edge e of G with ends $v_1, v_2 \notin U$.

Assume that $v_1 \neq v_2$. We claim that for every $s \in A_n$,

$$F_{G,U}(s) = F_{G/e,U}(s) - F_{G-e,U}(s). \tag{5.5}$$

Clearly, any $\varphi \in \Phi_{G-e,U}(s)$ can be considered as a flow from $\Phi_{G/e,U}(s)$, and any $\varphi \in \Phi_{G,U}(s)$ can be transformed to exactly one flow from $\Phi_{G/e,U}(s)$. On the other hand, any $\varphi \in \Phi_{G/e,U}(s)$ can be considered as an A-chain in $G - e$, which is a nowhere-zero A-flow in the network $(G - e, (v_1, v_2, u_1, \ldots, u_n))$, whence by Lemma 5.2.3, $\partial \varphi(v_1) + \partial \varphi(v_2) = 0$. If $\partial \varphi(v_1) = \partial \varphi(v_2) = 0$, then φ is from $\Phi_{G-e,U}(s)$; otherwise, φ can be extended to exactly one flow from $\Phi_{G,U}(s)$. In this way, we get a bijective mapping from $\Phi_{G/e,U}(s)$ to $\Phi_{G,U}(s) \cup \Phi_{G-e,U}(s)$. This implies (5.5) because $\Phi_{G,U}(s) \cap \Phi_{G-e,U}(s) = \emptyset$.

We have $|E(G)| > |E(G/e)|, |E(G - e)|$. Thus, by the induction hypothesis, there are integral vectors $\mathbf{x}_{G/e,U}$ and $\mathbf{x}_{G-e,U}$ such that for every $s \in A_n$, $F_{G/e,U}(s) = \chi_n(s) \cdot$

$\mathbf{x}_{G/e,U}$ and $F_{G-e,U}(s) = \chi_n(s) \cdot \mathbf{x}_{G-e,U}$, whence by (5.5), $F_{G,U}(s) = F_{G/e,U}(s)$ $- F_{G-e,U}(s) = \chi_n(s) \cdot (\mathbf{x}_{G/e,U} - \mathbf{x}_{G-e,U})$. Thus, the vector $\mathbf{x}_{G,U} = \mathbf{x}_{G/e,U} - \mathbf{x}_{G-e,U}$ satisfies the assumptions of theorem.

Let $v_1 = v_2$, that is, e be a loop of G. Then every nowhere-zero A-flow in $G - e$ can be extended to exactly $k - 1$ nowhere-zero A-flows in G, where k is the order of A. Thus, for every $s \in A_n$, $F_{G,U}(s) = (k - 1) \cdot F_{G-e,U}(s)$, whence the vector $\mathbf{x}_{G,U} = (k - 1)\mathbf{x}_{G-e,U}$ satisfies the assumptions of theorem. $\qquad\square$

5.3.2   First Reduction Principle

The symmetric group on $\{1, \ldots, n\}$ is the group of all permutations of elements $1, \ldots, n$. By a permutation group Γ on $\{1, \ldots, n\}$, we mean any subgroup of the symmetric group on $\{1, \ldots, n\}$. If $\gamma \in \Gamma$ and $Q \subseteq \{1, \ldots, n\}$, then

$$\gamma(Q) = \{\gamma(i); i \in Q\}.$$

If $P = \{Q_1, \ldots, Q_r\} \in \mathcal{P}_n$, then define $\gamma(P) = \{\gamma(Q_1), \ldots, \gamma(Q_r)\}$. We say that partitions P and $\gamma(P)$ are Γ-*equivalent*. Let $\mathcal{P}_{\Gamma,n}$ be the set of Γ-equivalence classes and $p_{\Gamma,n} = |\mathcal{P}_{\Gamma,n}|$. For each $P_\Gamma \in \mathcal{P}_{\Gamma,n}$ and $s \in A_n$, define

$$\chi(s, P_\Gamma) = \sum_{P \in P_\Gamma} \chi(s, P). \tag{5.6}$$

$\mathcal{P}_{\Gamma,n}$ is considered as an ordered $p_{\Gamma,n}$-tuple $(P_{\Gamma,1}, \ldots, P_{\Gamma,p_{\Gamma,n}})$. Then denote by $\chi_{\Gamma,n}(s)$ the integral vector $(\chi(s, P_{\Gamma,1}), \ldots, \chi(s, P_{\Gamma,p_{\Gamma,n}}))$.

If $\gamma \in \Gamma$ and $s = (a_1, \ldots, a_n) \in A_n$, then denote by $\gamma(s) = (a_{\gamma(1)}, \ldots, a_{\gamma(n)})$. Thus, $\chi(s, P) = \chi(\gamma(s), \gamma^{-1}(P))$ for each $P \in \mathcal{P}_n$, whence $\chi(s, P_\Gamma) = \chi(\gamma(s), P_\Gamma)$ for each $P_\Gamma \in \mathcal{P}_{\Gamma,n}$, and therefore for each $\gamma \in \Gamma$, we have

$$\chi_{\Gamma,n}(s) = \chi_{\Gamma,n}(\gamma(s)). \tag{5.7}$$

We say that s and $\gamma(s)$ are Γ-*equivalent* (this equivalence defined on A_n, thus it is formally different from those ones defined on the set of partitions $\mathcal{P}_n$; from the context, it will be always clear which operation we have in mind). Denote by $A_{\Gamma,n}$ the set of Γ-equivalence classes. Notice that $A_{\Gamma,n}$ is a partition of A_n. For each $s_\Gamma \in A_{\Gamma,n}$, define

$$\chi_{\Gamma,n}(s_\Gamma) = \chi_{\Gamma,n}(s), \tag{5.8}$$

where $s \in s_\Gamma$. This is well defined by (5.7). Furthermore, for each $s_\Gamma \in A_{\Gamma,n}$, define

$$F_{\Gamma,G,U}(s_\Gamma) = \sum_{\gamma \in \Gamma} F_{G,U}(\gamma(s)), \tag{5.9}$$

where $s \in s_\Gamma$. This is also well defined because Γ is a group. The following statement was proved in Kochol [22].

Theorem 5.3.2 *Let* (G, U), $U = (u_1, \ldots, u_n)$, *be a simple network and* Γ *be a permutation group on* $\{1, \ldots, n\}$. *Then there exists a vector* $\mathbf{y}_{\Gamma,G,U} \in \mathbb{Z}^{p_{\Gamma,n}}$ *such that for every* $s_\Gamma \in A_{\Gamma,n}$,

$$F_{\Gamma,G,U}(s_\Gamma) = \chi_{\Gamma,n}(s_\Gamma) \cdot \mathbf{y}_{\Gamma,G,U}.$$

Proof. If $P_i \in \mathcal{P}_n$, then there exists $f(i) \in \{1, \ldots, p_{\Gamma,n}\}$ such that $P_i \in P_{\Gamma,f(i)}$ (f is a surjective mapping from $\{1, \ldots, p_n\}$ to $\{1, \ldots, p_{\Gamma,n}\}$). Define $\Gamma_i = \{\gamma \in \Gamma; \gamma(P_i) = P_i\}$. Γ_i is a subgroup of Γ. For each $P' \in P_{\Gamma,f(i)}$, there exists $\gamma' \in \Gamma$ such that $P' = \gamma'(P_i)$ and $\{\gamma \in \Gamma; \gamma(P_i) = P'\} = \{\gamma'\gamma; \gamma \in \Gamma_i\}$, whence $|\Gamma| = |\Gamma_i| \cdot |P_{\Gamma,f(i)}|$. Thus, $|\Gamma|/|P_{\Gamma,f(i)}|$ is an integer and for each $s \in A_n$, we have

$$\sum_{\gamma \in \Gamma} \chi(s, \gamma(P_i)) = |\Gamma_i| \sum_{P' \in P_{\Gamma,f(i)}} \chi(s, P')$$

$$= \frac{|\Gamma|}{|P_{\Gamma,f(i)}|} \sum_{P' \in P_{\Gamma,f(i)}} \chi(s, P'),$$

whence by (5.6),

$$\sum_{\gamma \in \Gamma} \chi(s, \gamma(P_i)) = \frac{|\Gamma|}{|P_{\Gamma,f(i)}|} \chi(s, P_{\Gamma,f(i)}). \tag{5.10}$$

By the definitions of $\chi(s, P_i)$, $\gamma(s)$, and $\gamma(P_i)$, we have $\chi(\gamma(s), P_i) = \chi(s, \gamma(P_i))$ for each $s \in A_n$ and $\gamma \in \Gamma$. By Theorem 5.3.1, there exists an integral vector $(x_1, \ldots, x_{p_n})$ such that for every $s \in A_n$, $F_{G,U}(s) = \sum_{i=1}^{p_n} \chi(s, P_i)x_i$. If $s_\Gamma \in A_{\Gamma,n}$ and $s \in s_\Gamma$, then by (5.9) and (5.10),

$$F_{\Gamma,G,U}(s_\Gamma) = \sum_{\gamma \in \Gamma} F_{G,U}(\gamma(s)) = \sum_{\gamma \in \Gamma} \left(\sum_{i=1}^{p_n} \chi(\gamma(s), P_i)x_i \right)$$

$$= \sum_{i=1}^{p_n} \left(\sum_{\gamma \in \Gamma} \chi(\gamma(s), P_i)x_i \right) = \sum_{i=1}^{p_n} \left(\sum_{\gamma \in \Gamma} \chi(s, \gamma(P_i))x_i \right)$$

$$= \sum_{i=1}^{p_n} \frac{|\Gamma|}{|P_{\Gamma,f(i)}|} \chi(s, P_{\Gamma,f(i)}) \, x_i = \sum_{j=1}^{p_{\Gamma,n}} \chi(s, P_{\Gamma,j}) \frac{|\Gamma|}{|P_{\Gamma,j}|} \left(\sum_{i \in f^{-1}(j)} x_i \right).$$

Thus, setting $\mathbf{y}_{\Gamma,G,U} = (y_1, \ldots, y_{p_{\Gamma,n}}) \in \mathbb{Z}^{p_{\Gamma,n}}$ such that for $j = 1, \ldots, p_{\Gamma,n}$,

$$y_j = \frac{|\Gamma|}{|P_{\Gamma,j}|} \left(\sum_{i \in f^{-1}(j)} x_i \right), \tag{5.11}$$

we have by (5.8) $F_{\Gamma,G,U}(s_\Gamma) = \chi_{\Gamma,n}(s_\Gamma) \cdot \mathbf{y}_{\Gamma,G,U}$ for each $s_\Gamma \in A_{\Gamma,n}$. This proves the statement. $\square$

5.3.3   Planar Case

We identify a planar graph with its embedding in the plane. A network (G, U), $U = \{u_1, \ldots, u_n\}$, is called *planar*, if G is a planar graph and vertices $u_1, \ldots, u_n$ are in one face with clockwise or anticlockwise ordering. We give an example of a planar network.

Let C_n be the circuit of order n, that is, the graph having vertices $v_1, \ldots, v_n$ and edges $v_1 v_2, v_2 v_3, \ldots, v_n v_1$. Let H_n arise from C_n after adding new vertices $u_1, \ldots, u_n$ and edges $u_1 v_1, \ldots, u_n v_n$. Then the simple network (H_n, U_n), $U_n = (u_1, \ldots, u_n)$, is called an *n-wheel*. (H_n, U_n) is indicated in Figure 5.8. It is an example of a planar network. Note that the term wheel usually denotes a graph arising from H_n after identifying the vertices of U_n into one vertex. Also nowhere-zero A-flows in such graphs correspond to nowhere-zero A-flows in networks (H_n, U_n).

Let $P = \{Q_1, \ldots, Q_r\}$ be a partition of the set $\{1, \ldots, n\}$. Take the network (H_n, U_n) and for $j = 1, \ldots, r$, identify the set of vertices $\{u_i; i \in Q_j\}$ to a new vertex x_j. Suppose that the resulting graph has an embedding in the plane so that no vertex x_j, $j = 1, \ldots, r$, is inside the circuit C_n and no two edges intersect. Then P is called *planar*. Let $\overline{\mathcal{P}}_n$ be the set of nontrivial planar partitions of $\{1, \ldots, n\}$ and $\overline{p}_n = |\overline{\mathcal{P}}_n|$. Similarly, for each $s \in A_n$, denote by $\overline{\chi}_n(s)$ the integral vector $(\chi_{s,1}, \ldots, \chi_{s,\overline{p}_n})$ consisting of the coordinates of $\chi_n(s)$ corresponding to partitions from $\overline{\mathcal{P}}_n$.

We always assume that $\overline{\mathcal{P}}_n = \{P_{n,1}, \ldots, P_{n,\overline{p}_n}\}$, that is, the first $\overline{p}_n$ partitions from $\mathcal{P}_n$ are planar.

Clearly, contracting or deleting an edge in a planar network, we get again a planar network. Using this fact, we can simplify Theorems 5.3.1 and 5.3.2.

Theorem 5.3.3 *Let (G, U), $U = (u_1, \ldots, u_n)$, be a simple planar network and $|\overline{\mathcal{P}}_n| = \overline{p}_n$. Then there exists an integral vector $\mathbf{x}_{G,U} = (x_1, \ldots, x_{\overline{p}_n})$ such that for every $s \in A_n$, $F_{G,U}(s) = \overline{\chi}_n(s) \cdot \mathbf{x}_{G,U}$.*

Proof. If $|E(G)| = n$, then G is bipartite with partition sets $\{u_1, \ldots, u_n\}$ and $\{v_1, \ldots, v_r\}$. Thus, there exists a planar partition $P = \{Q_1, \ldots, Q_r\}$ of $\{1, \ldots, n\}$ such that $i \in Q_j$ if and only if u_i is adjacent to v_j ($i \in \{1, \ldots, n\}, j \in \{1, \ldots, r\}$). The rest of the proof runs in a similar way as in Theorem 5.3.1. $\square$

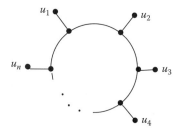

FIGURE 5.8:   Network (H_n, U_n).

A permutation γ on $\{1,\ldots,n\}$ is called *planar* if $\gamma(P)$ is planar for each $P \in \overline{\mathcal{P}}_n$. A permutation group Γ is called *planar* if it consists of planar permutations.

An example of a planar permutation group is the dihedral group Γ_n on $\{1,\ldots,n\}$, that is, the permutation group on $\{1,\ldots,n\}$ generated by permutations γ_n and γ'_n, where γ_n maps i to $i+1 \bmod n$ and γ'_n maps i to $n+1-i$ ($i = 1,\ldots,n$). Clearly, Γ_n has $2n$ elements of the form $(\gamma_n)^i(\gamma'_n)^j$, where $i = 0,1,\ldots,n-1, j = 0,1$. It corresponds with the group of automorphisms of C_n and H_n.

If Γ is a planar permutation group, then define by $\overline{\mathcal{P}}_{\Gamma,n}$ the set of Γ-equivalence classes that decompose $\overline{\mathcal{P}}_n$. It is considered as an $\bar{p}_n$-tuple $(P_{\Gamma,1},\ldots,P_{\Gamma,\bar{p}_{\Gamma,n}})$. Then denote by $\overline{\chi}_{\Gamma,n}(s)$ the integral vector $(\chi(s,P_{\Gamma,1}),\ldots,\chi(s,P_{\Gamma,\bar{p}_{\Gamma,n}}))$. For each $s_\Gamma \in A_{\Gamma,n}$, define

$$\overline{\chi}_{\Gamma,n}(s_\Gamma) = \overline{\chi}_{\Gamma,n}(s),$$

where $s \in s_\Gamma$. This is well defined by (5.7).

Theorem 5.3.4 *Let* (G,U), $U = (u_1,\ldots,u_n)$, *be a simple planar network and* Γ *be a planar permutation group on* $\{1,\ldots,n\}$. *Then there exists a vector* $\mathbf{y}_{\Gamma,G,U} \in \mathbb{Z}^{\bar{p}_{\Gamma,n}}$ *such that for every* $s_\Gamma \in A_{\Gamma,n}$, $F_{\Gamma,G,U}(s_\Gamma) = \overline{\chi}_{\Gamma,n}(s_\Gamma) \cdot \mathbf{y}_{\Gamma,G,U}$.

Proof. It runs analogously as in Theorem 5.3.2, after considering all permutations to be planar. $\qquad\square$

5.3.4 Numbers of Partitions

In this section, we evaluate recursive formulas for computing p_n and $\bar{p}_n$ as presented in Kochol [18].

We start with formula for evaluating p_n, the number of all nontrivial partitions from $\mathcal{P}_n$. If $\{Q_1,\ldots,Q_r\} \in \mathcal{P}_n$ and $n \in Q_j$, then $|Q_j| \in \{2,\ldots,n-2,n\}$. For $i = 2,\ldots,n-2$ (resp. $i = n$), $\mathcal{P}_n$ contains exactly $\binom{n-1}{i-1}p_{n-i}$ (resp. 1) partitions such that the element n is contained in an i-element subset. Hence, for any $n \geq 2$, we have

$$p_n = 1 + \sum_{i=2}^{n-2} \binom{n-1}{i-1}p_{n-i}. \tag{5.12}$$

We introduce the formula for evaluating $\bar{p}_n$, the number of all nontrivial planar partitions from $\overline{\mathcal{P}}_n$. Clearly, $\bar{p}_1 = 0$ and $\bar{p}_2 = \bar{p}_3 = 1$. It is natural to set $\bar{p}_0 = 1$. $\overline{\mathcal{P}}_n, n \geq 4$, can be partitioned into subsets $A_i, B_i, i = 1,\ldots,n-1$, such that a planar partition $P = \{Q_1,\ldots,Q_r\} \in \overline{\mathcal{P}}_n$ is from A_i (resp. B_i) if $n \in Q_1$, i is the smallest element from Q_1, and $|Q_1| = 2$ (resp. $|Q_1| \geq 3$). Note that we allow that some of the sets A_i, B_i are empty, for example, $B_{n-1} = A_2 = A_{n-2} = \emptyset$. Deleting set $\{i,n\}$ from $P \in A_i$, we get planar partitions of $\{1,\ldots,i-1\}$ and $\{i+1,\ldots,n-1\}$, and conversely, adding to such two partitions $\{i,n\}$, we get a partition from A_i. Thus, $|A_i| = \bar{p}_{i-1}\bar{p}_{n-i-1}$. Deleting element n from a partition $P \in B_i$, we get planar partitions of $\{1,\ldots,i-1\}$ and

$\{i, \ldots, n-1\}$. On the other hand, taking planar partitions P' and P'' of $\{1, \ldots, i-1\}$ and $\{i, \ldots, n-1\}$, respectively, and adding to the set from P'' containing i a new element n, we get a partition from B_i. Thus, $|B_i| = \overline{p}_{i-1}\overline{p}_{n-i}$. Therefore, supposing $\overline{p}_1 = 0$ and $\overline{p}_2 = \overline{p}_3 = \overline{p}_0 = 1$, we have for every $n \geq 4$

$$\overline{p}_n = \sum_{i=1}^{n-1} \overline{p}_{i-1}\left(\overline{p}_{n-i-1} + \overline{p}_{n-i}\right). \tag{5.13}$$

5.3.5   Flows on Wheels

For $i = 1, \ldots, n$, let x_i denote the arc of H_n directed from u_i to v_i and y_i denote the arc directed from v_i to v_{i+1} (considering the indices mod n). Define

$$C(A, n) = \{s \in A_n; F_{H_n, U_n}(s) \neq 0\},$$
$$c(A, n) = |C(A, n)|,$$
$$C_i(A, n) = \{s \in C(A, n); F_{H_n, U_n}(s) = i\}, \ i \geq 1,$$
$$c_i(A, n) = |C_i(A, n)|, \ i \geq 1.$$

The aim of this section is to find recursive formulas evaluating $c_i(A, n)$ and $c(A, n)$.

Lemma 5.3.1 *Let φ and φ' be nowhere-zero A-flows in (H_n, U_n) such that $\partial \varphi(U_n) = \partial \varphi'(U_n)$. Then there exists $a \in A$ such that $\varphi(y_i) = \varphi'(y_i) + a$ for every $i = 1, \ldots, n$.*

Proof. The fact that $\partial \varphi(U_n) = \partial \varphi'(U_n)$ implies $\varphi(x_i) = \varphi'(x_i)$ for $i = 1, \ldots, n$. Thus, by induction on i, $\varphi(y_i) = \varphi(y_1) - \sum_{j=2}^i \varphi(x_j)$ and $\varphi'(y_i) = \varphi'(y_1) - \sum_{j=2}^i \varphi'(x_j)$ for $i = 1, \ldots, n$, whence $\varphi(y_i) - \varphi'(y_i) = \varphi(y_1) - \varphi'(y_1)$. This implies the statement. $\square$

Before formulating the main theorem, we need some more notation. If φ is a nowhere-zero A-flow in (H_n, U_n) and $\varphi(x_{n-1}) + \varphi(x_n) = 0$ (resp. $\varphi(x_{n-1}) + \varphi(x_n) \neq 0$), then denote by $\overline{\varphi}$ the nowhere-zero A-flow in (H_{n-2}, U_{n-2}) (resp. (H_{n-1}, U_{n-1})) such that $\overline{\varphi}(x_i) = \varphi(x_i)$, $\overline{\varphi}(y_i) = \varphi(y_i)$ for $i = 1, \ldots, n-2$ (resp. $\overline{\varphi}(x_{n-1}) = \varphi(x_n) + \varphi(x_{n-1})$, $\overline{\varphi}(y_{n-1}) = \varphi(y_n)$, and $\overline{\varphi}(x_i) = \varphi(x_i)$, $\overline{\varphi}(y_i) = \varphi(y_i)$ for $i = 1, \ldots, n-2$).

If $(s_1, \ldots, s_n) \in A_n$ and $s_{n-1} + s_n = 0$ (resp. $s_{n-1} + s_n \neq 0$), then denote by $\overline{s} = (s_1, \ldots, s_{n-2}) \in A_{n-2}$ (resp. $\overline{s} = (s_1, \ldots, s_{n-2}, s_{n-1} + s_n) \in A_{n-1}$).

If ψ is a nowhere-zero A-flow in (H_{n-2}, U_{n-2}) and $a \in A \backslash \{0\}$, then denote by $\psi^{[a]}$ the A-flow in (H_n, U_n) such that $\psi^{[a]}(x_{n-1}) = -\psi^{[a]}(x_n) = a$, $\psi^{[a]}(y_n) = \psi(y_{n-2})$, $\psi^{[a]}(y_{n-1}) = \psi(y_{n-2}) - a$, and $\psi^{[a]}(x_i) = \psi(x_i)$, $\psi^{[a]}(y_i) = \psi(y_i)$ for $i = 1, \ldots, n-2$. Note that $\psi^{[a]}$ is nowhere-zero if and only if $\psi^{[a]}(y_{n-1}) = \psi(y_{n-2}) - a \neq 0$, that is, $a \neq \psi(y_{n-2})$. Furthermore, $\overline{\psi^{[a]}} = \psi$ for any $a \in A \backslash \{0\}$, $a \neq \psi(y_{n-2})$.

Similarly, if $t = (t_1, \ldots, t_{n-2}) \in A_{n-2}$ and $a \in A \backslash \{0\}$, denote $t^{[a]} = (t_1, \ldots, t_{n-2}, a, -a) \in A_n$. Clearly, $\overline{t^{[a]}} = t$.

If $a \in A \backslash \{0\}$, denote by

$$Y_a = \{(a_1, a_2); a_1 + a_2 = a, a_1 \neq 0 \neq a_2\}.$$

By the arithmetic in A, we have $|Y_a| = |A| - 2$. If ψ is a nowhere-zero A-flow in (H_{n-1}, U_{n-1}) and $(a_1, a_2) \in Y_{\psi(x_{n-1})}$, then denote by $\psi^{[a_1,a_2]}$ the A-flow in (H_n, U_n) such that $\psi^{[a_1,a_2]}(x_{n-1}) = a_1$, $\psi^{[a_1,a_2]}(x_n) = a_2$, $\psi^{[a_1,a_2]}(y_n) = \psi(y_{n-1})$, $\psi^{[a_1,a_2]}(y_{n-1}) = \psi(y_{n-2}) - a_1$ $(= \psi(y_{n-1}) + a_2$, because $a_1 + a_2 = \psi(x_{n-1}) = \psi(y_{n-2}) - \psi(y_{n-1}))$, and $\psi^{[a_1,a_2]}(x_i) = \psi(x_i)$, $\psi^{[a_1,a_2]}(y_i) = \psi(y_i)$ for $i = 1, \ldots,$ $n - 2$. Note that $\psi^{[a_1,a_2]}$ is nowhere-zero if and only if $\psi^{[a_1,a_2]}(y_{n-1}) = \psi(y_{n-2}) - a_1 \neq 0$, that is, $a_1 \neq \psi(y_{n-2})$. Furthermore, $\overline{\psi^{[a_1,a_2]}} = \psi$ for any $(a_1, a_2) \in Y_{\psi(x_{n-1})}$, $a_1 \neq \psi(y_{n-2})$.

Similarly, if $t = (t_1, \ldots, t_{n-1}) \in A_{n-1}$ and $(a_1, a_2) \in Y_{t_{n-1}}$, denote by $t^{[a_1,a_2]} = (t_1, \ldots, t_{n-2}, a_1, a_2) \in A_n$. Clearly, $\overline{t^{[a_1,a_2]}} = t$.

The following statement was proved in [24].

Theorem 5.3.5 *Let A be an Abelian group of order k. Then the following applies:*

(1) $c_i(A, 1) = 0$ *for every $i \geq 1$,*

(2) $c_{k-2}(A, 2) = k - 1$ *and $c_i(A, 2) = 0$ for every $i \geq 1$, $i \neq k - 2$,*

(3) *For every $i, n \geq 3$,*

$$c_i(A, n) = (k - i - 1)c_i(A, n - 2) + (i + 1)c_{i+1}(A, n - 2)$$
$$+ (k - i - 2)c_i(A, n - 1) + (i + 1)c_{i+1}(A, n - 1),$$

(4) $c_i(A, n) = 0$ *for every $i \geq k - 1$, $n \geq 1$,*

(5) $c(A, n) = \sum_{i=1}^{k-2} c_i(A, n)$ *for every $n \geq 1$.*

Proof. (1) follows from the fact that $A_1 = \emptyset$.

A_2 is the set of couples $(a, -a)$ where $a \in A \setminus \{0\}$, whence $|A_2| = k - 1$. A nowhere-zero A-chain φ in (H_2, U_2) belongs to $\Phi_{H_2, U_2}(a, -a)$ if and only if $\varphi(x_1) = a$, $\varphi(x_2) = -a$ and $(\varphi(y_1^{-1}), \varphi(y_2)) \in Y_a$. Since $|Y_a| = k - 2$, we have $F_{H_2, U_2}(a, -a) = k - 2$. This implies (2).

Now we prove (3). Let $s = (s_1, \ldots, s_n) \in C(A, n)$ and $\varphi \in \Phi_{H_n, U_n}(s)$. If $s_n + s_{n-1} = 0$, then $\overline{\varphi}$ is a nowhere-zero A-flow in (H_{n-2}, U_{n-2}), whence $\overline{s} \in C(A, n - 2)$ and s is of the form $t^{[a]}$ where $t \in C(A, n - 2)$. If $s_n + s_{n-1} \neq 0$, then $\overline{\varphi}$ is a nowhere-zero A-flow in (H_{n-1}, U_{n-1}), whence $\overline{s} \in C(A, n - 1)$ and s is of the form $t^{[a_1,a_2]}$ where $t \in C(A, n - 1)$.

Let $t = (t_1, \ldots, t_{n-2}) \in C_i(A, n-2)$ and $\Phi_{H_{n-2}, U_{n-2}}(t) = \{\psi_j; j = 1, \ldots, i\}$. Denote $Y = \{\psi_j(y_{n-2}); j = 1, \ldots, i\}$. By Lemma 5.3.1, $|Y| = i$. For each $a \in A \setminus \{0\}$, we have $\Phi_{H_n, U_n}(t^{[a]}) = \{\psi_j^{[a]}; j = 1, \ldots, i, a \neq \psi_j(y_{n-2})\}$ (because $\psi_j^{[a]}$ is nowhere-zero if and only if $a \neq \psi_j(y_{n-2})$). Hence, $t^{[a]} \in C_i(A, n)$ if $a \in A \setminus \{0\}$, $a \notin Y$, and $t^{[a]} \in C_{i-1}(A, n)$ if $a \in Y$. Thus, $C_i(A, n)$ has exactly $(k - i - 1)c_i(A, n - 2) + (i + 1)c_{i+1}(A, n - 2)$ elements $(s_1, \ldots, s_n)$ such that $s_n + s_{n-1} = 0$.

Let $t = (t_1, \ldots, t_{n-1}) \in C_i(A, n - 1)$ and $\Phi_{H_{n-1}, U_{n-1}}(t) = \{\psi_j; j = 1, \ldots, i\}$. Denote $Y' = \{(\psi_j(y_{n-2}), -\psi_j(y_{n-1})); j = 1, \ldots, i\}$. $Y' \subseteq Y_{t_{n-1}}$, because for

$j = 1, \ldots, i$, $\psi_j(y_{n-2}) - \psi_j(y_{n-1}) = \psi_j(x_{n-1}) = t_{n-1}$. But $|Y_{t_{n-1}}| = k - 2$ and, by Lemma 5.3.1, $|Y'| = i$, whence $|Y_{t_{n-1}} \setminus Y'| = k - i - 2$. For each $(a_1, a_2) \in Y_{t_{n-1}}$, we have $\Phi_{H_n, U_n}(t^{[a_1, a_2]}) = \{\psi_j^{[a_1, a_2]}; j = 1, \ldots, i, a_1 \neq \psi_j(y_{n-2})\}$ (because $\psi_j^{[a_1, a_2]}$ is nowhere-zero if and only if $a_1 \neq \psi_j(y_{n-2})$). Hence, $t^{[a_1, a_2]} \in C_i(A, n)$ if $(a_1, a_2) \in Y_{t_{n-1}} \setminus Y'$, and $t^{[a_1, a_2]} \in C_{i-1}(A, n)$ if $(a_1, a_2) \in Y'$. Thus, $C_i(A, n)$ has exactly $(k - i - 2)c_i(A, n-1) + (i+1)c_{i+1}(A, n-1)$ elements $(s_1, \ldots, s_n)$ such that $s_n + s_{n-1} \neq 0$. This proves (3).

(4) is a trivial consequence of Lemma 5.3.1 and the fact that $\varphi(y_1) \neq \varphi(x_1)$ for each nowhere-zero A-flow φ in (H_n, U_n).

(5) follows from (4). $\qquad\square$

Corollary 5.3.1 *If A and A' are two Abelian groups of the same cardinality, then $c_i(A, n) = c_i(A', n)$ and $c(A, n) = c(A', n)$ for every $i, n \geq 1$.*

Proof. The statement holds true for $n = 1, 2$ by items (1), (2), and (5) of Theorem 5.3.5. Hence, by induction on n and items (3) and (5) of Theorem 5.3.5, the statement holds true in general. $\qquad\square$

Note that Corollary 5.3.1 coincides with the classical result of Tutte [30,31] that the number of nowhere-zero A-flows in a graph does not depend on the structure of A, but only on the order of A.

By Corollary 5.3.1, we can use the notation $c(|A|, n)$ and $c_i(|A|, n)$ instead of $c(A, n)$ and $c_i(A, n)$, respectively.

The following statement is used in the sequel.

Corollary 5.3.2 $c(5, 1) = c_1(5, 1) = c_2(5, 1) = c_3(5, 1) = c_1(5, 2) = c_2(5, 2) = 0$, $c(5, 2) = c_3(5, 2) = 4$, *and for every* $n \geq 3$,

$$c_1(5, n) = 3c_1(5, n-2) + 2c_2(5, n-2) + 2c_1(5, n-1) + 2c_2(5, n-1),$$
$$c_2(5, n) = 2c_2(5, n-2) + 3c_3(5, n-2) + c_2(5, n-1) + 3c_3(5, n-1),$$
$$c_3(5, n) = c_3(5, n-2),$$
$$c(5, n) = c_1(5, n) + c_2(5, n) + c_3(5, n).$$

Proof. This follows from Theorem 5.3.5 after setting $A = \mathbb{Z}_5$. $\qquad\square$

5.4   Nowhere-Zero 5-Flows

In this section, we show that the smallest counterexample to the 5-flow conjecture must be cyclically 6-edge connected and has girth at least 11.

Throughout this section, we always consider $A = \mathbb{Z}_5$, because we deal with nowhere-zero $\mathbb{Z}_5$-flows on graphs.

5.4.1 Restrictions of Cyclical Edge Connectivity

Let $\mathcal{A}$ denote the automorphism group of $\mathbb{Z}_5$. The elements of $\mathcal{A}$ are $\alpha_0 = \mathrm{id}$, $\alpha_1 = (1, 2, 4, 3)$, $\alpha_2 = (1, 4)(2, 3)$, and $\alpha_3 = (1, 3, 4, 2)$. If $s = (s_1, \ldots, s_n) \in A_n$ and $\alpha \in \mathcal{A}$, then denote $\alpha(s) = (\alpha(s_1), \ldots, \alpha(s_n)) \in A_n$. Clearly, $\chi_n(s) = \chi_n(\alpha(s))$, whence by Theorem 5.3.1, $F_{G,U}(s) = F_{G,U}(\alpha(s))$ for every simple network (G, U) with n outer vertices.

The following statement was proved in Kochol [17].

Theorem 5.4.1 *Suppose that $\overline{G}$ is a counterexample to the 5-flow conjecture of minimal order. Then $\overline{G}$ is a cyclically 6-edge-connected snark.*

Proof. In Section 5.2.2, we have proved that $\overline{G}$ must be a cubic graph without a 3-edge coloring.

Suppose that $\overline{G}$ has a cyclic n-edge cut $\{f_1, \ldots, f_n\}$, $5 \geq n \geq 2$. Furthermore, we assume that cardinality of this cut is minimally possible. Then $G - \{f_1, \ldots, f_n\}$ has exactly two components G' and G'', which are cyclic and bridgeless. For $i = 1, \ldots, n$, let v_i' and v_i'' be the end of f_i contained in G' and G'', respectively. Add to G' (G'') n vertices $u_1, \ldots, u_n$ $(w_1, \ldots, w_n)$, and for $i = 1, \ldots, n$, join u_i with v_i' $(w_i$ with $v_i'')$. We get from G' (G'') a new graph X (Y). Consider the simple networks (X, U), $U = (u_1, \ldots, u_n)$, and (Y, W), $W = (w_1, \ldots, w_n)$.

$F_{X,U}(s), F_{Y,W}(s) \geq 0$ for every $s \in A_n$. If there exists $s \in A_n$ such that $F_{X,U}(s)$, $F_{Y,W}(s) > 0$, then (X, U) and (Y, W) have nowhere-zero A-flows φ_1 and φ_2, respectively, such that $\partial \varphi_1(U) = s$ and $\partial \varphi_2(W) = \alpha_2(s)$ (because $F_{Y,W}(\alpha_2(s)) = F_{Y,W}(s) > 0$), which can be pieced together into a nowhere-zero A-flow in G, a contradiction. Thus, $F_{X,U}(s) \cdot F_{Y,W}(s) = 0$ for every $s \in A_n$.

By Theorem 5.3.1, there exist integers x_i and y_i, $i = 1, \ldots, p_n$, such that for every $s \in A_n$, $F_{X,U}(s) = \sum_{i=1}^{p_n} \chi_{s,i} x_i$ and $F_{Y,W}(s) = \sum_{i=1}^{p_n} \chi_{s,i} y_i$ where $(\chi_{s,1}, \ldots, \chi_{s,p_n}) = \chi_n(s)$.

If $n \in \{2, 3\}$, then $\mathcal{P}_n$ contains exactly one partition and $\chi_n(s) = (1)$ for every $s \in A_n$. Thus, by Theorem 5.3.1, $F_{G,U}(s) = F_{G,U}(s')$ for every $s, s' \in A_n$ and every simple network (G, U) with n outer vertices. This implies that a smallest counterexample to the 5-flow conjecture must be cyclically 4-edge connected.

Let $n = 4$. By (5.12), $p_4 = 4$ and $\mathcal{P}_4$ contains the following four partitions:

$$P_{4,1} = \{\{1, 2, 3, 4\}\},$$
$$P_{4,2} = \{\{1, 2\}, \{3, 4\}\},$$
$$P_{4,3} = \{\{2, 3\}, \{4, 1\}\},$$
$$P_{4,4} = \{\{1, 3\}, \{2, 4\}\}.$$

Since $F_{X,U}(1, 1, 1, 2) = x_1$ and $F_{Y,W}(1, 1, 1, 2) = y_1$, we have $x_1, y_1 \geq 0$ and $x_1 \cdot y_1 = 0$. Suppose that $x_1 = 0$. Then $F_{X,U}(1, 4, 2, 3) = x_2 \geq 0$, $F_{X,U}(1, 2, 3, 4) = x_3 \geq 0$, $F_{X,U}(1, 2, 4, 3) = x_4 \geq 0$. If $x_1 = \ldots = x_4 = 0$, then identifying u_1 (u_3) with u_2 (u_4) in X and suppressing the vertices of valency 2, we get a smaller counterexample. Therefore, at least one x_i must be positive and without loss of generality, we can assume that $x_2 > 0$. Then $F_{X,U}(s) > 0$ and $F_{Y,W}(s) = 0$ if $\chi_4(s)$ has the second coordinate 1 ($s \in A_4$). Thus, identifying w_1 (w_3) with w_2 (w_4) in Y and suppressing the vertices of valency 2, we get a smaller counterexample. This implies that a smallest counterexample to the 5-flow conjecture must be cyclically 4-edge connected. Notice that this fact was already proved by Celmins [4].

Let $n = 5$. By (5.12), $p_5 = 11$ and $\mathcal{P}_5$ contains the following partitions:

$$P_{5,1} = \{\{1,2,3,4,5\}\},$$
$$P_{5,2} = \{\{1,2\},\{3,4,5\}\},$$
$$P_{5,3} = \{\{2,3\},\{4,5,1\}\},$$
$$P_{5,4} = \{\{3,4\},\{5,1,2\}\},$$
$$P_{5,5} = \{\{4,5\},\{1,2,3\}\},$$
$$P_{5,6} = \{\{5,1\},\{2,3,4\}\},$$
$$P_{5,7} = \{\{1,3\},\{2,4,5\}\},$$
$$P_{5,8} = \{\{2,4\},\{3,5,1\}\},$$
$$P_{5,9} = \{\{3,5\},\{4,1,2\}\},$$
$$P_{5,10} = \{\{4,1\},\{5,2,3\}\},$$
$$P_{5,11} = \{\{5,2\},\{1,3,4\}\}.$$

If $s = (s_1, \ldots, s_5) \in A_5$, then define $\pi(s) = (s_5, s_1, s_2, s_3, s_4)$. Let $\mathbf{e}_i$ denote the ith standard basis vector of $\mathbb{Q}^{11}$.

Now $F_{X,U}(1,1,1,1,1) = x_{11}$ and $F_{Y,W}(1,1,1,1,1) = y_1$. Thus, $x_1, y_1 \geq 0$ and $x_1 \cdot y_1 = 0$. Without loss of generality, we can assume that $x_1 = 0$.

Suppose that $x_i \geq 0$ for every $i = 1, \ldots, 10$. If $x_1 = \cdots = x_{11} = 0$, then $F_{X,U}(s) = 0$ for every $s \in A_5$. Identify u_1 with u_2 and u_3 with u_4, u_5 in X and suppress the vertex of valency 2. The resulting cubic graph is bridgeless, has order smaller than G, and does not have a nowhere-zero 5-flow (otherwise, $F_{X,U}(s) > 0$ for some $s \in A_5$), which contradicts the minimality of G. Hence, at least one x_i must be positive. We can choose the ordering of edges $f_1, \ldots, f_5$ so that $x_2 > 0$. Thus, $F_{X,U}(s) > 0$ and $F_{Y,W}(s) = 0$ if $s \in A_5$ and $\chi_5(s)$ has the second coordinate 1. Then identifying w_1 with w_2 and w_3 with w_4, w_5 in Y and suppressing the vertex of valency 2, we get a bridgeless cubic graph without a nowhere-zero 5-flow and of order smaller than G, a contradiction.

Therefore, at least one x_i is negative. We can choose the ordering of edges $f_1, \ldots, f_5$ so that $x_2 < 0$. Consider the following:

$$\begin{array}{lll}
p_1 = (1,4,1,2,2), & \chi_5(p_1) = \mathbf{e}_1 + \mathbf{e}_2 + \mathbf{e}_3, & \\
p_2 = (1,4,2,1,2), & \chi_5(p_2) = \mathbf{e}_1 + \mathbf{e}_2 + \mathbf{e}_8, & \\
p_3 = (1,4,2,2,1), & \chi_5(p_3) = \mathbf{e}_1 + \mathbf{e}_2 + \mathbf{e}_{11}, & \\
p_4 = (4,1,1,2,2), & \chi_5(p_4) = \mathbf{e}_1 + \mathbf{e}_2 + \mathbf{e}_7, & (5.14) \\
p_5 = (4,1,2,1,2), & \chi_5(p_5) = \mathbf{e}_1 + \mathbf{e}_2 + \mathbf{e}_{10}, & \\
p_6 = (4,1,2,2,1), & \chi_5(p_6) = \mathbf{e}_1 + \mathbf{e}_2 + \mathbf{e}_6. &
\end{array}$$

Since $F_{X,U}(p_i) \geq 0$ for $i = 1, \ldots, 6$, we have $x_3, x_6, x_7, x_8, x_{10}, x_{11} \geq -x_1 - x_{11} = -x_{11} > 0$. If one of x_4, x_5, x_9 is negative, we can choose the ordering of edges f_3, f_4, f_5

so that $x_4 < 0$. For $i = 1, \ldots, 6$, replacing p_i with $\pi^2(p_i)$ in (5.14) and using the fact that $F_{G,U}(\pi^2(p_i)) \geq 0$, we get $x_3, x_5, x_7, x_8, x_9, x_{10} \geq -x_4 - x_1 = -x_4 > 0$. Thus, without abuse of generality, we can assume that exactly one of the following cases occurs:

1. $x_2 < 0$, $x_3, x_6, x_7, x_8, x_{10}, x_{11} \geq -x_2$, and $x_4, x_5, x_9 \geq 0$;

2. $x_2, x_4 < 0$, $x_3, x_6, x_7, x_8, x_{10}, x_{11} \geq -x_2$, and $x_3, x_5, x_7, x_8, x_9, x_{10} \geq -x_4$.

Let S be the set of permutations $(s_1, s_2, s_3, s_4, s_5)$ of $1, 1, 1, 3, 4$. S is a proper subset of A_5 (for instance, $(1, 1, 1, 1, 1)$ and $(1, 1, 2, 2, 4)$ belong to A_5 but not to S). We claim that $F_{X,U}(s) > 0$ for every $s \in S$.

Let $s = (s_1, \ldots, s_5) \in S$. Then $\chi_5(s) = \mathbf{e}_a + \mathbf{e}_b + \mathbf{e}_c + \mathbf{e}_1$ and $F_{X,U}(s) = x_a + x_b + x_c$ where a, b, c are pairwise distinct elements from $\{2, \ldots, 11\}$ such that partitions $P_{5,a}, P_{5,b}, P_{5,c}$ are of the form $\{\{i_1, i_2\}, \{i_3, i_4, i_5\}\}$ where $s_{i_1} + s_{i_2} = 0$.

If $a = 2$, that is, $s_1 + s_2 = 0$, then $b, c \notin \{4, 5, 9\}$ (otherwise, at least one of the sums $s_3 + s_4$, $s_4 + s_5$, $s_3 + s_5$ equals 0, whence at least one of s_5, s_3, s_4 is 0, a contradiction). Since $x_3, x_6, x_7, x_8, x_{10}, x_{11} \geq -x_2$, we have $x_a + x_b + x_c \geq -x_2 > 0$.

Let $a, b, c \neq 2$ and case (1) occur. Then $\{4, 5, 9\} \neq \{a, b, c\}$ (otherwise, $s_3 + s_4 = s_4 + s_5 = s_3 + s_5 = 0$, whence $s_3 = s_4 = s_5 = 0$, a contradiction). Thus, x_a, x_b, x_c are nonnegative integers and at least one of them is positive, whence $x_a + x_b + x_c > 0$.

Let $a, b, c \neq 2$ and case (2) occur. If $4 \in \{a, b, c\}$, then we can choose the ordering of edges $f_1, \ldots, f_5$ so that we get the case $a = 2$. If $4 \notin \{a, b, c\}$, then x_a, x_b, x_c are positive, and so is $x_a + x_b + x_c$.

Therefore, for every $s \in S$, $F_{X,U}(s) > 0$ and $F_{Y,W}(s) = 0$. Consider

$$\begin{aligned}
p_7 &= (4, 3, 1, 1, 1), & \chi_5(p_7) &= \mathbf{e}_1 + \mathbf{e}_6 + \mathbf{e}_7 + \mathbf{e}_{10}, \\
p_8 &= (4, 1, 3, 1, 1), & \chi_5(p_8) &= \mathbf{e}_1 + \mathbf{e}_6 + \mathbf{e}_2 + \mathbf{e}_{10}, \\
p_9 &= (4, 1, 1, 3, 1), & \chi_5(p_9) &= \mathbf{e}_1 + \mathbf{e}_6 + \mathbf{e}_2 + \mathbf{e}_7.
\end{aligned} \tag{5.15}$$

Since $p_7, p_8, p_9 \in S$, we have

$$\begin{aligned}
0 &= F_{Y,W}(p_7) - F_{Y,W}(p_8) = y_7 - y_2, \\
0 &= F_{Y,W}(p_7) - F_{Y,W}(p_9) = y_{10} - y_2, \\
0 &= F_{Y,W}(p_8) - F_{Y,W}(p_9) = y_{10} - y_7.
\end{aligned} \tag{5.16}$$

Therefore, $y_2 = y_7 = y_{10}$. For $i = 1, 2, 3, 4$, replacing p_7, p_8, p_9 with $\pi^i(p_7)$, $\pi^i(p_8)$, $\pi^i(p_9)$, respectively, in (5.15) and (5.16), we get $y_3 = y_8 = y_{11}$, $y_4 = y_9 = y_7$, $y_5 = y_{10} = y_8$, $y_6 = y_{11} = y_9$, whence $y_2 = \cdots = y_{11}$. Furthermore, $F_{X,U}(1, 1, 2, 2, 4) = x_1 + x_6 + x_{11} > 0$ in both cases (1) and (2). Thus, $0 = F_{Y,W}(1, 1, 2, 2, 4) = y_1 + y_6 + y_{11}$ and $0 = F_{Y,W}(1, 1, 1, 3, 4) - F_{Y,W}(1, 1, 2, 2, 4) = y_9$. Therefore $y_1 = \cdots = y_{11} = 0$ and we get a smaller counterexample in a similar way as in the case $x_1 = \cdots = x_{11} = 0$. This proves the statement. $\qquad \square$

5.4.2   Forbidden Networks

Let (H, U), $U = (u_1, \ldots, u_n)$, be a simple network. (H, U) is called *quasicubic*, if every vertex of H has a degree of at most 3. By the *cubic order* of (H, U), denoted by $v_3(H, U)$, we mean the number of the vertices of H of degree 3. Denote by

$$A_{H,U} = \{s \in A_n; F_{H,U}(s) > 0\}$$

and by $V_{H,U}$ the linear hull of $\{\chi_n(s); s \in A_{H,U}\}$ in $\mathbb{R}^{p_n}$.

We say that (H, U) is a *forbidden network* if H cannot be a subgraph of a graph homeomorphic to a smallest counterexample to the 5-flow conjecture.

Assume that H is a subgraph of a graph G and (H', U'), $U' = (u_1', \ldots, u_n')$, is a simple network. Let G' arise from G after deleting the vertices from $V(H) \backslash U$ and identifying u_i with u_i' for $i = 1, \ldots, n$. We say that G' arises from G after *replacing* (H, U) by (H', U').

We say that (H, U) can be *regularly replaced* by (H', U') in a class of graphs $\mathcal{C}$, if for every graph G of $\mathcal{C}$, the graph G' arising from G after replacing (H, U) by (H', U') is always bridgeless.

Lemma 5.4.1 *Let (H, U), $U = (u_1, \ldots, u_n)$, and (H', U'), $U' = (u_1', \ldots, u_n')$, $n \geq 2$, be quasicubic networks such that $v_3(H, U) > v_3(H', U')$, $V_{H',U'} \subseteq V_{H,U}$, and (H, U) can be regularly replaced by (H', U') in the class of cyclically 6-edge-connected quasicubic graphs. Then (H, U) is a forbidden network.*

Proof. Let $\overline{G}$ be a counterexample to the 5-flow conjecture of the smallest possible order. Then by Theorem 5.4.1, $\overline{G}$ is a cyclically 6-edge-connected cubic graph. Suppose that F is homeomorphic with $\overline{G}$ and H is a subgraph of F. Without abuse of generality, we can assume that $u_1, \ldots, u_n$ have all degree 2 in F. Let F' be the graph arising after replacing (H, U) by (H', U'). By assumptions, F' is bridgeless and homeomorphic with a cubic graph G'. Since $v_3(H, U) > v_3(H', U')$, the order of G' is smaller than the order of G; therefore, G' and F' admit nowhere-zero 5-flows.

Let I (I') be the graph arising from F (F') after deleting the vertices from $V(H) \backslash U$ ($V(H') \backslash U'$). Then (I, U) and (I', U') are simple networks, and there is an isomorphism of I and I' that maps $u_1, \ldots, u_n$ to $u_1', \ldots, u_n'$, respectively. Thus, $F_{I,U}(s) = F_{I',U'}(s)$ for every $s \in A_n$.

If there exists $s \in A_n$ such that $F_{H,U}(s)$, $F_{I,U}(s) > 0$, then (H, U) and (I, U) have nowhere-zero $\mathbb{Z}_5$-flows φ_1 and φ_2, respectively, such that $\partial \varphi_1(U) = \partial \varphi_2(U) = s$ and the flows φ_1 and $-\varphi_2$ can be pieced together into a nowhere-zero $\mathbb{Z}_5$-flow in F, a contradiction. Thus, $F_{H,U}(s)F_{I,U}(s) = 0$ for every $s \in A_n$. Since $A_{H,U} = \{s \in A_n; F_{H,U}(s) > 0\}$, we have $F_{I,U}(s) = 0$ for every $s \in A_{H,U}$.

By Theorem 5.3.1, there exist integers $x_1, \ldots, x_{p_n}$ such that for every $s \in A_n$, $F_{I,U}(s) = \sum_{i=1}^{p_n} c_{s,i}x_i$ where $(c_{s,1}, \ldots, c_{s,p_n}) = \chi_n(s)$. Choose n-tuples $t_1, \ldots, t_r$ from $A_{H,U}$ so that $\chi_n(t_1), \ldots, \chi_n(t_r)$ form a basis in $V_{H,U}$. Suppose $s \in A_n$ such that $\chi_n(s) \in V_{H,U}$. Then there are numbers $y_{s,1}, \ldots, y_{s,r}$ such that $\chi_n(s) = \sum_{j=1}^{r} y_{s,j}\chi_n(t_j)$, and therefore,

$$F_{I,U}(s) = \sum_{i=1}^{p_n} c_{s,i} x_i = \sum_{i=1}^{p_n} \left(\sum_{j=1}^{r} y_{s,j} c_{t_j,i} \right) x_i$$

$$= \sum_{j=1}^{r} y_{s,j} \left(\sum_{i=1}^{p_n} c_{t_j,i} x_i \right) = \sum_{j=1}^{r} y_{s,j} F_{I,U}(t_j) = 0$$

(because $t_1, \ldots, t_r \in A_{H,U}$). Thus, if $s \in A_n$ and $\chi_n(s) \in V_{H,U}$, then $F_{I,U}(s) = 0$.

Since F' has a nowhere-zero $\mathbb{Z}_5$-flow, there exists $t \in A_{H',U'} \cap A_{I',U'}$, that is, $F_{H',U'}(t)$, $F_{I',U'}(t) > 0$. By assumptions, $V_{H',U'} \subseteq V_{H,U}$, whence $F_{I',U'}(t) = F_{I,U}(t) = 0$, which is a contradiction. This proves the statement. □

5.4.3 Restrictions of Girths

In this section, we use results presented in Kochol [19].

Consider a graph H_{n-2} and change the notation of its vertices by adding primes, that is, denote them by $v'_1, \ldots, v'_{n-2}, u'_1, \ldots, u'_{n-2}$. Similarly, change the notation of the arcs. Add new vertices $v'_{n-1}, v'_n, u'_{n-1}, u'_n$ and edges $v'_{n-1}u'_{n-1}, v'_n u'_n, v'_{n-1}v'_n$. Furthermore, let x'_{n-1}, x'_n, and z'_n denote the arcs of H'_n directed from u'_{n-1} to v'_{n-1}, from u'_n to v'_n, and from v'_{n-1} to v'_n, respectively. Then (H'_n, U'_n), $U'_n = (u'_1, \ldots, u'_n)$, is a simple network. It is indicated in Figure 5.9.

Lemma 5.4.2 *For $n \geq 6$, $v_3(H_n, U_n) > v_3(H'_n, U'_n)$ and (H_n, U_n) can be replaced by (H'_n, U'_n) regularly in the class of cyclically 6-edge-connected cubic graphs.*

Proof. $v_3(H_n, U_n) = n > n - 2 = v_3(H'_n, U'_n)$. Let G' arise from G after replacing (H_n, U_n) by (H'_n, U'_n). If G' has a bridge, then G is not cyclically 6-edge connected. □

Let V_n be the linear hull of $\{\chi_n(s); s \in A_n\}$ in $\mathbb{R}^{p_n}$.

Lemma 5.4.3 *If $V_n = V_{H_n, U_n}$, then (H_n, U_n) is a forbidden network. In particular, the smallest counterexample to the 5-flow conjecture cannot have a circuit of order n.*

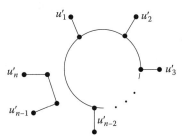

FIGURE 5.9:   Network (H'_n, U'_n).

Proof. Then $v_3(H_n, U_n) > v_3(H'_n, U'_n)$ and (H_n, U_n) can be regularly replaced by (H'_n, U'_n) in the class of cyclically 6-edge-connected quasicubic graphs. Furthermore, $V_{H'_n, U'_n} \subseteq V_n = V_{H_n, U_n}$. Thus, (H_n, U_n) is a forbidden network by Lemma 5.4.1.   □

Recall that $\mathcal{A}$ denotes the automorphism group of $\mathbb{Z}_5$ (with elements $\alpha_0 = $ id, $\alpha_1 = (1, 2, 4, 3)$, $\alpha_2 = (1, 4)(2, 3)$, and $\alpha_3 = (1, 3, 4, 2)$). Furthermore, if $s = (s_1, \ldots, s_n) \in A_n$ and $\alpha \in \mathcal{A}$, then $\alpha(s) = (\alpha(s_1), \ldots, \alpha(s_n)) \in A_n$, and $\chi_n(s) = \chi_n(\alpha(s))$ and $F_{U,G}(s) = F_{U,G}(\alpha(s))$ for any simple network (G, U) with n outer vertices. We say that s and $\alpha(s)$ are σ_n-equivalent.

Let $v_n = |A_n|/4$ be the number of σ_n-classes and $t_1, \ldots, t_{v_n}$ be pairwise non-σ_n-equivalent elements of A_n. Let M_n be the $v_n \times p_n$ matrix so that the ith row of M_n is $\chi_n(t_i)$, $i = 1, \ldots, v_n$. Note that the rowspace V_n of M_n does not depend on the choice of $t_1, \ldots, t_{v_n}$.

Let $c(n) = |A_{H_n, U_n}|/4$, that is, $c(n) = c(5, n)/4$. Without loss of generality, we can suppose that $t_1, \ldots, t_{c(n)}$ are elements of A_{H_n, U_n}. Thus, A_{H_n, U_n} contains all $s' \in A_n$ that are σ_n-equivalent to $s \in \{t_1, \ldots, t_{c(n)}\}$. Let M_{H_n, U_n} be the $c(n) \times p_n$ submatrix of M_n so that the ith row of M_{H_n, U_n} is $\chi_n(t_i)$, $i = 1, \ldots, c(n)$. Note that the rowspace V_{H_n, U_n} of M_{H_n, U_n} does not depend on the choice of $t_1, \ldots, t_{c(n)}$. Clearly, $V_{H_n, U_n} = V_n$ if and only if M_{H_n, U_n} and M_n have the same rank. Thus, the following statement is a more convenient reformulation of Lemma 5.4.3 in matrix terms.

Lemma 5.4.4 *If the matrices M_{H_n, U_n} and M_n have the same rank, then the smallest counterexample to the 5-flow conjecture has no circuit of order n.*

Clearly, $v_1 = 0$ and $v_2 = 1$. Furthermore, A_n has exactly $4|A_{n-2}|$ $(3|A_{n-1}|)$ n-tuples so that the sum of the last two elements is zero (nonzero). Thus, $|A_n| = 4|A_{n-2}| + 3|A_{n-1}|$, whence $v_n = 4v_{n-2} + 3v_{n-1}$ for any $n \geq 3$. We evaluate v_n for $n = 2, \ldots, 8$ and also p_n and v_{H_n} using (5.12) and Corollary 5.3.2, respectively,

n	1	2	3	4	5	6	7	8
$c(n)$	0	1	3	13	45	151	483	1513
v_n	0	1	3	13	51	205	819	3277
p_n	0	1	1	4	11	41	162	715

Thus, for $n = 2, 3, 4$, $c(n) = v_n$, whence matrices M_{H_n, U_n} and M_n have the same rank.

Using a computer, we can check that M_{H_n, U_n} equals the rank of M_n for $n = 5, \ldots, 8$. More precisely, the rank of M_n is 11, 40, 147, 568 if $n = 5, 6, 7, 8$, respectively. Thus, by Lemma 5.4.4, the following statement holds true.

Theorem 5.4.2 *The smallest counterexample of the 5-flow conjecture has girth at least 9.*

Notice that Theorem 5.4.1 implies that the smallest counterexample to the 5-flow conjecture has girth at least 6.

5.4.4 Reductions by Dihedral Groups

In this section, we use results presented in Kochol [22]. We use the principles from Section 5.3.2 for reduction of size of the matrices.

Let (G, U), $U = (u_1, \ldots, u_n)$, be a simple network and Γ be a permutation group on $\{1, \ldots, n\}$. We say that Γ *acts regularly* on (G, U) if for each $\gamma \in \Gamma$, there exists an automorphism of G that maps u_i to $u_{\gamma(i)}$ $(i = 1, \ldots, n)$.

Lemma 5.4.5 *If a permutation group Γ acts regularly on a simple network (G, U), $U = (u_1, \ldots, u_n)$, then $F_{G,U}(s) = F_{G,U}(\gamma(s))$ for each $s \in A_n$ and $\gamma \in \Gamma$.*

Proof. By the definition, for each $\gamma \in \Gamma$, there exists an automorphism $\tilde{\gamma}$ of G such that $\tilde{\gamma}(u_i) = u_{\gamma(i)}$, whence $F_{G,U}(s) = F_{G,U}(\gamma(s))$ for each $s \in A_n$. $\qquad\square$

Lemma 5.4.6 *Let (H, U), $U = (u_1, \ldots, u_n)$, $n \geq 2$, be a quasicubic network and Γ be a permutation group on $\{1, \ldots, n\}$ such that $F_{H,U}(s) > 0$ for every $s \in \cup \{s_\Gamma; s_\Gamma \in A_{\Gamma,H,U}\}$. Suppose there exists a quasicubic network (H', U'), $U' = (u'_1, \ldots, u'_n)$, such that $v_3(H, U) > v_3(H', U')$, $V_{\Gamma,H',U'} \subseteq V_{\Gamma,H,U}$, and (H, U) can be regularly replaced by (H', U') in the class of cyclically 6-edge-connected quasicubic graphs. Then (H, U) is a forbidden network.*

Proof. Let $\overline{G}$ be a counterexample to the 5-flow conjecture of the smallest possible order. Then by Theorem 5.4.1, $\overline{G}$ is a cyclically 6-edge-connected cubic graph. Suppose that G is homeomorphic with $\overline{G}$ and H is a subgraph of G. Without loss of generality, we can assume that $u_1, \ldots, u_n$ have in G degrees of 2. Let G' be the graph arising from G after replacing (H, U) by (H', U'). By assumptions, G' is bridgeless and homeomorphic with a cubic graph $\overline{G}'$. Since $v_3(H, U) > v_3(H', U')$, the order of $\overline{G}'$ is smaller than the order of $\overline{G}$; therefore, $\overline{G}'$ and G' admit nowhere-zero 5-flows.

Let I (I') be the graph arising from G (G') after deleting the vertices from $V(H)\backslash U$ ($V(H')\backslash U'$). Then (I, U) and (I', U') are simple networks, and there is an isomorphism of I and I' that maps $u_1, \ldots, u_n$ to $u'_1, \ldots, u'_n$, respectively. Thus, $F_{I,U}(s) = F_{I',U'}(s)$ for every $s \in A_n$.

If there exists $s \in A_n$ such that $F_{H,U}(s)$, $F_{I,U}(s) > 0$, then (H, U) and (I, U) have nowhere-zero $\mathbb{Z}_5$-flows φ_1 and φ_2, respectively, such that $\partial\varphi_1(U) = \partial\varphi_2(U) = s$ and the flows φ_1 and $-\varphi_2$ can be pieced together into a nowhere-zero $\mathbb{Z}_5$-flow in G, a contradiction. Thus, $F_{H,U}(s)F_{I,U}(s) = 0$ for every $s \in A_n$.

By Lemma 5.4.5, $F_{H,U}(s) = F_{H,U}(\gamma(s))$ for each $s \in A_n$ and $\gamma \in \Gamma$. Therefore, $s_\Gamma \in A_{\Gamma,H,U}$ if and only if $F_{H,U}(s) > 0$ for each $s \in s_\Gamma$. Hence, $F_{I,U}(s) = 0$ for every $s \in \cup \{s_\Gamma; s_\Gamma \in A_{\Gamma,H,U}\}$, and thus $F_{\Gamma,I,U}(s_\Gamma) = 0$ for every $s_\Gamma \in A_{\Gamma,H,U}$.

By Theorem 5.3.2, there exists a vector $\mathbf{y}_{\Gamma,I,U} \in \mathbb{Q}^{p_{\Gamma,n}}$, such that for every $s_\Gamma \in A_{\Gamma,n}$, $F_{\Gamma,I,U}(s_\Gamma) = \chi_{\Gamma,n}(s_\Gamma) \cdot \mathbf{y}_{\Gamma,I,U}$. Choose $t_{\Gamma,1}, \ldots, t_{\Gamma,r} \in A_{\Gamma,H,U}$ so that $\chi_{\Gamma,n}(t_{\Gamma,1}), \ldots, \chi_{\Gamma,n}(t_{\Gamma,r})$ form a basis in $V_{\Gamma,H,U}$. We know that $F_{\Gamma,I,U}(t_{\Gamma,i}) = 0$ for $i = 1, \ldots, r$. Suppose $s_\Gamma \in A_{\Gamma,n}$ such that $\chi_{\Gamma,n}(s_\Gamma) \in V_{\Gamma,H,U}$. Then there are numbers $z_1, \ldots, z_r$ such that $\chi_{\Gamma,n}(s_\Gamma) = \sum_{i=1}^{r} z_i \chi_{\Gamma,n}(t_{\Gamma,i})$ whence

$$F_{\Gamma,I,U}(s_\Gamma) = \chi_{\Gamma,n}(s_\Gamma) \cdot y_{\Gamma,I,U} = \left(\sum_{i=1}^{r} z_i \chi_{\Gamma,n}(t_{\Gamma,i}) \right) \cdot y_{\Gamma,I,U}$$

$$= \sum_{i=1}^{r} z_i \left(\chi_{\Gamma,n}(t_{\Gamma,i}) \cdot y_{\Gamma,I,U} \right) = \sum_{i=1}^{r} z_i F_{\Gamma,I,U}(t_{\Gamma,i}) = \sum_{i=1}^{r} z_i 0 = 0.$$

Thus, if $s_\Gamma \in A_{\Gamma,n}$ such that $\chi_{\Gamma,n}(s_\Gamma) \in V_{\Gamma,H,U}$, then $F_{\Gamma,I,U}(s_\Gamma) = 0$.

Since G' has a nowhere-zero $\mathbb{Z}_5$-flow, there exists $s \in A_n$ such that $F_{H',U'}(s)$, $F_{I',U'}(s) > 0$. Thus, $s \in s_\Gamma \in A_{\Gamma,H',U'} \cap A_{\Gamma,I',U'}$, that is, $F_{\Gamma,I',U'}(s_\Gamma) > 0$. By assumptions, $V_{\Gamma,H',U'} \subseteq V_{\Gamma,H,U}$, whence $\chi_{\Gamma,n}(s_\Gamma) \in V_{\Gamma,H,U}$, and we have proved that then $F_{\Gamma,I,U}(s_\Gamma) = 0$; thus also $F_{\Gamma,I',U'}(s_\Gamma) = 0$, a contradiction with the fact that $F_{\Gamma,I',U'}(s_\Gamma) > 0$. This proves the statement. □

Note that in Lemma 5.4.6, we do not need that Γ acts regularly on (G, U). It suffices that $F_{G,U}(\gamma(s)) = 0$ for each $\gamma \in \Gamma$ whenever $F_{G',U'}(s) > 0$ and $F_{G,U}(s) = 0$.

By Lemma 5.4.6, to show that a smallest counterexample to the 5-flow conjecture has no circuit of length n, it suffices to prove that $V_{\Gamma,H'_n,U'_n} \subseteq V_{\Gamma,H_n,U_n}$ for a permutation subgroup Γ on (H_n, U_n).

Recall that Γ_n denote the dihedral group on $\{1, \ldots, n\}$, that is, the permutation group on $\{1, \ldots, n\}$ generated by permutations γ_n and γ'_n, where γ_n maps i to $i + 1$ mod n and γ'_n maps i to $n + 1 - i$ ($i = 1, \ldots, n$). Γ_n has $2n$ elements of the form $(\gamma_n)^i (\gamma'_n)^j$, where $i = 0, 1, \ldots, n - 1, j = 0, 1$, and Γ_n acts regularly on (H_n, U_n). It corresponds with the group of automorphisms of C_n and H_n.

By using computers, we have proved the following statement. More details about the proof are discussed in the following section.

Lemma 5.4.7 $V_{\Gamma_9,H'_9,U'_9} \subseteq V_{\Gamma_9,H_9,U_9}$ *and* $V_{\Gamma_{10},H'_{10},U'_{10}} \subseteq V_{\Gamma_{10},H_{10},U_{10}}$.

Theorem 5.4.3 *A smallest counterexample to the 5-flow conjecture has girth at least 11.*

Proof. Let $\overline{G}$ be the smallest counterexample to the 5-flow conjecture. By Theorems 5.4.1 and 5.4.2, $\overline{G}$ is a cyclically 6-edge-connected cubic graph with girth at least 9. By Lemmas 5.4.6 and 5.4.7, $\overline{G}$ cannot have circuits of orders 9 and 10. Thus, $\overline{G}$ has girth at least 11. □

Let us note that there are no known snarks that are cyclically 6-edge connected and have girth at least 12 (see [14]).

5.4.5   Discussion about Computations

In this section, we continue to follow results from Kochol [22].

Recall that $\mathcal{A}$ denotes the automorphism group of $\mathbb{Z}_5$. Also now, it suffices to consider only elements from $A'_n = \{(1, s_2, \ldots, s_n) \in A_n\}$ instead of A_n.

Let Γ be a permutation group on $\{1, \ldots, n\}$. Then $\chi_{\Gamma,n}(s) = \chi_{\Gamma,n}$ $(\alpha(s))$ and $\alpha(\gamma(s)) = \gamma(\alpha(s))$ for every $s \in A_n$, $\alpha \in \mathcal{A}$, and $\gamma \in \Gamma$. Thus, if $s_\Gamma \in A_{\Gamma,n}$, then either $\alpha(s_\Gamma) = \{\alpha(s); s \in s_\Gamma\}$ coincides with s_Γ or these two sets are distinct. But $\chi_{\Gamma,n}(s_\Gamma) = \chi_{\Gamma,n}(\alpha(s_\Gamma))$. Therefore, it suffices to consider a minimal set $A'_{\Gamma,n}$ so that $A_{\Gamma,n} = \{\alpha(s); s \in A'_{\Gamma,n}\}$ instead of $A_{\Gamma,n}$. Note that $A'_{\Gamma,n}$ is not uniquely defined, but every set with this property has the same number of elements.

For any network (G, U), $U = (u_1, \ldots, u_n)$, we can consider $A'_{G,U} = A_{G,U} \cap A'_n$ and $A'_{\Gamma,G,U} = A_{\Gamma,G,U} \cap A'_{\Gamma,n}$ instead of $A_{G,U}$ and $A_{\Gamma,G,U}$.

For example, we discuss the case when $n = 5$ and $\Gamma = \Gamma_5$ (the dihedral group on $\{1, \ldots, 5\}$). By 5.12, $p_5 = 11$ and $\mathcal{P}_5$ contains the following partitions:

$$P_{5,1} = \{\{1, 2, 3, 4, 5\}\},$$
$$P_{5,2} = \{\{1, 2\}, \{3, 4, 5\}\},$$
$$P_{5,3} = \{\{2, 3\}, \{4, 5, 1\}\},$$
$$P_{5,4} = \{\{3, 4\}, \{5, 1, 2\}\},$$
$$P_{5,5} = \{\{4, 5\}, \{1, 2, 3\}\},$$
$$P_{5,6} = \{\{5, 1\}, \{2, 3, 4\}\},$$
$$P_{5,7} = \{\{1, 3\}, \{2, 4, 5\}\},$$
$$P_{5,8} = \{\{2, 4\}, \{3, 5, 1\}\},$$
$$P_{5,9} = \{\{3, 5\}, \{4, 1, 2\}\},$$
$$P_{5,10} = \{\{4, 1\}, \{5, 2, 3\}\},$$
$$P_{5,11} = \{\{5, 2\}, \{1, 3, 4\}\}.$$

We evaluate $\chi_5(t_i)$, $i = 1, \ldots, 7$, where

$$
\begin{aligned}
t_1 &= (1, 1, 1, 1, 1), & \chi_5(t_1) &= (1, 0, 0, 0, 0, 0, 0, 0, 0, 0, 0), \\
t_2 &= (1, 1, 2, 4, 2), & \chi_5(t_2) &= (1, 0, 0, 0, 0, 0, 0, 1, 0, 1, 0), \\
t_3 &= (1, 4, 1, 2, 2), & \chi_5(t_3) &= (1, 1, 1, 0, 0, 0, 0, 0, 0, 0, 0), \\
t_4 &= (1, 1, 4, 2, 2), & \chi_5(t_4) &= (1, 0, 1, 0, 0, 0, 1, 0, 0, 0, 0), \\
t_5 &= (1, 2, 1, 4, 2), & \chi_5(t_5) &= (1, 0, 0, 1, 0, 0, 0, 0, 0, 1, 0), \\
t_6 &= (1, 1, 1, 3, 4), & \chi_5(t_6) &= (1, 0, 0, 0, 0, 1, 0, 0, 1, 0, 1), \\
t_7 &= (1, 1, 3, 1, 4), & \chi_5(t_7) &= (1, 0, 0, 0, 1, 1, 0, 0, 0, 0, 1).
\end{aligned}
$$

Furthermore, $\mathcal{P}_{\Gamma_5,5} = (P_{\Gamma_5,1}, P_{\Gamma_5,2}, P_{\Gamma_5,3})$, where

$$P_{\Gamma_5,1} = \{P_{5,1}\},$$
$$P_{\Gamma_5,2} = \{P_{5,i}; i = 2, \ldots, 6\},$$
$$P_{\Gamma_5,3} = \{P_{5,i}; i = 7, \ldots, 11\}.$$

Thus, $p_{\Gamma_{5,5}} = 3$. We get $\chi_{\Gamma_{5,5}}(t_i)$ from $\chi_5(t_i)$ (see (5.6) and (5.8)), and setting $t_{\Gamma_{5,i}} = \{\gamma(t_i); \gamma \in \Gamma_5\}$, we can evaluate $|t_{\Gamma_{5,i}}|$ for $i = 1, \ldots, 7$,

$$
\begin{array}{lll}
t_1 = (1,1,1,1,1), & \chi_{\Gamma_{5,5}}(t_1) = (0,0,1), & |t_{\Gamma_{5,1}}| = 1, \\
t_2 = (1,1,2,4,2), & \chi_{\Gamma_{5,5}}(t_2) = (1,0,2), & |t_{\Gamma_{5,2}}| = 5, \\
t_3 = (1,4,1,2,2), & \chi_{\Gamma_{5,5}}(t_3) = (1,2,0), & |t_{\Gamma_{5,3}}| = 5, \\
t_4 = (1,1,4,2,2), & \chi_{\Gamma_{5,5}}(t_4) = (1,1,1), & |t_{\Gamma_{5,4}}| = 10, \\
t_5 = (1,2,1,4,2), & \chi_{\Gamma_{5,5}}(t_5) = (1,1,1), & |t_{\Gamma_{5,5}}| = 10, \\
t_6 = (1,1,1,3,4), & \chi_{\Gamma_{5,5}}(t_6) = (1,1,2), & |t_{\Gamma_{5,6}}| = 10, \\
t_7 = (1,1,3,1,4), & \chi_{\Gamma_{5,5}}(t_7) = (1,2,1), & |t_{\Gamma_{5,7}}| = 10.
\end{array}
$$

Hence, $S'' = \cup_{i=1}^{7} t_{\Gamma_{5,i}}$ is a set of cardinality 51 containing pairwise non-σ_5-equivalent elements from A_5. We know that $|A_5'| = 51$. Thus, every element of S'' is σ_5-equivalent with exactly one element from A_5' and vice versa. Therefore, we can choose $A_{\Gamma_{5,5}}' = \{t_{\Gamma_{5,i}}; i = 1, \ldots, 7\}$. We can check that $t_1, t_2 \notin A_{H_5, U_5}$ and $t_3, \ldots, t_7 \in A_{H_5, U_5}$. Hence, $A_{\Gamma_5, H_5, U_5}' = \{t_{\Gamma_{5,i}}; i = 3, \ldots, 7\}$ and V_{Γ_5, H_5, U_5} is the linear hull of vectors $\chi_{\Gamma_{5,5}}(t_i)$, $i = 3, \ldots, 7$, that is, $V_{\Gamma_5, H_5, U_5} = \mathbb{Q}^3$.

Now we discuss the computations implying Lemma 5.4.7. Let M_n' be a matrix of size $|A_{\Gamma_n, H_n, U_n}'| \times p_{\Gamma_{n,n}}$, whose rows are vectors of the form $\chi_{\Gamma_{n,n}}(s_\Gamma)$ where $s_\Gamma \in A_{\Gamma_n, H_n, U_n}'$. Adding to that matrix rows of the form $\chi_{\Gamma_{n,n}}(s_\Gamma)$ where $s_\Gamma \in A_{\Gamma_n, H_n', U_n'}' \backslash A_{\Gamma_n, H_n, U_n}'$, we get a matrix M_n. The rank of M_n' is the dimension of V_{Γ_n, H_n, U_n}, and if M_n and M_n' have the same rank, then $V_{\Gamma_n, H_n', U_n'} \subseteq V_{\Gamma_n, H_n, U_n}$.

By using computers, we have checked that $p_{\Gamma_9, 9} = 238$, $|A_{\Gamma_9, H_9, U_9}'| = 262$, and $|A_{\Gamma_9, H_9', U_9'}' \backslash A_{\Gamma_9, H_9, U_9}'| = 168$. Similarly, for parameter 10, we have checked that $p_{\Gamma_{10}, 10} = 1\,079$, $|A_{\Gamma_{10}, H_{10}, U_{10}}'| = 792$, and $|A_{\Gamma_{10}, H_{10}', U_{10}'}' \backslash A_{\Gamma_{10}, H_{10}, U_{10}}'| = 623$.

Thus, M_9', M_9, M_{10}', and M_{10} have size 262×238, 430×238, 792×1079, and 1415×1079, respectively. By using computers, we have verified that M_9 and M_9' have rank 151. We used Maple programming language to evaluate them, and the program runs few minutes on a personal computer. In a similar way, we have verified that M_{10} and M_{10}' have rank 539. This program runs 1 day. The fact that M_n' and M_n have the same rank for $n = 9, 10$ implies Lemma 5.4.7.

In order to stress how much we can reduce the size of matrices using Theorem 5.3.2, we discuss what happens if we use the trivial permutation group (containing only identical permutation) instead of Γ_n. Then the matrix corresponding to M_n has p_n columns. Clearly, $p_1 = 0$ and $p_2 = p_3 = 1$. By 5.12, $p_9 = 3\,425$ and $p_{10} = 17\,722$.

Let $c(n) = |A_{H_n, U_n}'| = c(5, n)/4$. By Corollary 5.3.2, $c(9) = |A_{H_9, U_9}'| = 4\,665$ and $c(10) = |A_{H_{10}, U_{10}}'| = 14\,251$. In the following section (see Lemma 5.5.4), we prove that $|A_{H_n', U_n'}' \backslash A_{H_n, U_n}'| = c_1(n - 2)$. Therefore, $|A_{H_9', U_9'}' \backslash A_{H_9, U_9}'| = c_1(7) = 420$ and $|A_{H_{10}', U_{10}'}' \backslash A_{H_{10}, U_{10}}'| = c_1(8) = 1\,386$.

Therefore, replacing Γ_n by the trivial permutation group on $\{1, \ldots, n\}$, instead of matrices M_9', M_9, M_{10}', and M_{10}, we must deal with matrices of size $4\,665 \times 3\,425$, $5\,085 \times 3\,425$, $14\,251 \times 17\,722$, and $15\,637 \times 17\,722$, respectively. We were not able to evaluate the ranks of these matrices by personal computers.

5.4.6 Open Problems

By Lemma 5.4.3, the affirmative solution of the following problem implies the 5-flow conjecture.

Conjecture 5.4.1 $V_{H_n,U_n} = V_n$ *for every* $n \geq 2$.

We have verified Conjecture 5.4.1 for $n \leq 8$ in Theorem 5.4.2. With respect to Lemma 5.4.6, it is enough to consider weaker conjectures.

Conjecture 5.4.2 $V_{\Gamma_n,H'_n,U'_n} \subseteq V_{\Gamma_n,H_n,U_n}$ *for every* $n \geq 2$.

Conjecture 5.4.3 *For each* $n \geq 2$, *there exists a quasicubic network* (H'', U''), $U'' = (u''_1, \ldots, u''_n)$, *such that* $v_3(H_n, U_n) > v_3(H'', U'')$, (H_n, U_n) *can be regularly replaced by* (H'', U'') *in the class of cyclically* 6-*edge-connected quasicubic graphs, and* $V_{\Gamma_n,H'',U''} \subseteq V_{\Gamma_n,H_n,U_n}$.

Really, by Lemma 5.4.6, if one of Conjectures 5.4.1 through 5.4.3 is true, then the smallest counterexample to the 5-flow conjecture cannot have a circuit of any order, whence the 5-flow conjecture holds true.

In fact, it suffices to consider any subgroup of Γ_n, because each such subgroup acts regularly on (H_n, U_n) and, furthermore, the following statement holds true.

Lemma 5.4.8 *Let* (H, U), $U = (u_1, \ldots, u_n)$, $n \geq 2$, *be a quasicubic network and* Γ *be a permutation group on* $\{1, \ldots, n\}$. *Suppose there exists a quasicubic network* (H', U'), $U' = (u'_1, \ldots, u'_n)$, *and a subgroup* Γ' *of* Γ *such that* $V_{\Gamma',H',U'} \subseteq V_{\Gamma',H,U}$. *Then* $V_{\Gamma,H',U'} \subseteq V_{\Gamma,H,U}$.

Proof. Since Γ' is a subgroup of Γ, then for each $P_{\Gamma',i} \in \mathcal{P}_{\Gamma',n}$, there exists $g(i) \in \{1, \ldots, p_{\Gamma,n}\}$ such that $P_{\Gamma',i} \subseteq P_{\Gamma,g(i)}$. Clearly, g is a surjective mapping form $\{1, \ldots, p_{\Gamma',n}\}$ to $\{1, \ldots, p_{\Gamma,n}\}$. For each $s \in A_n$, $\chi(s, P_{\Gamma',i}) = \sum_{P \in P_{\Gamma',i}} \chi(s, P)$ and

$$\chi(s, P_{\Gamma,j}) = \sum_{P \in P_{\Gamma,j}} \chi(s, P) = \sum_{i \in g^{-1}(j)} \chi(s, P_{\Gamma',i}). \qquad (5.17)$$

Choose $t_{\Gamma',1}, \ldots, t_{\Gamma',r} \in A_{\Gamma',H,U}$ so that $\chi_{\Gamma',n}(t_{\Gamma',1}), \ldots, \chi_{\Gamma',n}(t_{\Gamma',r})$ form a basis in $V_{\Gamma',H,U}$. Choose $t_i \in t_{\Gamma',i}$ for $i = 1, \ldots, r$. Suppose $s_\Gamma \in A_{\Gamma,H',U'}$. Then there exists $s \in s_\Gamma$ such that $F_{H',U'}(s) > 0$; thus, $s \in s_{\Gamma'} \in A_{\Gamma',H',U'}$, whence, by assumptions and (5.8), $\chi_{\Gamma',n}(s) = \chi_{\Gamma',n}(s_{\Gamma'})$ is a linear combination of vectors $\chi_{\Gamma',n}(t_1), \ldots, \chi_{\Gamma',n}(t_r)$. But by (5.17), we get that then $\chi_{\Gamma,n}(s_\Gamma) = \chi_{\Gamma,n}(s)$ is a linear combination of vectors $\chi_{\Gamma,n}(t_1), \ldots, \chi_{\Gamma,n}(t_r)$. Thus, $V_{\Gamma,H',U'} \subseteq V_{\Gamma,H,U}$. $\qquad \square$

In view of Lemma 5.4.8, if we prove that $V_{\Gamma',H'_n,U'_n} \subseteq V_{\Gamma',H_n,U_n}$, where Γ' is a subgroup of Γ_n, then it implies that $V_{\Gamma_n,H'_n,U'_n} \subseteq V_{\Gamma_n,H_n,U_n}$. But in order to prove the first formula, we need to deal with matrices of larger size then in the second formula. Furthermore, the second formula can be true though the first one could be false. Thus, the most suitable choice is to deal with group Γ_n, that is, the maximal permutation group that acts regularly on (H_n, U_n).

5.5   Second Reduction Principle

In this section, we use ideas introduced by Kochol et al. [25] and present another way how to decrease the size of matrices. This is an alternative reduction to those presented in Section 5.3.2. Unfortunately, in contrasts with the dihedral group applications presented in Section 5.4.4, the principles used here can be applied only for excluding girth 7 from the smallest counterexample to the 5-flow conjecture.

5.5.1   Reductions by Superproper Permutations

A proper partition $P = \{Q_1, \ldots, Q_r\}$ of the set $\{1, \ldots, n\}$, $n \geq 2$, is called *superproper* if n and $n - 1$ are contained in the same set from $Q_1, \ldots, Q_r$. Let $\mathcal{P}'_n$ denote the set of superproper partitions of $\{1, \ldots, n\}$. (e.g., $\{\{1, 2\}, \{3, 4, 5\}\} \in \mathcal{P}'_5$.) We consider $\mathcal{P}'_n$ as an p'_n-tuple $(P_{n,1}, \ldots, P_{n,p'_n})$. For any $s \in A_n$, denote by $\chi'_n(s)$ the integral vector $(c_{s,1}, \ldots, c_{s,p'_n})$ so that $c_{s,i} = 1$ $(c_{s,i} = 0)$ if $P_{n,i}$ is (is not) compatible with s, $i = 1, \ldots, p'_n$. Vector $\chi'_n(s)$ contains the coordinates of $\chi_n(s)$ corresponding with the superproper partitions from $\mathcal{P}_n$.

Lemma 5.5.1 *If s and t are θ_n-equivalent elements of A_n, then $\chi'_n(s) = \chi'_n(t)$.*

Proof. Follows from the definitions of $\mathcal{P}'_n$ and χ'_n.   □

Clearly, $p'_n = p_n = 1$ for $n = 2, 3$. If $n \geq 4$, then $\mathcal{P}'_n$ contains exactly p_{n-2} partitions such that n and $n - 1$ are contained in a 2-element subset and exactly p_{n-1} partitions such that n and $n - 1$ are contained in at least 3-element subset (in this case, we can delete n and get all partitions from $\mathcal{P}_{n-1}$). Thus, for $n \geq 4$,

$$p'_n = p_{n-1} + p_{n-2}. \tag{5.18}$$

The main idea standing behind the reductions presented here is that instead of vectors $\chi_n(s)$, we consider vectors of the form $\chi_n(s) - \chi_n(t)$ where $s \equiv t(\theta_n)$. By Lemma 5.5.1, vectors of this form have all coordinates corresponding to superproper partitions equal to 0. Thus, instead of p_n-dimensional vectors, we deal with $p_n - p'_n = p_n - p_{n-1} - p_{n-2}$-dimensional vectors. For the case $n = 7$, we get reduction from $p_7 = 162$ to $p_7 - p_6 - p_5 = 162 - 41 - 11 = 110$.

Lemma 5.5.2 *Let φ_1, φ_2 be nowhere-zero $\mathbb{Z}_5$-flows in (H_n, U_n) such that $\partial \varphi_1(U_n) = \partial \varphi_2(U_n)$. Then either $\varphi_1 = \varphi_2$ or $\varphi_1(y_i) \neq \varphi_2(y_i)$ for $i = 1, \ldots, n$.*

Proof. Follows from the fact that for $i = 1, \ldots, n$, $\varphi_1(y_i) - \varphi_1(y_1) = \sum_{j=2}^{i} \varphi_1(x_j) = \sum_{j=2}^{i} \varphi_2(x_j) = \varphi_2(y_i) - \varphi_2(y_1)$.   □

Let

$$C(5, n) = \{s \in A_n; F_{H_n, U_n}(s) \neq 0\},$$
$$C_i(5, n) = \{s \in C(5, n); F_{H_n, U_n}(s) = i\},$$
$$C'(5, n) = \{s \in A_n; F_{H'_n, U'_n}(s) \neq 0\}, \tag{5.19}$$
$$C''(5, n) = C'(5, n) \backslash C(5, n).$$

Using the notation from Section 5.3, $|C(5, n)| = c(5, n)$ and $|C_i(5, n)| = c_i(5, n)$. By Theorem 5.3.5, $C(5, n) = C_1(5, n) \cup C_2(5, n) \cup C_3(5, n)$ for every $n \geq 2$. Before formulating another lemma, we introduce some more technical notation.

If φ is a nowhere-zero $\mathbb{Z}_5$-flow in (H_{n-2}, U_{n-2}) and $a \in \mathbb{Z}_5 \backslash \{0, \varphi(y_{n-2})\}$, then denote by $\varphi^{[a]}$ the nowhere-zero $\mathbb{Z}_5$-flow in (H_n, U_n) such that $\varphi^{[a]}(x_n) = -\varphi^{[a]}(x_{n-1}) = a$, $\varphi^{[a]}(y_n) = \varphi(y_{n-2})$, $\varphi^{[a]}(y_{n-1}) = \varphi(y_{n-2}) - a$, and $\varphi^{[a]}(x_i) = \varphi(x_i)$, $\varphi^{[a]}(y_i) = \varphi(y_i)$ for $i = 1, \ldots, n-2$. (Note that writing $\varphi(x_i)$, $\varphi(y_i)$, we consider the arcs x_i, y_i to be from H_{n-2}, and writing $\varphi^{[a]}(x_i)$, $\varphi^{[a]}(y_i)$, we consider the arcs x_i, y_i to be from H_n.)

If φ is a nowhere-zero $\mathbb{Z}_5$-flow in (H_n, U_n) and $\varphi(x_{n-1}) + \varphi(x_n) = 0$ (resp. $\varphi(x_{n-1}) + \varphi(x_n) \neq 0$), then denote by $\overline{\varphi}$ the nowhere-zero $\mathbb{Z}_5$-flow in (H_{n-2}, U_{n-2}) (resp. (H_{n-1}, U_{n-1})) such that $\overline{\varphi}(x_i) = \varphi(x_i), \overline{\varphi}(y_i) = \varphi(y_i)$ for $i = 1, \ldots, n-2$ (resp. $\overline{\varphi}(x_{n-1}) = \varphi(x_n) + \varphi(x_{n-1}), \overline{\varphi}(y_{n-1}) = \varphi(y_n)$, and $\overline{\varphi}(x_i) = \varphi(x_i), \overline{\varphi}(y_i) = \varphi(y_i)$ for $i = 1, \ldots, n-2$).

If $(s_1, \ldots, s_n) \in A_n$ and $s_{n-1} + s_n = 0$ (resp. $s_{n-1} + s_n \neq 0$), then denote by $\overline{s} = (s_1, \ldots, s_{n-2}) \in A_{n-2}$ (resp. $\overline{s} = (s_1, \ldots, s_{n-2}, s_{n-1} + s_n) \in A_{n-1}$).

Lemma 5.5.3 *Let $s = (s_1, \ldots, s_n) \in A_n$. Then $s \in C''(5, n)$ if and only if $s_{n-1} + s_n = 0$, $\overline{s} \in C_1(5, n-2)$, and $s_n = \varphi(y'_{n-2})$ where $\varphi \in \Phi_{H_{n-2}, U_{n-2}}(\overline{s})$.*

Proof. Suppose $\varphi' \in \Phi_{H'_n, U'_n}(s)$ and φ'' be the restriction of φ' to $D(H_{n-2})$. Then $s_n = \varphi'(x'_n) = -\varphi'(z'_n) = -\varphi'(x'_{n-1}) = -s_{n-1}$ and $\partial \varphi''(u'_1, \ldots, u'_{n-2}) = \overline{s}$. If $s_n \neq \varphi''(y'_{n-2})$, then $\varphi''^{[s_n]} \in \Phi_{H_n, U_n}(s)$. Thus, if $s \in C''(5, n)$, then s must satisfy the assumptions.

If s satisfies the assumptions, then $s \in C'(5, n)$ and we can choose $\varphi' \in \Phi_{H'_n, U'_n}(s)$. If also $s \in C(5, n)$, take $\varphi \in \Phi_{H_n, U_n}(s)$. Then $\overline{\varphi}$ must be the restriction of φ' to $D(H_{n-2})$ (because $\overline{s} \in C_1(5, n-2)$), whence $\varphi(x_n) = s_n = \varphi(y_{n-2}) = \varphi(y_n) = -\varphi(y_n^{-1})$, which is not possible because it requires $\varphi(y_{n-1}) = 0$. Thus, $s \in C''(5, n)$. $\square$

Denote by $[s]\theta_n = \{t \in A_n; t \equiv s(\theta_n)\}$ for any $s \in A_n$. By the arithmetic in the group $\mathbb{Z}_5$, if $(s_1, \ldots, s_n) \in A_n$ and $s_{n-1} + s_n = 0$ ($s_{n-1} + s_n \neq 0$), then $|[s]\theta_n| = 4$ ($|[s]\theta_n| = 3$). Let us note that $C(5, n) = C_1(5, n) \cup C_2(5, n) \cup C_3(5, n)$ for every $n \geq 2$.

Lemma 5.5.4 *Let $s = (s_1, \ldots, s_n) \in A_n$.*

(1) *If $s_{n-1} + s_n = 0$ and $\overline{s} \in C_1(5, n-2)$, then $|[s]\theta_n \cap C(5, n)| = 3$ and $|[s]\theta_n \cap C''(5, n)| = 1$.*

(2) *If $s_{n-1} + s_n = 0$ and $\bar{s} \in C_2(5, n - 2) \cup C_3(5, n - 2)$, then $|[s]\theta_n \cap C(5, n)| = 4$ and $|[s]\theta_n \cap C''(5, n)| = 0$.*

(3) *If $s_{n-1} + s_n \neq 0$ and $\bar{s} \in C_1(5, n - 1)$, then $|[s]\theta_n \cap C(5, n)| = 2$ and $|[s]\theta_n \cap C''(5, n)| = 0$.*

(4) *If $s_{n-1} + s_n \neq 0$ and $\bar{s} \in C_2(5, n - 1) \cup C_3(5, n - 1)$, then $|[s]\theta_n \cap C(5, n)| = 3$ and $|[s]\theta_n \cap C''(5, n)| = 0$.*

(5) *If neither of the assumptions from (1) through (4) occurs, then $|[s]\theta_n \cap C(5, n)| = 0$ and $|[s]\theta_n \cap C''(5, n)| = 0$.*

Proof. Let $s_{n-1} + s_n = 0$ and $\bar{s} \in C(5, n - 2)$. Then $\Phi_{H_{n-2}, U_{n-2}}(\bar{s}) \neq \emptyset$. If $\bar{s} \in C_1(5, n - 2)$ ($\bar{s} \in C_2(5, n - 2) \cup C_3(5, n - 2)$), then by Lemma 5.5.3, $|[s]\theta_n \cap C''(5, n)| = 1$ ($|[s]\theta_n \cap C''(5, n)| = 0$). Using $\psi^{[a]}$ for all $\psi \in \Phi_{H_{n-2}, U_{n-2}}(\bar{s})$ and all $a \in \mathbb{Z}_5 \setminus \{0, \psi(y_{n-2})\}$, we get that $|[s]\theta_n \cap C(5, n)| = 3$ ($|[s]\theta_n \cap C(5, n)| = 4$). This proves (1) and (2).

Let $s_{n-1} + s_n \neq 0$. Without loss of generality we can assume that $s_{n-1} + s_n = 1$.

Assume that $\bar{s} \in C_1(5, n - 1)$ and $\varphi' \in \Phi_{H_{n-1}, U_{n-1}}(\bar{s})$. Then $\varphi'(x_{n-1}) = 1$ and $\varphi'(y_{n-2})$, $\varphi'(y_{n-1})$ can be either 1,2 or 2,3 or 3,4, respectively. In all three cases, there exist exactly two nowhere-zero $\mathbb{Z}_5$-flows φ_1, φ_2 in (H_n, U_n) such that $\varphi_1(x_n) \neq \varphi_2(x_n)$ and $\overline{\varphi_1} = \overline{\varphi_2} = \varphi'$. In particular, if $\varphi'(y_{n-2})$, $\varphi'(y_{n-1})$ are 1,2 (2,3 or 3,4), then $\varphi_1(x_{n-1})$, $\varphi_1(y_{n-1})$, $\varphi_1(x_n)$, $\varphi_2(x_{n-1})$, $\varphi_2(y_{n-1})$, $\varphi_2(x_n)$ are 2, 3, 4, 3, 4, 3 (2, 4, 4, 4, 1, 2 or 3, 1, 3, 4, 2, 2), respectively. Thus, $|[s]\theta_n \cap C(5, n)| = 2$. By Lemma 5.5.3, $|[s]\theta_n \cap C''(5, n)| = 0$. This proves (3).

Assume that $\bar{s} \in C_2(5, n-1) \cup C_3(5, n-1)$ and $\varphi'_1, \varphi'_2 \in \Phi_{H_{n-1}, U_{n-1}}(\bar{s})$, $\varphi'_1 \neq \varphi'_2$. Then we can choose the notation of φ'_1, φ'_2 so that $\varphi'_1(y_{n-2})$, $\varphi'_1(y_{n-1})$, $\varphi'_2(y_{n-2})$, $\varphi'_2(y_{n-1})$ are either 1,2,2,3 or 1,2,3,4 or 2,3,3,4, respectively. In all three cases, there exist nowhere-zero $\mathbb{Z}_5$-flows $\varphi_1, \varphi_2, \varphi_3$ in (H_n, U_n) such that $\varphi_1(x_n)$, $\varphi_2(x_n)$, $\varphi_3(x_n)$ are pairwise different and $\overline{\varphi_1}, \overline{\varphi_2}, \overline{\varphi_3} \in \{\varphi'_1, \varphi'_2\}$. In particular, if $\varphi'_1(y_{n-2})$, $\varphi'_1(y_{n-1})$, $\varphi'_2(y_{n-2})$, $\varphi'_2(y_{n-1})$ are 1,2,2,3 (1,2,3,4 or 2,3,3,4), then $\varphi_1(x_{n-1})$, $\varphi_1(y_{n-1})$, $\varphi_1(x_n)$, $\varphi_2(x_{n-1})$, $\varphi_2(y_{n-1})$, $\varphi_2(x_n)$, $\varphi_3(x_{n-1})$, $\varphi_3(y_{n-1})$, $\varphi_3(x_n)$ can be 2, 3, 4, 3, 4, 3, 4, 1, 2 (2, 3, 4, 3, 4, 3, 4, 2, 2 or 2, 4, 4, 4, 1, 2, 3, 1, 3), respectively. Thus, $|[s]\theta_n \cap C(5, n)| = 3$. By Lemma 5.5.3, $|[s]\theta_n \cap C''(5, n)| = 0$. This proves (4).

Let $s_{n-1} + s_n = 0$ ($s_{n-1} + s_n \neq 0$) and $\bar{s} \notin C(5, n - 2)$ ($\bar{s} \notin C(5, n - 1)$). Suppose there exists $t \in [s]\theta_n \cap C(5, n)$. Then there exists $\varphi \in \Phi_{H_n, U_n}(t)$, whence $\overline{\varphi} \in \Phi_{H_{n-2}, U_{n-2}}(\bar{s})$ ($\overline{\varphi} \in \Phi_{H_{n-1}, U_{n-1}}(\bar{s})$), because $\bar{s} = \bar{t} \in A_{n-2}$ ($\bar{s} = \bar{t} \in A_{n-1}$). Thus, $\bar{s} \in C(5, n-2)$ ($\bar{s} \in C(5, n-1)$), a contradiction. Therefore, $|[s]\theta_n \cap C(5, n)| = 0$. Furthermore, $[s]\theta_n \cap C''(5, n) = \emptyset$; otherwise, from Lemma 5.5.3, we get that $s_{n-1} + s_n = 0$ and $\bar{s} \in C_1(5, n - 2)$, a contradiction. This implies (5). $\qquad\square$

5.5.2 Application

Lemma 5.5.5 *Let $\alpha_1, \ldots, \alpha_r, \beta_1, \ldots, \beta_{2s+r}$, $r < s$, be not necessarily different vectors from $\mathbb{R}^p$ and $\alpha_1 - \beta_1$, ..., $\alpha_r - \beta_r$ be contained in the linear hull of $\beta_{r+1} - \beta_{r+2}$,...., $\beta_{2s+r-1} - \beta_{2s+r}$. Then $\alpha_1, \ldots, \alpha_r$ are contained in the linear hull of $\beta_1, \ldots, \beta_{2s+r}$.*

Proof. This follows immediately from properties of linear dependence in linear spaces. □

Let $\mathcal{A}$ denote the automorphism group of $\mathbb{Z}_5$. We do not need to consider all elements from A_n, but only non-σ_n-equivalent representatives of the σ_n-equivalence classes (each of them has exactly four elements). Thus, we consider only elements from $A_n' = \{(1, s_2, \ldots, s_n) \in A_n\}$ instead of A_n.

Similarly, denote by $C(n)$, $C'(n)$, $C''(n)$, and $C_i(n)$, $i = 1, 2, 3$ the sets of elements from $C(5, n)$, $C'(5, n)$, $C''(5, n)$, and $C_i(5, n)$, $i = 1, 2, 3$, respectively, of the form $(1, s_2, \ldots, s_n)$. Let $c(n) = |C(5, n)| = c(5, n)/4$, $c'(n) = |C'(n)|$, $c''(n) = |\tilde{C}''(n)|$, and $c_i(n) = |C_i(n)| = c_i(5, n)/4$, $i = 1, 2, 3$.

Theorem 5.5.1 $V_{H_7', U_7'} \subseteq V_{H_7, U_7}$.

Proof. Following Lemma 5.5.4, we can denote the elements from $C(n) \cup C'(n)$ as $s^{[i,j,k]}$ where $1 \le i \le 4$ (which corresponds to item (i) of Lemma 5.5.4), $1 \le j \le b_i(n)$, $1 \le k \le r_i(n)$,

$$
\begin{aligned}
b_1(n) &= c_1(n-2), & b_2(n) &= c_2(n-2) + c_3(n-2), \\
b_3(n) &= c_1(n-1), & b_4(n) &= c_2(n-1) + c_3(n-1), \\
r_1(n) &= r_2(n) = 4 & r_3(n) &= 2, \quad r_4(n) = 3,
\end{aligned}
$$

assuming that $\{s^{[i,j,k]}; k = 1, \ldots, r_i(n)\}$, $i = 1, \ldots, 4$, $j = 1, \ldots, b_i(n)$ are pairwise θ_n-equivalent and satisfy the condition (i) from Lemma 5.5.4. If $i = 1$, then we also assume that $s^{[i,j,1]} \in \tilde{C}''(n)$ for $j = 1, \ldots, b_1(n)$.

Define

$$
\begin{aligned}
A_n &= \Big\{ \gamma^{[i,j,k]} = \chi_n(s^{[i,j,k]}) - \chi_n(s^{[i,j,r_i(n)]}); \\
& \quad 1 \le i \le 4, \ 1 \le j \le b_i(n), \ 1 \le k < r_i(n) \Big\}, \\
B_n &= \Big\{ \gamma^{[1,j,1]}; \ 1 \le j \le b_1(n) \Big\}.
\end{aligned}
$$

Then $B_n \subseteq A_n$. Let W_n be the linear hull of $A_n \backslash B_n$ in $\mathbb{R}^{p_n}$. If $B_n \subseteq W_n$, then by Lemma 5.5.5, vectors $\chi_n(s^{[1,j,1]})$, $1 \le j \le b_1(n)$ (which correspond to $\alpha_1, \ldots, \alpha_r$ from Lemma 5.5.5) are contained in the linear hull of the vectors of the form $\chi_n(s^{[i,j,k]})$ where $s^{[i,j,k]} \in C(n)$ (which correspond to $\beta_1, \ldots, \beta_{2s+r}$ from Lemma 5.5.5). Hence, by Lemma 5.5.4, all vectors of the form $\chi_n(s'')$ where $s'' \in C''(n)$ are contained in the linear hull of vectors of the form $\chi_n(s)$ where $s \in \tilde{C}(n)$, which means $V_{H_n', U_n'} \subseteq V_{H_n, U_n}$.

Using computers, we have verified that $B_7 \subseteq W_7$. Thus, $V_{H_7', U_7'} \subseteq V_{H_7, U_7}$. □

By Lemmas 5.4.1 and 5.4.2 and Theorem 5.5.1, the smallest counterexample to the 5-flow conjecture cannot have a circuit of order 7.

Now we discuss the computations mentioned in the proof of Theorem 5.5.1. Let M_n be the matrix whose rows are the vectors from A_n'. Furthermore, we assume that the first $|A'_n \setminus B_n|$ rows correspond with the elements from $A'_n \setminus B_n$ and denote the submatrix composed from these rows by M'_n. We can also assume that M_n does not contain the columns corresponding with the superproper partitions from $\mathcal{P}_n$ (because all these entries are 0 by Lemma 5.5.1). Thus, M_n has $q_n = p_n - p'_n$ columns and $a_n = 3b_1(n) + 3b_2(n) + b_3(n) + 2b_4(n)$ rows, and M'_n has $a'_n = a_n - b_1(n)$ rows.

We can evaluate $c(n)$ and $c_i(n)$ ($i = 1, 2, 3$) by Corollary 5.3.2.

Let $v_n = |A'_n|$. We have already shown that $v_2 = v_3 = 1$ and $v_n = 3v_{n-2} + 4v_{n-1}$ for $n \geq 4$. Using (5.12) and (5.18), we can evaluate q_n, a_n and a'_n for $1 \leq n \leq 8$.

n	1	2	3	4	5	6	7	8
$c_1(n)$	0	0	0	6	30	120	420	1386
$c_2(n)$	0	0	3	6	15	30	63	126
$c_3(n)$	0	1	0	1	0	1	0	1
$c(n)$	0	1	3	13	45	151	483	1513
v_n	0	1	3	13	51	205	819	3277
p_n	0	1	1	4	11	41	162	715
q_n	0	0	0	2	6	26	110	512
a_n	0	0	2	9	29	99	317	999
a'_n	0	0	2	9	29	93	287	879

To check whether $B_n \subseteq W_n$, we need to apply Gauss elimination to a matrix M_n and check whether the nonzero entries are only in the rows of submatrix M'_n. We have applied this for $n = 7$, and we got that M'_7 and M_7 have the same rank. Note that M_7 and M'_7 have size 317×110 and 287×110, respectively. This presents certain improvement comparing the computation based only on ideas from Lemma 5.4.3.

Using computers, we have applied the same approach also for $n = 8$. But in this case, we get that $B_8 \not\subseteq W_8$. Thus, our method cannot be applied for excluding girth 8.

5.6   Edge Colorings of Cubic Graphs

The four-color theorem, proved by Appel and Haken [2], is that every planar graph is 4-colorable. This was a long-standing open problem in mathematics. By Theorem 5.2.3, this statement is equivalent with the fact that every bridgeless planar graph has a nowhere-zero 4-flow.

A graph is internally k-connected if it is $k - 1$-connected and every separating vertex set of order $k - 1$ divides the graph into exactly one component and one single vertex. By Birkhoff [3], a smallest counterexample to the four-color theorem must be an internally 6-connected graph. Furthermore, we can assume it is a triangulation. Thus, the four-color theorem suffices to prove for internally 6-connected planar triangulations.

A planar graph is called a *simple 5-cut graph* if it is cyclically 5-edge connected and the edges of every cyclic 5-edge cut are incident to a circuit of order 5. Clearly, a planar triangulation is internally 6-connected if and only if its dual is a simple 5-cut graph. In this section, we present a proof of the dual version of Birkhoff's [3] result. We prove the following statement.

Theorem 5.6.1 *Let $\overline{\mathcal{G}}$ be the class of planar bridgeless graphs without a nowhere-zero $\mathbb{Z}_2 \times \mathbb{Z}_2$-flow. If this class in nonempty, then it contains a planar simple 5-cut snark.*

Proof. Let $\overline{G}$ be a graph from $\overline{\mathcal{G}}$ with the smallest possible number of vertices, and among graphs with this property, it has the smallest possible number of edges. Clearly, $\overline{G}$ has no loops, no multiple edges, and no vertex of valency 1 or 2. Suppose that $\overline{G}$ has an n-edge cut $C = \{e_1, \ldots, e_n\}$, $n \geq 2$, so that $\overline{G} - C$ does not contain an isolated vertex (edge cuts with this property are called *nontrivial*). Then $\overline{G} - C$ has exactly two connected components G' and G''. Furthermore, assume that C has the smallest possible cardinality (i.e., n is the smallest possible). We can choose the ordering of edges $e_1, \ldots, e_n$ so that there exists an embedding of $\overline{G}$ in the plane where a simple Jordan curve intersects $e_1, \ldots, e_n$ in cyclic order and no other edge or vertex of the graph. For $i = 1, \ldots, n$, let v_i' and v_i'' be the end of e_i contained in G' and G'', respectively. Add to G' (G'') n new pairwise different vertices $u_1, \ldots, u_n$ ($w_1, \ldots, w_n$) and, for $i = 1, \ldots, n$, join u_i with v_i' (w_i with v_i''). We get from G' (G'') a new graph $X(Y)$. (X, U), $U = (u_1, \ldots, u_n)$, and (Y, W), $W = (w_1, \ldots, w_n)$ are simple planar networks.

$F_{X,U}(s)$, $F_{Y,W}(s) \geq 0$ for every $s \in A_n$. If there exists $s \in A_n$ such that $F_{X,U}(s)$, $F_{Y,W}(s) > 0$, then (X, U) and (Y, W) have nowhere-zero A-flows φ_1 and φ_2, respectively, such that $\partial\varphi_1(U) = s$ and $\partial\varphi_2(W) = (s)$ that can be pieced together into a nowhere-zero A-flow in G, a contradiction. Thus, for every $s \in A_n$, we have

$$F_{X,U}(s) \cdot F_{Y,W}(s) = 0. \tag{5.20}$$

By Theorem 5.3.3, there exist integers x_i and y_i, $i = 1, \ldots, \overline{p}_n$, such that for every $s \in A_n$, $F_{X,U}(s) = \sum_{i=1}^{\overline{p}_n} \chi_{s,i} x_i$ and $F_{Y,W}(s) = \sum_{i=1}^{\overline{p}_n} \chi_{s,i} y_i$ where $(\chi_{s,1}, \ldots, \chi_{s,\overline{p}_n}) = \overline{\chi}_n(s)$.

If $n \in \{2, 3\}$, then $\overline{\mathcal{P}}_n$ contains exactly one partition $P_{n,1} = \{\{1, \ldots, n\}\}$ and $\overline{\chi}_n(s) = (1)$ for every $s \in A_n$. Thus, by Theorem 5.3.3, $F_{G,U}(s) = F_{G,U}(s')$ for every $s, s' \in A_n$ and every simple network (G, U) with n outer vertices. Then either $F_{X,U}(s) = 0$ or $F_{Y,W}(s) = 0$. If $F_{X,U}(s) = 0$ ($F_{Y,W}(s) = 0$), then $(X, U)_{P_{n,1}}$ $((Y, W)_{P_{n,1}})$ is a graph from $\overline{\mathcal{G}}$ having a smaller order than $\overline{G}$, a contradiction. Therefore, $n \geq 4$.

Suppose that G' (resp. G'') has a bridge e'. Then there exists $1 \leq r < n$ such that $C' = \{e_1, \ldots, e_r, e'\}$ and $C'' = \{e', e_{r+1}, \ldots, e_n\}$ are edge cuts of $\overline{G}$ and $n > \min\{|C'|, |C''|\}$, a contradiction with the choice of C. Thus, G' and G'' are bridgeless.

Suppose that $\overline{G}$ has a vertex v of valency ≥ 4. Consider two edges $e_1 = vv_1$ and $e_2 = vv_2$ contained in one face. Let H arise from $\overline{G}$ after deleting e_1, e_2 and adding a new edge v_1v_2. Then H is bridgeless; otherwise, we get a smaller edge cut with the aforementioned property, a contradiction with the minimality of $n \geq 4$. Each nowhere-zero $\mathbb{Z}_2 \times \mathbb{Z}_2$-flow in H indicates such a flow in $\overline{G}$, whence H is from $\overline{\mathcal{G}}$, a contradiction with minimality of $\overline{G}$. Thus, $\overline{G}$ is cubic (note that this fact also follows from the results by Fleischner [6]).

Let $n = 4$. By (5.13), $\overline{p}_4 = 3$ and $\overline{\mathcal{P}}_4$ contains the following partitions:

$$P_{4,1} = \{\{1, 2, 3, 4\}\},$$
$$P_{4,2} = \{\{1, 2\}, \{3, 4\}\},$$
$$P_{4,3} = \{\{2, 3\}, \{4, 1\}\}.$$

We have $F_{X,U}(1, 2, 1, 2) = x_1$, $F_{Y,W}(1, 2, 1, 2) = y_1$, whence $x_1, y_1 \geq 0$ and, by (5.20), $x_1 \cdot y_1 = 0$. Suppose that $x_1 = 0$. Then $F_{X,U}(1, 1, 2, 2) = x_2 \geq 0$ and $F_{X,U}(1, 2, 2, 1) = x_3 \geq 0$. If $x_1 = x_2 = x_3 = 0$, then identifying u_1 (u_3) with u_2 (u_4) in X and suppressing the vertices of valency 2, we get from (X, U) a graph of $\overline{G}$ having a smaller order than $\overline{G}$. Therefore, at least one x_i must be positive and without loss of generality, we can assume that $x_2 > 0$. Then $F_{X,U}(s) > 0$ and $F_{Y,W}(s) = 0$ if $\chi_4(s)$ has the second coordinate 1 ($s \in A_4$). Thus, identifying w_1 (w_3) with w_2 (w_4) in Y and suppressing the vertices of valency 2 we get from (Y, W) a graph of $\overline{G}$ having a smaller order than $\overline{G}$. This implies that $n = 5$.

Let $n = 5$. By (5.13), $p_5 = 6$ and $\overline{\mathcal{P}}_5$ contains the following partitions:

$$P_{5,1} = \{\{1, 2, 3, 4, 5\}\},$$
$$P_{5,2} = \{\{1, 2\}, \{3, 4, 5\}\},$$
$$P_{5,3} = \{\{2, 3\}, \{4, 5, 1\}\},$$
$$P_{5,4} = \{\{3, 4\}, \{5, 1, 2\}\},$$
$$P_{5,5} = \{\{4, 5\}, \{1, 2, 3\}\},$$
$$P_{5,6} = \{\{5, 1\}, \{2, 3, 4\}\}.$$

Consider s_i and $\overline{\chi}_5(s_i)$, $i = 1, \ldots, 10$, as indicated in Table 5.1.

We claim that

$$\text{either } F_{X,U}(s_i) > 0 \text{ for all } i \in \{6, \ldots, 10\},$$
$$\text{or } F_{Y,W}(s_i) > 0 \text{ for all } i \in \{6, \ldots, 10\}. \tag{5.21}$$

TABLE 5.1:   Values s_i and $\overline{\chi}_5(s_i)$ for $i = 1, \ldots, 10$

$s_1 = (1, 1, 2, 1, 3)$,	$\overline{\chi}_5(s_1) = (1, 1, 0, 0, 0, 0)$,
$s_2 = (3, 1, 1, 2, 1)$,	$\overline{\chi}_5(s_2) = (1, 0, 1, 0, 0, 0)$,
$s_3 = (1, 3, 1, 1, 2)$,	$\overline{\chi}_5(s_3) = (1, 0, 0, 1, 0, 0)$,
$s_4 = (2, 1, 3, 1, 1)$,	$\overline{\chi}_5(s_4) = (1, 0, 0, 0, 1, 0)$,
$s_5 = (1, 2, 1, 3, 1)$,	$\overline{\chi}_5(s_5) = (1, 0, 0, 0, 0, 1)$,
$s_6 = (1, 1, 1, 2, 3)$,	$\overline{\chi}_5(s_6) = (1, 1, 1, 0, 0, 0)$,
$s_7 = (3, 1, 1, 1, 2)$,	$\overline{\chi}_5(s_7) = (1, 0, 1, 1, 0, 0)$,
$s_8 = (2, 3, 1, 1, 1)$,	$\overline{\chi}_5(s_8) = (1, 0, 0, 1, 1, 0)$,
$s_9 = (1, 2, 3, 1, 1)$,	$\overline{\chi}_5(s_9) = (1, 0, 0, 0, 1, 1)$,
$s_{10} = (1, 1, 2, 3, 1)$,	$\overline{\chi}_5(s_{10}) = (1, 1, 0, 0, 0, 1)$.

To prove the claim, we need to check several cases. Notice that $(X, U)_{P_{n,i}}$ and $(Y, W)_{P_{n,i}}$ are bridgeless for each $i = 1, \ldots, 6$, because any bridge would indicate a nontrivial 2- or 3-edge cut in $\overline{G}$, a contradiction with the minimality of n.

Case 1. $F_{X,U}(s_i) = 0$ for $i = 1, \ldots, 5$. Then by Table 5.1, $x_1 = -x_j$ for $j = 2, \ldots, 6$, and $F_{X,U}(s_6) = x_1 + x_2 + x_3 \geq 0$, whence $x_1 \leq 0$.

Case 1.1. $x_1 = 0$. Then $x_2 = \cdots = x_6 = 0$, whence $F_{X,U}(s) = 0$ for each $s \in A_5$. Identifying u_1, u_2 (u_3, u_4, u_5) into a new vertex u' (u'') and suppressing u', we get from (X, U) a graph of $\overline{\mathcal{G}}$ having a smaller order than $\overline{G}$, a contradiction.

Case 1.2. $x_1 < 0$. Then $x_2 = \ldots = x_6 = -x_1 > 0$, whence (5.21) holds true.

Case 2. $F_{Y,W}(s_i) = 0$ for $i = 1, \ldots, 5$. It suffices to interchange the role of (X, U) and (Y, W), and we get the situation from Case 1.

Case 3. Suppose that there exist $i, j \in \{1, \ldots, 5\}$ such that $F_{X,U}(s_i) > 0$ and $F_{Y,W}(s_j) > 0$. Then by (5.20), there are three indices $i_1, i_2, i_3 \in \{1, \ldots, 5\}$ such that either $F_{X,U}(s_{i_j}) = 0$ for $j = 1, 2, 3$ or $F_{Y,W}(s_{i_j}) = 0$ for $j = 1, 2, 3$. Without abuse of generality, we can assume that the latter case occurs. Then at least two of the indices are consecutive, and we can choose the notation so that $i_1 = 1$ and $i_2 = 2$. Hence, $F_{Y,W}(s_1) = F_{Y,W}(s_2) = 0$, that is, $y_1 + y_2 = y_1 + y_3 = 0$. Furthermore, $F_{Y,W}(s_6) = y_1 + y_2 + y_3 \geq 0$, whence $y_1 \leq 0$.

Case 3.1. If $y_1 = 0$, then (since $y_1 + y_2 = y_1 + y_3 = 0$) we have $y_1 = y_2 = y_3 = 0$. We need to check three subcases.

Case 3.1.1. $i_3 = 3$. Then $F_{Y,W}(s_3) = 0 = y_1 + y_4$ and $y_4 = 0$. Identifying w_1, w_2 (w_3, w_4, w_5) into a new vertex w' (w''), we get from (Y, W) a graph Y'. Y' does not have a nowhere-zero A-flow, because $F_{Y,W}(s_k) = 0$ for $k \in \{2, 6, 8\}$. Thus, suppressing w', we get from Y' a graph of $\overline{\mathcal{G}}$ having a smaller order than $\overline{G}$, a contradiction.

Case 3.1.2. $i_3 = 5$. We get the same situation as in Case 3.1.1. after reordering W into $(w_5, w_1, \ldots, w_4)$.

Case 3.1.3. $i_3 = 4$. Then $F_{Y,W}(s_4) = 0 = y_1 + y_5$ and $y_5 = 0$. Identifying w_1, w_5 (w_3, w_4) into a new vertex w' (w''), add new edges $w'w_2$, $w''w_2$. Denote by Y'' the resulting graph. Y'' does not have a nowhere-zero A-flow, because $F_{Y,W}(s_k) = 0$ for $k \in \{1, 2, 4, 6\}$. Thus, Y'' is a graph from $\overline{\mathcal{G}}$ having a smaller order than $\overline{G}$, a contradiction.

Case 3.2. $y_1 < 0$. Then $y_2, \ldots, y_6 \geq -y_1$ (because $F_{X,U}(s_i) \geq 0$ for $k = 1, \ldots, 5$), and (5.21) holds true.

Thus, (5.21) holds true, and without abuse of generality, we can assume that $F_{X,U}(s_i) > 0$ for $i = 6, \ldots, 10$. Thus, by (5.20), $F_{Y,W}(s_i) = 0$ for $i = 6, \ldots, 10$. Identifying the vertices of U by a circuit of order 5, we get from X a planar graph $X' \in \overline{\mathcal{G}}$. This implies that Y must be a circuit, and the statement holds true. $\qquad \square$

It is a known consequence of Euler's theorem that every planar cubic graph has girth at most 5 (see, e.g., Diestel [5]). Thus, improving Theorem 5.6.1 so that $\overline{G}$ contains a graph of girth at least 6 gives a proof of the four-color theorem.

By the four-color theorem and Theorem 5.6.1, there is no planar simple 5-cut snark. In [14, Section 10.2], we construct an infinite family of nonplanar simple 5-cut snarks.

Theorem 5.6.2 *Suppose there exists planar graph without a 4-coloring. Then there exists an internally 6-connected planar triangulation without a 4-coloring.*

Proof. Follows from Theorems 5.6.1 and 5.2.3.   □

Acknowledgments

This chapter was partially supported by grants VEGA 2/0118/10 and 2/0017/14.

References

1. M. Aigner, *Combinatorial Theory*, Springer-Verlag, Berlin, Germany (1979).

2. K. Appel and W. Haken, *Every Planar Map Is Four Colorable*, Contemporary Mathematics, Vol. 98, American Mathematical Society, Providence, RI (1989).

3. G. D. Birkhoff, The reducibility of maps, *American Journal of Mathematics* **35** (1913) 115–128.

4. U. A. Celmins, On cubic graphs that do not have an edge-3-colouring, PhD thesis, Department of Combinatorics and Optimization, University of Waterloo, Waterloo, Ontario, Canada (1984).

5. R. Diestel, *Graph Theory*, Springer-Verlag, New York (1997).

6. H. Fleischner, Eine gemeinsame Basis für die Theorie der eulerschen Graphen und den Satz von Petersen, *Monatshefte für Mathematik* **81** (1976) 267–278.

7. M. Gardner, Mathematical games: Snarks, Boojums and other conjectures related to the four-color-map theorem, *Scientific American* **234** (April 1976) 126–130.

8. F. Jaeger, Flows and generalized coloring theorems in graphs, *Journal of Combinatorial Theory*, Series B **26** (1979) 205–216.

9. F. Jaeger, Nowhere-zero flow problems, in: *Selected Topics in Graph Theory 3* (L. W. Beineke and R. J. Wilson, Eds.), Academic Press, New York (1988), pp. 71–95.

10. T. R. Jensen and B. Toft, *Graph Coloring Problems*, Wiley, New York (1995).

11. P. A. Kilpatrick, Tutte's first colour-cycle conjecture, PhD thesis, Cape Town, South Africa (1975).

12. M. Kochol, Snarks without small cycles, *Journal of Combinatorial Theory*, Series B **67** (1996) 34–47.

13. M. Kochol, Hypothetical complexity of the nowhere-zero 5-flow problem, *Journal of Graph Theory* **28** (1998) 1–11.

14. M. Kochol, Superposition and constructions of graphs without nowhere-zero *k*-flows, *European Journal of Combinatorics* **23** (2002) 281–306.

15. M. Kochol, Polynomials associated with nowhere-zero flows, *Journal of Combinatorial Theory*, Series B **84** (2002) 260–269.

16. M. Kochol, Tension polynomials of graphs, *Journal of Graph Theory* **40** (2002) 137–146.

17. M. Kochol, Reduction of the 5-flow conjecture to cyclically 6-edge-connected snarks, *Journal of Combinatorial Theory*, Series B **90** (2004) 139–145.

18. M. Kochol, Decomposition formulas for the flow polynomial, *European Journal of Combinatorics* **26** (2005) 1086–1093.

19. M. Kochol, Restrictions on smallest counterexamples to the 5-flow conjecture, *Combinatorica* **26** (2006) 83–89.

20. M. Kochol, Polyhedral embeddings of snarks in orientable surfaces, *Proceedings of the American Mathematical Society* **137** (2009) 1613–1619.

21. M. Kochol, *Superposition, Snarks and Flows*, EDIS—Žilina University Press, Žilina, Slovakia (2009).

22. M. Kochol, Smallest counterexample to the 5-flow conjecture has girth at least eleven, *Journal of Combinatorial Theory*, Series B **100** (2010) 381–389.

23. M. Kochol, Linear algebraic approach to an edge-coloring result, *Journal of Combinatorial Optimization*, in press.

24. M. Kochol, N. Krivoňáková, S. Smejová, and K. Šranková, Counting nowhere-zero flows on wheels, *Discrete Mathematics* **308** (2008) 2050–2053.

25. M. Kochol, N. Krivoňáková, S. Smejová, and K. Šranková, Matrix reduction in a combinatorial computation, *Information Processing Letters* **111** (2011) 164–168.

26. A. Schrijver, *Combinatorial Optimization—Polyhedr a and Efficiency*, Springer-Verlag, Berlin, Germany (2003).

27. P. D. Seymour, Nowhere-zero 6-flows, *Journal of Combinatorial Theory*, Series B **30** (1981) 130–135.

28. P. G. Tait, Remarks on the colouring of maps, *Proceedings of the Royal Society of Edinburgh* **10** (1880) 729.

29. C. Thomassen, The weak 3-flow conjecture and the weak circular flow conjecture, *Journal of Combinatorial Theory*, Series B **102** (2012) 521–529.

30. W. T. Tutte, On the imbedding of linear graphs in surfaces, *Proceedings of the London Mathematical Society*, Second Series **51** (1950) 474–483.

31. W. T. Tutte, A contribution to the theory of chromatic polynomials, *Canadian Journal of Mathematics* **6** (1954) 80–91.

32. W. T. Tutte, A class of Abelian groups, *Canadian Journal of Mathematics* **8** (1956) 13–28.

33. W. T. Tutte, On the algebraic theory of graph colorings, *Journal of Combinatorial Theory* **1** (1966) 15–50.

34. D. J. A. Welsh, *Complexity: Knots, Colourings and Counting*, London Mathematical Society Lecture Notes Series 186, Cambridge University Press, Cambridge, U.K. (1993).

35. D. H. Younger, Integer flows, *Journal of Graph Theory* **7** (1983) 349–357.

36. C.-Q. Zhang, *Integral Flows and Cycle Covers of Graphs*, Marcel Dekker, New York (1997).

37. C.-Q. Zhang, *Cycle Double Covers of Graphs*, London Mathematical Society Lecture Note Series (No. 399), Cambridge University Press, Cambridge, U.K. (2012).

Chapter 6

Width-Measures for Directed Graphs and Algorithmic Applications

Stephan Kreutzer and Sebastian Ordyniak

Contents

6.1	Introduction	182
6.2	Preliminaries	185
	6.2.1 Graph Theory	185
	6.2.2 Complexity Theory	186
6.3	Cops and Robber Games	187
	6.3.1 Cops and Robber Games on Undirected Graphs	188
	6.3.2 Monotonicity in Cops and Robber Games	189
	6.3.3 Directed Cops and Robber Games	189
	6.3.4 Monotonicity in Directed Cops and Robber Games	190
6.4	Directed Tree Width	191
	6.4.1 Directed Tree Decompositions and Directed Tree Width	191
	6.4.2 Computing Directed Tree Decompositions	193
	6.4.3 Obstructions: Havens, Brambles, and Well-Linked Sets	197
	6.4.3.1 Havens in Digraphs	197
	6.4.3.2 Brambles in Directed Graphs	198
	6.4.3.3 Well-Linked Sets	200
	6.4.3.4 Putting Things Together	201
6.5	D-Width	202
6.6	DAG Width	204
	6.6.1 Computing DAG Decompositions	206
6.7	Kelly Width	206
	6.7.1 Definition and Characterizations of Kelly Width	207
	6.7.2 Computing Kelly Decompositions	210
6.8	Directed Path Width	211
6.9	Clique Width and Bi-Rank Width	212
6.10	Further Proposals of Directed Decompositions	215
6.11	Nowhere Crownful Classes of Digraphs	215
6.12	Disjoint Paths and General Linkage Problems	219
6.13	Query Evaluation in Graph Databases	223
6.14	Foundations of Model Checking	224
6.15	Boolean Networks	225
6.16	Directed Dominating Sets and Similar Problems	225
6.17	Conclusion	226
References		227

6.1   Introduction

Graphs are among the most fundamental mathematical models in computer science and many practical problems in routing, scheduling, and other areas have natural graph theoretical formalizations. It has, however, long been realized that a large class of important algorithmic problems seems to evade all attempts of efficient solvability, leading to the theory of NP-hardness as model for computational intractability.

In the study of hard algorithmic problems on graphs, methods derived from structural graph theory have proved to be a valuable tool. The rich theory of special classes of graphs developed in this area has been used to identify classes of graphs on which many computationally hard problems can be solved efficiently. Most of these classes are defined by some structural property of graphs and this structural information can be exploited algorithmically. The motivation for this is twofold: On the one hand, instances of problems arising in applications often have much more structure than just being general graphs. For instance, road or railway maps are nearly planar and computer or communication networks are relatively sparse, and this information can be used in the design of efficient algorithms. Thus, algorithms that work for restricted classes of graphs may be enough for many applications. On the other hand, by analyzing structural properties of input instances that allow for these problems to be solved efficiently, our understanding of the nature of these problems is advanced and their boundary of tractability can be explored.

Of particular importance in this context is the concept of tree width introduced by Robertson and Seymour as part of their celebrated graph minor project [61]. Tree width associates with every undirected graph a natural number, its *tree width*, which measures similarity of the graph to being a tree. A huge number of otherwise intractable problems can be solved efficiently on graph classes with a fixed upper bound on their tree width, and thousands of papers have been written using algorithmic aspects of tree width in a wide range of application areas (see, e.g., [11] for a survey). Besides applications in the theory of algorithms, tree width has also found applications in very practical areas. For instance, it has been observed that various standard algorithms for probabilistic inference in causal or Bayesian networks run in time exponential in the tree width of the (morality graph of the) network but only polynomial in the size of the network (see, e.g., [19]).

Graphs of bounded tree width are examples of sparse graphs-graphs with moderate edge density. Besides tree width, many other structural properties of sparse graphs have been studied, which can be used to design more efficient algorithms for otherwise hard problems (see, e.g., [13] for a survey of graph classes). Among the most important are planar graphs or, much more generally, graph classes excluding a fixed minor (see, e.g., [20]). A relatively new addition to the family of graph parameters studied with algorithmic applications in mind are nowhere dense classes of graphs [30,31,54], which further generalize classes with excluded minors.

The structural parameters discussed previously relate to undirected graphs. However, in many applications in computer science, directed graphs are a more natural model. Given the enormous success width parameters had for problems defined on undirected graphs, it is natural to ask whether they can also be used to analyze the complexity of NP-hard problems on digraphs. While in principle it is possible to apply the structure theory for undirected graphs to directed graphs by ignoring the direction of edges, this implies a significant information loss. Hence, for computational problems whose instances are directed graphs, methods based on the structure theory for undirected graphs may be less useful.

Reed [60] and Johnson et al. [36] initiated the development of a decomposition theory for directed graphs with the aim of defining a directed analogue of undirected tree width. They introduced the concept of *directed tree width* and showed that the k-disjoint path problem (see the following) and more general linkage problems can be solved in polynomial time on classes of digraphs of bounded directed tree width. Following this initial proposal, several alternative notions of width measures for sparse classes of digraphs have been introduced, for instance, *directed path width* (see [6], initially proposed by Robertson et al.), *D-width* [63], *DAG width*, [7,8,55] *Kelly width* [34], or *entanglement* [9]. For each of these, some algorithmic applications were given, mostly in relation to linkage problems or a form of combinatorial games known as *parity games*.

A different approach is taken in a width measure known as *directed clique width* [17] and $\mathbb{F}$- or *bi-rank width* [39,40]. Directed clique width was introduced in [15] by means of graph grammars. Later, a decomposition-based variant of it was defined in [17]. Generalizing the concept of rank width from undirected graphs to digraphs, Kanté et al. defined the concepts of $\mathbb{F}$- or *bi-rank width* [39,40], which provide an equivalent characterization of clique width inspired by the matroid theory. Using this approach, a rich structure theory for classes of digraphs of bounded directed clique width was established [39,40], which we explain in more detail in the following. Directed clique width with its equivalent representation by bi-rank width provides a very interesting width measure for digraphs. It is somewhat different in spirit than the other width measures discussed so far. It is a measure meant for very dense graphs and, for instance, it is not preserved under subgraphs (i.e., a subgraph can have much higher clique width than the graph itself) and is not a measure for connectivity but for *homogeneity* in some sense. Also, if a class $\mathcal{C}$ of digraphs has bounded directed clique width, then the class of underlying undirected graphs, that is, the class of undirected graphs obtained from digraphs in $\mathcal{C}$ by forgetting the edge orientation, has bounded undirected clique width. This is very different from the other width measures where for classes of digraphs of bounded directed path width, for example, the class of underlying undirected graphs does not necessarily have bounded tree or clique width.

All of these differences are not surprising, given that bi-rank width was not intended as a connectivity measure exceeding tree or clique width. Algorithmically, the main advantage of directed clique width/bi-rank width is that every digraph property definable in *monadic second-order logic* (MSO) can be decided in polynomial time (in fact, linear time plus the time needed to construct the decomposition) on any class of digraphs of bounded directed clique width. This makes bi-rank width by far the directed width measure with the broadest algorithmic applications.

Encouraged by initial successes of digraph width measures for disjoint paths problems, several papers investigated how rich the algorithmic theory of classes of digraphs of bounded width with respect to these measures is (see, e.g., [27,28,48,50]). Unfortunately, for many standard graph algorithmic problems other than those related to disjoint paths, strong intractability results have been obtained showing that the algorithmic applicability of the existing directed width measures on general digraphs is very limited.

However, directed width measures have successfully been used in various application areas. For instance, in [66], Tamaki studied the problem of finding attractors in Boolean networks. His experimental results showed that on Boolean networks of small directed path width, he was able to compute all attractors even for networks of size where traditional methods did not succeed. Another example is the evaluation of regular simple path queries, which in [4] were shown to be solvable in polynomial time on graph databases of bounded directed tree width. Similar results can be obtained in the theory of verification logics. For instance, in [12], it was shown that model checking for the modal μ-calculus, a prominent logic in the theory of verification, is fixed-parameter tractable on classes of transition systems of bounded Kelly width, whereas it is not known if the problem is fixed-parameter tractable in general. The results mentioned previously for parity games point into the same direction. Hence, despite the several intractability results for standard graph algorithmic problems, width measures for digraphs can be useful in concrete application areas.

Furthermore, much better algorithmic results can be obtained if in addition to limiting the widths of digraphs with respect to some directed width measure, additional constraints are imposed. In particular, on tournaments, directed path width has been used very successfully to show that directed edge-disjoint path problems become fixed-parameter tractable [24]. Results like these show the way to overcome some of the limitations of existing directed width measures.

The width measures mentioned so far are all based on a concept of decomposition of the digraph into smaller pieces. An entirely different approach is taken in [49] where classes of digraphs are defined based on a concept of directed minors. It can be shown that on such classes, problems become tractable, which remain hard for all the other width measures mentioned earlier. On the other hand, disjoint path problems are as yet not known to become easier on such classes than on general digraphs.

In conclusion, while directed structure properties and width measures exhibit a promising approach to algorithms on digraphs, much more research is needed to fully assess their potential. In particular, more research is needed to fully understand the cases, where these measures will prove algorithmically useful, and to develop the algorithmic tools necessary to exploit their structure.

In this chapter, we provide an overview over the existing width measures and their applications. In the first part, we cover the most common width measures, state their definition, comment on ways to compute the associated decompositions and known algorithmic applications, and finally review structural properties of digraphs related to the width measure. The aim of these sections is to review the state of the art in this line of research.

In the second part, we then state possible application areas that may benefit from directed width measures and thereby give an extra motivation for studying these measures. However, we also demonstrate their limitations on general digraphs and show possible ways to overcome these problems.

Organization: After fixing notation and introducing basic concepts of (directed) graphs (Section 6.2) and cops and robber games (Section 6.3), the first part of this chapter deals with the theory of directed width measures. Here, we introduce and survey directed width measures based on cops and robber games, such as directed tree width, D-width, DAG width, Kelly width, and directed path width (Sections 6.4 through 6.8); directed width measures based on algebraic decompositions, such as clique width and bi-rank width (Section 6.9); and finally directed width measures based on minors in Section 6.11.

The second part of this chapter explores the algorithmic applications (and limitations) of the directed width measures studied so far. After showing the usefulness of directed tree width for a very general form of the directed disjoint path problem in Section 6.12, we survey two practical applications of directed path width for graph databases and Boolean networks in Sections 6.13 and 6.15, we show applications of DAG width and Kelly width in the theory of model checking in Section 6.14, and we show the usefulness of directed width measures based on minors in Section 6.16. In Section 6.17, we conclude the chapter with a discussion about the algorithmic limitations of the known digraph width measures and provide an outlook on future directions in research on directed width measures.

6.2 Preliminaries

6.2.1 Graph Theory

We use standard notation from graph theory as can be found in, for example, [20]. All graphs and directed graphs in this paper are finite and simple.

Let G be a (directed) graph. We denote the vertex set of G by $V(G)$ and the edge set of G by $E(G)$. We denote by $e \sim v$ that the edge $e \in E(G)$ is incident to the vertex $v \in V(G)$. For $X \subseteq V(G)$, we denote by $G[X]$ the subgraph of G induced by X and by $G - X$ the subgraph of G induced by $V(G) - X$. Similarly for $Y \subseteq E(G)$, we set $G - Y$ to be the subgraph of G obtained by deleting all edges in Y. If G is a digraph, $d \in \mathbb{N}$ and $v \in V(G)$, we denote by $N^+(v)$ the out-neighborhood of v in G, that is, the set $\{u : (v, u) \in E(G)\}$ and by $N_d^+(v)$ the set of vertices that can be reached by a directed path of length at most d from v. The *underlying undirected graph* of a digraph G is the graph with vertex set $V(G)$ and edge set $\{\{u, v\} : (u, v) \in E(G)\}$.

For a set X and $k \in \mathbb{N}$, we denote by $\mathcal{P}(X)$ the power set of X and by $[X]^{\leq k}$ and $[X]^{<k}$ the set of all subsets of X of cardinality at most k or less than k, respectively. If M is a set and $\beta : M \to N$ is a function defined on M, then for every $S \subseteq M$,

we write $\beta(S) := \bigcup_{s \in S} \beta(s)$. For an element x, we sometimes write x to denote the set $\{x\}$.

We often consider tree or DAG-like decompositions of graphs. A tree is an undirected connected and acyclic graph and a directed acyclic graph (or DAG) is a directed graph without cycles. A directed tree is a directed graph whose underlying undirected graph is a tree with a designated root where all edges are directed away from the root. Directed trees and DAGs naturally define a partial ordering on their vertices, which we denote by $\preceq_G$ for a directed tree or DAG G, that is, for two vertices u and v of G, $u \preceq_G v$ if G contains a directed path from u to v. For a directed tree or DAG G and a vertex $v \in V(G)$, we denote by G_v the subgraph of G induced by the vertices in $\{u : v \preceq_T u\}$.

Let G be a graph. A set S of vertices of G is called a *separator* if the graph $G-S$ has at least two components. S is called an (A, B)-*separator* (or a separator for A and B) if the graph $G - S$ contains no path between a vertex in A and a vertex in B. If G is a digraph, we distinguish between (directed) separators and strong separators. Whereas a *directed (A, B)-separator* hits all directed paths from a vertex in A to a vertex in B, a *strong (A, B)-separator* ensures that no strong component of $G - S$ contains both a vertex from A and a vertex from B. We note here that a *strongly connected component*, or just component or strong component, of a digraph is a maximal set of vertices that are pairwise connected by a directed path.

The following theorem is due to Menger and provides a connection between separators and connecting paths in an undirected graph.

Theorem 6.2.1 *Let G be an undirected graph and A and B two sets of vertices of G. Then, the size of a minimal (A, B)-separator is equal to the maximum number of vertex-disjoint paths between A and B.*

We will also need the following generalization of Menger's theorem for directed graphs.

Theorem 6.2.2 *Let G be a digraph and A and B two sets of vertices of G. Then, the size of a minimal (A, B)-separator is equal to the maximum number of vertex-disjoint paths from A to B.*

6.2.2   Complexity Theory

We also assume familiarity with basic concepts from complexity theory, in particular the concept of NP-completeness. See, for example, [3]. In the study of computational problems on classes of graphs defined by some width parameter, a more refined complexity analysis than provided by classical complexity is often required. We will therefore use concepts from parameterized complexity theory; see [22,23].

In parameterized complexity, we study *parameterized problems*. An input to a parameterized problem is a pair (w, k), where w is a normal input to a computational problem and $k \in \mathbb{N}$ is the parameter. For instance, the parameterized dominating set problem for undirected graphs is the problem

p-DOMINATING SET
Input: Graph G, $k \in \mathbb{N}$
Parameter: k
Problem: Does G contain a dominating set of size $\leq k$

where a dominating set in a graph G is a set $X \subseteq V(G)$ such that for all $u \in V(G) \setminus X$ there is a $v \in X$ with $\{u, v\} \in E(G)$.

A parameterized problem is called *fixed-parameter tractable* (fpt) if there is a constant $c \in \mathbb{N}$ and a computable function $f : \mathbb{N} \rightarrow \mathbb{N}$ such that on input (w, k) the problem can be solved in time $f(k) \cdot |w|^c$. The class FPT of all fixed-parameter tractable problems is the parameterized analogue of PTIME in classical complexity as model for tractability. There is also a corresponding concept of intractability, playing the same rôle as NP in classical complexity. However, in parameterized complexity, there is a whole hierarchy, known as the *W-hierarchy* with classes $W[1], W[2], \ldots$, which takes on this rôle. Another parameterized complexity class we will use in the chapter is XP. A parameterized problem is in XP if there is a computable function $f : \mathbb{N} \rightarrow \mathbb{N}$ such that on input (w, k) the problem can be solved in time $|w|^{f(k)}$.

It is well known that the dominating set problem is (presumably) not fixed-parameter tractable in general. In fact, it is complete for the class $W[2]$. On the other hand, the *vertex cover* problem is fixed-parameter tractable. See [22,23] for a comprehensive treatment of this subject.

For our purposes here, it suffices to know the basic definitions. Note that the complexity of a parameterized problem depends crucially on its parameter. For instance, if we take the size of the input as parameter, then FPT simply means computable. Usually, the parameter is the solution size or, in our context, the width of the input graph with respect to some width measure.

6.3 Cops and Robber Games

Cops and robber games are a type of pursuit-evasion games played by two players on a (directed) graph. The minimal number of cops needed to catch a robber in a particular game variant yields a natural and intuitive graph invariant or width measure, and it has been shown that on undirected graphs, several prominent width measures such as tree or path width can be characterized by a corresponding cops and robber game. The main advantage of this characterization of width measures is that the games provide an intuitive explanation for the abstract concepts behind width measures, for instance, in terms of decompositions corresponding to winning strategies for the cops, whereas obstructions such as havens or brambles correspond naturally to winning strategies for the robber.

In this section, we briefly review the main concepts of cops and robber games as far as they are relevant for the digraph width measures discussed in this chapter.

6.3.1   Cops and Robber Games on Undirected Graphs

Cops and robber games are a form of one- or two-player combinatorial games played on graphs or digraphs. In a nutshell, in a cops and robber game, a number of cops (or searchers) try to catch a robber (or fugitive) hiding on the vertices of a graph or digraph. The aim of the cops is to capture the robber with as few cops as possible. See [25,47] for surveys on the type of cops and robber games relevant for this chapter.

As an example, we consider the following game, known as *visible node-searching game*, or *visible cops and robber game*, played on undirected graphs. Let G be an undirected graph. A position in the game is described by a pair (C, r), where $C \subseteq V(G)$ is the current position of the cops and $r \in V(G)$ is the current robber position. Initially, there is no cop on the graph and in the first round of the game, the robber can choose a position $r_0 \in V(G)$ to start from. The first cop position is $C_0 = \emptyset$. In round $i+1$, with (C_i, r_i) being the current position, the game is then played as follows. The cops propose their new position $C_{i+1} \subseteq V(G)$. The robber can then choose any vertex r_{i+1} as his new position, as long as r_{i+1} is reachable from r_i in the graph $G - (C_i \cap C_{i+1})$. A graphic illustration is the idea that the cops move from C_i to C_{i+1} by helicopters, where the cops in $C_i \cap C_{i+1}$ stay where they are. While the other cops fly to their new positions, the robber can move to his new position avoiding the cops that did not move.

If $r_i \in C_i$, after some round i, then the cops capture the robber and win the play. Otherwise, that is, if the play never ends, then the robber wins. Clearly, on any finite graph G, the cops can capture the robber by simply putting a cop on every node of the graph. However, by employing a more sophisticated strategy, they may be able to capture the robber using fewer cops, where the number of cops used in a play is the maximum size of any set C_i during the play. The minimum number of cops needed to capture a robber on a graph G defines a connectivity measure on undirected graphs G.

By varying the exact rules for the cops or the robber to move from one position to the next, a number of different cops and robber games can be defined. In this way, cops and robber games provide a range of natural and intuitive connectivity measures for graphs or digraphs. For instance, a well-studied variant of the previous game is obtained by making the robber invisible. That is, the cops have to move without knowing the current robber position, which essentially means that the game is a one-player game. This variant is known as *invisible node searching* or the *invisible cops and robber game*.

Several of the most common width measures for undirected graphs, in particular *path width* and *tree width*, can be characterized exactly by cops and robber games. More precisely, for any graph G, the minimal number of cops required to capture a robber in the *invisible node-searching game* or the *visible node-searching game* is exactly the path or tree width of the graph, respectively.

In relation to digraph width measures, an interesting feature of cops and robber games is that they have relatively straightforward and natural generalizations to directed graphs, while it is not clear what the right generalization of tree width or path width to digraphs is. For this reason, many of the proposed width measures for digraphs are based on a variant of cops and robber games for digraphs.

6.3.2 Monotonicity in Cops and Robber Games

So far, we have not imposed any restrictions on the strategies used by the cops. In particular, it is possible that cops move to positions that were already occupied earlier on in the game or that they move in a way that the robber can suddenly reach parts of the graph that were unreachable to him before. These types of general strategies are called *nonmonotone*. Nonmonotone strategies can exhibit a rather erratic behavior. For the purpose of graph decompositions, therefore, more systematic strategies are more useful. Systematic strategies are captured by the concept of *monotonicity*.

A strategy for the cops is called *cop monotone* if the cops never move twice to the same vertex, that is, if they place a cop on a vertex v and then move this cop elsewhere, they are not allowed to move back to v.

A different form of monotonicity is *robber monotonicity*. To define it properly, we first need the concept of robber space. For any game variant and any position (C_i, r_i) in that game, let R_i be the set of nodes reachable by the robber from position r_i if he were allowed to move with all cops remaining on C. A strategy for the cops is *robber monotone* if in every play $(C_0, r_0), \ldots, (C_i, r_i), \ldots$ the robber space is nonincreasing, that is, $R_{i+1} \subseteq R_i$ for all i.

A cops and robber game is called *robber monotone* if whenever k cops suffice to catch a robber on a graph G, then k cops have a robber-monotone winning strategy. And the game is called *cop monotone* if the analogous condition holds for cop-monotone winning strategies.

Among the most important problems for any cops and robber game variant is the monotonicity problem, that is, the problem whether a cops and robber game is robber or cop monotone. For the two games considered previously, this is indeed the case as shown in [10,64].

6.3.3 Directed Cops and Robber Games

Cops and robber games have natural generalizations to directed graphs. Recall that in the node-searching variants considered earlier, a robber can move as follows: if the current game position is (C, r) and the cops move to C', then the robber can move to any vertex v reachable from r by a path in the graph $G - (C \cap C')$. In other words, r and v must be in the same component of $G - (C \cap C')$. On undirected graphs, being in the same component and being reachable by a path are the same concepts. For directed graphs, however, these two concepts generalize differently. The first can be translated into the robber moving along directed paths, whereas the second translates to the robber staying in a strongly connected component of $G - (C \cap C')$. These two lead to different directed variants of the cops and robber, which we describe next.

Strong component searching. In the *(strong) component* search variant, the rules for the robber are as follows. Suppose the current position on a digraph G is (C, r) and the cops propose to move to C'. Then the robber can move to any position $r' \in V(G)$ in the same strongly connected component of $G - (C \cap C')$ as r. Again, we can consider the visible and invisible robber variant of the game.

Reachability searching: In the *reachability* search variant, the rules for the robber are as follows. Suppose the current position on a digraph G is (C, r) and the cops propose to move to C'. Then, the robber can move to any position $r' \in V(G)$ such that there is a directed path in $G - (C \cap C')$ from r to r'. Again, we can consider the visible and invisible robber variant of the game.

We briefly mention another variation known as *inert robber*. In this variant, if (C, r) is the current position in a game played on a digraph G and C' is the proposed new cop position, then the robber can only move if $r \in C'$. That is, the robber is lazy in the sense that he only moves if he is in immediate danger of being caught by the cops. Inertness can be combined with reachability and strong component searching and also with visible and invisible robber. However, in most cases, inertness has only been considered with an invisible robber.

Theorem 6.3.1 *Let G be an undirected graph and let G' be the directed graph obtained from G by replacing each undirected edge $\{u, v\} \in E(G)$ by two directed edges (u, v) and (v, u) in opposite directions. Then, the number of cops needed to catch a robber in the cops and robber game on G is the same as the number of cops needed to catch the robber in the reachability or strong component game on G'.*

6.3.4   Monotonicity in Directed Cops and Robber Games

Cops and robber games played on directed graphs have the same concepts of monotonicity as their undirected variants. However, whereas on undirected graphs many game variants turn out to be monotone, that is, robber and cop monotone, for directed graphs, the situation is quite different.

It was shown by Johnson et al. [36] that the strong component game with visible robber is not cop monotone and a counterexample was given. Adler [1] improved this result by showing that the game is also not robber monotone. However, it has bounded robber monotonicity, because whenever k cops have a winning strategy in the game, then $3k$ cops have a robber-monotone winning strategy. This follows from [36] and is a consequence of the results we present in Section 6.4. See the following for details. For cop monotonicity, an example showing that the number of cops needed for a cop-monotone strategy cannot be bounded by any function in the number of cops needed with any unrestricted strategy has recently been announced [38].

For reachability games with a visible robber, Kreutzer and Ordyniak showed that the game is neither cop nor robber monotone [48]. However, it is unknown whether the monotonicity cost can be bounded by a constant or, indeed, any function. The same situation occurs in the inert reachability game; see again [48].

Part I: The Theory—Directed Width-Measures

In this first part of the chapter, we present several directed width measures and other structural parameters for directed graphs. The focus of this part is on the structural aspects, the connection to graph searching games, and the methods for computing associated decompositions and obstructions, that is, information one can derive about a graph if its width is high with respect to some of these width parameters. Algorithmic applications are the topic of the second part of this chapter.

6.4 Directed Tree Width

In this section, we present the concept of *directed tree width*. Directed tree width was introduced by Reed [60] and Johnson et al. [36], and it was the first width measure for directed graphs developed explicitly as an attempt to generalize tree width to digraphs. In this sense, the work on directed tree width has initiated the research on tree-width variants and directed width measures. While today there is a wider variety of width measures for digraphs, directed tree width is to date the best understood among these, both algorithmically and from a structural perspective.

6.4.1 Directed Tree Decompositions and Directed Tree Width

We use a version of directed tree width introduced in [57]. Note that this differs from the original definition in [36] but only in a constant factor.

Definition 6.4.1 *Let G be a digraph. A* directed tree decomposition *of G is a tuple $\mathcal{T} := (T, \beta, \gamma)$, where*

- *T is a directed tree*

- *β is function $\beta : V(T) \to \mathcal{P}(V(G))$ such that $(\beta(t))_{t \in V(T)}$ is a partition of $V(G)$*

- *γ is a function $\gamma : E(T) \to \mathcal{P}(V(G))$*

such that

1. *For all $e = (s,t) \in E(T)$, the set $\beta(T_t) := \bigcup_{t \leq_T t'} \beta(t')$ is a strong component of $G - \gamma(e)$*

2. *$\bigcup_{t <_T t'} \beta(t') \cap \bigcup_{e \sim t} \gamma(e) = \emptyset$ for every $t \in V(T)$*

The width $w(\mathcal{T})$ *of $\mathcal{T} := (T, \beta, \gamma)$ is $\max \{ |\Gamma(t)| : t \in V(T) \}$, where $\Gamma(t) := \beta(t) \cup \bigcup_{e \sim t} \gamma(e)$. The* directed tree width $\mathrm{dtw}(G)$ *of G is the minimum width of any of its directed tree decompositions. A class $\mathcal{C}$ of digraphs has* bounded directed tree width *if there is a $k \in \mathbb{N}$ such that $\mathrm{dtw}(G) \leq k$ for all $G \in \mathcal{C}$.*

We introduce some notation that will frequently be used in the following. Let $\mathcal{T} := (T, \beta, \gamma)$ be a directed tree decomposition of a graph G. For a subtree $S \subseteq T$, we define $\beta(S) := \bigcup_{s \in V(S)} \beta(s)$. By definition of directed tree decompositions, if $S, S' \subseteq T$ are disjoint subtrees, then $\beta(S) \cap \beta(S') = \emptyset$.

Example 6.4.2 *We give some examples illustrating directed tree-decompositions.*

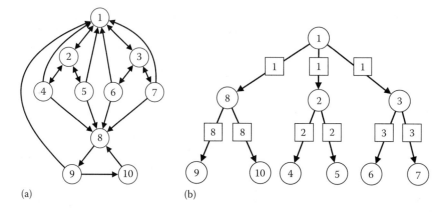

(a)   (b)

FIGURE 6.1:   (a) A sample digraph and (b) a corresponding directed tree decomposition of width 2.

As a first example, consider the graph in Figure 6.1a and its directed tree decomposition in Figure 6.1b. Here, the bags $\beta(t)$ are illustrated by circles and the guards $\gamma(e)$ by square boxes. The width of the decomposition is 2.

As a second example, it is easily seen that DAGs have directed tree-width 1. In fact, a simple directed path in which every bag contains exactly 1 vertex and the vertices appear on this path in any topological order of the DAG is a valid directed tree decomposition.

Finally, as a last example, it is not very hard to see that if G is an undirected graph and G' is the digraph obtained from G by replacing each undirected edge by two directed edges oriented in opposite directions, then the directed tree width of G' equals the tree width of G.

Directed tree width is broadly characterized by the visible strong cops and robber game described in Section 6.3. We will make this statement precise below (see Corollary 6.4.16), but essentially, the minimal number of cops required to catch the robber in these games is within a small constant factor of its directed tree width.

The next lemma, which follows immediately from the definition, establishes some simple connections between directed tree decompositions and strong separators in digraphs.

Lemma 6.4.3 *Let $\mathcal{T} := (T, \beta, \gamma)$ be a directed tree decomposition of a digraph G.*

1. *For every $e \in E(T)$, $\gamma(e)$ is a strong separator in G. More precisely, if S_1, S_2 are the two components of $T - e$, then every strongly connected component of $G - \gamma(e)$ is either contained in $\beta(S_1)$ or $\beta(S_2)$.*

2. *If $t \in V(T)$ and $T_1, \ldots, T_s$ are the components of $T - t$, then every strong component of $G - \Gamma(t)$ is contained in exactly one $\beta(T_i)$ for some i.*

We now establish a property of directed tree decompositions that will prove to be useful for the design of algorithms based on small directed tree width.

Lemma 6.4.4 *Let G be a digraph and $\mathcal{T} := (T, \beta, \gamma)$ be a directed tree decomposition of G. For every $t \in V(T)$, there is an ordering $<_t$ on the successors $s_1, \ldots, s_k$ of t in T so that if $s_i <_t s_j$, then G does not contain any edge $e = (u, v) \in E(G)$ with $u \in \beta(T_{s_i})$ and $v \in \beta(T_{s_j})$.*

Proof. Let $t \in V(T)$ and let $s_1, \ldots, s_k$ be its children and $e_i = (t, s_i)$, for $1 \leq i \leq k$, be the connecting edges. By definition, for all $1 \leq i \leq k$, $\beta(T_{s_i})$ is a strong component of $G - \gamma(e_i)$ and, as $\beta(T_{s_i}) \cap \bigcup_{1 \leq j \leq k} \gamma(e_j) = \emptyset$, for all i, a strong component of $G - \Gamma(t)$.

Now let H be the graph with vertex set $\{s_1, \ldots, s_k\}$ and a directed edge from s_i to s_j if G contains an edge $e = (u, v)$ with $u \in \beta(T_{s_i})$ and $v \in \beta(T_{s_j})$. As $\beta(T_{s_i})$ and $\beta(T_{s_j})$ are strongly connected and pairwise disjoint whenever $i \neq j$, H is acyclic. Hence, there exists a topological ordering $<$ on $V(H)$ such that all edges in H are directed toward smaller elements with respect to this ordering. Clearly, this ordering satisfies the requirements of the lemma. □

6.4.2 Computing Directed Tree Decompositions

An important step in using decomposition concepts algorithmically is to be able to compute the decomposition efficiently. However, it follows immediately from the NP-completeness of undirected tree width [2] that computing the directed tree width of a digraph is NP-hard. On the other hand, the problem is in NP as any directed tree decomposition can at most have a linear number of nodes, and hence its size is polynomial in the size of the digraph.

Theorem 6.4.5 *The problem to decide for a given digraph G and $k \in \mathbb{N}$ whether G has directed tree width at most k is NP-complete.*

An immediate consequence of the previous theorem is that unless $P = NP$, there cannot be a polynomial time algorithm for computing directed tree decompositions of optimal width. However, in this section, we will show that given a digraph G of directed tree width k, we can compute an approximate directed tree decomposition of width $3k + 5$ by means of a parameterized algorithm.

The algorithm for constructing directed tree decompositions is based on the concept of balanced separators.

Definition 6.4.6 *Let G be a digraph and $W \subseteq V(G)$. A balanced W-separator is a set $S \subseteq V(G)$ such that every strong component of $G - S$ contains at most $\frac{|W|}{2}$ vertices of W. The order of the separator is $|S|$. A set $W \subseteq V(G)$ is k-linked if G does not contain a balanced W-separator of order k.*

We show first that a digraph of directed tree width at most k does not contain a $k + 1$-linked set.

Lemma 6.4.7 *Let G be a digraph of directed tree width at most k. Then every set* $W \subseteq V(G)$ *has a balanced W-separator S of order at most k.*

Proof. Let $\mathcal{T} := (T, \beta, \gamma)$ be a directed tree decomposition of G of width k and let $W \subseteq V(G)$. We orient every edge $e \in E(T)$ as follows. Let $C_1, \ldots, C_l$ be the strong components of $G - \gamma(e)$ containing an element of W. If no C_i contains more than $\frac{|W|}{2}$ elements of W, then $\gamma(e)$ is a balanced W-separator and we are done. Otherwise, there is a (unique) component C_i containing more than half of the elements of W. Let T_1 and T_2 be the two components of $T - e$. By Lemma 6.4.3, either $C_i \subseteq \beta(T_1) - \gamma(e)$ or $C_i \subseteq \beta(T_2) - \gamma(e)$. In the first case, we orient e toward T_1; otherwise, it is oriented toward T_2.

Now suppose all edges have been oriented in this way. As T is a tree, there must be a node $t \in V(T)$ such that all incident edges are oriented toward t. We claim that $\Gamma(t)$ is a balanced W-separator. Toward a contradiction, suppose $\Gamma(t)$ is not a balanced W-separator and let C be the strong component of $G - \Gamma(t)$ containing more than $\frac{|W|}{2}$ of the elements of W.

Let $e, e_1, \ldots, e_s$ be the edges of T incident to t where e is the edge from the predecessor of t and $e_1, \ldots, e_s$ are the edges toward the children $t_1, \ldots, t_s$ of t. As e is oriented toward t, $V(C) \subseteq \beta(T_t) - \gamma(e)$. Let $U := \beta(T_t) - \gamma(e)$. Again by Lemma 6.4.3, as C is a strong component of $G - \Gamma(t)$, C must be contained in $\beta(T_{t_i})$ for some successor t_i of t. But this contradicts the fact that the edge between t and t_i was oriented toward t, as $\gamma(e_i) \subseteq \Gamma(t)$. $\qquad\square$

We will show next that every digraph G either has directed tree width at most $3k + 2$ or contains a set W that is k-linked and hence, by the previous lemma, has directed tree width at least k.

Theorem 6.4.8 ([36]) *Every digraph G either has directed tree width at most $3k + 2$ or contains a set W that is k-linked and is a witness that G has directed tree width at least k.*

Proof. To prove the theorem, we inductively construct a directed tree decomposition (T, β, γ) of G, maintaining the following invariant:

1. For every inner vertex $t \in V(T)$, $|\Gamma(t)| \leq 3k + 2$.

2. For every edge $e \in E(T)$, $|\gamma(e)| \leq 2k + 1$.

Either this process will succeed and therefore produce a directed tree decomposition of the required width or it will fail at some point at which we obtain a k-linked set.

We initialize the construction by the trivial directed tree decomposition $\mathcal{T} := (T, \beta, \gamma)$, where T is the tree with one node r and $\beta(r) := V(G)$. Clearly, this satisfies the previously mentioned invariant.

Now suppose $\mathcal{T} = (T, \beta, \gamma)$ has already been constructed. If $\mathcal{T}$ does not contain a leaf $t \in V(T)$ with $|\Gamma(t)| > 3k+2$, then we are done. So let $t \in V(T)$ be such a leaf.

Let e be the edge incident with t. By construction, $|\gamma(e)| \leq 2k + 1$. If $\gamma(e)$ is k-linked, we are done. Otherwise, let S be a balanced $\gamma(e)$-separator of order at most k. Let $v \in \beta(t)$ be an arbitrary vertex and let $X := S \cup \{v\}$. By construction, $|X| \leq k+1$, $X \cap \beta(t) \neq \emptyset$ and every strong component C of $G - X$ contains at most $\frac{|\gamma(e)|}{2} \leq k$ elements of $\gamma(e)$. Let $C_1, \ldots, C_s$ be the strong components of $G - (X \cup \gamma(e))$. By the definition of a directed tree decomposition, either $C_i \subseteq \beta(t)$ or $C_i \cap \beta(t) = \emptyset$, for all $1 \leq i \leq s$. Let $D_1, \ldots, D_l$ be the components among $\{C_1, \ldots, C_s\}$ with $C_i \subseteq \beta(t)$. For each D_i, let D_i' be the component of $G - X$, such that $V(D_i) \subseteq V(D_i')$ and let $W_i = \left(\gamma(e) \cap V(D_i')\right) \cup X$. Then

$$|W_i| \leq |\gamma(e) \cap V(D_i')| \cup |X| \leq k + k + 1 = 2k + 1$$

and D_i is also strong component of $G - W_i$.

We extend $\mathcal{T}$ as follows to obtain a new decomposition $\mathcal{T}' := (T', \beta', \gamma')$: add new nodes $t_1, \ldots, t_l$ and edges $e_i := (t, t_i)$ to T, for all $1 \leq i \leq l$, and set $\beta'(t) := X \cap \beta(t)$, $\beta'(t_i) := V(D_i)$, and $\gamma'(e_i) := W_i$. For all other nodes t and edges e, we set $\beta'(t) := \beta(t)$ and $\gamma'(e) := \gamma(e)$. It is easily seen that $\mathcal{T}'$ is a directed tree decomposition of G maintaining the previous invariant. In particular, $|\beta'(t)| \leq |X| \leq k + 1$ and $|\gamma'(e_i)| \leq 2k + 1$. Furthermore, $\gamma'(e_i) \subseteq X \cup \gamma(e)$ and thus $\Gamma'(t) = \beta'(t) \cup \gamma'(e) \cup \bigcup\{\gamma'(e_i) : 1 \leq i \leq s\} \subseteq X \cup \gamma(e)$. It follows that $|\Gamma'(t)| \leq k + 1 + 2k + 1 = 3k + 2$. Furthermore, as $D_1, \ldots, D_l$ are strong components of $G - (X \cup \gamma(e))$, Conditions (1) and (2) of Definition 6.4.1 are still satisfied.   □

The proof of the previous theorem is essentially algorithmic. To turn it into an algorithm for computing directed tree decompositions of approximate width, we need to be able to compute balanced separators. However, there are no efficient algorithms for this problem available at the moment. We therefore relax the notion of balanced separators slightly to obtain weakly balanced separators that can be computed efficiently.

Definition 6.4.9 *Let G be a digraph and $W \subseteq V(G)$. A weakly balanced W-separation is a triple (X, S, Y) of pairwise disjoint sets $X, Y \subseteq W$ and $S \subseteq V(G)$ such that*

1. *$W = X \cup (S \cap W) \cup Y$*

2. *There is no directed path from X to Y in $G - S$*

3. *$0 < |X|, |Y| \leq \frac{3}{4}|W|$*

The order *of the separation is $|S|$. A set $W \subseteq V(G)$ is weakly k-linked if G does not contain a weakly balanced W-separator of order k.*

Lemma 6.4.10 *Let G be a digraph and $W \subseteq V(G)$ such that $|W| \geq 2k + 2$. If $\mathrm{dtw}(G) \leq k$, then there exists a weakly balanced W-separation of order at most k.*

Proof. As $\mathrm{dtw}(G) \leq k$, Lemma 6.4.7 implies that there is a balanced W-separator S of order at most k. Let $C_1, \ldots, C_m$ be the components of $G - S$ ordered in a way such that there is no edge from a vertex in C_i to a vertex in C_j, for any i, j with $1 \leq j < i \leq m$. We distinguish between two cases.

Suppose first that $|W \cap V(C_i)| > \frac{1}{4}|W|$ for some $1 \leq i \leq m$. Let

$$X' := \bigcup_{1 \leq j < i} V(C_j) \cap W \text{ and } Y' := \bigcup_{i+1 \leq j \leq m} V(C_j) \cap W.$$

If $|X'| \leq |Y'|$, we set $X := X' \cup (W \cap V(C_i))$ and $Y := Y'$, otherwise we set $X := X'$ and $Y := Y' \cup (W \cap V(C_i))$. In the first case, we have $|X| > 0$ and $|Y| < \frac{3}{4}|W|$. Furthermore, $|Y| \geq \frac{|W-S|-|W \cap V(C_i)|}{2} \geq \frac{|W|/2+1-|W|/4}{2} > 0$ and because S is a balanced W separator and therefore $W \cap V(C_i)$ cannot contain more than half of the elements of W, we obtain $|X| \leq \frac{|W-S|-|W \cap V(C_i)|}{2} + |W \cap V(C_i)| \leq \frac{|W|+|W \cap V(C_i)|}{2} \leq \frac{3}{4}|W|$. The proof for the case $|X'| > |Y'|$ is analogous.

Now suppose that $|C_i| \leq \frac{1}{4}|W|$ for all $1 \leq i \leq m$. As $|W - S| > \frac{1}{2}|W|$, there is an $1 \leq i \leq m$ such that $\sum_{j=1}^{j=i} |V(C_j) \cap W| > \frac{1}{4}|W|$. Let i be minimal subject to this condition. We set $X = \bigcup_{j=1}^{j=i} V(C_j) \cap W$ and $Y = \bigcup_{j=i+1}^{m} V(C_j) \cap W$. Then $|X| > 0$ and $|Y| < \frac{3}{4}|W|$. Furthermore, it follows from the minimality of i that $|X| < \frac{1}{4}|W| + \frac{1}{4}|W| = \frac{1}{2}|W| \leq \frac{3}{4}|W|$ and hence $|Y| \geq |W - S| - |X| \geq |W|/2 + 1 - |W|/2 > 0$, as required.   □

Lemma 6.4.11 *There is an algorithm running in time $\mathcal{O}(3^{2k+2}k|E(G)|)$, which, on input a digraph G, a number $k \geq 1$, and a set $W \subseteq V(G)$ of size $2k + 2$, computes a weakly balanced W-separation of order at most k if such a separation exists.*

Proof sketch. The algorithm is an adaptation of the algorithm in [23, Corollary 11.22] to the directed setting. For every two disjoint subsets $X, Y \subseteq W$ with $0 < |X|, |Y| \leq \frac{3}{4}|W|$, the algorithm computes a minimal directed separator from X to Y in $G - S$, where $S := W - (X \cup Y)$. If there are such subsets X and Y, such that there is a directed separator S' from X to Y with $|S' \cup S| \leq k$, then the algorithm outputs $S' \cup S$. Otherwise, the algorithm outputs that G does not have a weakly balanced W-separation of order at most k. Because there are at most 3^{2k+2} possible such subsets X and Y, and for each choice of X and Y a minimal directed separator from X to Y can be computed in time $O(k|E(G)|)$ (e.g., via a standard flow algorithm), the running time of the algorithm is at most $\mathcal{O}(3^{2k+2}k|E(G)|)$, as required.   □

Using the previous lemma, we can now prove the following algorithmic result.

Theorem 6.4.12 *There is an algorithm with running time $\mathcal{O}(3^{2k+2} \cdot k \cdot |E(G)| \cdot |V(G)|)$ which, on input G and $k \geq 1$, either computes a directed tree decomposition of G of width at most $5k + 10$ or a weakly k-linked set W.*

Proof. It is easy to see that the proof of Theorem 6.4.12 can be adapted to use weakly W-separators instead of W-balanced separators. Doing this, the width of the constructed directed tree decomposition increases to $(4k + 1) + (k + 1) = 5k + 2$. The running time follows from Lemma 6.4.11 and the fact that a directed tree decomposition has at most $|V(G)|$ nodes.   □

Furthermore, the directed tree width of a digraph can be approximated up to a logarithmic factor.

Theorem 6.4.13 ([44]) *There exists a polynomial time approximation algorithm that, given a digraph G, computes a directed tree decomposition of G, whose width is at most $\mathcal{O}(\log^{\frac{3}{2}} |V(G)| \cdot \text{dtw}(G))$.*

6.4.3 Obstructions: Havens, Brambles, and Well-Linked Sets

The existence of a directed tree decomposition of a digraph G is a witness for its directed tree width being small. We will see algorithmic applications exploiting directed tree decompositions later on, in Section 6.17. In this section, we study several *obstructions* for directed tree width, that is, structural information we can derive from the fact that the directed tree width of a digraph G is large. See [60] and [36] where most results of this section were proved originally.

6.4.3.1 Havens in Digraphs

The first type of obstructions we study are *havens* defined in [36].

Definition 6.4.14 *Let G be a digraph. A* haven *of G of order k is a function $h:[V(G)]^{<k} \to \mathcal{P}(V(G))$ assigning to every set X of fewer than k vertices a strong component of $G - X$ such that if $Y \subseteq X \subseteq V(G)$ with $|X| < k$, then $h(X) \subseteq h(Y)$.*

We show next that any digraph containing a haven of large order must have large directed tree width.

Theorem 6.4.15 ([36])

1. *If G is a digraph of tree width at most k, then G does not contain a haven of order $k + 1$.*

2. *Conversely, if G does not contain a haven of order $k + 1$, then G has tree-width $< 3k + 2$.*

Proof. The proof of Part (1) is essentially the same as for Lemma 6.4.7. If G contains a haven h of order $k + 1$ but has a directed tree decomposition (T, β, γ) of width k, we can orient the edges of T as follows. We orient an edge $e = (u, v)$ in the direction containing $h(\gamma(e))$. Hence, there must be a node t such that all incident edges are oriented toward t. It follows that for all incident edges e, $h(\gamma(e)) \subseteq \beta(t)$. Now consider $\Gamma(t) = \beta(t) \cup \bigcup_{e \sim t} \gamma(e)$. Clearly, $\gamma(e) \subseteq \Gamma(t)$ for all $e \sim t$ and hence, $h(\Gamma(t)) \subseteq h(\gamma(e))$. But as $h(\Gamma(t)) \cap \beta(t) = \emptyset$, this yields a contradiction.

Toward Part (2), it is easily seen that if G contains a k-linked set W, for some $k \geq 1$, then it contains a haven of order $k + 1$. For, let W be a k-linked set. We define a haven $h : [V(G)]^{\leq k} \to \mathcal{P}(V(G))$ by setting, for each $X \in [V(G)]^{\leq k}$, $h(X) := C$ for the (unique) strong component of $G - X$ containing more than half of the elements of W. As W is k-linked and hence has no balanced separator of order k, this component always exists. It is easily checked that this is indeed a haven.

Now, by Theorem 6.4.8, if G has directed tree-width $3k + 2$, then it contains a k-linked set W and hence a haven of order $k + 1$.   □

A direct consequence of this result is that directed tree width can, up to a small constant factor, be characterized by the visible strong cops and robber game, which was established in [36].

Corollary 6.4.16 *Let G be a digraph and $k \in \mathbb{N}$.*

1. *If has directed tree width at most k, then k cops have a robber-monotone winning strategy in the visible strong cops and robber game on G.*

2. *Conversely, if k cops have a winning strategy in this game on G, monotone or not, then G has tree width less than $3k + 2$.*

3. *If k cops have a winning strategy in the visible strong cops and robber game on G, then $3k + 2$ cops have a robber-monotone winning strategy on G.*

Proof. It is easily seen that a directed tree decomposition of width k of a digraph G immediately yields a robber-monotone winning strategy for the cops. They simply need to follow the tree from the root to a leaf in the direction of the bag containing the robber. This establishes Part (1).

On the other hand, any haven h of order k immediately yields a winning strategy for the robber, where for any cop position C, the robber always remains in $h(C)$. Hence, if k cops have any winning strategy on G, then this implies that G cannot contain a haven of order $k + 1$ and hence, by Theorem 6.4.15, the directed tree width of G is at most $3k + 2$. This proves Part (2).

Part (3) follows immediately from Part (1) and (2).   □

6.4.3.2   Brambles in Directed Graphs

We now define a sequence of other obstructions for directed tree width, originally defined in [60].

Definition 6.4.17 *A bramble in a digraph G is a set $\mathcal{B}$ of strongly connected subgraphs of G such that for any pair $B, B' \in \mathcal{B}$, either $V(B) \cap V(B') \neq \emptyset$ or there are edges e, e' linking B and B' in both directions. A bramble $\mathcal{B}$ is strong if $V(B) \cap V(B') \neq \emptyset$ for all $B, B' \in \mathcal{B}$.*

A cover, or hitting set, of $\mathcal{B}$ is a set $X \subseteq V(G)$ such that $X \cap V(B) \neq \emptyset$ for all $B \in \mathcal{B}$. The order of $\mathcal{B}$ is the minimum size of a cover of $\mathcal{B}$.

Example 6.4.18 *Figure 6.2 illustrates a bramble $\mathcal{B}$ of order 2 in the digraph of Figure 6.1a. A minimal cover of $\mathcal{B}$ is the set $\{1, 8\}$. Obviously, there is no cover of $\mathcal{B}$ of order 1 as $\mathcal{B}$ contains two disjoint elements. The bramble is not strong, as the subgraph $G[\{1, 2, 4\}] \in \mathcal{B}$ has an empty intersection with the subgraph $G[\{8, 9, 10\}] \in \mathcal{B}$.*

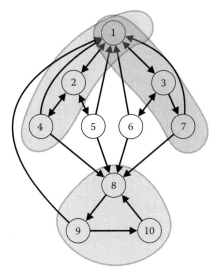

FIGURE 6.2: A bramble of order 2 in the digraph of Figure 6.1a.

Lemma 6.4.19 *Let G be a digraph. If G contains a k-linked set, then it contains a strong bramble of order k.*

Proof. Let W be a k-linked set. By definition, for every $X \subseteq V(G)$ with $|X| \leq k$, there is a strong component $C(X)$ of $G - X$ containing more than half the vertices of W. We define $\mathcal{B} := \{C(X) : X \subseteq V(G), |X| \leq k\}$. By construction, each $B \in \mathcal{B}$ is connected. Furthermore, $V(B) \cap V(B') \neq \emptyset$ for every $B, B' \in \mathcal{B}$, as both contain more than half the vertices of W and therefore must have a common vertex. Hence, $\mathcal{B}$ is a strong bramble. The order of $\mathcal{B}$ is at least k as any set X of fewer than k vertices avoids the element $C(X) \in \mathcal{B}$. □

Strong brambles and havens are closely related, as the following lemma shows.

Lemma 6.4.20 *Let G be a digraph. If G contains a bramble of order k, then G contains a haven of order k + 1. Furthermore, if G contains a haven of order k + 1, then it contains a strong bramble of order $\frac{k}{2}$.*

Proof. Let $\mathcal{B}$ be a bramble in G of order k. We define a function $h : [V(G)]^{\leq k} \to \mathcal{P}(V(G))$ as follows. For every $X \in [V(G)]^{\leq k}$, we define $h(X)$ as the strong component of $G - X$ containing $\bigcup\{B \in \mathcal{B} : V(B) \cap X = \emptyset\}$. As the union of any set of bramble elements is strongly connected and each $X \in [V(G)]^{\leq k}$ avoids some element of $\mathcal{B}$, $h(X)$ is well defined. It is now easily seen that if $Y \subseteq X \subseteq V(G)$ with $|X| \leq k$, then $h(X) \subseteq h(Y)$. This shows the first part of the lemma.

Toward the second part, let h be a haven in G of order $k + 1$. We define $\mathcal{B} := \{h(X) : X \subseteq V(G), |X| \leq \frac{k}{2}\}$. By definition of h, each $B \in \mathcal{B}$ is strongly connected. Furthermore, every set $X \subseteq V(G)$ of at most $\frac{k}{2}$ elements avoids an element of $\mathcal{B}$; hence, the order of $\mathcal{B}$ is at least $\frac{k}{2}$. We need to show that any $B, B' \in \mathcal{B}$ share a vertex.

So let $B = h(X)$ and $B' = h(X')$, for some $X, X' \in [V(G)]^{\leq \frac{k}{2}}$, be two elements of $\mathcal{B}$. Then $h(X \cup X') \subseteq B \cap B'$ and hence $B \cap B' \neq \emptyset$. □

6.4.3.3   Well-Linked Sets

The last obstruction we consider are well-linked sets [60].

Definition 6.4.21 *Let G be a digraph. A set $W \subseteq V(G)$ is* well linked *if for any $X, Y \subseteq W$ with $|X| = |Y|$ there are $|X| = |Y|$ pairwise vertex disjoint paths from X to Y in $G - W$. The* order *of a well-linked set W is $|W|$.*

Lemma 6.4.22 *Let G be a digraph and let $\mathcal{B}$ be a bramble of G of order k. Then G contains a well-linked set of order k.*

Proof. Let W be a minimum cover of $\mathcal{B}$. Hence, $|W| = k$. We claim that W is well linked.

Toward a contradiction, assume that W is not well linked. Hence, there are $X, Y \subseteq W$ of size $|X| = |Y| = r$ such that there are no r pairwise vertex disjoint paths from X to Y. W.l.o.g. we assume that $X \cap Y = \emptyset$.

By Menger's theorem (see Theorem 6.2.2), there is a set $S \subseteq V(G)$ of size less than r such that there is no path from X to Y in $G - S$. We claim that either $W_X := (W - X) \cup S$ or $W_Y := (W - Y) \cup S$ is a cover of $\mathcal{B}$ of order less than $|W|$, contradicting the minimality of W.

Toward a contradiction, suppose that neither W_X nor W_Y are a cover of $\mathcal{B}$. Hence, there are $B_X \in \mathcal{B}$ and $B_Y \in \mathcal{B}$ such that $W_X \cap B_X = \emptyset$ and $W_Y \cap B_Y = \emptyset$. Because W is a cover of $\mathcal{B}$, there must be an $x \in X \cap V(B_X)$ and an $y \in Y \cap V(B_Y)$. Furthermore, because $\mathcal{B}$ is a bramble and $(V(B_X) \cup V(B_Y)) \cap S = \emptyset$, the union of B_X and B_Y is strongly connected in $G - S$. It follows that x and y are in the same strongly connected component of $G - S$, contradicting our assumption that S separated X from Y. □

Lemma 6.4.23 *Let G be a digraph. If G contains a well-linked set of order $4k + 1$, then G contains a k-linked set.*

Proof. Let W be a well-linked set of order $4k + 1$. We claim that W is k-linked. Toward this aim, let $S \subseteq V(G)$ be a set of size k. We show that there is a strong component of $G - S$ containing more than one-half of the elements of W.

W.l.o.g. we assume that $S \cap W = \emptyset$. If not, we apply the analysis to $W - S$ and $S - W$ instead. Clearly, $W - S$ is still well linked and $4|S - W| + 1 \leq |W - S|$. Hence, we can apply the following argument to these sets instead.

Let $(C_1, \ldots, C_m)$ be a topological ordering of the strong components of $G - S$, that is, for every i and j with $1 \leq i < j \leq m$, there is no edge from a vertex in C_j to a vertex in C_i. Assume first that there is a component C_j that contains more than k elements of W. Let $X := W \cap \left(\bigcup_{i=1}^{j-1} C_i \right)$ and $Y := W \cap \left(\bigcup_{i=j+1}^{m} C_i \right)$. Then, because $|W \cap C_j| \leq \frac{|W|}{2}$, either $|X| \geq \frac{|W|}{4} = \frac{4k+1}{4} > k$ or $|Y| \geq \frac{|W|}{4} = \frac{4k+1}{4} > k$. If $|X| > k$, then because W is well linked, there must be more than k paths from C_i

to X in G, and hence $G - S$ must contain at least one path from C_i to X. However, this cannot be the case since $(C_1, \ldots, C_m)$ is a topological ordering. The argument for the case that $|Y| > k$ is similar, because now G contains more than k paths from Y to C_i. So suppose there is no such component C_j and let j be minimal such that the set $X := |W \cap (\bigcup_{i=1}^{j} C_i)|$ has more than k elements. Then, $|X| \le 2k$ and hence $Y := |W \cap (\bigcup_{i=j+1}^{m} C_i)| > k$. Again, because W is well linked, there is at least one path from Y to X in $G - S$, contradicting our assumption that $(C_1, \ldots, C_m)$ is a topological ordering. $\qquad\square$

6.4.3.4 Putting Things Together

Each of the previous types of obstructions defines its own width parameter of digraphs, and the results presented earlier show that all these parameters are within a constant factor of each other. More precisely, the *linkedness* of a digraph G is the maximal k for which G contains a k-linked set. The *well-linkedness* of G is the maximal order of a well-linked set in G. Similarly, the *(monotone) visible strong cops and robber number* of G is the minimal number k of cops that have a (robber-monotone) winning strategy on G. Finally, the *haven number* and the *(strong) bramble number* of G is the maximal order of a haven or a (strong) bramble in G, respectively. Then the previous results establish the next corollary.

Corollary 6.4.24 *For every digraph G, the directed tree width, the linkedness, the well-linkedness, the monotone and nonmonotone visible strong cops and robber number, the haven number, and the (strong) bramble number of G are all within a (small) constant factor of each other.*

The obstructions considered so far provide valuable insight into the structure of digraphs of large directed tree width. However, all types of obstructions seen previously are in themselves not very *well structured*. A type of obstruction that has much more order are directed, or cyclic, grids, illustrated in Figure 6.3.

It was conjectured by Reed [59] and Johnson et al. [36] that any digraph of sufficiently high directed tree width should contain a large cylindrical grid as a butterfly minor. We refer the reader to [36] for a formal definition of directed grids and butterfly minors. An illustration of the two considered types of directed grids is also given in Figure 6.3.

Conjecture 6.4.25 ([36,59]) *Every digraph of high directed tree width contains a large directed grid, that is, there is a function $f : \mathbb{N} \to \mathbb{N}$ such that every digraph of directed tree width at least $f(k)$ contains a cyclic grid of order k as a butterfly minor.*

The conjecture was confirmed for planar graphs in [37]. In [42], the authors generalize this to classes of digraphs excluding an undirected minor.

Theorem 6.4.26 ([42]) *Let $\mathcal{C}$ be a class of directed graphs excluding a fixed undirected graph H as a minor. There is a computable function $f : \mathbb{N} \to \mathbb{N}$ such that for all $G \in \mathcal{C}$ and all $k \in \mathbb{N}$, if the directed tree width of G is at least $f(k)$, then G contains a cylindrical grid of order k as a butterfly minor.*

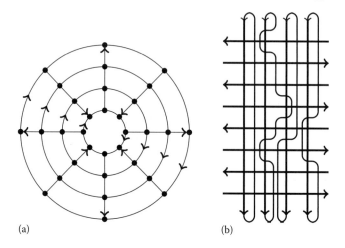

FIGURE 6.3:   (a) Cylindrical grid G_4 and (b) half-integral grid H_4.

More recently, a weaker form of the conjecture was established in [35]. We refer to [35] for a definition of half-integral grids.

Theorem 6.4.27 ([35]) *There is a computable function $f : \mathbb{N} \to \mathbb{N}$ such that for all digraphs G and all $k \in \mathbb{N}$, if the directed tree width of G is at least $f(k)$, then G contains a half-integral grid of order k as a butterfly minor.*

However, the full version of the conjecture remains open to date.

6.5   D-Width

In the previous section, we have considered directed tree width, a width measure based on a decomposition that yields a strong control over strongly connected sets in directed graphs. Algorithmically, a problem with directed tree decompositions is that while the bags form a partition of the vertex set and therefore every vertex of G only occurs in one bag, the guards do not satisfy any such kind of monotonicity condition. For algorithmic purposes, this is sometimes inconvenient.

Employing the connection to cops and robber games, the problem is that directed tree width is based on a robber-monotone game variant but not on a cop-monotone variant, which usually yield nicer decompositions. In [63], Safari therefore introduced a variant of directed tree width, called *D-width*, which appeared to be closer to the cop-monotone game.

Here, we present a slightly different decomposition than the one originally proposed in [63]. However, the two only differ by a factor of at most 2 and hence define the same classes of digraphs of bounded D-width.

Definition 6.5.1 (DS-Decompositions and D-Width) *Let G be a directed graph. A DS-decomposition of G is a pair (T, β), where T is a tree and $\beta : V(T) \to \mathcal{P}(V(G))$ is a function such that*

1. *For every vertex $v \in V(G)$, the set $\beta^{-1}(v) := \{t \in V(T) : v \in \beta(t)\}$ is nonempty and connected in T*

2. *For every edge $e = \{u, v\} \in V(T)$, the set $\beta(u) \cap \beta(v)$ is a strong separator of G, that is, there is no strongly connected set $C \subseteq G - (\beta(u) \cap \beta(v))$, which contains a vertex in $\beta(T_v) \setminus (\beta(u) \cap \beta(v))$ and a vertex in $\beta(T_u) \setminus (\beta(u) \cap \beta(v))$, where T_u, T_v are the two components of $T - e$.*

The width of (T, β) is $\max\{|\beta(t)| : t \in V(T)\}$ and the D-width of G, denoted by D-width(G), is the minimum width of any of its DS-decompositions. A class $\mathcal{C}$ of digraphs has bounded D-width if there is a $k \in \mathbb{N}$ such that D-width(G) $\leq k$ for all $G \in \mathcal{C}$.

Example 6.5.2 *As an example, consider again the digraph G in Figure 6.1a. Figure 6.4 illustrates a DS-decomposition of G of width 2. Conceptually, the DS-decomposition appears simpler than, for instance, the corresponding directed tree decomposition of G in Example 6.4.2.*

As noted previously, D-width appears to be similar to the cop-monotone visible strong cops and robber game. In particular, it is easily seen that if a digraph G has a DS-decomposition of width k, then k cops have a cop-monotone winning strategy in the visible strong component cops and robber game: given a DS-decomposition (T, β), the cops start by placing k cops on the root r of T, where any node r can be chosen as root. From this position, they move to a neighbor t of r in T such that the current robber position is in $\beta(T_t)$. As $\beta(r) \cap \beta(t)$ is a strong separator of G, the robber cannot escape from T_t while the cops move from $\beta(r)$ to $\beta(t)$. Continuing in this

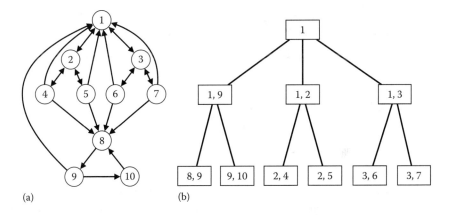

(a)   (b)

FIGURE 6.4: (a) The digraph from Figure 6.1 and (b) a corresponding DS-decomposition of width 2.

way, the cops will eventually capture the robber in a leaf of the tree T. Condition 1 of Definition 6.5.1 ensures that this strategy is cop monotone. This shows the following lemma.

Lemma 6.5.3 *On every digraph G of D-width k, k cops have a cop-monotone winning strategy in the visible strong component cops and robber game on G.*

Whether the converse is true is still open and in fact we conjecture that the converse fails.*

Conjecture 6.5.4 *There is a family $\mathcal{C}$ of digraphs of unbounded D-width and a constant $c \in \mathbb{N}$ such that c cops have a cop-monotone winning strategy for the visible strong component cops and robber game on every $G \in \mathcal{C}$.*

6.6   DAG Width

DAG width is a proposal for a digraph width measure introduced in [7,8,55] to overcome certain problems with directed tree decompositions. In particular, for every node t of a directed tree decomposition, $\beta(T_t)$ is a strong component of $G - \gamma(e)$, where e is the incoming edge of t, but there might still be edges of G going out of $\beta(T_t)$ or going into $\beta(T_t)$ and the guarding set $\gamma(e)$ does not yield any control over these. It only certifies that if we follow a path emerging from $\beta(T_t)$ in G, then it will hit $\gamma(e)$ before coming back into $\beta(T_t)$. For several applications such as directed disjoint paths, this level of control is sufficient, but for other problems, this level of control is too weak.

For DAG width, therefore, a more direct concept of guarding was used.

Definition 6.6.1 *Let G be a digraph and $V' \subseteq V(G)$. A set $W \subseteq V(G)$ guards V' if for all $(u, v) \in E(G)$, if $u \in V'$ then $v \in V' \cup W$.*

We are now ready to define the concept of DAG width.

Definition 6.6.2 (DAG decomposition) *Let G be a digraph. A DAG-decomposition of G is a pair (D, β), where D is a DAG and $\beta : V(D) \to \mathcal{P}(V(G))$ is a function such that*

1. $\bigcup_{d \in V(D)} \beta(d) = V(G)$

2. $\beta(d) \cap \beta(d'') \subseteq \beta(d')$, *for all nodes $d \preceq_D d' \preceq_D d''$*

3. *For all edges $(d, d') \in E(D)$, $\beta(d) \cap \beta(d')$ guards $\beta(D_{d'}) \setminus \beta(d)$, where $D_{d'}$ stands for $\bigcup_{d' \preceq_D d''} \beta(d'')$. For any source d, $\beta(D_d)$ is guarded by $\emptyset$*

The width of a DAG decomposition (D, β) is defined as $\max\{|\beta(d)| : d \in V(D)\}$. The DAG width of a digraph G, denoted by $\mathrm{dagw}(G)$, is defined as the minimal width of any of its DAG decompositions. A class $\mathcal{C}$ of digraphs has bounded DAG width if there is a $k \in \mathbb{N}$ such that $\mathrm{dagw}(G) \leq k$ for all $G \in \mathcal{C}$.

---

* A positive solution to this conjecture has recently been announced but is as yet unpublished, see [38].

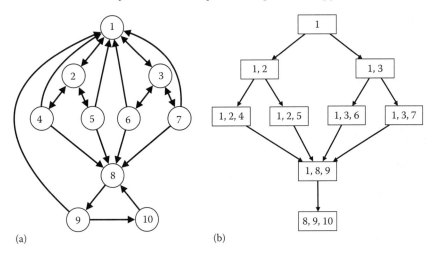

FIGURE 6.5:   (a) The digraph from Figure 6.1 and (b) a corresponding DAG decomposition.

It may be worth pointing out that in a DAG decomposition (D, β) of a digraph G, every directed path in D induces a directed path decomposition of a subgraph of D. See Section 6.8 for a definition of directed path width. The next example illustrates the concept of DAG decompositions.

Example 6.6.3 *As an example, consider again the digraph G in Figure 6.1a. Figure 6.5 illustrates a DAG decomposition of G of width 3.*

Note the difference to directed tree decompositions as here the underlying structure is indeed a DAG. This is necessary as if we were to delete, for instance, the edge from the node labeled $1, 3, 7$ to the node $1, 8, 9$; then 8 and subsequently also 9 and 10 would need to be added to $1, 3, 7$, increasing the width.

It can be shown, for instance, using the characterization of DAG width by a cops and robber game in the following, that G has no DAG decomposition of order 2. Hence, the DAG width of G is higher than its directed tree width or D-width.

As mentioned previously, DAG width is based on a monotone variant of a directed cops and robber game.

Theorem 6.6.4 ([8]) *A digraph G has DAG width k if, and only if, k cops have a monotone winning strategy in the visible reachability cops and robber game on G.*

As an immediate consequence of this theorem and the fact that tree width is characterized exactly by the corresponding cops and robber game on undirected graphs (see [64]), we get the following.

Corollary 6.6.5 *Let G be an undirected graph and let G' be the directed graph obtained from G by replacing each undirected edge $\{u, v\} \in E(G)$ by two directed edges (u, v) and (v, u) in opposite directions.*

Then the DAG width of G' equals the tree width of G.

Recall that directed tree width was characterized by the visible strong cops and robber game. Clearly, if k cops can catch a visible robber in the reachability game, they can also catch the robber in the strong game, as in the latter the robber has less vertices to move to. The converse is not true, however, as shown in [8]. In fact, let T be a directed binary tree of height k where we add for every node t and every predecessor s of t the edge (t, s). The nodes $\{1, 2, 3, 4, 5, 6, 7\}$ in the digraph in Figure 6.1 induce such a tree of height 2. It is not hard to see that on this tree, k cops are needed to catch the robber in the reachability game, but only 2 are needed in the strong component game. Hence, the class of such trees has unbounded DAG-width but directed tree-width 2. It is this observation that also explains why the DAG width of the digraph in Figure 6.1 is 3 but its directed tree width is only 2.

As a consequence, we get the next corollary.

Corollary 6.6.6 ([8]) *Every class of digraphs of bounded DAG width also has bounded directed tree width. However, there are classes of digraphs of bounded directed tree width but unbounded DAG width.*

6.6.1   Computing DAG Decompositions

We have seen that computing the directed tree width of a digraph is NP-complete but that directed tree decompositions of approximate width can be computed efficiently by parameterized algorithms. Corollary 6.6.5 immediately implies that deciding DAG width is NP-hard as well.

Corollary 6.6.7 *Deciding the DAG width of a directed graph is NP-hard.*

However, unlike directed tree decompositions, the number of nodes in a DAG decomposition is not necessarily linearly bounded in the number n of vertices in the digraph and could in principle be as high as n^k, where k is the DAG width. Hence, it is not clear whether computing the DAG width is in NP. However, for every fixed k, a DAG decomposition of optimal width can be computed in polynomial time.

Theorem 6.6.8 ([8]) *Given a digraph G of DAG width at most k, a DAG decomposition of G of width at most k can be computed in time $|G|^{\mathcal{O}(k)}$.*

6.7   Kelly Width

In the previous section, we have considered the concept of DAG width. As we have seen, it facilitates a very strong concept of guarding compared to directed tree width and D-width. On the other hand, with the exception of a cops and robber game, no further characterizations of DAG width by other structural parameters are known, and furthermore the size of a DAG decomposition is not necessarily linearly (or fixed polynomially) bounded in the order n of the digraphs. In fact, it can be as large as n^k, where k is the DAG width.

We now present a width measure for digraphs known as *Kelly width*,* which in many ways is similar to DAG width, especially in its concept of guarding, but where decompositions have linear size. Furthermore, Kelly width can be characterized in a number of seemingly unrelated ways making it a robust and natural measure.

6.7.1   Definition and Characterizations of Kelly Width

As the other width measures so far, Kelly width can be defined in terms of a decomposition.

Definition 6.7.1 (Kelly decomposition and Kelly width [34]) *A Kelly decomposition of a digraph G is a triple* $\mathcal{D} := (D, \beta, \gamma)$ *so that*

1. *D is a DAG*

2. $\beta : V(D) \rightarrow \mathcal{P}(V(G))$ *is a function such that* $\big(\beta(t)\big)_{t \in V(D)}$ *partitions* $V(G)$

3. $\gamma : V(D) \rightarrow \mathcal{P}(V(G))$ *is a function and for all* $t \in V(D)$, $\gamma(t)$ *guards* $\beta(V(D_t)) := \bigcup_{t' \geq_D t} \beta(t')$

4. *For all* $s \in V(D)$, *there is a linear order* $<_s$ *on its children* $t_1, \ldots, t_p$ *so that for all* $1 \leq i \leq p$,

$$\gamma(t_i) \subseteq \beta(s) \cup \gamma(s) \cup \bigcup_{j <_s i} \beta(V(D_{t_j})).$$

Similarly, there is a linear order $<_r$ *on the roots such that*

$$\gamma(r_i) \subseteq \bigcup_{j <_r i} \beta(V(D_{r_j})).$$

The width *of* $\mathcal{D}$ *is* $\max\{|\beta(t) \cup \gamma(t)| : t \in V(D)\}$. *The* Kelly width kellyw(G) *of* G *is the minimal width of any of its Kelly decompositions. A class* $\mathcal{C}$ *of digraphs has* bounded Kelly width *if there is a* $k \in \mathbb{N}$ *such that* kellyw$(G) \leq k$ *for all* $G \in \mathcal{C}$.

Example 6.7.2 *As an example, consider again the digraph G in Figure 6.1a. Figure 6.6 illustrates a Kelly decomposition of G of width 3. Here, the circled labels are the bags* $\beta(t)$, *whereas the guards* $\gamma(t)$ *are drawn inside the rectangles. The order among the children of a node is always from left to right as drawn in the figure. In the same way as the DAG decomposition of this graph in Example 6.6.3, the underlying structure of the Kelly decomposition is again a DAG.*

Example 6.7.3 *As a second example, consider the digraph and associated Kelly decomposition in Figure 6.7. Condition 4 of the previous definition states that if we*

---

* Named after a famous Australian outlaw in reference to the cops and robber game characterization established in the following.

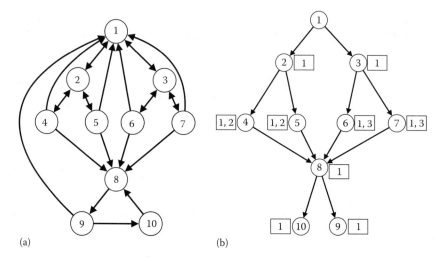

FIGURE 6.6: (a) The digraph from Figure 6.1 and (b) a corresponding Kelly decomposition.

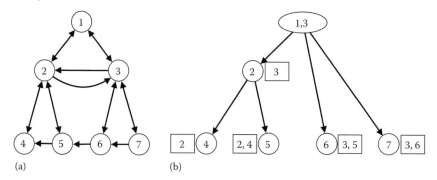

FIGURE 6.7: (a) The digraph from Figure 6.1 and (b) an associated Kelly decomposition.

consider a node s of a Kelly decomposition of a digraph G and its children $t_1, \ldots, t_p$, then edges that start at a vertex occurring in the subdag rooted at any t_i can have their endpoint only in the same subdag, in subdags to the left of t_i, or in the bag or guard of its parent node s. For instance, the edge from 6 to 5 implies that 5 must occur in the guard of the node labeled by 6 in the example. However, as the order among the nodes labeled by (2), (6) and (7), respectively, is from left to right, the edge in G from 2 to 3 cannot be guarded this way, that is, we cannot simply add 3 to the guard of 2 in the Kelly decomposition but have to add the vertex 3 to the root node of the decomposition also.

The order among the children of a node in a Kelly decomposition yields a natural direction in which algorithms can proceed through a Kelly decomposition so that edges of G only point to parts of the graph that have already been processed. Even though it is not explicitly stated in [34], there is a linear order $<$ on $V(G)$ such that for each $s \in V(D)$ the order $<_s$ on its children is consistent with $<$. This is particularly useful for nodes with multiple parents.

Besides its definition by an explicit decomposition, Kelly width can be characterized in many different ways. We refer to [34] for proofs of the following equivalences. The next characterization we present extends the idea of vertex elimination to digraphs. Vertex elimination is a process of systematically removing vertices from a graph but adding edges to preserve reachability. The complexity measure we are interested in is the maximum out-degree of eliminated vertices.

Definition 6.7.4 (Directed elimination ordering [34]) *Let G be a digraph. A directed elimination ordering $\lhd$ is a linear ordering on $V(G)$. Given an elimination ordering $\lhd := (v_0, v_1, \ldots, v_{n-1})$ of G, we define $G_0^\lhd := G$ and $G_{i+1}^\lhd$ to be the graph obtained from $G_i^\lhd$ by deleting v_i and adding (if necessary) new edges (u, v) if $(u, v_i), (v_i, v) \in E(G_i^\lhd)$, and $u \neq v$. $G_i^\lhd$ is the directed elimination graph at step i according to $\lhd$.*

The width of an elimination ordering is the maximum over all i of the out-degree of v_i in $G_i^\lhd$.

Example 6.7.5 *A directed elimination ordering of the digraph in Figure 6.1a is $10, 9, 8, 4, 5, 2, 6, 7, 3, 1$. The width of this ordering is 2. For instance, at the time 8 gets eliminated, we have added the edge $(8, 1)$ and this is the only outgoing edge of 8 when it is eliminated. On the other hand, when 10 gets eliminated, the edge $(9, 8)$ is added to the graph so that when we eliminate 9, its out-degree is 2.*

The next characterization we present is based on a cops and robber game, called the *inert invisible reachability game*. The game is played in the same way as the invisible reachability game, that is, the robber can move along unguarded directed paths and the cops cannot see his current position. However, in this game, the robber is *inert*. This means that if the cops move from position C to C', then the robber can only move from his current position r if $r \in C'$, that is, if the cops would otherwise capture the robber. See [34] for details.

We will see in the following that the number of cops needed to catch a robber in this game in a robber-monotone strategy is exactly its Kelly width.

The last characterization we consider is a generalization, called partial k-DAGs, of partial k-trees [62], a concept in undirected graph theory equivalent to tree width. The class of k-trees can be viewed as a class of graphs generated by a generalization of how one might construct a tree. In the same way, k-DAGs are a class of digraphs generated by a generalization of how one might construct a DAG in a top-down manner.

Definition 6.7.6 ((Partial) k-DAG [34]) *The class of k-DAGs is defined recursively as follows:*

1. A complete digraph with k vertices is a k-DAG.

2. *A k-DAG with n+1 vertices can be constructed from a k-DAG $\mathcal{H}$ with n vertices by adding a vertex v and edges satisfying the following:*

   a. *At most k edges from v to $\mathcal{H}$.*

   b. *If X is the set of endpoints of the edges added in the previous subcondition, an edge from $u \in V(H)$ to v if $(u, w) \in E(H)$ for all $w \in X \setminus \{u\}$. Note that if $X = \emptyset$, this condition is true for all $u \in V(H)$.*

A partial k-DAG is a subgraph of a k-DAG.

The main structural result of this section is that the various measures introduced earlier are equivalent on digraphs.

Theorem 6.7.7 ([34]) *Let G be a digraph. The following are equivalent:*

1. *G has Kelly width at most $k + 1$.*

2. *G has a directed elimination ordering of width $\leq k$.*

3. *$k+1$ cops have a monotone winning strategy to capture an inert invisible robber.*

4. *G is a partial k-DAG.*

Finally, we mention a few further characterizations of Kelly width given in the literature. In [51], a concept of chordal digraphs was introduced and it was shown that Kelly-width k can be characterized by a natural subclass of chordal digraphs. Finally, in [45], the authors study forbidden minors for Kelly width. It was already proved in [34] that Kelly width is invariant under butterfly minors. In [45], a different form of directed minors is introduced and several excluded minors for graph classes of small Kelly width are given.

6.7.2   Computing Kelly Decompositions

In the same way that DAG width and directed tree width coincide with undirected tree width on undirected graphs, Kelly width equals the tree width on any undirected graph. Hence, deciding the Kelly width of a digraph is NP-hard. However, unlike DAG width, a Kelly decomposition is a polynomially sized object, and therefore to decide whether a digraph G has Kelly width k, we can simply guess a Kelly decomposition of this width. Hence, the problem is in NP.

Theorem 6.7.8 *The problem to decide for a given digraph G and $k \in \mathbb{N}$ whether G has Kelly width at most k is NP-complete.*

It is an open problem whether deciding Kelly width is fixed-parameter tractable. In fact, it is not even known whether for each fixed k there is a polynomial time algorithm deciding whether a digraph G has Kelly-width k, that is, whether the problem is in the parameterized complexity class XP. However, explicit algorithms for Kelly width are known.

Theorem 6.7.9 ([34]) *The Kelly width of a graph with n vertices can be determined in time $\mathcal{O}^*(2^n)$ and space $\mathcal{O}^*(2^n)$, or in time $\mathcal{O}^*(4^n)$ and polynomial space, where $\mathcal{O}^*(f(n))$ means that polynomial factors are suppressed.*

Furthermore, the Kelly width of a digraph can be approximated up to a logarithmic factor.

Theorem 6.7.10 ([44]) *There exists a polynomial time approximation algorithm that, given a digraph G, computes a Kelly decomposition of G, whose width is $\mathcal{O}\left(\log^{\frac{3}{2}} n \cdot \mathrm{kellyw}(G)\right)$.*

Finally, for small values of k, efficient and explicit algorithms for deciding the Kelly width and computing corresponding decompositions were given, for example, in [52].

6.8 Directed Path Width

Around the same time as directed tree width, Robertson et al. also introduced the concept of directed path width. Directed path width is a straightforward generalization of the concept of undirected path width to digraphs.

Definition 6.8.1 *Let G be a digraph. A* directed path decomposition *is a pair (P, β), where $P = (v_1, \ldots, v_s)$ is a directed path and $\beta : V(P) \to \mathcal{P}(V(G))$ such that*

1. *For every $v \in V(G)$, the set $\beta^{-1}(v) := \{v_i \in V(P) : v \in \beta(v_i)\}$ is connected in P and nonempty*

2. *For every edge $e = (u, v) \in E(G)$, there are $i \leq j$ such that $u \in \beta(v_j)$ and $v \in \beta(v_i)$*

The width *of (P, β) is $\max\{|\beta(v_i)| : 1 \leq i \leq s\}$ and the* directed path-width $\mathrm{dpw}(G)$ *of a digraph is the minimum width of any of its directed path decompositions. A class $\mathcal{C}$ of graphs has bounded directed path width if there is a constant $k \in \mathbb{N}$ such that $\mathrm{dpw}(G) \leq k$ for all $G \in \mathcal{C}$.*

It is convenient to represent a directed path decomposition (P, β), where $P = (v_1, \ldots, v_k)$, simply by the sequence $(\beta(v_1), \ldots, \beta(v_k))$.

Example 6.8.2 *Consider again the digraph in Example 6.4.2 illustrated in Figure 6.1. A corresponding directed path decomposition of width 3 is shown in Figure 6.8.*

Intuitively, whereas directed tree width is characterized by the visible strong component game, directed path width is characterized by the invisible reachability game. This game is monotone, that is, the same number of cops are needed for robber- and cop-monotone strategies as for general strategies.

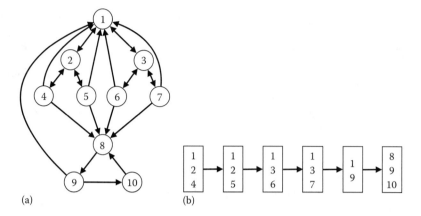

(a)   (b)

FIGURE 6.8:   (a) The digraph from Figure 6.1 and (b) an associated directed path decomposition.

Theorem 6.8.3 ([6,33]) *The invisible reachability cops and robber game on digraphs is monotone and the number of cops required to catch the robber on a digraph equals its directed path width.*

The problem of computing directed path decompositions has been considered in [46,67].

Theorem 6.8.4 ([46,67])

1. *There is an algorithm that, given a digraph G and $k \in \mathbb{N}$ as input, computes a directed path decomposition of G of width k, if it exists, in time $\mathcal{O}(|G|^{k+1} \cdot |E(G)|)$.*

2. *There is an algorithm computing a directed path decomposition of a digraph G of optimal width in time $\mathcal{O}^*(1.89^n)$, where $\mathcal{O}^*$ means that polynomial factors are suppressed.*

It is still open whether computing optimal path decompositions is fixed-parameter tractable.

6.9   Clique Width and Bi-Rank Width

Clique width is a width measure introduced in [15,17], first as a type of graph grammars and then as an explicit decomposition-based concept. Already in its original definition, it was defined for directed graphs. However, unlike all other measures considered so far, it is a measure that also applies to very dense graphs and digraphs and as a consequence is not closed under taking subgraphs. This makes it a very different type of width measure than what we have seen earlier.

Definition 6.9.1 (Labeled graph) *Let Γ be a finite set of colors. A Γ-labeled digraph is a digraph G together with a function $\gamma : V(G) \to \Gamma$ associating with every vertex a color.*

Definition 6.9.2 (Γ-expression) *Let Γ be a finite set of colors. The set of* clique expressions over Γ, *or Γ-expressions, is inductively defined as follows:*

 1. $\mathbf{i}$ *is a Γ-expression for all $i \in \Gamma$.*

 2. *If $i \neq j \in \Gamma$ and φ is a Γ-expression, then so are* $\mathrm{edge}_{i \to j}(\varphi)$ *and* $\mathrm{rename}_{i \to j}(\varphi)$.

 3. *If φ_1, φ_2 are Γ-expressions, then so is $(\varphi_1 \oplus \varphi_2)$.*

A Γ-expression φ generates a digraph $G(\varphi)$ colored by colors from Γ as follows. The k-expression $\mathbf{i}$ generates a digraph with one vertex colored by the color i and no edges.

The expression $\mathrm{edge}_{i \to j}$ is used to add edges. If φ is a Γ-expression generating the colored digraph $G := G(\varphi)$, then $\mathrm{edge}_{i \to j}(\varphi)$ defines the digraph H with $V(H) := V(G)$ and

$$E(H) := E(G) \cup \big\{ (u, v) : u \text{ has color } i \text{ and } v \text{ has color } j \big\}.$$

Hence, $\mathrm{edge}_{i \to j}(\varphi)$ adds edges between all vertices with color i and all vertices with color j.

The operation $\mathrm{rename}_{i \to j}(\varphi)$ recolors the digraph. Given the digraph G generated by φ, the Γ-expression $\mathrm{rename}_{i \to j}(\varphi)$ generates the digraph obtained from G by giving all vertices that have color i in G the color j in H. All other vertices keep their color.

Finally, if φ_1, φ_2 are Γ-expressions generating colored digraphs G_1, G_2, respectively, then $(\varphi_1 \oplus \varphi_2)$ defines the disjoint union of G_1 and G_2.

A digraph G has directed clique width k if there is a set Γ of size k and a Γ-expression that generates the colored digraph G, γ for some coloring $\gamma : V(G) \to \Gamma$. As before, a class $\mathcal{C}$ of digraphs has bounded clique width if there is a constant k such that the clique width of all $G \in \mathcal{C}$ is at most k.

Example 6.9.3 *Consider again the digraph depicted in Figure 6.1. A clique expression over the colors $\Gamma := \{\bot, a, b, c, d\}$ is given in Figure 6.9, where we have made some obvious abbreviations to reduce the size of the figure.*

For undirected graphs, Seymour and Oum [58] introduced a concept of *rank decompositions* and an associated *rank width* of undirected graphs and showed that this, algebraically defined, measure is equivalent to clique width in the sense that a class of undirected graphs has bounded clique width if, and only if, it has bounded rank width. Rank width is closely related to submodular functions, and using algorithms for minimizing submodular functions, explicit algorithms for computing rank decompositions and therefore also clique expressions have been developed. The best algorithm runs in cubic time and is due to [32].

Kanté et al. [39,40] generalized the concepts of rank width and introduced a concept of $\mathbb{F}$-rank width and $\mathbb{F}$-bi-rank width, for a field $\mathbb{F}$, which specialize to digraphs.

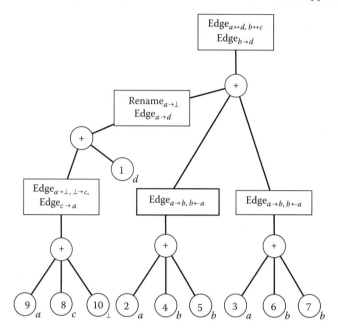

FIGURE 6.9:   A clique expression of width 5 generating the digraph of Figure 6.1.

With these concepts, they were able to generalize characterizations of rank width by vertex minors to directed graphs and also translate the cubic time recognition algorithm to directed graphs.

Theorem 6.9.4 ([40]) *For fixed k, there exists a cubic time algorithm that, for a directed graph G, either outputs a $\mathbb{F}$-rank decomposition of width k or confirms that the $\mathbb{F}$-rank width of G is larger than k.*

Directed clique width with its equivalent representation by bi-rank width provides a very interesting width measure for digraphs. As explained in the introduction, it is somewhat different in spirit than the other width measures discussed so far. For instance, it is not preserved under subgraphs (i.e., a subgraph can have much higher clique width than the graph itself) and is not a measure for connectivity but for *homogeneity* in some sense. Also, if a class $\mathcal{C}$ of digraphs has bounded directed clique width, then the class of underlying undirected graphs, that is, the class of undirected graphs obtained from digraphs in $\mathcal{C}$ by forgetting the edge orientation, has bounded clique width. This is very different from the other width measures where for classes of digraphs of bounded directed path width, for example, the class of underlying undirected graphs do not necessarily have bounded tree or clique width.

All of these differences are not surprising, given that bi-rank width was not introduced as a connectivity measure exceeding tree or clique width. Algorithmically, the main advantage of directed clique width/bi-rank width is that every digraph property definable in *monadic second-order logic* can be decided in polynomial time (in fact, linear time plus the time needed to construct the decomposition) on any class of digraphs of bounded directed clique width. This makes bi-rank width by far the directed width measure with the broadest algorithmic applications.

Theorem 6.9.5 ([16]) *Let C be a class of digraphs of bounded directed clique width. Then every graph property definable in MSO can be decided on C in polynomial time, in fact in linear time plus the time (currently cubic) to compute a decomposition.*

In other words, model checking for MSO is fixed-parameter tractable with the parameter being the size of the formula and the clique width.

Another interesting property of $\mathbb{F}$-rank width and bi-rank width is that they are preserved under directed vertex minors defined in [40]. As directed vertex minors form a well-quasi-ordering, this implies that the classes of bi-rank width k can be characterized by a finite set of forbidden vertex minors. This yields a strong structural description of bi-rank width that is not known for any other decomposition-based directed width measure.

6.10   Further Proposals of Directed Decompositions

We briefly mention some other width measures considered in the literature. In [28], the authors systematically study width parameters with the aim of assessing their algorithmic applications. Toward this goal, they introduce a range of further measures mostly for technical purposes. We refer to [28] for details. An old width measure studied in [28] is the concept of *cycle rank* going back to Eggan and Büchi in the 1960s. This measure is particularly useful in formal language theory and has recently been applied again in this context. We refer to [28] and references therein for details.

6.11   Nowhere Crownful Classes of Digraphs

All width measures defined in the previous sections are measures based on a concept of decomposition of digraphs and are often characterized by a cops and robber game. Furthermore, with the exception of clique width and bi-rank width, the class of acyclic digraphs has small width with respect to all these measures, usually 0 or 1. As we will see in the following, this limits the algorithmic applications of these measures quite dramatically for certain types of problems such as directed dominating sets and others of this form.

In general, the assumption that acyclic digraphs are *simple* and therefore should have small width seems obvious enough when interested in problems where strong connectivity is an important factor, but for many other purposes, this assumption is debatable. For instance, it is not hard to see that the directed dominating set and other problems are intractable even on DAGs (see, e.g., [28]). To motivate the next structural parameter for classes of directed graphs, we present the reduction for directed dominating sets.

Definition 6.11.1 (Directed Dominating Set) *Let G be a digraph. A directed dominating set in G is a set $X \subseteq V(G)$ such that for all $u \in V(G) \setminus X$ there is a $v \in X$ with $(v, u) \in E(G)$.*

The DIRECTED DOMINATING SET *problem is the problem to decide for a given digraph G and $k \in \mathbb{N}$ whether G contains a directed dominating set of size $\leq k$.*

It is well known that for undirected graphs, the dominating set problem is NP-complete and hard for the parameterized complexity class $W[2]$. The following reduction shows that the same hardness results apply to the directed case. For any undirected graph G, let G_d be the digraph with vertex set $V(G_d) := \{r\} \,\dot\cup\, \{v, v' : v \in V(G)\}$ and edge set

$$E(G_d) := \{(r, v), (v, v') : v \in V(G)\} \cup \{(v, u') : \{v, u\} \in E(G)\}.$$

Now, a set $X \subseteq V(G)$ is a dominating set in G if, and only if, the set $X' := \{r\} \cup \{v : v \in X\}$ is a directed dominating set in G_d. As a consequence, we obtained the following theorem.

Theorem 6.11.2 *The directed dominating set problem is NP-complete and $W[2]$-hard even for DAGs.*

An *oriented bipartite graph* is a directed graph $G := (A \dot\cup B, E)$ whose vertex set is partitioned into two sets A and B and $E \subseteq A \times B$. The previous construction shows that the directed dominating set problem is $W[2]$-hard even on nearly bipartite digraphs, as $G_d - r$ is an oriented bipartite graph. Similar constructions can be used to show hardness for various other problems on DAGs and nearly bipartite DAGs. See Section 6.17 as well as [18,29,41,48] and references therein. As, with the exception of directed clique width, all previously discussed width measures have small width on the class of DAGs, these width measures are of little use for problems such as directed dominating sets.

Motivated by these examples, in [49], an entirely different approach for defining classes of digraphs is pursued. Instead of trying to generalize undirected tree width to directed graphs, the classes defined in [49] generalize the concepts of classes defined by excluded minors or more, generally, nowhere dense classes of undirected graphs introduced in [54] to directed graphs. See [30] for a recent survey of nowhere dense classes and [53] for a comprehensive treatment of the subject.

The classes of digraphs introduced in [49], called *nowhere crownful*, are based on excluding *crowns* as *directed minors*. Contrary to undirected graphs, where there is a universally accepted notion of minor, for digraphs, no such general concept is known. Several proposals of minors have been made in the literature, including butterfly minors in [36] and other proposals, for instance, in [43,45]. In [49], the following concept of a minor is used, which generalizes the definition of undirected minors based on *models*.

Definition 6.11.3 *A digraph H has a* directed model *in a digraph G if there is a function* δ *mapping vertices v* ∈ *V(H) of H to subgraphs* δ(v) ⊆ *G and edges e* ∈ *E(H) to edges* δ(e) ∈ *E(G) such that*

- *If v* ≠ *u, then* δ(v) ∩ δ(u) = ∅

- *If e* := *uv, then* δ(e) *has its start point in* δ(u) *and its end in* δ(v)

For v ∈ *V(H), we set*

$$\text{in}(\delta(v)) := V(\delta(v)) \cap \bigcup_{e:=(u,v)\in E(H)} V(\delta(e))$$

and

$$\text{out}(\delta(v)) := V(\delta(v)) \cap \bigcup_{e:=(v,w)\in E(H)} V(\delta(e))$$

and require that

- *There is a directed path in* δ(v) *from any u* ∈ in(δ(v)) *to every u′* ∈ out(δ(v))

- *There is at least one source vertex* s_v ∈ δ(v) *that reaches every element of* out(δ(v))

- *There is at least one sink vertex* t_v ∈ δ(v) *that can be reached from every element of* in(δ(v))

Note that the last two items are only needed in case in(δ(v)) *or* out(δ(v)) *are empty. We write H* $\preceq^d$ *G if H has a directed model in G and say H is a* directed minor *of G. We call the sets* δ(v) *for v* ∈ *V(H) the* branch-sets *of the model.*

To define nowhere crownful classes of digraphs, the minor relation has to be parameterized by the radius of the graphs used in the model.

Definition 6.11.4 *A digraph H is a* (shallow) depth r minor *of a digraph G, denoted as H* $\preceq_r^d$ *G, if there exists a directed model* δ *of H in G in which for every v* ∈ *V(H) the maximal length of any path from a vertex in* in(δ(v)) *to a vertex in* out(δ(v)) *is bounded by r.*

Hence, a depth 0 minor is simply a subgraph and any minor of G is a depth n minor. This is analogous to the definition of depth r minors in undirected graphs, denoted by H $\preceq_r$ G.

The last concept we need to define nowhere crownful classes of digraphs are special digraphs called *crowns*.

Definition 6.11.5 *A* crown of order q, *for q* > 0, *is the graph* S_q *with*

- $V(S_q) := \{v_1, \ldots, v_q\} \dot\cup \{u_{i,j} : 1 \le i < j \le q\}$

- $E(S_q) := \{(u_{i,j}, v_i), (u_{i,j}, v_j) : 1 \leq i < j \leq q\}$

We call the vertices $v_1, \ldots, v_n$ *the* principal vertices *of the crown.*

We can now define nowhere crownful classes of digraphs.

Definition 6.11.6 *A class* C *of digraphs is* nowhere crownful *if for every* $r \in \mathbb{N}$, *there exists a* $q \in \mathbb{N}$ *so that for all* $G \in C$, $S_q \not\preceq_r^d G$.

Nowhere crownful classes of digraphs can also be characterized by another property, which is algorithmically very useful.

Definition 6.11.7 *Let* G *be a digraph and* $d \in \mathbb{N}$. *A set* $U \subseteq V(G)$ *is* d-scattered *if there is no* $v \in V(G)$ *and* $u_1 \neq u_2 \in U$ *with* $u, u' \in N_d^+(v)$.

Note that as $N_0^+(v) = \{v\}$, any subset of $V(G)$ is 0-scattered.

Definition 6.11.8 *A class* C *of directed graphs is* uniformly quasi-wide *if there are functions* $s : \mathbb{N} \to \mathbb{N}$ *and* $N : \mathbb{N} \times \mathbb{N} \to \mathbb{N}$ *such that for every* $G \in C$ *and all* $d, m \in \mathbb{N}$ *and* $W \subseteq V(G)$ *with* $|W| > N(d, m)$ *there is a set* $S \subseteq V(G)$ *with* $|S| \leq s(d)$ *and* $U \subseteq W$ *with* $|U| = m$ *such that* U *is* d-scattered in $G - S$. s, N *are called the* margin *of* C.
If s *and* N *are computable, then we call* C effectively uniformly quasi-wide.

Theorem 6.11.9 *A class* C *of digraphs is nowhere crownful if, and only if, it is directed uniformly quasi-wide.*

Nowhere crownful classes of digraphs are very general in that the graphs in them can be quite dense. However, the class of DAGs is not nowhere crownful and hence nowhere crownful classes of digraphs and digraph classes defined by any of the other width measures in this chapter are incomparable. The fact that nowhere crownful classes do not include the class of DAGs implies that the hardness results mentioned earlier for directed dominating sets do not apply to nowhere crownful classes. And indeed, as we will see in the following, these problems become tractable in nowhere crownful classes of digraphs.

Part II: Algorithmic Applications of Directed Width-Measures

In the previous part, we have introduced a range of different width measures or structure properties for directed graphs. In this part of the chapter, we will focus on corresponding algorithmic applications. In the individual sections, we focus on different types of problems. In Section 6.17, finally, we mention strong intractability results that have been obtained in the literature and comment on future directions in algorithmic applications of digraph structure parameters.

6.12   Disjoint Paths and General Linkage Problems

In this section, we explore algorithmic applications of directed width measures for problems related to routing or linkages in digraphs. In particular, we will show that problems such as the directed Hamiltonian path problem or the k-disjoint paths problem can be solved in polynomial time on classes of digraphs of bounded directed tree width. This was first shown by Johnson et al. in [36] and was the first significant application of the directed width measures considered previously.

The k-disjoint path problem on directed and undirected graphs is well known to be NP-complete. But whereas on undirected graphs the problem is fixed-parameter tractable, it is NP-complete on directed graphs even for $k = 2$, as shown by Fortune et al. [26].

Theorem 6.12.1 ([26]) *The following problem is NP-complete for all $k \geq 2$:*

> DIRECTED k-DISJOINT PATHS
> *Input:*   A digraph G and terminals $s_1, t_1, s_2, t_2, \ldots, s_k, t_k$.
> *Problem:* Find k pairwise internally vertex disjoint paths $P_1, \ldots, P_k$ such that P_i is from s_i to t_i for $i = 1, \ldots, k$.

Furthermore, as shown by Slivkins [65], the directed k-disjoint path problem is already $W[1]$-hard on DAGs. But it can be solved in polynomial time for every fixed value of k on any fixed class $\mathcal{C}$ of digraphs of bounded directed tree width. We prove the theorem here for weighted graphs. An *(edge-)weighted graph* is a pair (G, ω), where G is a digraph and $\omega : E(G) \to \mathbb{R}_{\geq 0}$.

Definition 6.12.2 *A* linkage *in a weighted digraph (G, ω) is a set $\mathcal{L}$ of pairwise internally vertex disjoint directed paths. The* order *$|\mathcal{L}|$ is the number of paths in $\mathcal{L}$; its* size *is $|V(\mathcal{L})|$, where $V(\mathcal{L}) := |\bigcup_{P \in \mathcal{L}} V(P)|$; and its* weight *is the sum of the edge weights of its paths. Let $\sigma := \{(s_1, t_1), \ldots, (s_k, t_k)\}$ be a set of k pairs of vertices in G. A σ-linkage is a linkage $\mathcal{L} := \{P_1, \ldots, P_k\}$ of order k such that P_i links s_i to t_i.*

The first algorithmic result we establish is the following theorem.

Theorem 6.12.3 ([36]) *Let (G, ω) be a weighted digraph and $\mathcal{T} := (T, \beta, \gamma)$ be a directed tree decomposition of G of width w. Let $k, l \geq 1$, $r \in \mathbb{R}$, and σ be a set of k pairs of vertices in G. Then it can be decided in time $\mathcal{O}(|V(D)|^{6(k+w)+3}(k+w)^{k+w})$ whether G contains a σ-linkage of size l and weight at most r.*

The theorem immediately implies the following corollary.

Corollary 6.12.4 *The Hamiltonian cycle, the Hamiltonian path, and, for all k, the k-disjoint path problem can be solved in polynomial time on any class $\mathcal{C}$ of digraphs of bounded directed tree width.*

Note that in the terminology of parameterized complexity, the previous result shows that the k-disjoint paths is in XP with parameter $k + w$. Unless FPT=$W[1]$, this cannot be improved to FPT in the parameter k for every fixed width w, as Slivkins [65] showed that the disjoint path problem is $W[1]$-hard already on DAGs.

To prove the theorem, we first establish a few simple observations. Let $\mathcal{L}$ be a linkage in a digraph G and let $A \subseteq V(G)$. We write $\mathcal{L}[A]$ for the linkage $\mathcal{L}$ restricted to the digraph $G[A]$, that is, $\mathcal{L}[A]$ contains one path for every maximal subpath of a path in $\mathcal{L}$ that lies completely in $G[A]$. A set $A \subseteq V(G)$ is *guarded* by a set $X \subseteq V(G)$ if A is the union of strong components of $G - X$.

Definition 6.12.5 *Let G be a digraph, $k, w \in \mathbb{N}$, and $Z \subseteq V(G)$. We say a set $S \subseteq V(G) - Z$ is*

1. *Z-normal if every directed path from and to a vertex in S that is not completely contained in $G[S]$ uses at least one vertex from Z*

2. *w-protected if there is a set $Z \subseteq V(G)$ with $|Z| \leq w$, such that S is Z-normal*

We also say that a linkage $\mathcal{L}$ of G is (k, w)-limited in S if $V(\mathcal{L}) \subseteq S$, and for every w-protected set $S' \subseteq S$, the linkage $\mathcal{L}[S']$ has order at most $k + w$.

Lemma 6.12.6 *Let G be a digraph, $k \in \mathbb{N}$, and $S \subseteq V(G)$, and let $\mathcal{L}$ be a linkage of order k in $G[S]$. Then for every $w \in \mathbb{N}$, $\mathcal{L}$ is (k, w)-limited in S.*

Proof. Let $P_1, \ldots, P_k$ be the paths belonging to $\mathcal{L}$ and let $S' \subseteq S$ be w-protected. Then, there is a set $Z \subseteq V(G)$, such that $|Z| \leq w$ and S' is Z-normal. It follows that if $P_i[S']$ has j weak components, then $|V(P_i) \cap Z| \geq j - 1$. Hence, $\mathcal{L}[S']$ has order at most $k + w$. □

The following two lemmas show that a directed tree decomposition of width w decomposes a digraph G into sets S of vertices, such that every linkage of order at most k in G has order at most $k + w$ in S.

Lemma 6.12.7 *Let G be a digraph and $T := (T, \beta, \gamma)$ be a directed tree decomposition of G of width w. Then, for every $t \in V(T)$, the set $S := \beta(T_t) \setminus \beta(t)$ is w-protected.*

Proof. Let $Z = \Gamma(t)$. We show that S is Z-normal. Because $|Z| \leq w$ and $Z \cap S = \emptyset$ (because of Property 2 of a directed tree decomposition), the result follows. So suppose for a contradiction that S is not Z-normal. Then, G has a directed path from and to a vertex in S that contains a vertex $v \in V(G) - S - Z$. Because v exists, t cannot be the root of T. Hence, there is a unique edge e with head t in T. Because $\beta(T_t)$ is a strong component of $G - \gamma(e)$ (because of Property 1 of a directed tree decomposition), $\gamma(e) \subseteq Z$, and $S = \beta(T_t) \setminus Z$, such a path cannot exist. □

Lemma 6.12.8 *Let G be a digraph; $A, B \subseteq V(G)$ be two disjoint sets of vertices, such that there is no edge from a vertex in B to a vertex in A; and $\mathcal{L}$ be a linkage in G with $V(\mathcal{L}) \subseteq A \cup B$ of order at most k. Then, $\mathcal{L}[A]$ and $\mathcal{L}[B]$ have order at most k.*

Proof. Toward a contradiction, suppose that one of $\mathcal{L}[A]$ or $\mathcal{L}[B]$ has order larger than k. Then, there is a directed path P in $\mathcal{L}$, which decomposes into more than one path in $\mathcal{L}[A]$ or $\mathcal{L}[B]$, respectively. But this is not possible, because $V(\mathcal{L}) \subseteq (A \cup B)$ and there are no edges from B to A in G. $\square$

Definition 6.12.9 *Let* (G, ω) *be a weighted digraph,* $S \subseteq V(G)$, *and* $k, w \in \mathbb{N}$. *We denote by* $\mathrm{Link}_{k,w}^{(G,\omega)}(S)$ *a set of triples* (σ, l, r), *where* σ *is a set of at most* $k + w$ *pairs of vertices in* S, *l is a natural number, and r is a real number, such that*

- *For every* σ *and* l *(as previously defined),* $\mathrm{Link}_{k,w}^{(G,\omega)}(S)$ *contains at most one triple* (σ, l, r)

- *If* $(\sigma, l, r) \in \mathrm{Link}_{k,w}^{(G,\omega)}(S)$, *then there is a* σ-*linkage in* $G[S]$ *of size* l *and weight* r

- *If* r *is the minimum weight of a* (k, w)-*limited* σ-*linkage* $\mathcal{L}$ *in* $G[S]$ *of size* l, *then* $(\sigma, l, r) \in \mathrm{Link}_{k,w}^{(G,\omega)}(S)$

The following lemma shows that for certain subsets $A, B \subseteq V(G)$, the set $\mathrm{Link}_{k,w}^{(G,\omega)}(A \cup B)$ can be efficiently computed from the sets $\mathrm{Link}_{k,w}^{(G,\omega)}(A)$ and $\mathrm{Link}_{k,w}^{(G,\omega)}(B)$.

Lemma 6.12.10 *Let* G *be a digraph,* $k, w \in \mathbb{N}$, *and* A *and* B *two disjoint sets of vertices of* G, *such that the sets* A *and* $A \cup B$ *are* w-*protected and either* B *is* w-*protected or* $|B| \leq w$. *Then,* $\mathrm{Link}_{k,w}^{(G,\omega)}(A \cup B)$ *can be computed from* $\mathrm{Link}_{k,w}^{(G,\omega)}(A)$ *and* $\mathrm{Link}_{k,w}^{(G,\omega)}(B)$ *in time* $\mathcal{O}(|A \cup B|^{2(k+w)+1}|A|^{2(k+w)+1}|B|^{2(k+w)}$ $(k+w)^{k+w}) = \mathcal{O}((|V(G)|)^{6(k+w)+2}(k+w)^{k+w})$.

Proof. To compute $\mathrm{Link}_{k,w}^{(G,\omega)}(A \cup B)$, we have to find the (k, w)-limited σ-linkage $\mathcal{L}$ of minimum weight in $G[A \cup B]$ for every set σ of at most $k + w$ pairs of vertices in $A \cup B$ and every natural number l with $1 \leq l \leq |A \cup B|$.

Because A and $A \cup B$ are w-protected, every such (k, w)-limited linkage consists of at most $k + w$ paths in A and $A \cup B$. The same holds for B, because either B is also w-protected or $|B| \leq w$. Hence, to compute $\mathrm{Link}_{k,w}^{(G,\omega)}(A \cup B)$, it is sufficient to consider linkages in $G[A \cup B]$, $G[A]$, and $G[B]$ of order at most $k + w$.

Let σ, σ_A, and σ_B be sets of at most $k + w$ pairs of vertices in $A \cup B$, A, and B, respectively. For $C \in \{A, B\}$, we denote by $s(\sigma_C)$ the set of vertices that are sources of σ_C but not sources of σ and by $t(\sigma_C)$ the set of vertices that are targets of σ_C but not targets of σ. We say that σ can be decomposed into σ_A and σ_B if every source of σ is either a source of σ_A or a source of σ_B and every target of σ is either a target of σ_A or a target of σ_B, and furthermore $|t(\sigma_A)| = |s(\sigma_B)|$ and $|t(\sigma_B)| = |s(\sigma_A)|$. In the following, assume that σ can be decomposed into σ_A and σ_B.

Let $\varphi_{A \to B}$ be a one-to-one mapping from $t(\sigma_A)$ to $s(\sigma_B)$ and let $\varphi_{B \to A}$ be a one-to-one mapping from $t(\sigma_B)$ to $s(\sigma_A)$, such that for every $t \in t(\sigma_A)$, it holds that $(t, \varphi_{A \to B}(t)) \in E(G)$, and for every $t \in t(\sigma_B)$, it holds that $(t, \varphi_{B \to A}(t)) \in E(G)$. We denote by $\omega(\sigma, \sigma_A, \sigma_B, \varphi_{A \to B}, \varphi_{B \to A})$ the sum of the edges in between

A and B, that is, $\omega(\sigma, \sigma_A, \sigma_B, \varphi_{A \to B}, \varphi_{B \to A})$ is equal to $\sum_{t \in t(\sigma_A)} \omega(t, \varphi_{A \to B}(t)) + \sum_{t \in t(\sigma_B)} \omega(t, \varphi_{B \to A}(t))$. Furthermore, we denote by $\sigma(\sigma_A, \sigma_B, \varphi_{A \to B}, \varphi_{B \to A})$ the set of all pairs (s, t) such that

1. s is a source of σ and t is a target of σ.

2. There is a sequence of pairs $(s_1, t_1), \ldots, (s_n, t_n)$, such that

    a. $s_1 = s$ and $t_n = t$

    b. For every i with $1 \le i \le n$, it holds that $(s_i, t_i) \in \sigma_A \cup \sigma_B$

    c. For every i with $1 \le i < n$, if $t_i \in A$, then $\varphi_{A \to B}(t_i) = s_{i+1}$, and similarly if $t_i \in B$, then $\varphi_{B \to A}(t_i) = s_{i+1}$

We denote by $\omega(\sigma, \sigma_A, \sigma_B)$ the smallest weight $\omega(\sigma, \sigma_A, \sigma_B, \varphi_{A \to B}, \varphi_{B \to A})$ over any $\varphi_{A \to B}$ and $\varphi_{B \to A}$ such that $\sigma = \sigma(\sigma_A, \sigma_B, \varphi_{A \to B}, \varphi_{B \to A})$.

It is now straightforward to check that there is a $(\sigma, l, r) \in \text{Link}_{k,w}^{(G,\omega)}(A \cup B)$ if and only if r is the smallest weight such that there is a $(\sigma_A, l_A, r_A) \in \text{Link}_{k,w}^{(G,\omega)}(A)$ and a $(\sigma_B, l - l_A, r - r_A - \omega(\sigma, \sigma_A, \sigma_B)) \in \text{Link}_{k,w}^{(G,\omega)}(B)$ such that σ can be decomposed into σ_A and σ_B. Hence, $\text{Link}_{k,w}^{(G,\omega)}(A \cup B)$ can be computed by going over all choices of σ, l, σ_A, l_A, σ_B, $\varphi_{A \to B}$, and $\varphi_{B \to A}$.

The running time of the algorithm follows immediately from the earlier claim, because there are at most $|A \cup B|^{2(k+w)}$, $|A \cup B|$, $|A|^{2(k+w)}$, $|B|^{2(k+w)}$, $|A|$, $(k+w)^{k+w}$, and $(k+w)^{k+w}$ possible choices for σ, l, σ_A, l_A, σ_B, $\varphi_{A \to B}$, and $\varphi_{B \to A}$, respectively. □

We are now ready to prove Theorem 6.12.3.

Proof. Let $T = (T, \beta, \gamma)$ be the given directed tree decomposition of width at most w.

The algorithm computes a set $\text{Link}_{k,w}^{(G,\omega)}(\beta(T_t))$ for every $t \in V(T)$ in a bottom-up manner starting with the leaves of T. Eventually, the set $\text{Link}_{k,w}^{(G,\omega)}(V(G))$ is computed for the root r of T, and because of Lemma 6.12.6, it contains the smallest weight of any linkage in G of order at most k and size l.

Let $t \in V(T)$. If t is a leaf, a set $\text{Link}_{k,w}^{(G,\omega)}(\beta(T_t))$ can be computed in time depending only on w (because $|\beta(T_t)| \le w$). Hence, let t be an inner node of T with children $t_1, \ldots, t_d$ in T and suppose that the sets $\text{Link}_{k,w}^{(G,\omega)}(\beta(T_{t_i}))$ have already been computed for every $1 \le i \le d$. Because of Lemma 6.4.4, we can assume that the children of t are ordered in such a way that there is no edge from a vertex in $\beta(T_{t_i})$ to a vertex in $\beta(T_{t_j})$ for $i > j$. Because of Lemmas 6.12.7 and 6.12.8, the sets $\bigcup_{i=1}^{j} \beta(T_{T_i})$ are w-protected for every j with $1 \le j \le d$, and because of the properties of a directed tree decomposition, this also holds for the sets $\beta(T_{t_i})$. By repeated application of Lemma 6.12.10, the sets $\text{Link}_{k,w}^{(G,\omega)}\left(\bigcup_{i=1}^{j} \beta(T_{T_i})\right)$ for every $1 \le j \le d$ can be computed one by one in time $\mathcal{O}(d \cdot |V(D)|^{6(k+w)+2}(k+w)^{k+w})$.

Furthermore, because $|\beta(t)| \le w$, we can compute a set $\text{Link}_{k,w}^{(G,\omega)}(\beta(t))$ in time depending only on w. Again using Lemma 6.12.10, we can compute $\text{Link}_{k,w}^{(G,\omega)}(\beta(T_t))$

from $\text{Link}_{k,w}^{(G,\omega)}\left(\bigcup_{j=1}^{d}\beta(T_{t_j})\right)$ and $\text{Link}_{k,w}^{(G,\omega)}(\beta(t))$ in time $\mathcal{O}(|V(D)|^{6(k+w)+2}$ $(k+w)^{k+w})$. The overall running time of the algorithm then follows because $|V(T)| \leq |V(G)|$. □

As mentioned earlier, by Slivkins' result [65], there is no hope (unless FPT= W[1]) to improve the previous result to fixed-parameter tractability with parameter k (for fixed classes $\mathcal{C}$ of bounded directed tree width). However, using dynamic programming over directed path decompositions of tournaments and corresponding obstructions in case of tournaments of large directed path width, Fomin and Pilipczuk [24] showed that the k-edge disjoint path problem is fixed-parameter tractable in tournaments. In fact, they showed much more general results.

We close this section by mentioning another result that relies on Theorem 6.12.3 and Theorem 6.4.27.

Theorem 6.12.11 ([35]) *For every fixed $k \geq 1$, there is a polynomial time algorithm for deciding the following problem.*

QUARTER-INTEGRAL DISJOINT PATHS
  Input: A digraph G and terminals $s_1, t_1, s_2, t_2, \ldots, s_k, t_k$.
Problem:

- Find k paths $P_1, \ldots, P_k$ such that P_i is from s_i to t_i for $i = 1, \ldots, k$ and every vertex in G is in at most four of the paths, or

- Conclude that G does not contain disjoint paths $P_1, \ldots, P_k$ such that P_i is from s_i to t_i for $i = 1, \ldots, k$.

6.13   Query Evaluation in Graph Databases

In this section, we very briefly mention an application of directed width measures in the context of databases, taken from [4].

Let Σ be an alphabet. A *graph database* is essentially a labeled directed graph, where in this example labels are from Σ and occur on the edges of the graph. Let $\mathcal{L} \subseteq \Sigma^*$ be a language over Σ.

Let $\mathcal{D}$ be a graph database. A *regular path query* (RPQ) selects pairs of vertices connected by a (possibly not simple) path whose edge labels form a word in $\mathcal{L}$. Graph databases and RPQs have been studied intensively in the literature on databases. See references in [4]. A particularly interesting type of regular path queries are *simple regular path queries* (SRPQ), which select pairs of vertices connected by a simple path whose edge labels form a word in $\mathcal{L}$. Hence, whereas in the path for an RPQs a vertex may appear multiple times, this is forbidden for SRPQs. RPQs are computable in time polynomial in both query and data complexity. On the other hand, the evaluation of SRPQs is NP-complete even for fixed basic languages such as $(aa)^*$ or a^*ba^*. However, the problem can be solved in polynomial time for database graphs of bounded directed tree width.

Theorem 6.13.1 ([4]) *Let $k \geq 0$ and let C be a class of graph databases with directed tree width at most k. Then for every language $\mathcal{L} \subseteq \Sigma^*$, there is a polynomial time algorithm for evaluating the SRPQ defined by $\mathcal{L}$ in graphs from C.*

6.14   Foundations of Model Checking

In this section, we briefly mention some applications of directed width measures in the theory of verification. An important tool for hard- and software verification is an approach known as *model checking* [5,14]. In this approach, a process, for example, the description of a circuit, communication protocol, or even software, is modeled as a labeled transition system, that is, a labeled directed graph. The property that one wants to verify in the system is formalized by a temporal logic formula. The main algorithmic problem is then to evaluate the logical formulas in the transition system.

The logics commonly used in this area are *linear time temporal logic* (LTL) or branching time logics such as CTL and CTL*. Another prominent logic in this context is the modal μ-calculus L_μ introduced by Kozen in 1982. The μ-calculus is a particularly interesting logic for the theoretical foundations of model checking as it encompasses most other logics, in particular the logics LTL, CTL, CTL* mentioned earlier, but on the other hand seems to exhibit a good balance between expressive power and evaluation complexity. It is known that the problem to verify whether an L_μ-formula is true in a transition system is in NP ∩ coNP. It is a long-standing open problem whether the model-checking problem for L_μ is in P. L_μ model checking is polynomial time equivalent to solving a form of combinatorial games played on digraphs known as *parity games*. Again, the problem whether these games can be solved in polynomial time is an open problem.

However, as parity games are simply labeled directed graphs, directed width measures apply to parity games as well, and for special classes of parity games, polynomial time algorithms are known.

Theorem 6.14.1 ([8,34,56]) *For every class C of parity games of bounded DAG width, Kelly width, or directed clique width, there is a polynomial time algorithm for solving all parity games in C.*

Note that the previous result does not imply that L_μ model checking is polynomial time solvable on classes of transition systems of bounded DAG width as the translation from L_μ model checking to parity games increases the width of the transition system for all width measures studied here. Only very recently, more efficient algorithms for L_μ model checking on specific classes of transition systems defined by width measures have been developed.

Theorem 6.14.2 ([12]) L_μ *model checking is fixed-parameter tractable on any class of graphs of bounded Kelly width, where the parameter is the width and the size of the formula, provided the Kelly decomposition is part of the input.*

6.15 Boolean Networks

The last example of algorithmic applications of directed width measures we present is computing attractors in *Boolean networks*.

For our purpose here, a Boolean network consists of a set V of nodes and a transition function $f : 2^V \to 2^V$. A Boolean network describes a dynamical system where the state set is 2^V. That is, any combination of nodes that can hold at any given time describes a possible state of the network and the function f describes the transition from one state to the next.

Boolean networks were originally introduced by Kauffman in the study of metabolic systems but have since then received significant attention. An important algorithmic problem is to compute attractors in such a network. A state $s \in 2^V$ is called *recurrent* if there is some $k > 0$ with $f^k(s) = s$. If s is a recurrent state and k is minimal with $f^k(s) = s$, then once a network reaches s, the next k states repeat forever. This cyclic sequence of states is called an attractor.

Computing the set of all attractors of a network is computationally challenging. In [66], Tamaki studied Boolean networks of bounded directed path width and conducted experimental studies for computing attractors. For small values of w, he was able to compute the set of all attractors of Boolean networks of directed path width at most w of size exceeding by far the network sizes that traditional approaches can handle.

6.16 Directed Dominating Sets and Similar Problems

We close the part on algorithmic applications of directed width measures or directed structure properties by considering some classical graph algorithmic problems. In Section 6.11, we have already mentioned that the directed dominating set problem remains intractable on DAGs. However, it turns out that this problem and many other problems of the same kind become fixed-parameter tractable on nowhere crownful classes of digraphs. We refer to [49] for details.

Theorem 6.16.1 ([49]) *Let C be a class of directed graphs that is nowhere crownful. Then the directed (independent or unrestricted) dominating set problem, the dominating outbranching problem, and the independent set problem as well as their distance-d-versions are fixed-parameter tractable on C.*

As mentioned earlier, on any class of bounded directed clique width, every MSO definable problem is tractable.

Theorem 6.16.2 ([16]) *Let* C *be a class of digraphs of bounded directed clique width. Then every problem definable in* MSO *can be decided in polynomial time on* C.

As a consequence, problems such as directed dominating sets and a very wide range of other computational problems are fixed-parameter tractable on classes C of bounded directed clique width. However, the result in [16] is even more powerful. In fact, for a given formula $\varphi(X)$ of MSO with a free set variable, we can compute a minimal set in a digraph $G \in C$ satisfying the formula. Hence, the directed dominating set problem is not only fixed-parameter tractable but in fact solvable in polynomial time on classes C of bounded clique width.

6.17 Conclusion

In the previous sections, we have seen a range of algorithmic applications for directed width measures and nowhere crownful classes of digraphs. These examples show that some important but computationally intractable problems can be solved efficiently on classes of digraphs with some structural restriction such as bounded width with respect to specific directed width measures.

Encouraged by the initial results such as the polynomial time solvability of the k-disjoint paths problem (Theorem 6.12.3), several authors conducted systematic studies into the extent to which directed width measures can be applied to standard algorithmic digraph problems. See for example, [18,29,41,48]. The results obtained showed that none of the width measures such as directed tree width, DAG width, Kelly width, or even directed path width had very broad algorithmic applications comparable to undirected tree width, as many problems such as directed dominating sets, Kernels, and various disjoint paths problems remained fixed-parameter intractable on DAGs or digraphs that are close to DAGs. However, fixed-parameter algorithms for edge-disjoint paths problems on tournaments [24] or for dominating set problems on nowhere crownful classes show that these negative results may not be the final answer.

In fact, currently, the situation is that width measures such as directed tree width are very useful in the analysis of linkage and other *reachability* problems, whereas proposals such as nowhere crownful classes yield a key to attack dominating set problems. It would be worthwhile to combine the two approaches and, for instance, study directed tree decomposition of nowhere crownful graphs, or much simpler on planar digraphs.

We believe this to be a very promising direction in directed structure theory. For instance, the model-checking problem for the model μ-calculus we studied in Section 6.14 can easily be reduced to the model-checking problem for L_μ on planar digraphs. Hence, if one wanted to use an approach based on directed tree width and their obstructions in terms of excluded grids (see Section 6.4) to show that the modal μ-calculus is fixed-parameter tractable in general, then it would be enough to study directed tree width for planar digraphs. The same is true for the study of disjoint path problem, for instance, on planar digraphs or digraphs of bounded genus. In each of these cases, it would be enough to combine directed tree width and directed tree decompositions with other structural restrictions such as planarity. Results such as in [21] (for undirected graphs) show that one might be able to obtain much better decompositions on, say, planar digraphs, than in general.

References

1. I. Adler. Directed tree-width examples. *J. Comb. Theory Ser. B*, 97(5):718–725, 2007.

2. S. Arnborg, D. G. Corneil, and A. Proskurowski. Complexity of finding embeddings in a *k*-tree. *SIAM J. Algebr. Discr. Methods*, 8:277–284, 1987.

3. S. Arora and B. Barak. *Computational Complexity*. Cambridge University Press, 2009.

4. G. Bagan, A. Bonifati, and B. Groz. A trichotomy for regular simple path queries on graphs. In *Proceedings of the 32nd Symposium on Principles of Database Systems (PODS)*, pp. 261–272, 2013.

5. C. Baier and J.-P. Katoen. *Principles of Model Checking*. MIT Press, 2008.

6. J. Barát. Directed path-width and monotonicity in digraph searching. *Graphs Comb.*, 22(2):161–172, 2006.

7. D. Berwanger, A. Dawar, P. Hunter, and S. Kreutzer. DAG-width and parity games. In *Symposium on Theoretical Aspects of Computer Science (STACS)*, 2006.

8. D. Berwanger, A. Dawar, P. Hunter, S. Kreutzer, and J. Obržálek. DAG-width and parity games. *J. Comb. Theory Ser. B*, 102(4):900–923, 2012.

9. D. Berwanger, E. Grädel, L. Kaiser, and R. Rabinovich. Entanglement and the complexity of directed graphs. *Theor. Comput. Sci.*, 463:2–25, 2012.

10. D. Bienstock and P. Seymour. Monotonicity in graph searching. *J. Algorith.*, 12(2):239–245, 1991.

11. H. L. Bodlaender. Treewidth: Characterizations, applications, and computations. In F. V. Fomin, ed., *Graph-Theoretic Concepts in Computer Science*, vol. 4271 of *Lecture Notes in Computer Science*, pp. 1–14. Springer, 2006.

12. M. Bojanczyk, C. Dittmann, and S. Kreutzer. Decomposition theorems and model-checking for the modal μ-calculus. *ACM/IEEE. Symposium on Logic in Computer Science (LICS)*, 2014.

13. A. Brandstädt, V. B. Le, and J. Spinrad. *Graph Classes: A Survey*. SIAM Monographs on Discrete Mathematics and Applications. SIAM, 1999.

14. E. M. Clarke, O. Grumberg, and D. A. Peled. *Model Checking*. MIT Press, 1999.

15. B. Courcelle, J. Engelfriet, and G. Rozenberg. Handle-rewriting hypergraphs grammars. *J. Comput. Syst. Sci.*, 46:218–270, 1993.

16. B. Courcelle, J. Makowsky, and U. Rotics. Linear time solvable optimization problems on graphs of bounded clique-width. *Theory Comput. Syst.*, 33(2): 125–150, 2000.

17. B. Courcelle and S. Olariu. Upper bounds to the clique width of graphs. *Discr. Appl. Math.*, 1–3:77–114, 2000.

18. P. Dankelmann, G. Gutin, and E. J. Kim. On complexity of minimum leaf out-branching problem. *Discr. Appl. Math.*, 157(13):3000–3004, 2009.

19. R. Dechter. Bucket elimination: A unifying framework for probabilistic inference. In *Proceedings of the 12th Annual Conference on Uncertainty in Artificial Intelligence (UAI)*, pp. 211–219, 1996.

20. R. Diestel. *Graph Theory*, 3rd edn. Springer-Verlag, 2005.

21. F. Dorn, E. Penninkx, H. Bodlaender, and F. Fomin. Efficient exact algorithms on planar graphs: Exploiting sphere cut branch decompositions. In *ESA*, pp. 95–106, 2005.

22. R. G. Downey and M. Fellows. *Parameterized Complexity*. Springer, 1998.

23. J. Flum and M. Grohe. *Parameterized Complexity Theory*. Springer, 2006.

24. F. V. Fomin and M. Pilipczuk. Jungles, bundles, and fixed parameter tractability. In *Proceedings of the 24th Annual ACM-SIAM Symposium on Discrete Algorithms (SODA)*, pp. 396–413, 2013.

25. F. V. Fomin and D. M. Thilikos. An annotated bibliography on guaranteed graph searching. *Theor. Comput. Sci.*, 399(3):236–245, 2008.

26. S. Fortune, J. E. Hopcroft, and J. Wyllie. The directed subgraph homeomorphism problem. *Theor. Comput. Sci.*, 10:111–121, 1980.

27. R. Ganian, P. Hlinený, J. Kneis, D. Meister, and J. Obdrzálek, P. Rossmanith, and S. Sikdar. Are there any good digraph width measures? In *Parameterized and Exact Computation (IPEC)*, pp. 135–146, 2010.

28. R. Ganian, P. Hlinený, J. Kneis, A. Langer, J. Obdrzálek, and P. Rossmanith. On digraph width measures in parameterized algorithmics. In *IWPEC*, pp. 185–197, 2009.

29. R. Ganian, P. Hliněný, J. Kneis, A. Langer, J. Obržálek, and P. Rossmanith. On digraph width measures in parameterized algorithmics. Available on arxiv.org.

30. M. Grohe, S. Kreutzer, and S. Siebertz. Characterisations of nowhere dense graphs (invited talk). In *FSTTCS*, pp. 21–40, 2013.

31. M. Grohe, S. Kreutzer, and S. Siebertz. Deciding first-order properties of nowhere dense graphs. *CoRR*, abs/1311.3899, 2013.

32. P. Hlinený and S. Oum. Finding branch-decompositions and rank-decompositions. *SIAM J. Comput.*, 38(3):1012–1032, 2008.

33. P. Hunter. Losing the +1 or directed path-width games are monotone. Unpublished manuscript, Available from the authors webpage.

34. P. Hunter and S. Kreutzer. Digraph measures: Kelly decompositions, games, and ordering. *Theor. Comput. Sci. (TCS)*, 399(3), 2008.

35. K. Kawarabayashi, Y. Kobayashi, and S. Kreutzer. An excluded half-integral grid theorem for digraphs and the directed disjoint paths problem. In *Proceedings of the ACM Symposium on Theory of Computing (STOC)*, 2014.

36. T. Johnson, N. Robertson, P. D. Seymour, and R. Thomas. Directed tree-width. *J. Comb. Theory Ser. B*, 82(1):138–154, 2001.

37. T. Johnson, N. Robertson, P. D. Seymour, and R. Thomas. Excluding a grid minor in digraphs. Unpublished manuscript, 2001.

38. Ł. Kaiser, S. Kreutzer, R. Rabinovich, and S. Siebertz. Directed width measures and monotonicity of directed graph searching. Unpublished, 2014.

39. M. M. Kanté and M. Rao. Directed rank-width and displit decomposition. In *Workshop on Graph-Theoretical Concepts in Computer Science (WG)*, pp. 214–225, 2009.

40. M. M. Kanté and M. Rao. The rank-width of edge-coloured graphs. *Theory Comput. Syst.*, 52(4):599–644, 2013.

41. G. Kaouri, M. Lampis, and V. Mitsou. On the algorithmic effectiveness of digraph decompositions and complexity measures. *Discr. Optim.* 8(1):129-138, 2011.

42. K. Kawarabayashi and S. Kreutzer. An excluded grid theorem for digraphs with forbidden minors. In *ACM/SIAM Symposium on Discrete Algorithms (SODA)*, 2014.

43. I. Kim and P. D. Seymour. Tournament minors. ArXiv e-prints, arXiv:1206.3135, 2012.

44. S. Kintali, N. Kothari, and A. Kumar. Approximation algorithms for digraph width parameters. ArXiv preprint, arXiv:1107.4824, 2011.

45. S. Kintali and Q. Zhang. Forbidden directed minors and Kelly-width. ArXiv e-prints, arXiv:1308.5170, 2013.

46. K. Kitsunai, Y. Kobayashi, K. Komuro, H. Tamaki, and T. Tano. Computing directed pathwidth in o(1.89 n) time. In D. M. Thilikos and G. J. Woeginger, eds., *Parameterized and Exact Computation*, vol. 7535 of *Lecture Notes in Computer Science*, pp. 182–193. Springer, Berlin, Germany, 2012.

47. S. Kreutzer. Graph searching games. In K. R. Apt and E. Grädel, eds., *Lectures in Game Theory for Computer Scientists*, Chapter 7, pp. 213–263. Cambridge University Press, 2011.

48. S. Kreutzer and S. Ordyniak. Digraph decompositions and monotonocity in digraph searching). In *34th International Workshop on Graph-Theoretic Concepts in Computer Science (WG)*, 2008.

49. S. Kreutzer and S. Tazari. Directed nowhere dense classes of graphs. In *Proc. of the 23rd ACM-SIAM Symposium on Discrete Algorithms (SODA)*, 2012.

50. M. Lampis, G. Kaouri, and V. Mitsou. On the algorithmic effectiveness of digraph decompositions and complexity measures. *Discr. Optim.*, 8(1):129–138, 2011.

51. D. Meister and J. A. Telle. Chordal digraphs. In *35th International Workshop on Graph-Theoretic Concepts in Computer Science (WG)*, pp. 273–284, 2009.

52. D. Meister, J. A. Telle, and M. Vatshelle. Characterization and recognition of digraphs of bounded Kelly-width. In *33th International Workshop on Graph-Theoretic Concepts in Computer Science (WG)*, pp. 270–279, 2007.

53. J. Nešetřil and P. O. de Mendez. *Sparsity*. Springer, 2012.

54. J. Nešetřil and P. Ossona de Mendez. On nowhere dense graphs. *Eur. J. Comb.*, 32(4):600–617, 2011.

55. J. Obdržálek. Dag-width: Connectivity measure for directed graphs. In *Symposium on Discrete Algorithms (SODA)*, pp. 814–821, 2006.

56. J. Obdrzalek. Clique-width and parity games. In *Computer Science Logic (CSL)*, pp. 54–68, 2007.

57. S. Ordyniak. Gerichtete Graphen: Zerlegungen und Algorithmen. Master's thesis, Diplomarbeit (Master's thesis) Humboldt Universität zu Berlin, 2006.

58. S.-I. Oum and P. D. Seymour. Approximating clique-width and branch-width. *J. Comb. Theory Ser. B*, 96:514–528, 2006.

59. B. Reed. Tree width and tangles: A new connectivity measure and some applications. In R.A. Bailey, ed., *Surveys in Combinatorics*, pp. 87–162. Cambridge University Press, 1997.

60. B. Reed. Introducing directed tree-width. *Electron. Notes Discr. Math.*, 3:222–229, 1999.

61. N. Robertson and P. D. Seymour. Graph minors I–XXIII, 1982–2010. Appearing in *J. Comb. Theory Ser. B* from 1982 till 2010.

62. D. J. Rose. Triangulated graphs and the elimination process. *J. Math. Anal. Appl.*, 32:597–606, 1970.

63. M. A. Safari. D-width: A more natural measure for directed tree width. In *Proceedings of Mathematical Foundations of Computer Science (MFCS)*, vol. 3618 in *Lecture Notes in Computer Science*, pp. 745–756, 2005.

64. P. D. Seymour and R. Thomas. Graph searching and a min-max theorem for tree-width. *J. Comb. Theory Ser. B*, 58(1):22–33, 1993.

65. A. Slivkins. Parameterized tractability of edge-disjoint paths on directed acyclic graphs. In *European Symposium on Algorithms*, pp. 482–493, 2003.

66. H. Tamaki. A directed path-decomposition approach to exactly identifying attractors of Boolean networks. In *2010 International Symposium on Communications and Information Technologies (ISCIT)*, pp. 844–849, October 2010.

67. H. Tamaki. A polynomial time algorithm for bounded directed pathwidth. In P. Kolman and J. Kratochvíl, eds., *Graph-Theoretic Concepts in Computer Science*, vol. 6986 of *Lecture Notes in Computer Science*, pp. 331–342. Springer, Berlin, Germany, 2011.

Chapter 7

Betweenness Centrality in Graphs

Silvia Gago, Jana Coroničová Hurajová, and Tomáš Madaras

Contents

7.1 Introduction..233
7.2 General Properties...234
    7.2.1 Definition, Analogues, and Related Graph Invariants....................235
    7.2.2 Possible Values of the Betweenness Centrality and Relations with
             Graph Parameters...237
    7.2.3 Relations with Graph Operations......................................242
7.3 Betweenness-Uniform Graphs..244
7.4 Spectral Relations...251
7.5 Conclusion..254
References..255

7.1 Introduction

The central paradigm of network analysis relies on the fact that, the essential information to explain the functionality and the behavior of individuals within a complex network or the whole network itself is contained in its abstract structure, that is, the underlying graph model. Based on this assumption and reflecting the recent advances in graph theory and the development of specialized computer software, the contemporary analysis of complex networks (see [33] for a general introduction) uses a wide variety of quantitative methods for modeling, measuring their characteristics and visualization. Along with determining statistical properties, the global structure, or the detection of network communities, one of the basic objectives is the identification of the most important objects within a network. The common way to express the importance of network objects is to quantify it by evaluating a specific centrality index on the vertices of the graph representing a given network, where the vertices with the higher values of centrality are perceived as being the more important. The centrality index, as a real-valued function on the vertices of a graph, is a structural descriptor, that is, it is invariant under graph isomorphism.

In practice, the choice of a specific centrality index depends on our interpretation of importance. For instance in graphs representing social networks, where edges represent relations of personal knowledge or friendship, the vertex degree is an appropriate quantitative measure as highly important actors usually have many direct ties to others. However, for locating the best place for a fire station in a city, one should look for a vertex in the city-graph with the highest reciprocal of the eccentricity, and, the reciprocal of total sum of all distances from a fixed vertex (known as the closeness centrality) is suitable for best locations of facilities which have to be as close as possible to every building. In many situations, the network to be analyzed consists on objects whose mutual relations reflect an ability of objects to communicate in direct way; among the typical examples, one can mention mobile networks or the Internet. The most important objects in such networks are those ones lying on many communication paths; as the information spreading within a network propagates mainly through routes which are as short as possible, the importance of vertices in a network graph relies on their occurrence at geodesic paths. This approach yields a centrality index known as the *betweenness centrality*. It was first introduced by Freeman [15] (and independently by Anthonisse [1]) and, since then, it became one of the most popular and frequently used centrality index. Despite the success of betweenness centrality in network analyses and developing progress in algorithmic aspects of quantitative computation for this invariant (see, e.g., [2,5,19]), its graph-theoretical properties started to be systematically investigated only recently. Our aim is to give a detailed overview of results on betweenness centrality and related indices. In Section 7.2, we concentrate on general properties of betweenness centrality and its relation to other graph invariants as well as on its estimates; we also describe the most common analogues of betweenness centrality and their connections with the original betweenness centrality concept. Section 7.3 deals with properties of graphs whose vertices have the same value of betweenness centrality. We present several constructions of such graphs (mostly based on the local join operation) as well as wide graph families whose members do have this property (namely, distance-regular and strongly regular graphs). The final Section 7.4 reviews the results on the relations of betweenness centrality and graph spectra.

7.2   General Properties

The graph terminology used in the chapter is taken from the book [32]. For a graph G, $V(G) = V$ and $E(G) = E$ denote its vertex and edge set, respectively; $n = |V(G)|$ is the order and $m = |E(G)|)$ is the size of the graph; and the set of all 2-element subsets of V is denoted by $\binom{V}{2}$. Given two vertices $u, v \in V$, the distance $d(u, v)$ is the length of the shortest path from u to v (if u, v belong to different components of G, we set $d(u, v) := +\infty$); the maximum distance between two vertices of G is the diameter of G and is denoted by $D = \mathrm{diam}(G)$. The set of vertices of G at distance k from a vertex $x \in V$ is denoted by $N_k(x) = \{v \in V : d(u, x) = k\}$, for any $1 \leq k \leq D$. The total number of vertices at distance 1 from a vertex $x \in V$, $|N_1(x)|$, is called the degree of the vertex, and it is denoted by $\deg(x)$, and $\Delta = \Delta(G)$ and $\delta = \delta(G)$ stand for the minimum and the maximum degree of the graph G.

The complete graph on p vertices and the complete bipartite graph with parts having p and q vertices are denoted as K_p and $K_{p,q}$, respectively. The graph $K_{1,r}$ is referred as an r-star. The symbol $\overline{G}$ stands for the complement of a graph G.

7.2.1 Definition, Analogues, and Related Graph Invariants

The original definition of vertex betweenness centrality given by Freeman in [15] is the following: given an undirected graph $G = (V, E)$ and three distinct vertices u, $v, x \in V$, let $\sigma_{u,v} \neq 0$ be the number of shortest paths between u and v in G, and let $\sigma_{u,v}(x)$ be the number of shortest u–v paths containing x, and then the betweenness centrality of the vertex x is

$$B(x) = \sum_{\{u,v\} \in \binom{V}{2}} \frac{\sigma_{u,v}(x)}{\sigma_{u,v}}. \tag{7.1}$$

The ratio $\delta_{u,v}(x) = \sigma_{u,v}(x)/\sigma_{u,v}$ is called the *pairwise dependency* of x with respect to u, v.

(Note that the definition of vertex betweenness centrality can be extended also to disconnected graphs by adopting the convention that vertices from different components do not contribute to the previously mentioned sum.)

The majority of authors prefer to use the summation over unordered pairs of vertices in (7.1); but sometimes, as in [17] or [18], the ordered pairs are considered. The extension of the definition of betweenness centrality for digraphs (where the values $\sigma_{u,v}$ and $\sigma_{v,u}$ may differ) has been considered in [34]. However, we must point out that for undirected graphs, it yields just twice the value given by (7.1). In the following, the results of [17] and [18] that refer to betweenness centrality values are adjusted to reflect, in the commonly used definition of betweenness centrality, the summation over unordered pairs of vertices.

From this definition of betweenness centrality for a vertex, other parameters are also studied: the *average vertex betweenness centrality* of the graph, defined as

$$\overline{B}(G) = \frac{1}{n} \sum_{x \in V} B(x), \tag{7.2}$$

and the *maximum vertex betweenness centrality*, defined as

$$B^{max} = \max\{B(x), \ x \in V\}. \tag{7.3}$$

Girvan and Newman [20] describe an edge analogue of this definition and introduce the edge betweenness centrality of an edge as the fraction of shortest paths between pairs of vertices that run along it, that is, the *betweenness for an edge* $e \in E$ is the sum

$$B(e) = \sum_{\{u,v\} \in \binom{V}{2}} \frac{\rho_{u,v}(e)}{\sigma_{u,v}}. \tag{7.4}$$

where $\rho_{u,v}(e)$ denotes the number of shortest u–v paths containing e. Similar definitions can be made for the *average edge betweenness centrality of a graph G of size m*, defined as

$$\overline{B}_e(G) = \frac{1}{m} \sum_{e \in E} B(e), \qquad (7.5)$$

and the *maximum edge betweenness centrality*

$$B_e^{max} = \max_{e \in E} B(e),$$

related with other expansion parameters of the graph (see Section 7.4).

Besides of the classical vertex betweenness centrality definition, some authors (see [10]) deal rather with a related vertex centrality, the *adjusted betweenness centrality* whose definition is based on the edge betweenness centrality concept. The adjusted betweenness centrality of a vertex x is the sum of edge betweenness centralities of edges incident with x:

$$B^*(x) = \sum_{x \in e} B(e). \qquad (7.6)$$

The standard and adjusted vertex betweenness centralities are mutually related by the formula

$$B^*(x) = 2B(x) + n - 1.$$

There are several centrality indices derived from the original betweenness centrality that take into account a reduction of the influence of long paths. Everett and Borgatti [14] consider the variant where the betweenness centrality sum concerns only pairs of vertices with bounded distance; in this way, they define *k-betweenness centrality* of a vertex x in G as the sum

$$B_{G;k}(x) = \sum_{\substack{\{u,v\} \in \binom{V}{2} \\ d(u,v) \le k}} \frac{\sigma_{u,v}(x)}{\sigma_{u,v}}. \qquad (7.7)$$

The contribution of long paths to betweenness centrality can be also reduced by penalizing the pairwise dependencies by reciprocals of vertex distances; in this spirit, Everett and Borgatti define *length-scaled betweenness centrality* of x in G as

$$B_{G;dist}(x) = \sum_{\{u,v\} \in \binom{V}{2}} \frac{1}{d(u,v)} \cdot \frac{\sigma_{u,v}(x)}{\sigma_{u,v}}. \qquad (7.8)$$

Another scaled variant, mentioned in [6], is *linearly scaled betweenness centrality* which counts, as a penalization factor, the relative distance from the source of a geodesic path to a vertex whose betweenness centrality is considered:

$$B_{G;lin}(x) = \sum_{\{u,v\} \in \binom{V}{2}} \frac{d(x,v)}{d(u,v)} \cdot \frac{\sigma_{u,v}(x)}{\sigma_{u,v}}. \qquad (7.9)$$

The vertex and edge betweenness centrality values are also closely related to another global distance-based invariants, as the *Wiener index* $W(G) = \sum_{\{u,v\} \in \binom{V}{2}} d(u,v)$ and the *mean distance* $\bar{l}(G) = W(G)/\binom{n}{2}$. Namely, it holds

$$\sum_{u \in V} B^*(u) = 2W(G). \tag{7.10}$$

Furthermore, for the average vertex betweenness centrality, it holds

$$\bar{B}(G) = \frac{(n-1)}{2}(\bar{l}(G) - 1). \tag{7.11}$$

(Note that, if the sum in the betweenness centrality is taken over an ordered pair of vertices, the average betweenness centrality is twice this value; see [11].)

7.2.2 Possible Values of the Betweenness Centrality and Relations with Graph Parameters

Firstly, we study possible values of the betweenness centrality for vertices, as it is clear that for any vertex $x \in V$ of a graph G of order n, the betweenness centrality of x lies between 0 and $\binom{n-1}{2}$, where the value 0 is reached if and only if all neighbors of x induce a clique in G, and the value $\binom{n-1}{2}$ exactly in the case when $G \cong K_{1,n-1}$ and x is its central vertex.

It is straightforward that, for a given positive rational number p/q, we can construct a graph with a vertex with betweenness centrality p/q: take a copy of K_p, a copy of K_{q-1} and an edge uv, and add new edges between vertices of K_p and K_{q-1}; also, join u with all vertices of K_p and similarly v with all vertices of K_{q-1} (see Figure 7.1). In the resulting graph, the betweenness centrality of u is equal to p/q.

On the other hand, the problem to determine, for a given integer n, all possible values of betweenness centrality of vertices of graphs of order n, is open. Furthermore,

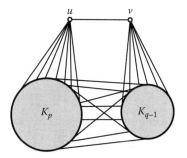

FIGURE 7.1: A construction of graph with prescribed betweenness centrality of its vertex.

it seems that the characterization of graphical betweenness centrality sequences (analogues of degree sequences) is very hard.

Next, we state general lower and upper bounds for vertex betweenness centrality in terms of other graph characteristics. First, we describe estimates based on properties of vertex neighborhoods. Let u, v be nonadjacent vertices of a graph G with a common neighbor x, let $M_{u,v} = \{y \in V : uy \notin E \vee vy \notin E\}$ be the set of all vertices that are not common neighbors of u and v (note that $u, v \in M_{u,v}$); and let $N_{u,v}(x)$ be the set of all vertices of $M_{u,v}$ that are neighbors of x; denote $\nu_{u,v}(x)$ the cardinality of $N_{u,v}(x)$.

Lemma 7.2.1 $B(x) \geq \displaystyle\sum_{u,v\in N(x),uv\notin E} \dfrac{1}{n - \nu_{u,v}(x)}.$

For graphs of diameter $D = 2$, we can obtain precise values:

Lemma 7.2.2 *[18] If $D = 2$ then* $B(x) = \displaystyle\sum_{\substack{u,v\in N(x) \\ uv\notin E(G)}} \dfrac{1}{\sigma_{u,v}}.$

In the next, we consider various non-local estimates. The following are formulae and bounds concerning cut-vertices:

Theorem 7.2.1 *[22] Let x be a cut vertex of a graph G, and let $n_1, \ldots, n_k$ be the orders of the components of $G - x$. If $k = d(x)$, then $B(x) = \sum_{i<j} n_i n_j$.*

Theorem 7.2.2 *[22] Let x be a cut vertex of a graph G of order n, and let $n_1, \ldots, n_k$ be the orders of the components of $G - x$. Then*

$$B(x) \leq \binom{k - r}{2}\left(\frac{n - 1 - r}{k}\right)^2 + \binom{r}{2}\left(\frac{n - 1 - r}{k} + 1\right)^2$$
$$+ r(k - r)\frac{n - 1 - r}{k}\left(\frac{n - 1 - r}{k} + 1\right)$$

where $r = n - 1 - k\left\lfloor\dfrac{n-1}{k}\right\rfloor$.

The bound is attained if and only if $G - x$ has $k - r$ components of the same order $\frac{n-1-r}{k}$ and r components of order $\frac{n-1-r}{k} + 1$.

Theorem 7.2.3 *[13] If G is a bipartite graph with bipartitions V_1, V_2 of order n_1 and n_2, respectively, then, for every vertex $x \in V_1$, it is satisfied*

$$B(x) \leq \frac{1}{2}\left(n_2^2(p + 1)^2 + n_2(p + 1)(2r - p - 1) - r(2p - r + 3)\right),$$

where p and r are positive integers, $r < p$, such that $n_1 - 1 = pn_2 + r$.

The upper bound is attained if and only if $G - x$ has n_2 components and their orders differ by at most 1.

Recently, Majstorović and Caporossi considered the estimates of adjusted betweenness centrality in trees; their results can be summarized as follows:

Theorem 7.2.4 *[26] If T is a tree of order n, then, for each vertex $x \in V(T)$ of degree $d(x)$, it is satisfied*

$$(2n - 1)d(x) - d^2(x) - n + 1 \leq B^*(x) \leq n(n - 1) - \frac{(n - 1)^2}{d(x)}.$$

Moreover, the lower bound is attained if and only if x is adjacent with at least $d(x) - 1$ pendant vertices, and the upper bound is attained if and only if all the components of $T - x$ are of the same order.

Theorem 7.2.5 *[26] Let T be a tree of order n with p pendant vertices, and let $p_1(x)$ be the number of pendant neighbors of a vertex x of T. Then, for any nonpendant vertex x,*

$$B^*(x) \leq (n - 1)^2 - p + p_1(x) - \frac{(p - p_1(x))^2}{d(x) - p_1(x)}.$$

The upper bound is attained if and only if $p - p_1(x) \equiv 0 \pmod{(d(x) - p_1(x))}$ and each component of $T - x$ is either a single vertex or $\frac{p - p_1(x)}{d(x) - p_1(x)}$-star whose central vertex is adjacent to x in T.

Theorem 7.2.6 *[26] If T is a tree of order n having p pendant vertices, then, for every nonpendant vertex x of T of degree $d(x)$, it is satisfied*

$$B^*(x) \leq \begin{cases} (n - 1)^2 - 6(p - d(x)), & \text{if } p \leq 3d(x), \\ (n - 1)^2 - p - \dfrac{p^2}{d(x)}, & \text{if } p > 3d(x). \end{cases}$$

For $p \leq 3d(x)$, the upper bound is attained if and only if x is a central vertex of a star or each component of $T - x$ is a 3-star or a 4-star with central vertex adjacent to x in T. For $p > 3d(x)$, the upper bound is attained for a vertex x of extremal trees of Theorem 7.2.5 such that $d_1(x) = 0$.

They also considered mutual relations of adjusted betweenness centrality and the *vertex transmission*: for a graph G and a vertex $x \in V$, it is defined as the sum

$$t(x) = \sum_{y \in V(G)} d(x, y). \tag{7.12}$$

Notice that $t(x)$ is also known as the vertex status, the farness, or total distance. For describing the extremal trees concerning these relations, the following notation is used: A *starlike tree* $S(n_1, \ldots, n_\Delta)$ is a tree having exactly one vertex v of degree at least three and the maximal paths starting at v have lengths $n_1, \ldots, n_\Delta$. A starlike tree $S(n - \Delta, 1, \ldots, 1)$ is called a *comet* $C_{n,\Delta}$ and its pendant vertex adjacent to a degree-two vertex is a *tail vertex*. A *caterpillar* is a tree whose removal of all pendant vertices results in a path; if the resulting path has length 1, the caterpillar is called a *double star*. A double star whose central vertices have equal degrees is an *equilibrated double star*.

Theorem 7.2.7 *[26] If T is a tree of order n with p pendant vertices, then, for every vertex x of T it is satisfied*

$$B^*(x) + 2t(x) \leq \begin{cases} n^2 - 1 + p - \dfrac{p^2}{d(x)}, & \text{if } x \text{ is not pendant,} \\ n^2 - p^2 + 3p - 3, & \text{if } x \text{ is pendant.} \end{cases}$$

For a pendant vertex x, the equality holds for a tail vertex of a comet and pendant vertices of a path and a star, respectively.

For a nonpendant vertex x, the equality holds if and only if x is either the central vertex of a starlike tree or x is a vertex of a path or else each component of $T - x$ is either a comet (with the tail vertex adjacent to x in T) or a star (with the central vertex adjacent to x in T) with the same numbers of pendant vertices.

Theorem 7.2.8 *[26] If T is a tree of order n with p pendant vertices and $\Delta(T) = \Delta > 2$, then, for every vertex x of T, it is satisfied*

$$B^*(x) + 2t(x) \leq \begin{cases} n^2 - 1 - \dfrac{\Delta - 1}{\Delta - 2} \cdot \dfrac{(p - d(x))^2}{d(x)}, & \text{if } x \text{ is not pendant,} \\ n^2 - 1 - \dfrac{\Delta - 1}{\Delta - 2}(p - 2)^2, & \text{if } x \text{ is pendant.} \end{cases}$$

For pendant vertex x, the equality holds if and only if x is the tail vertex of a comet, or a pendant vertex of a star, or else a pendant vertex of a caterpillar whose nonpendant vertices are either of degree 2 or Δ such that, starting from x, all nonpendant vertices are in increasing order.

For nonpendant vertex x, the equality holds if and only if x is either the central vertex of a starlike tree or x is a vertex of a path, or each component of $T - x$ is either a comet with maximum degree Δ (with the tail vertex adjacent to x in T) or a $(\Delta - 1)$-star (with the central vertex adjacent to x in T), or else all components of $T - x$ are caterpillars with the same number of pendant vertices such that all other vertices have degree 2 or Δ and, starting from x, in each caterpillar, the degrees of nonpendant vertices are in increasing order.

The following theorem gives us a lower bound relating the adjusted betweenness centrality and the vertex transmission.

Theorem 7.2.9 *[26] If T is a tree of order n, then, for every vertex $x \in V(T)$ it is satisfied*

$$B^*(x) \geq n - 1 + (2n - 3)t(x) - t^2(x).$$

The lower bound is attained exactly in the case when x belongs to the center of a double star or x is a pendant vertex of a star.

Regarding the estimates of average vertex betweenness centrality, we point out that they can be obtained (using the relation (7.11)) from estimates for the average graph distance of the graph. Sometimes, the opposite approach is used: for estimating the average distance, we use an estimation of average betweenness centrality, as in the next theorem.

Theorem 7.2.10 *[17] Let G be a graph of order n, size m, and diameter D > 1. Then*

$$\bar{B}(G) \leq 1 + \frac{(n-4)D}{2} - \frac{2(m-n+1)}{n} - \frac{D^2 - 6D^2 + h(D)}{6n(n-1)},$$

where $h(D) = \begin{cases} 6 - D, & \text{if } n - D \text{ is odd,} \\ 2D, & \text{if } n - D \text{ is even} \end{cases}$.

Theorem 7.2.11 *[17] From a tree of order n, T, of maximum degree Δ,*

$$\bar{B}(T) \leq \frac{(\Delta - 1)(n - \Delta)(n - 1)^2}{2n\Delta}$$

The upper bound is attained for an $(n - 1)$*-star.*

Theorem 7.2.12 *[3] Let G be a graph of order n and size m. Then*

$$\frac{n-1}{2} - \frac{m}{n} \leq \bar{B}(G) \leq \frac{\binom{C}{2} + C\binom{P+1}{2} + P(C - \alpha) + \binom{P+1}{3}}{n} - \frac{n-1}{2}$$

where $C = \left\lfloor \dfrac{3 + \sqrt{9 + 8m - 8n}}{2} \right\rfloor$, $P = n - C$, *and* $\alpha - m - P + 1 - \binom{C}{2}$.

We further review several known as well as unpublished results for edge betweenness centrality:

Lemma 7.2.3 *Let G be a graph of diameter D = 2. Then for every edge e = uv ∈ E(G),*

$$B(e) = \sum_{x,y \in V} \frac{1}{\sigma_{xy}} = \sum_{y \in N(v) \setminus (N(u) \cup \{u\})} \frac{1}{\sigma_{uy}} + \sum_{x \in N(u) \setminus (N(v) \cup \{v\})} \frac{1}{\sigma_{vx}} + 1.$$

Proof. By definition,

$$B(e) = \sum_{x,y \in V} \frac{\sigma_{xy}(e)}{\sigma_{xy}}.$$

The diameter of G is 2; therefore, $\sigma_{xy}(e) > 0$ if and only if one of the following three situations sets in:

- $x = u$ and $y = v$. Then $\frac{\sigma_{uv}(e)}{\sigma_{uv}} = 1$,

- $x = u$, $uy \notin E(G)$, and $vy \in E(G)$. Then $y \in N(v)\setminus(N(u) \cup \{u\})$,

- $y = v$, $xv \notin E(G)$, and $ux \in E(G)$. In this case, $x \in N(u)\setminus(N(v)\setminus\{v\})$.

In any other case, $\sigma_{xy}(e) = 0$. Hence,

$$B(e) = \sum_{x,y \in V} \frac{\sigma_{xy}(e)}{\sigma_{xy}} = \sum_{x,y \in V} \frac{1}{\sigma_{xy}} = \sum_{y \in N(v)\setminus(N(u) \cup \{u\})} \frac{1}{\sigma_{uy}} + \sum_{x \in N(u)\setminus(N(v) \cup \{v\})} \frac{1}{\sigma_{vx}} + 1.$$

$\square$

Theorem 7.2.13 *[3] Let G be a graph of order n and size m. Then the average edge betweenness centrality $\overline{B}_e(G)$ of G is equal to $\frac{1}{m}\binom{n}{2}\overline{l}(G)$. Furthermore,*

$$\frac{n(n-1)}{m} - 1 \le \overline{B}_e(G) \le \frac{\binom{C}{2} + C\binom{P+1}{2} + P(C - \alpha) + \binom{P+1}{3}}{m}.$$

where $C = \left\lfloor \dfrac{3 + \sqrt{9 + 8m - 8n}}{2} \right\rfloor$, $P = n - C$, and $\alpha - m - P + 1 - \binom{C}{2}$.

The graph-theoretical properties of other variants of betweenness centrality are still little explored; we list selected known results:

Theorem 7.2.14 *[24] For a graph G of order n, its average scaled vertex betweenness centrality is*

$$\overline{B}_{dist}(G) \le \frac{n-1}{2}\left(1 - \frac{1}{\overline{l}(G)}\right)$$

Theorem 7.2.15 *[24] For a graph G of order n, average degree $\overline{d}$, and diameter D,*

$$\frac{n-1-\overline{d}}{4} \le \overline{B}_{dist}(G) \le \frac{n-1-\overline{d}}{2}\left(1 - \frac{1}{D}\right)$$

Theorem 7.2.16 *[24] For a graph G of order n and diameter D,*

$$\frac{\overline{B}(G)}{D} \le \overline{B}_{dist}(G) \le \frac{\overline{B}(G)}{2}.$$

7.2.3 Relations with Graph Operations

As mentioned before, the graph-theoretic properties of betweenness centrality are just getting to be studied in a deeper detail, and it seems that the influence of graph operations on betweenness centrality changes is not explored. The individual values of vertices are hard to predict; thus, attention was focused on changes of average betweenness centrality according to selected graph operations; some results are listed in [17].

Lemma 7.2.4 *[17] If G is a graph of order n and G′ = G + uv is the graph obtained from G by joining two vertices u, v of distance d > 1 with a new edge, then*

$$\overline{B}(G') \leq \overline{B}(G) - \frac{d-1}{n}$$

Theorem 7.2.17 *[17] If H is a spanning subgraph of a graph G and r is the number of edges of G that are not contained in G′, then*

$$\overline{B}(G) \leq \overline{B}(H) - \frac{r}{n}.$$

Corollary 7.2.1 *[17] If T is a spanning tree of a graph G of order n and size m, then*

$$\overline{B}(G) \leq \overline{B}(T) - \frac{m-n+1}{n}.$$

Theorem 7.2.18 *[17] If G is a graph of order n and G′ is a graph obtained from G by connecting a new pendant vertex to a vertex x, then*

$$\overline{B}(G') = \frac{n}{n+1}\overline{B}(G) + \frac{2}{n+1}\sum_k kn_k(x)$$

where $n_k(x)$ is the number of vertices at distance k from x.

Theorem 7.2.19 *[17] If G is a graph of order n and G′ is a graph obtained from G by connecting a new vertex to two vertices at distance at most 2, then*

$$\overline{B}(G') \geq \frac{1}{n+1}(n\overline{B}(G) + n - 2).$$

Similar results can be obtained also for edge betweenness centrality:

Theorem 7.2.20 *Let G be a graph of order n and size m, and let H be the graph obtained from G by connecting two vertices u, v ∈ V(G) at distance d(u, v) > 1 with a new edge. Then*

$$\overline{B}_e(H) \leq \frac{m}{m+1}\overline{B}_e(G) - \frac{d-1}{m+1}.$$

Theorem 7.2.21 *Let G be a Hamiltonian graph of order n and size m. Then*

$$\overline{B}_e(G) \leq \frac{n}{m}\overline{B}_e(C_n) - \frac{m-n}{m}.$$

Note that selected graph operations (e.g., graph join) yield graphs where individual values of vertex betweenness centrality can be expressed by exact formula (e.g., the one by Theorem 7.2.2).

7.3   Betweenness-Uniform Graphs

When considering the distribution of values for centrality indices of vertices of a graph, two extremal cases are possible: The first one concerns highly diverse networks with the maximum possible number of distinct values, and the second one refers to homogeneous networks having vertices of the same centrality. For example, if we take, as the centrality index, the vertex degree, then heterogeneous networks are precisely antiregular graphs (i.e., graphs with exactly two vertices of the same degree; see [4,30]), while the homogeneous ones are regular graphs. Similarly, the homogeneous graphs with respect to vertex eccentricity are self-centered graphs (see [8]). Here, we will concentrate on graphs whose vertices possess the same value of betweenness centrality; we will call them *betweenness-uniform graphs*.

There are several general conditions that betweenness-uniform graphs have to satisfy. In [18], it was proved that each betweenness-uniform graph is 2-connected. This stems from the fact that if a graph G contains a cut vertex X, then it is quite easy to find another vertex v whose betweenness centrality differs from $B(X)$: We take a block B_1 of G with the smallest number of vertices, say $n_1 + 1$, and then we take the vertex $v \in B_1$ at the maximum distance from X. Then $B(v)$ is determined only by the vertices from the same block; hence, $B(v) \leq \binom{n_1}{2}$. On the other hand, to compute the betweenness centrality of X, we have to count at least each pair of vertices from different blocks of G; thus, $B(X) \geq \sum_{i<j}^{k} n_i n_j \geq (n_1)^2 > B(v)$, where k is the number of components of $G - X$.

Note that, unlike the eccentricity or total distance where the vertices forming the graph center (i.e., the set of vertices having the minimum centrality) lie in a single block (see [9]), neither the vertices of minimum betweenness centrality nor the ones with the maximum betweenness centrality need to be contained in the same block of a graph. As an example, one can take the graph obtained a double star by subdividing its central edge: The vertices with the zero betweenness centrality are pendant ones, and the maximum betweenness centrality is realized on two nonadjacent vertices of maximum degree.

The maximum degree has also an influence on the uniformity of betweenness centrality—if a betweenness-uniform graph contains a universal vertex (i.e., a vertex adjacent to each other vertex), it is isomorphic to a complete graph [18]. Furthermore, it is easy to show that each betweenness-uniform graph G with n vertices having a subuniversal vertex (i.e., a vertex of degree $n - 2$) has diameter 2 [18]; on the other hand, the families of betweenness-uniform graphs with subuniversal vertices are not explored well and show many diverse members. More generally, we conjecture that each betweenness-uniform n-vertex graph G with $\Delta = n - k$ has $D \leq k$.

Betweenness-uniform graphs of diameter $D = 2$ are a quite plausible graph family, as their betweenness centrality can be computed by Lemma 7.2.2, being equal to $n - 1 - 2m/n$. This is frequently used in various constructions developed in [18].

In [18], it was proved that for a betweenness-uniform graph G, the common value of vertex betweenness centralities of G is either 0 or at least $1/2$. It is an open question whether, for any rational number $p/q \in (1/2, \infty)$, there exists a betweenness-uniform graph whose vertices reach the betweenness centrality $B(x) = p/q$, for any $x \in V$.

According to the general definition of centrality index (being a real-valued function on the vertex set of a graph, which is invariant under graph isomorphism), for each vertex-transitive graph, its vertices have the same value of the given index, including betweenness centrality (note, however, that the possible values of betweenness centrality of vertex-transitive graphs were not systematically studied). Hence, the attention was focused on exploring nontransitive betweenness-uniform graphs, of which an important family is the family of *distance-regular graphs*, that is, the family of graphs such that, for any pair u, v of their vertices, the number of vertices at distance j from u and the number of vertices at distance $d(u, v) - j$ from v are the same. To each distance-regular graph G of diameter D, we can assign a formal matrix (called intersection array of G):

$$
i(G) = \begin{Bmatrix} - & c_1 & \dots & c_{D-1} & c_D \\ a_0 & a_1 & \dots & a_{D-1} & a_D \\ b_0 & b_1 & \dots & b_{D-1} & - \end{Bmatrix}
$$

with $a_k = |N_k(u) \cap N(v)|$, $b_k = |N_{k+1}(u) \cap N(v)|$, and $c_k = |N_{k-1}(u) \cap N(v)|$ for any pair u, v of vertices of G and $0 \le k \le D$. It can be shown (see [18]) that, for any vertex x of a distance-regular graph G and any pair u, v of vertices such that $d(u, x) = l, d(u, v) = k$, the pairwise dependency

$$
\delta_{u,v}(x) = \frac{\prod_{i=1}^{l} c_i}{\prod_{i=k-l}^{k} c_i},
$$

hence, it does not depend on x; consequently, $B(x)$ is the same for all $x \in V(G)$.

Distance-regular graphs of diameter $D = 2$ are called *strongly regular graphs* (equivalently, they are k-regular graphs such that each pair of adjacent vertices has λ common neighbors while each pair of nonadjacent vertices has μ common neighbors). Thus, they are betweenness-uniform with the common value of vertex betweenness centrality $\frac{k}{\mu}(k - \lambda - 1)$. Moreover, by [29], for each finite group A, there exists a strongly regular graph such that its automorphism group is isomorphic to A; therefore, one can construct betweenness-uniform graphs with prescribed automorphism groups.

In the following, we present some constructions of [18], which yield to nonregular betweenness-uniform graphs. Several of these constructions are based on the graph cloning, an operation where the vertices are replaced by the graphs. More precisely, having a graph G with a vertex set $V(G) = \{v_1, v_2, \dots, v_n\}$ and graphs $H_1, H_2, \dots, H_n$ used for the replacement, the graph cloning operation replaces each vertex v_i by its substitute H_i and performs graph join on H_i and H_j whenever v_i, v_j are adjacent in G (see Figure 7.2). The obtained graph is denoted as $G[H_1, H_2, \dots, H_n]$, or simply $G[H]$ for $H_1 = H_2 = \dots = H_n = H$ (note that $G[H]$ is the lexicographic product).

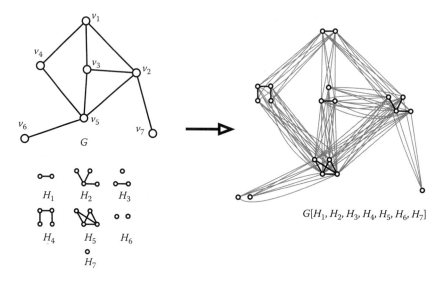

FIGURE 7.2: An example of graph cloning.

Theorem 7.3.1 *[18] For every integer m and every betweenness-uniform graph G, the graph $G[K_m]$ is betweenness-uniform.*

Theorem 7.3.2 *[18] For every regular betweenness-uniform graph G and H = $\bigcup_{i=1}^{\ell} n_i K_{r_i}$ being the disjoint union of n_i complete graphs on r_i vertices, $i = 1, \ldots, \ell$, the graph $G[H]$ is betweenness-uniform.*

The latter construction was used in [18] to determine the lower bound for the maximum possible number of distinct degrees (the *irregularity index*; see [28]) of betweenness-uniform graphs. Having $G \cong K_2$ and $H \cong \bigcup_{i=1}^{s} K_i$, $G[H]$ is betweenness-uniform graph of order $s(s + 1)$ with irregularity index s. Thus, there are n-vertex betweenness-uniform graphs with $(\sqrt{4n + 1} - 1)/2 = \Omega(\sqrt{n})$ distinct degrees. An upper bound $n - 3$ on irregularity index of n-vertex betweenness-uniform graphs follows from the fact that the minimum degree is at least 2 and the maximum degree is, due to [4] and the aforementioned exclusion of universal vertices, at most $n - 2$.

Another observation based on this construction is that there exist superpolynomially many nonisomorphic nontransitive betweenness-uniform graphs on n vertices. To show this, consider a number partition $k = k_1 + \cdots + k_l$ and take $G := K_2$, H being the disjoint union of complete graphs of orders k_i, $i = 1, \ldots, l$. The graph $G[H]$ is betweenness-uniform; moreover, different number partitions of k correspond to nonisomorphic cloned graphs. Therefore, the number of nonisomorphic betweenness-uniform graphs on $2k$ vertices is at least the number of all number partitions of k, which is, by Hardy–Ramanujan formula, asymptotically equal to $(1/4k\sqrt{3})e^{\sqrt{2k/3}}$, which is superpolynomial in k. The majority of these graphs (in the sense that their number is also superpolynomial in k) is nontransitive due to the fact that the number

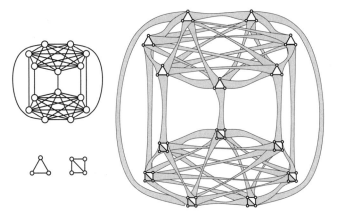

FIGURE 7.3: An example of construction with different substitutes in two complete halves.

of partitions of k with equal parts is sublinear in k and any other partitions yields a graph clone, which is not transitive.

The former two constructions have one common feature—each vertex is replaced by the same copy of complete graph or the same union of complete graphs. This can be further weakened in the following way: start with two copies of K_s and add new edges between these copies so that they form a matching. Now, replacing vertices of the first copy by K_r and the rest of the vertices with K_q, we obtain a graph that is betweenness-uniform (see Figure 7.3).

Theorem 7.3.3 *Let* u, v *be two vertices of* K_2 *and let* $H_1 = \cup_{i=1}^{s} K_{a_i}$ *and* $H_2 = \cup_{i=1}^{t} K_{b_i}$. *Let the graph* G *be obtained from* K_2 *by replacing* u, v *with* H_1 *and* H_2, *respectively, and, after adding new edges between any two vertices* x, y, *where* $x \in V(H_1)$ *and* $y \in V(H_2)$. *Then* G *is betweenness-uniform if and only if*

$$\sum_{k=1}^{s} a_k \sum_{i=1}^{s} \sum_{j>i}^{s} a_i a_j = \sum_{k=1}^{t} b_k \sum_{i=1}^{t} \sum_{j>i}^{t} b_i b_j.$$

The corollary of the previous theorem is that cloning the cycle on four vertices with K_a, K_b, K_c, K_d gives betweenness-uniform graph if and only if $ac(a+c) = bd(b+d)$; on the other hand, the clone $C_5[K_a, K_b, K_c, K_d, K_e]$ is betweenness-uniform if and only if $a = b = c = d = e$. Other examples are graphs $K_{2,3}[K_1, K_4; K_1, K_1, K_2]$ and $K_{3,7}[K_1, K_1, K_4; K_1, K_1, K_1, K_1, K_1, K_1, K_2]$.

The aforementioned results show that the graph cloning operation results, in many cases, in betweenness-uniform graphs even starting from graphs that are not betweenness-uniform. Hence, the following conjecture might be of some interest:

Conjecture 7.3.1 *For any n-vertex graph* G, *there exist graphs* $H_1, \ldots, H_n$ *such that* $G[H_1, \ldots, H_n]$ *is betweenness-uniform.*

Up to now, the presented constructions were based on the local join; we present further some other unpublished constructions.

Theorem 7.3.4 *Given an integer n, let $k_1, \ldots, k_n \geq 4$ be integers. Then the graph $K_n[\overline{C}_{k_1}, \ldots, \overline{C}_{k_n}]$ is betweenness-uniform.*

Proof. It is easy to check that the diameter of $K_n[\overline{C}_{k_1}, \ldots, \overline{C}_{k_n}]$ is 2. Set $p = \sum_{j=1}^{n} k_i$ and let $G_i = \overline{C}_{k_i}$ be ith substitute in $K_n[\overline{C}_{k_1}, \ldots, \overline{C}_{k_n}]$ (here, we order vertices such that $C_{k_i} = x_1^i \cdots x_{k_i}^i$). Then, for each $x = x_l^i \in V(G_i)$,

$$B(x) = \sum_{u,v \in V(G_i)} \frac{\sigma_{u,v}(x)}{\sigma_{u,v}} + \sum_{j \neq i} \sum_{u,v \in V(G_j)} \frac{\sigma_{u,v}(x)}{\sigma_{u,v}} + \sum_{j \neq i} \sum_{u \in V(G_i), v \in V(G_j)} \frac{\sigma_{u,v}(x)}{\sigma_{u,v}}$$

For the first sum, observe that only $k_i - 4$ pairs of nonadjacent vertices of G_i provide nonzero contribution (excluded are the pairs $\{x_l^i, x_{l+1}^i\}$, $\{x_{l+1}^i, x_{l+2}^i\}$, $\{x_{l-1}^i, x_l^i\}$, $\{x_{l-2}^i, x_{l-1}^i\}$; indices are taken modulo k_i). For each such a pair, there is $k_i - 4 + p - k_i$ geodetic paths of length 2, of which only one passes through x. Hence, the first sum is equal to $(k_i - 4)/((k_i - 4) + p - k_i)$. In a similar way, it can be shown that the second sum is $\sum_{j \neq i} k_j/((k_j - 4) + p - k_j)$. The third sum is equal to zero because, in the resulting graph, all vertices of G_i and G_j are joined by edges of local join. Thus, the betweenness centrality of x is

$$\frac{k_i - 4}{(k_i - 4) + p - k_i} + \sum_{j \neq i} \frac{k_j}{(k_j - 4) + p - k_j} + 0 = 1$$

We conclude that the betweenness centrality of each vertex of $K_n[\overline{C}_{k_1}, \ldots, \overline{C}_{k_n}]$ is equal to 1. □

Theorem 7.3.5 *Let s, t be positive integers, $s = t - 1$, and let H_1 be the graph obtained from the join of two disjoint copies K^1, K^2 of K_t on vertex sets $V_1 = \{u_1, \ldots, u_t\}$ and $V_2 = \{v_1, \ldots, v_t\}$, respectively, by deleting edges $u_i v_i$ for each $i = 1, \ldots, t$. Let $H_2 \cong K_{2s}$ where $V(H_2) = V_3 \cup V_4$, $|V_3| = |V_4| = s$, and $V_3 \cap V_4 = \emptyset$. Then the graph G obtained from H_1, H_2 by adding new edges xy where $x \in V_1, y \in V_3$ or $x \in V_2, y \in V_4$ is betweenness-uniform.*

Proof. It is easy to see that G is of diameter 2. Let $x \in V(H_1)$ and $y \in V(H_2)$ (without loss of generality, take $x = u_j \in V_1, y \in V_3$). Then

$$B(x) = \sum_{u,v \in V(H_1)} \frac{\sigma_{u,v}(x)}{\sigma_{u,v}} + \sum_{u,v \in V(H_2)} \frac{\sigma_{u,v}(x)}{\sigma_{u,v}} + \sum_{u \in V(H_1), v \in V(H_2)} \frac{\sigma_{u,v}(x)}{\sigma_{u,v}}$$

$$= 0 + \frac{1}{2} + \frac{s(t-1)}{s+t-1}$$

The first sum is 0 because vertices in H_2 do not contribute to the sum (they are mutually connected by edges); the second one is 1/2 because, for every two nonadjacent vertices u, v of H_1, there is a unique u-v path of length 2 passing through x

(any $u - v$-path with all internal vertices from H_2 has length at least 3). For the third sum, the nonzero contribution may come only from pairs $\{u, v\}$ where $u \in V_3, v \in V_2$. The pairs $\{u, v_j\}$ do not contribute and each other pair contributes $1/(s + t - 1)$; hence, the third sum is $(st - s)/(s + t - 1)$.

Similarly,

$$B(y) = \sum_{u,v \in V(H_1)} \frac{\sigma_{u,v}(y)}{\sigma_{u,v}} + \sum_{u,v \in V(H_2)} \frac{\sigma_{u,v}(y)}{\sigma_{u,v}} + \sum_{u \in V(H_1), v \in V(H_2)} \frac{\sigma_{u,v}(y)}{\sigma_{u,v}}$$

$$= 0 + 0 + \frac{st}{s + t - 1}$$

since nonadjacent vertices u, v of H_1 do not contribute (again, any $u - v$-path with internal vertices from H_2 is of length at least 3) and the vertices of H_2 are mutually adjacent. The value of the third sum comes from the fact that the only pairs $\{u, v\}$ with nonzero contribution are those ones with $u \in V_1, v \in V_4$; there are st such pairs, each of them giving $s + t - 1$ geodetic paths of length 2.

Thus, all vertices of H_1 have the same betweenness centrality; the same holds for H_2. As $s = t - 1$, we have $\frac{s}{s+t-1} = \frac{1}{2}$, which yields $\frac{st}{s+t-1} = \frac{1}{2} + \frac{s(t-1)}{s+t-1}$, that is, $B(x) = B(y)$. This proves the claim (Figure 7.4). $\qquad\square$

Theorem 7.3.6 *Let s, t be integers (s being even) such that $3s(s - 1) = t(t - 1)$, and let H_1 be the graph obtained from the join of two disjoint copies K^1, K^2 of K_t on vertex sets $V_1 = \{u_1, \dots, u_t\}$ and $V_2 = \{v_1, \dots, v_t\}$, respectively, by deleting edges $u_i v_i$ for each $i = 1, \dots, t$. Let H_2 be the graph obtained from the join of two disjoint copies K^3, K^4 of the complete graph K_s without a perfect match (assume that K^3, K^4 have vertex sets $V_3 = \{x_1, \dots, x_s\}$ and $V_4 = \{y_1, \dots, y_s\}$, respectively). Then the graph G obtained from H_1, H_2 by adding new edges xy such that $x \in V_1, y \in V_3$, or $x \in V_2, y \in V_4$ is betweenness-uniform.*

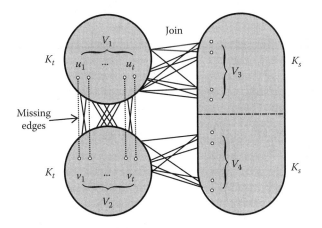

FIGURE 7.4: The construction for Theorem 7.3.5.

Proof. Again, it is easy to see that diam$(G) = 2$. Pick $u = u_i \in V(H_1)$ and $x = x_j \in V(H_2)$ (note that H_1 and H_2 are vertex transitive and, by the construction of G, all vertices of $V(H_1)$ have the same betweenness centrality in G; the same holds for vertices of $V(H_2)$). Then

$$B(u) = \sum_{v,w \in V(H_1)} \frac{\sigma_{v,w}(u)}{\sigma_{v,w}} + \sum_{v,w \in V(H_2)} \frac{\sigma_{v,w}(u)}{\sigma_{v,w}} + \sum_{v \in V(H_1), w \in V(H_2)} \frac{\sigma_{v,w}(u)}{\sigma_{v,w}}$$

$$= \frac{1}{2} + \frac{s}{2} \cdot \frac{1}{2s - 2 + t} + \frac{s(t-1)}{s+t-1}$$

and

$$B(x) = \sum_{v,w \in V(H_1)} \frac{\sigma_{v,w}(x)}{\sigma_{v,w}} + \sum_{v,w \in V(H_2)} \frac{\sigma_{v,w}(x)}{\sigma_{v,w}} + \sum_{v \in V(H_1), w \in V(H_2)} \frac{\sigma_{v,w}(x)}{\sigma_{v,w}}$$

$$= 0 + \frac{s-1}{2s - 2 + t} + \frac{st}{s+t-1}.$$

The first sum of $B(u)$ is equal to $\frac{t-1}{2(t-1)} = \frac{1}{2}$; the reason is that, for two nonadjacent vertices of H_1 in G, there is no geodetic path passing through any vertex of H_2. This also means that the first sum of $B(x)$ is zero. On the other hand, for each pair of nonadjacent vertices of H_2, there are $2s - 2 + t$ geodetic paths and exactly t of them contain some vertex of H_2. The number of such pairs of vertices with nonzero contribution for u is $s/2$ and, for x, it is $s - 1$. This explains the second sums. Moreover, u is adjacent only with vertices of the vertex set V_3, so its contribution is $s(t-1)/(s+t-1)$. Similarly, the third sum of $B(x)$ is $st/(s+t-1)$ because only the missing edges between the vertices of the vertex set V_1 and V_4 create nonzero contribution for this vertex.

We see that $B(u)$ as well as $B(x)$ does not depend on the choice of u or x. Since $3s(s-1) = t(t-1)$, we get that $B(u) = B(x)$ and it shows that G is betweenness-uniform (Figure 7.5). □

The centrality uniformity can be considered also for other variants of betweenness centrality; inspired by results of [18], we found their analogues for edge betweenness centrality:

Lemma 7.3.1 *Let G be an edge of betweenness-uniform graph of diameter $D = 2$. Then for every edge $e \in E(G)$,*

$$B(e) = \frac{n(n-1)}{m} - 1.$$

Proof. As $D = 2$ and $\overline{B}(e) = \frac{1}{m} \sum_{s,t \in V} d(s,t)$, we obtain

$$B(e) = \frac{1}{m} \sum_{s,t \in V} d(s,t) = \frac{1}{m} \left(\sum_{s,t \in V, st \in E} 1 + \sum_{s,t \in V, st \notin E} 2 \right)$$

$$= \frac{1}{m} \left(m + 2 \left(\frac{n(n-1)}{2} - m \right) \right) = 1 + \frac{n(n-1)}{m} - 2 = \frac{n(n-1)}{m} - 1. \quad □$$

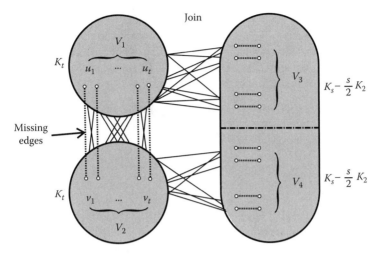

FIGURE 7.5: The construction for Theorem 7.3.6.

Theorem 7.3.7 *Every distance-regular graph is edge betweenness-uniform.*

Proof. For a distance-regular graph G, take $s, t, u, v \in V$ such that $d(s,t) = k$, $d(s,u) = l$, $d(v,t) = k - l - 1$, and $e = uv \in E$. Then

$$\sigma_{st}(uv) = \sigma_{su} \cdot \sigma_{vt} = \prod_{i=1}^{l} c_i \prod_{i=1}^{k-l-1} c_i,$$

and besides,

$$b_{st}(uv) = \frac{\sigma_{st}(uv)}{\sigma_{st}} = \frac{\sigma_{su} \cdot \sigma_{vt}}{\sigma_{st}} = \frac{\prod_{i=1}^{l} c_i \prod_{i=1}^{k-l-1} c_i}{\prod_{i=1}^{k} c_i} = \frac{\prod_{i=1}^{l} c_i}{\prod_{i=k-l}^{k} c_i}.$$

As a consequence, we obtain that each strongly regular graph with parameters (n, k, λ, μ) is edge betweenness-uniform with the common value of edge betweenness centrality equal to $2(k - \lambda - 1)\frac{1}{\mu} + 1$. □

7.4 Spectral Relations

In the last section of the chapter, we review some spectral bounds obtained in [11] for the vertex and edge betweenness centrality of the graph by means of the Laplacian eigenvalues. Furthermore, using the relations among them and other invariants of graphs, as the isoperimetric number or the vertex-forwarding index, a bound for the diameter of the graph can be found.

Consider the Laplacian matrix of the graph G, that is, the matrix $L = D - A$, where D is the diagonal matrix of vertex degrees of G and A is the adjacency matrix of the graph. In the following, we obtain some bounds for the betweenness centrality of a vertex $x \in V$ using the Laplacian eigenvalues. So let $spec(L) = \{0, \theta_2, \ldots, \theta_n\}$ be the nondecreasing sequence of Laplacian eigenvalues.

Proposition 7.4.1 *Let G be a connected graph of order n and $u, v, x \in V$ such that $u, v \in V$ are two vertices at distance 2 and x is a common neighbor, $d(u, x) = 1 = d(x, v)$. Let Δ be the maximum degree of G and let θ_n be the largest eigenvalue of the Laplacian matrix; then*

$$\frac{n}{(n-1)\theta_n} \leq \frac{1}{\Delta} \leq \delta_{u,v}(x).$$

Proposition 7.4.2 *Let G be a graph of order n, let $u, v, x \in V$ be vertices of G such that $d(u, v) = d > 2$, and suppose that $x \in N_i(u) \cap N_{d-i}(v)$ for some $1 \leq i \leq d - 1$. Let $n_i = |N_i(u) \cap N_{d-i}(v)|$, let Δ be the maximum degree of the graph, and let θ_2 be the second largest eigenvalue of the Laplacian; then*

$$\left(\frac{\theta_2}{n\Delta}\right)^{d-1} \leq \delta_{u,v}(x) \leq 1 - (n_i - 1)\left(\frac{\theta_2}{n\Delta}\right)^{d-1}.$$

The following results are some bounds for the maximum edge betweenness centrality of a graph and an improved bound for its diameter. Firstly, let us study the relation among the edge betweenness centrality with other parameters and invariants of a graph like the edge expansion index and the isoperimetric number.

Recall that for a graph G, $B(e)$ is the betweenness centrality of an edge $e \in E$, $\overline{B}_e(G)$ is the average edge betweenness centrality of the graph, and B_e^{max} is the maximum edge betweenness centrality of G. The following lemma provides us some basic properties of these parameters (Figure 7.6).

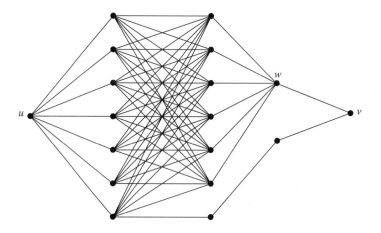

FIGURE 7.6: The pairwise dependency $\delta_{u,v}(w)$ approaches the upper bound of Proposition 7.4.2.

Lemma 7.4.1 *Let G be a graph and let e ∈ E be an edge with end vertices u, v ∈ V;*
then

1. $\rho_{u,v}(e) = 1 = \rho_{v,u}(e)$.

2. $1 \le B(e) \le n^2/4$ *if n is even and* $1 \le B(e) \le (n^2 - 1)/4$ *if n is odd.*

3. $B(e) \le \max\{B(u) + 1, B(v) + 1\}$.

4. $B(e) = n - 1$ *if one of the endvertices of e has degree* 1.

5. $B_e^{max} \le B^{max} + 1$.

6. $\overline{B}_e(G) \le \frac{m}{n}(\overline{B}(G) + 1)$.

Lemma 7.4.2 *[11] Let G be a graph of order n, then*

- *If* $e \in E$ *is an edge connecting two components* $G_1 \subset V$ *and* $G \backslash G_1$, *with*
  $|V(G_1)| = n_1$, *then* $B(e) = n_1(n - n_1)$.

- *If C is a cut-set of edges of the graph G, connecting two sets of vertices X and*
  $V(G) \backslash X$ *and* $|X| = n_x$, *then* $\sum_{e \in C} B(e) = n_x(n - n_x)$.

The edge betweenness centrality is also related with other known expansion parameters in graphs, as the so-called *edge expansion index* introduced in [31] as follows:

$$\beta = min \left\{ \frac{|\partial X|}{(|X||\overline{X}|)} : X \subset V, 1 \le |X| \le n - 1 \right\}.$$

where
    X is a proper set of vertices of V
    $\overline{X}$ is its complement in V
    $|\partial X|$ is the number of edges connecting X with $\overline{X}$.

Another expansion parameter of a graph is the *isoperimetric number*, introduced by
Mohar in [27] and defined as:

$$i(G) = min \left\{ \frac{|\partial X|}{|X|} : X \subset V, 1 \le |X| \le \frac{n}{2} \right\}.$$

As small isoperimetric number means that the graph can be easily disconnected in two
components. Related with both parameters is the *edge-forwarding index*, introduced
by Heydemann et al. [23] as an extension to edges of the vertex-forwarding index
defined by Chung et al. [12]. Let a route R be the set of $n(n - 1)$ paths connecting the
vertices of a graph, and let $R(e)$ be the number of paths of the route R that go through
the edge e, and then the edge-forwarding index is defined as

$$\Pi = min_{\mathbb{R}} max_{e \in E} R(e).$$

The relation among these parameters can be found in [31]: $i(G) \ge \beta n/2$ and $\Pi\beta \ge 2$.

Proposition 7.4.3 *[11] Let G be a graph of order n, edge expansion index β, and isoperimetric number i(G), then*

$$B^{max} + 2 \geq B_e^{max} \geq \frac{2}{\beta},$$

$$B^{max} + 2 \geq B_e^{max} \geq \frac{n}{i(G)}.$$

Observe from the last proposition that for graphs with poor expansion properties, that is, with small β, or graphs with a small isoperimetric number, B^{max} will be large.

By using the relations among these parameters obtained in [27] we get the relation with the second largest eigenvalue of the Laplacian, θ_2, in the next corollary.

Corollary 7.4.1 *[11] Let G be a graph of order n, maximum degree Δ and second eigenvalue of the Laplacian matrix θ_2, then*

$$\frac{n}{2\sqrt{\theta_2(2\Delta - \theta_2)}} \leq B_e^{max} \leq B^{max} + 1.$$

Finally, following the proof of Theorem 2.3 in [31] we use our results to improve a bound for the diameter of the graph *D* given there as Corollary 2.4. We point out that this bound also improves a former result from Mohar [27].

Corollary 7.4.2 *Let G be a graph of order n, maximum degree Δ, and diameter D; then*

$$D \leq \left\lceil \frac{\ln(n/2)}{\ln \frac{B_e^{max}\Delta + n}{B_e^{max}\Delta - n}} \right\rceil \leq \left\lceil \frac{\ln(n/2)}{\ln \frac{\Pi\Delta + n}{\Pi\Delta - n}} \right\rceil \leq \left\lceil \frac{\ln(n/2)}{\ln \frac{\Delta + i(G)}{\Delta - i(G)}} \right\rceil.$$

Solé [31]   Mohar [27]

We point out that the last bound for the diameter is easier to compute than the previous ones, as the maximum edge betweenness centrality is less difficult to compute than the edge-forwarding index or the isoperimetric number of a graph.

7.5   Conclusion

In this chapter, we have surveyed various results on betweenness centrality from the point of view of its graph-theoretic properties; we have described variants of betweenness centrality in unoriented graphs, and presented lower and upper bounds for vertex betweenness centrality and several related characteristics. We have also discussed the properties and constructions of graphs whose vertices have the same betweenness centralities. Most of the results covered here are taken from the literature; several results extend the previously published results by new constructions of betweenness-uniform graphs (Theorems 7.3.4 through 7.3.6) and new facts on edge betweenness centralities (Lemmas 7.2.3 and 7.3.1 and Theorems 7.2.20, 7.2.21, and 7.3.7).

Our survey does not touch the problematics of betweenness centrality in directed graphs (see [21,34]) as well as in weighted networks [16]. We also have not covered the discussion on comparison of use of betweenness centrality over other centrality indices (see, e.g., [25]), and the algorithmic aspects of betweenness centrality and similar indices; for exact computation, the paper [6] brings descriptions of fast algorithms for many betweenness centrality variants. The approximation algorithms are proposed, for example, in [2,7], or [19]. As it seems that, within the last decade, one may observe an increasing interest in mathematics of betweenness centrality index, it is possible to expect the increase of betweenness-related results.

References

1. J.M. Anthonisse, The rush in a graph, Technical Report BN 9/71, Stichting Mathematish Centrum, Amsterdam, the Netherlands (1971).

2. D.A. Bader, S. Kintali, K. Madduri, and M. Mihail, Approximating betweenness centrality, *The Fifth Workshop on Algorithms and Models for the Web-Graph (WAW2007)*, San Diego, CA, December 11–12 (2007) pp. 124–137.

3. D. Barmpoutis and R.M. Murray, Extremal properties of complex networks, arXiv:1104.5532v1 [q-bio.MN] 29 April 2011.

4. M. Behzad and G. Chartrand, No graph is perfect, *American Mathematical Monthly* **74** (1967) 962–963.

5. U. Brandes, A faster algorithm for betweenness centrality, *Journal of Mathematical Sociology* **25**(2) (2001) 163–177.

6. U. Brandes, On variants of shortest-path betweenness centrality and their generic computation, *Social Networks* **30**(2) (2008) 136–145.

7. U. Brandes and C. Pich, Centrality estimation in large networks, *International Journal of Bifurcation and Chaos* **17**(7) (2007).

8. F. Buckley, Self-centered graphs, *Annals of the New York Academy of Sciences* **576**(1) (1989) 71–78.

9. F. Buckley and F. Harary, *Distance in Graphs,* Addison-Wesley Publishing Company, Redwood City, CA (1990).

10. G. Caporossi, M. Paiva, D. Vukičević, and M. Segatto, Centrality and betweenness: Vertex and edge decomposition of the Wiener index, *MATCH: Communications in Mathematical and In computer Chemistry* **68** (2012) 293–302.

11. F. Comellas and S. Gago, Spectral bounds for the betweenness of a graph, *Linear Algebra and Its Applications* **423** (2007) 74–80.

12. F.R.K. Chung, E.G. Coman, M.I. Reiman, and B.E. Simon, The forwarding index of communication networks, *IEEE Transactions on Information Theory* **33** (1987) 224–232.

13. M.G. Everett, P. Sinclair, and P.A. Dankelmann, Some centrality results new and old, *Journal of Mathematical Sociology*, **28**(4) (2004) 215–227.

14. M.G. Everett and S.P. Borgatti, A graph-theoretic perspective on centrality, *Social Networks* **28** (2005) 466–484.

15. L.C. Freeman, Centrality in social networks: Conceptual clarification, *Social Networks* **1** (1978/1979) 215–239.

16. L.C. Freeman, S.P. Borgatti, and D.R. White, Centrality in valued graphs: A measure of betweennness based on network flow, *Social Networks* **13** (1991) 141–154.

17. S. Gago, J. Hurajová, and T. Madaras, Notes on the betweenness centrality of a graph, *Mathematica Slovaca* **62** (2012) 1–12.

18. S. Gago, J. Hurajová, and T. Madaras, On betweenness-uniform graphs, *Czechoslovak Mathematical Journal* **63**(138) (2013) 629–642.

19. R. Geisberger, P. Sanders, and D. Schultes, Better approximation of betweenness Centrality, *Proceedings of the 10th Workshop on Algorithm Engineering and Experiments (ALENEX 2008)*, SIAM, Philadelphia, PA (2008) pp. 90–100.

20. M. Girvan and M.E.J. Newman, Community structure in social and biological networks, *Proceedings of the National Academy of Sciences of the United States of America* **99** (2002) 7821–7826.

21. R.V. Gould, Measures of betweenness in non-symmetric networks, *Social Networks* **9** (1987) 277–282.

22. R. Grassi, R. Scapellato, S. Stefani, and A. Torriero, Betweenness centrality: Extremal values and structural properties, *Networks, Topology and Dynamics. Lecture Notes in Economics and Mathematical Systems* **613** (2009) 161–175.

23. M.C. Heydemann, J.C. Meyer, and D. Sotteau, On forwarding indices of networks, *Discrete Applied Mathematics* **23** (1987) 103–123.

24. J. Hurajová, Indexy centrality grafu, Master thesis, P.J. Šafárik University in Košice, Košice, Slovakia (2010).

25. A. Landherr, B. Friedl, and J. Heidemann, A critical review of centrality measures in social networks, *Business & Information Systems Engineering* **2**(6) (2010) 371–385.

26. S. Majstorović and G. Caporossi, Bounds and relations involving adjusted centrality of the vertices of a tree, submitted.

27. B. Mohar, Isoperimetric numbers of graphs, *Journal of Combinatorial Theory Series B* **47** (1989) 274–291.

28. S. Mukwembi, A note on diameter and the degree sequence of a graph, *Applied Mathematics Letters* **25** (2012) 175–178.

29. K. Phelps, Latin square graphs and their automorphism groups, *Ars Combinatoria* **7** (1979) 273–299.

30. J. Sedláček, Perfect and quasiperfect graphs, *Časopis pro pěstování matematiky* **100** (1975) 135–141.

31. P. Solé, Expanding and forwarding, *Discrete Applied Mathematics* **58** (1995) 67–78.

32. D. West, *Introduction to Graph Theory,* 2nd edn., Prentice Hall, Upper Saddle River, NJ (2001).

33. S. Wasserman and K. Faust, *Social Network Analysis. Methods and Applications*, Cambridge, U.K. (1994).

34. D.R. White and S.P. Borgatti, Betweenness centrality measures for directed graphs, *Social Networks* **16** (1994) 335–346.

Chapter 8

On a Variant Szeged and PI* Indices of Thorn Graphs

Mojgan Mogharrab and Reza Sharafdini

Contents

8.1 Introduction and Notions...259
8.2 Preliminaries..266
8.3 Main Results...269
8.4 Summary and Conclusion...275
References..276

8.1 Introduction and Notions

We present some of the basic definitions, notation, and terminology used in this chapter. Other terminologies will be introduced as they naturally occur in the text, and those concepts not defined can be found in [20].

In this chapter, all graphs are finite, undirected, connected, and without loops and multiple edges. The vertex and edge sets of a graph G are denoted by $V(G)$ and $E(G)$, respectively. We show an edge $e \in E(G)$ adjacent to the vertices $u, v \in V(G)$, by $e = uv$. A path in G is an alternating sequence of distinct vertices and edges beginning and ending with vertices in which each edge joins the vertex before it to the one following it. Under distance $d_G(u, v)$ between vertices $u, v \in V(G)$, we mean the standard distance of the simple graph G, that is, the number of edges on a shortest path connecting these vertices in G [4]. The degree of a vertex $v \in V(G)$, denoted by $\deg_G(v)$, is defined to be the number of vertices in $V(G)$ of distance one from v. A graph is called *d-regular* if all its vertices have the same degree d.

Two graphs G and H are said to be isomorphic if there exists a bijection ψ : $V(G) \to V(H)$ such that $uv \in E(G)$ if and only if $\psi(u)\psi(v) \in E(H)$.

---

* PI stands for "Padmakar–van". Padmakar is the first name of P.V. Khadikar, the inventor of the PI index [22], and Ivan comes from Ivan Gutman, one of the leading researchers in the field of theoretical chemistry [12–19].

A *tree* is a connected graph with no cycles. For a graph T, the following are equivalent:

1. T is a tree.

2. T is connected and has $|V(T)| - 1$ edges.

3. T has $|V(T)| - 1$ edges and no cycles.

4. For each pair of vertices of T, there exists a unique path between them.

In the class of trees on $n + 1$ vertices, there exists a unique (up to isomorphism) tree with exactly n vertices of degree one. It is called the *star* on $n+1$ vertices and denoted by $K_{1,n}$.

The reason why we have considered the concepts of graphs and specially trees is due to an obvious analogy existing between the structural formulas used in chemistry and graph theory. A *molecular graph* is a graph in which the vertices are atoms and the edges are chemical bonds of a given molecule. The degree of each vertex in such graphs is at most four. Hydrogen atoms add very little to the overall size of the molecule due to their small size, so hydrogen-suppressed molecular graphs are the commonly used representation in chemical graph theory.

Historically, Cayley [6] was the first scholar who considered the concept of molecular graphs. He introduced two types of such graphs, naming them kenograms and plerograms. According to Cayley, hydrogen-suppressed molecular graph is called kenogram (see, e.g., Figure 8.1), whereas if every atom in a molecule is represented by a vertex, then what we get is called a plerogram.

Graph theory has successfully provided chemists with a variety of useful tools [1,19], among which are the topological indices or molecular graph-based structure descriptors [33–35]. A *topological index* of a (molecular) graph G is a quantity $T(G)$, which is invariant under graph isomorphism. Indeed, $T(G) = T(H)$ for each graph H being isomorphic to G. Topological indices are nowadays extensively used in theoretical chemistry for the design of the so-called *quantitative structure-property relations*

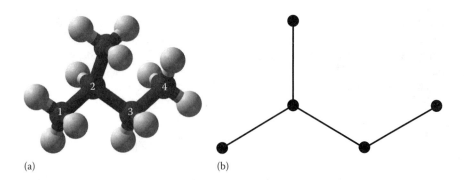

(a)   (b)

FIGURE 8.1:   (a) The plerogram of isopentane and (b) the kenogram of isopentane.

(QSPRs) and *quantitative structure–activity relations (QSARs)*, where under *property* are meant the physicochemical properties and under *activity* the pharmacologic and biological activities of the respective chemical compounds. For example, the Wiener index of a (molecular) graph G, denoted by $W(G)$, is defined as the sum of all distances between the vertices of a graph [41], that is,

$$W(G) = \sum_{\{u,v\} \subseteq V(G)} d_G(u, v).$$

The Wiener index found its first, simplest, and most straightforward applications within modeling of the properties of acyclic molecules, the so-called *alkanes* (in older times also called *paraffins*). In fact, let $\mathcal{A}$ be an alkane with the chemical formula C_nH_{2n+2} and consider the respective kenogram G. In 1974, Wiener [41] proved that the boiling point of $\mathcal{A}$, denoted by $t_B(\mathcal{A})$, is computed by the linear formula

$$t_B(\mathcal{A}) = aW(G) + bP(G) + c,$$

where a, b, and c are constants related to the corresponding isomeric group and $P(G)$ is the polarity of G defined as follows:

$$P(G) = \left| \left\{ \{u, v\} \subseteq V(G) \mid d_G(u, v) = 3 \right\} \right|.$$

In this chapter, we are concerned with some of these topological indices, which recently attracted much attention and found noteworthy chemical applications.

The *vertex-PI* index seems to be first considered by Khalifeh et al. [24] and is defined as

$$PI_v(G) = \sum_{e=uv \in E(G)} \left[n_u(e, G) + n_v(e, G) \right].$$

Here, the sum is taken over all edges of G, and for a given edge $e = uv \in E(G)$, the quantity $n_u(e, G)$ denotes the number of vertices of G closer to u than to v, and the quantity $n_v(e, G)$ is defined analogously. In fact, we can define the sets

$$N_u(e, G) = \left\{ x \in V(G) \mid d_G(x, u) < d_G(x, v) \right\},$$

$$N_v(e, G) = \left\{ x \in V(G) \mid d_G(x, v) < d_G(x, u) \right\},$$

and

$$N_0(e, G) = \left\{ x \in V(G) \mid d_G(x, v) = d_G(x, u) \right\},$$

by means of which we have:

$$n_u(e, G) = \left| N_u(e, G) \right|, \qquad n_v(e, G) = \left| N_v(e, G) \right|, \qquad n_0(e, G) = \left| N_0(e, G) \right|,$$

and

$$|V(G)| = n_u(e, G) + n_v(e, G) + n_0(e, G),$$

since

$$V(G) = N_u(e, G) \bigcup N_v(e, G) \bigcup N_0(e, G). \tag{8.1}$$

Given an edge $e = uv \in E(G)$, the distance between the edge $f = xy$ and the vertex u in the graph G, denoted by $d_G(f, u)$, is defined as

$$d_G(f, u) = \min \left\{ d_G(x, u), d_G(y, u) \right\}.$$

We can now introduce the sets

$$M_u(e, G) = \left\{ f \in E(G) \mid d_G(f, u) < d_G(f, v) \right\},$$
$$M_v(e, G) = \left\{ f \in E(G) \mid d_G(f, v) < d_G(f, u) \right\},$$

and

$$M_0(e, G) = \left\{ f \in E(G) \mid d_G(f, v) = d_G(f, u) \right\},$$

by means of which we have:

$$m_u(e, G) = \left| M_u(e, G) \right|, \qquad m_v(e, G) = \left| M_v(e, G) \right|, \qquad m_0(e, G) = \left| M_0(e, G) \right|,$$

and

$$|E(G)| = m_u(e, G) + m_v(e, G) + m_0(e, G),$$

due to the fact that

$$E(G) = M_u(e, G) \bigcup M_v(e, G) \bigcup M_0(e, G). \tag{8.2}$$

The *edge-PI* index is defined as [22]

$$PI_e(G) = \sum_{e = uv \in E(G)} \left[m_u(e, G) + m_v(e, G) \right].$$

Since this edge version was introduced first, the subscript e is usually omitted and the index is referred to simply as *PI* index. More details on the *PI* index are found in the review [23] and the references cited therein.

The *vertex-Szeged* index of a graph G is denoted by $Sz(G)$ and defined as [12]

$$Sz(G) = \sum_{e = uv \in E(G)} n_u(e, G) n_v(e, G). \tag{8.3}$$

It is obvious that an end-vertex of any edge is closer to itself than to the other end-vertex of that edge. Therefore, the product $n_u(e, G) n_v(e, G)$ is always positive. For more details on the Szeged index, see the review [17] and the references cited therein.

The following theorem is due to Dobrynin and Gutman [9].

Theorem 8.1.1 *Let G be a graph satisfying the following conditions:*

(a) *The shortest path between any two vertices of G is unique;*

(b) *For each edge $e = uv \in E(G)$, if $x \in N_u(e, G)$ and $y \in N_v(e, G)$, then, and only then, the shortest path between x and y contains the edge e.*

Then, $W(G) = Sz(G)$.

A graph G is called *bipartite* if $V(G)$ can be partitioned into two subsets A and B such that each edge of G connects a vertex in A to a vertex in B. It is well known that a graph G is bipartite if and only if it does not have an odd cycle. This implies that G is bipartite if and only if $n_0(e, G) = 0$ for all $e \in E(G)$.

The λ-*variable Wiener* indices of a tree T are defined as [37]

$$_\lambda W(T) = \frac{1}{2} \sum_{e=uv \in E(T)} \left[|V(T)|^\lambda - n_u(e, T)^\lambda - n_v(e, T)^\lambda \right].$$

Note that λ-variable Wiener indices can be defined for any connected graph. In this case, for $\lambda = 2$, we have

$$Sz(G) =_2 W(G) - \frac{1}{2}|E(G)||V(G)|^2 + \frac{1}{2} \sum_{e=uv \in E(G)} [n_u(e, G) + n_v(e, G)]^2,$$

and $Sz(G) =_2 W(G)$ if and only if G is bipartite.

The Szeged index does not take into account the contributions of the vertices that are equidistance from the endpoints of an edge. The problem occurs when the graph is bipartite. Therefore, Randić [30] introduced the *revised vertex-Szeged* index of a graph G, which shows to be a better descriptor for structure–property relationships for cyclic molecules. It is denoted by $Sz^\star(G)$ and defined as

$$Sz^\star(G) = \sum_{e=uv} \left(n_u(e, G) + \frac{n_0(e, G)}{2} \right) \left(n_v(e, G) + \frac{n_0(e, G)}{2} \right). \tag{8.4}$$

The *edge-Szeged* index is obtained by replacing $n_u(e, G)\, n_v(e, G)$ in Equation 8.3 by $m_u(e, G)\, m_v(e, G)$. Hence, the edge version of the Szeged index is given by [16]

$$Sz_e(G) = \sum_{e=uv \in E(G)} m_u(e, G)\, m_v(e, G).$$

The *revised edge-Szeged* index of a graph G is denoted by $Sz_e^\star(G)$ and defined as [11]

$$Sz_e^\star(G) = \sum_{e=uv} \left(m_u(e, G) + \frac{m_0(e, G)}{2} \right) \left(m_v(e, G) + \frac{m_0(e, G)}{2} \right). \tag{8.5}$$

The *edge-vertex-Szeged* index of a graph G is denoted by $Sz_{ev}(G)$ and defined as [25]

$$Sz_{ev}(G) = \frac{1}{2} \sum_{e=uv \in E(G)} \left[n_u(v, G)m_v(u, G) + n_v(u, G)m_u(v, G) \right].$$

The *total Szeged* index of a graph G is denoted by $Sz_t(G^*)$ and defined as

$$Sz_t(G^*) = \sum_{e=uv} \left(m_u(e, G) + n_u(e, G) \right) \left(m_v(e, G) + n_v(e, G) \right). \qquad (8.6)$$

It is straightforward to check that

$$Sz_t(G) = Sz(G) + Sz_e(G) + Sz_{ev}(G).$$

Given a graph G, the thorn graph $G^* = G^* \left(p_u \mid u \in V(G) \right)$ $(p_u \geq 0$ for all $u \in V(G))$ of G is obtained from G by attaching to each vertex $u \in V(G)$, p_u new vertices of degree one. The p_u new vertices attached to the vertex u are called thorns of u and G is called the parent of G^*(see Figure 8.2).

Special cases of thorn graphs attracting much attention are those in which for each vertex u of the parent graph G,

$$p_u := a \deg_G(u) + b,$$

where a and b are constant real numbers. Such thorn graphs are denoted by $G^*_{a,b}$. Thorn graphs in chemistry may be viewed as the hydrogen-completed graphs (or plerogram) of parent H-deleted graphs (or kenogram). For example, if T is a tree, then $T^*_{-1,b}$ is a special case of thorn graphs having a chemical relevance. For example, let T be a tree depicted in Figure 8.3, and then $T^*_{-1,4}$ is isomorphic to the molecular

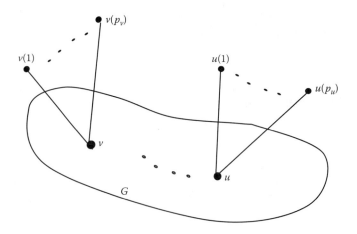

FIGURE 8.2:   The thorn graph $G^* \left(p_u \mid u \in V(G) \right)$.

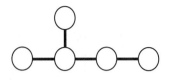

FIGURE 8.3: The parent graph (*H*-deleted graph or plerogram) of isopentane and a polyeneoid.

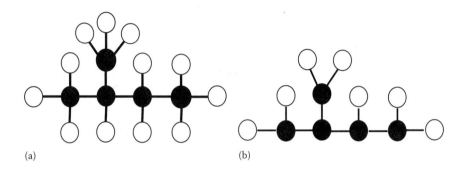

(a)   (b)

FIGURE 8.4: (a) The molecular graph (or kenogram) of isopentane and (b) the molecular graph (or kenogram) of a polyeneoid.

graph of an alkane (isopentane), while $T^*_{-1,3}$ is isomorphic to the molecular graph of a polyeneoid (see Figure 8.4).

Now let us describe a kind of composite graph using the concept of thorn graphs. The cluster $G\{H\}$ is obtained by taking one copy of G and $|V(G)|$ copies of a rooted graph H and by identifying the root of the ith copy of H with the ith vertex of G, $1 \le i \le |V(G)|$. The composite graph $G\{H\}$ was studied by Schwenk [32]; the name cluster is proposed here for the first time in [42]. If $p_u = p$ for all $u \in V(G)$, then the thorn graph $G^*\left(p_u \mid u \in V(G)\right)$ is isomorphic to $G\{K_{1,p}\}$, where $G\{K_{1,p}\}$ is isomorphic to $G^*_{0,p}$.

The classes of alkanes and polyeneoid compounds have only two types of vertices: nonterminal vertices of degree d and terminal vertices of degree 1. In the graph theory, such graphs are referred to as *proper graph* [20], while in the concept of thorn graphs, it is the so-called *d-thorn* graphs [5]. Let G be a d-thorn graph and let H be a graph obtained from G by deleting all its terminal vertices. It is easy to verify that G is isomorphic to $H^*_{-1,d}$, since $d = \deg_H(u) + p_u$ where p_u denotes the number of terminal adjacent vertices of u in G. Another class of d-thorn graphs is dendrimers, denoted by $D(n,d)$; they are by definition central trees with a center u_0 and terminal vertices of distance n from u_0 and all nonterminal vertices are of degree d. Note that $D(n+1,d)$ is isomorphic to $D(n,d)^*_{-1,d}$ (see Figure 8.5).

Special cases of thorn graphs have been already considered by Cayley [6] and later by Pólya, while they were trying to establish a fundamental enumeration theory

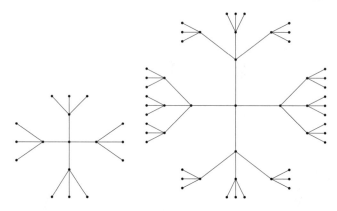

FIGURE 8.5:   Dendrimers $D(2,4)$ and $D(3,4)$.

to enumerate the number of distinct isomers of structural formula C_nH_{2n+2} [28,29]. Years later, Gutman [14] established relations between the Wiener indices of G^* and G. Since then, this concept has attracted much attention and several investigations of different graph invariants of general and some particular thorn graphs like Wiener number [3,5,45], modified Wiener index [36,43], variable Wiener index [44], altered Wiener index [39], terminal Wiener index [18,21], Hosoya polynomial [26,40], Zagreb polynomials [27], Schultz index [38], eccentric connectivity index [8], and augmented eccentric connectivity index [7] have already been considered.

    The aim of this article is to contribute to the theory of the previously described topological indices by showing how these can be computed in the case of thorn graphs.

8.2   Preliminaries

    Let G be a graph $G^* = G^*(p_u \mid u \in V(G))$ and let us denote the number of thorns attached to G by $T(G^*)$, that is, $T(G^*) = \sum_{u \in V(G)} p_u$. It follows from the definition of G^* that

$$V(G^*) = V(G) \bigcup \bigcup_{u \in V(G)} \left\{ u(i) \mid 1 \leq i \leq p_u \right\} \tag{8.7}$$

$$E(G^*) = E(G) \bigcup \bigcup_{u \in V(G)} \left\{ uu(i) \mid 1 \leq i \leq p_u \right\}. \tag{8.8}$$

Consequently,

$$|V(G^*)| = |V(G)| + T(G^*) \qquad |E(G^*)| = |E(G)| + T(G^*).$$

For each $e = uv \in E(G)$, we define

$$T_u(e, G^*) := \sum_{x \in N_u(e,G)} p_x, \qquad T_0(e, G^*) := \sum_{x \in N_0(e,G)} p_x.$$

It follows from Equation (8.1) that

$$T_u(e, G^*) + T_v(e, G^*) = T(G^*) - T_0(e, G^*).$$

For each $e = uv \in E(G)$, we define

$$\alpha_u(e, G) := \sum_{x \in N_u(e,G)} \deg_G(x), \qquad \alpha_0(e, G) := \sum_{x \in N_0(e,G)} \deg_G(x).$$

In this case,

$$2|E(G)| = \sum_{x \in V(G)} \deg_G(x) = \alpha_u(e, G) + \alpha_v(e, G) + \alpha_0(e, G).$$

It is obvious that any cycle of G^* is a cycle of G and vice versa. It follows that G is bipartite if and only if so is G^*. Let us summarize our discussion in the following lemma.

Lemma 8.2.1 *The following conditions are equivalent:*

(i) *G is bipartite.*

(ii) *$PI_v(G) = |E(G)||V(G)|.$*

(iii) *$T(G^*) = T_u(e, G^*) + T_v(e, G^*)$ for all $e \in E(G)$.*

(iv) *G^* is bipartite.*

(v) *$PI_v(G^*) = \big(|E(G)| + T(G^*)\big)\big(|V(G)| + T(G^*)\big).$*

(vi) *$2|E(G)| = \alpha_u(e, G) + \alpha_v(e, G).$*

We need to recall the following observation to discuss a special case of thorn graphs $G^*_{a,b}$.

The degree analogue of the Wiener index is defined as [31]

$$D(G) = \sum_{\{x,y\} \subseteq V(G)} \Big(\deg_G(x) + \deg_G(y)\Big) d_G(x, y),$$

which is known as the *Schultz index* [10,13]. Gutman in [13] examined another degree analogue of the Wiener index,

$$D^*(G) = \sum_{\{x,y\} \subseteq V(G)} \Big(\deg_G(x) \deg_G(y)\Big) d_G(x, y),$$

and later in [15] called it the *modified Schultz index*, but for which the name *Gutman index* has also been used [33]. As a generalization of Theorem 8.1.1, the following hold [2]:

Theorem 8.2.1 *Let G be a graph satisfying the conditions (a) and (b) of Theorem 8.1.1. Then,*

(i) $D(G) = \displaystyle\sum_{e=uv \in E(G)} \left[n_u(e, G)\alpha_v(e, G) + n_v(e, G)\alpha_u(e, G) \right].$

(ii) $D^*(G) = \displaystyle\sum_{e=uv \in E(G)} \left[\alpha_u(e, G)\alpha_v(e, G) \right].$

Lemma 8.2.2 *Given a graph G, the following hold for the thorn graph $G^*_{a,b}$:*

(i) $T_u(e, G^*_{a,b}) = a\alpha_u(e, G) + bn_u(e, G)$ *and for each $e = uv \in E(G)$,*

$$T_0(e, G^*_{a,b}) = a\alpha_0(e, G) + bn_0(e, G).$$

(ii) $T(G^*_{a,b}) = 2a|E(G)| + b|V(G)|.$

(iii) $\displaystyle\sum_{e=uv \in E(G)} \left[T_u(e, G^*_{a,b}) + T_v(e, G^*_{a,b}) \right] = bPI_v(G) + 2a|E(G)|^2 -$

$a \displaystyle\sum_{e=uv \in E(G)} \alpha_0(e, G).$

(iv) *If G satisfies the conditions (a) and (b) of Theorem 8.1.1, then*

$$\sum_{e=uv \in E(G)} T_u(e, G^*_{a,b})T_v(e, G^*_{a,b}) = a^2 D^*(G) + abD(G) + b^2 Sz(G),$$

$$\sum_{e=uv \in E(G)} \left[n_u(e, G)T_v(e, G^*_{a,b}) + n_v(e, G)T_u(e, G^*_{a,b}) \right] = aD(G) + bPI_v(G).$$

We present the vertices attached to the vertex $u \in V(G)$ by $u(1), \ldots, u(p_u)$. For all $u, v \in V(G)$, $1 \leq i \leq p_u$, and $1 \leq j \leq p_v$, we have the following:

$$d_{G^*}(u, v) = d_G(u, v), \qquad d_{G^*}\left(u(i), v(j) \right) = d_G(u, v) + 2.$$

It is easy to see that if $e = uu(t) \in E(G^*)$ such that $1 \leq t \leq p_u$, then

$$N_u(e, G^*) = V(G) \bigcup_{u \neq v \in V(G)} \left\{ v(i) \mid \leq i \leq p_v \right\} \bigcup \left\{ u(i) \mid 1 \leq i \leq p_u, i \neq t \right\},$$

$$N_{u(t)}(e, G^*) = \left\{ u(t) \right\},$$

$$M_u(e, G^*) = E(G) \bigcup_{u \neq v \in V(G)} \left\{ vv(i) \mid \leq i \leq p_v \right\} \bigcup \left\{ uu(i) \mid 1 \leq i \leq p_u, i \neq t \right\}.$$

And also if $e = uv \in E(G)$, then

$$N_u(e, G^*) = N_u(e, G) \bigcup \bigcup_{x \in N_u(e,G)} \left\{ x(i) \mid 1 \leq i \leq p_x \right\},$$

$$N_0(e, G^*) = N_0(e, G) \bigcup \bigcup_{x \in N_0(e,G)} \left\{ x(i) \mid 1 \leq i \leq p_x \right\},$$

$$M_u(e, G^*) = M_u(e, G) \bigcup \bigcup_{x \in N_u(e,G)} \left\{ xx(i) \mid 1 \leq i \leq p_x \right\}.$$

Therefore the following hold:

$$n_u(e, G^*) = \begin{cases} |V(G)| + T(G^*) - 1 & \text{if} \quad e = uu(i), 1 \leq i \leq p_u \\ n_u(e, G) + T_u(e, G^*) & \text{if} \quad e = uv \in E(G) \end{cases} ; \qquad (8.9)$$

$$n_0(e, G^*) = n_0(e, G) + T_0(e, G^*), \qquad \text{where} \quad e = uv \in E(G); \qquad (8.10)$$

$$n_{u(i)}\left(uu(i), G^* \right) = 1, \qquad 1 \leq i \leq p_u; \qquad (8.11)$$

$$n_0\left(uu(i), G^* \right) = 0, \qquad 1 \leq i \leq p_u; \qquad (8.12)$$

$$m_u(e, G^*) = \begin{cases} |E(G)| + T(G^*) - 1 & \text{if} \quad e = uu(i), 1 \leq i \leq p_u \\ m_u(e, G) + T_u(e, G^*) & \text{if} \quad e = uv \in E(G); \end{cases} ; \qquad (8.13)$$

$$m_0(e, G^*) = m_0(e, G) + T_0(e, G^*); \qquad e = uv \in E(G); \qquad (8.14)$$

$$m_{u(i)}\left(uu(i), G^* \right) = 0, \qquad 1 \leq i \leq p_u; \qquad (8.15)$$

$$m_0\left(uu(i), G^* \right) = 1, \qquad 1 \leq i \leq p_u. \qquad (8.16)$$

8.3 Main Results

Proposition 8.3.1

$$PI_v(G^*) = PI_v(G) + \sum_{e=uv \in E(G)} \left[T_u(e, G^*) + T_v(e, G^*) \right] + \left(|V(G)| + T(G^*) \right) T(G^*).$$

Besides,

$$PI_v(G^*_{a,b}) = (b+1)PI_v(G) + 2a|E(G)|^2 - a \sum_{e=uv \in E(G)} \alpha_0(e, G)$$

$$+ \left((b+1)|V(G)| + 2a|E(G)| \right) \left(2a|E(G)| + b|V(G)| \right).$$

Proof. It follows from Equations 8.8 and 8.9 that

$$PI_v(G^*) = \sum_{e=uv\in E(G^*)} \left[n_u(e, G^*) + n_v(e, G^*) \right] = \sum_{e=uv\in E(G)} \left[n_u(e, G^*) + n_v(e, G^*) \right]$$

$$+ \sum_{u\in V(G)} \sum_{i=1}^{p_u} \left[n_u(uu_i, G^*) + n_{u(i)}(uu_i, G^*) \right]$$

$$= \sum_{e=uv\in E(G)} \left[[n_u(e, G) + T_u(e, G^*)] + [n_v(e, G) + T_v(e, G^*)] \right]$$

$$+ \sum_{u\in V(G)} \sum_{i=1}^{p_u} \left[|V(G)| + T(G^*) - 1 + 1 \right]$$

$$= PI_v(G) + \sum_{e=uv\in E(G)} \left[T_u(e, G^*) + T_v(e, G^*) \right] + \left(|V(G)| + T(G^*) \right) T(G^*).$$

The second part is done by Lemma 8.2.2 and the first part.   □

Corollary 8.3.1

$$PI_v(G^*_{a,b}) = (b+1)PI_v(G) + 2a|E(G)|^2 - a \sum_{e=uv\in E(G)} \alpha_0(e, G)$$

$$+ \left((b+1)|V(G)| + 2a|E(G)| \right)\left(2a|E(G)| + b|V(G)| \right).$$

Besides, if G is bipartite, then

$$PI_v(G^*_{a,b}) = (b+1)PI_v(G) + 2a|E(G)|^2$$

$$+ \left((b+1)|V(G)| + 2a|E(G)| \right)\left(2a|E(G)| + b|V(G)| \right).$$

Corollary 8.3.2 *If T is a tree, then*

$$PI_v(T^*_{a,b}) = (b+1)|V(T)||E(T)| + 2a|E(G)|^2$$

$$+ \left((2a+b+1)|E(G)| + b + 1 \right)\left((2a+b)|E(T)| + b \right).$$

Corollary 8.3.3

$$PI_v(G\{K_{1,p}\}) = (p+1)PI_v(G) + |V(G)|^2 p(p+1).$$

Proposition 8.3.2

$$Sz(G^*) = Sz(G) + \sum_{e=uv\in E(G)} \left[n_u(e, G)T_v(e, G^*) + n_v(e, G)T_u(e, G^*) \right]$$

$$+ \sum_{e=uv\in E(G)} \left[T_u(e, G^*)T_v(e, G^*) \right] + \left[|V(G)| + T(G^*) - 1 \right] T(G^*).$$

Proof. It follows from Equations 8.8 and 8.9 that

$$Sz(G^*) = \sum_{e=uv\in E(G^*)} n_u(e, G^*)n_v(e, G^*)$$

$$= \sum_{e=uv\in E(G)} \left[n_u(e, G^*)n_v(e, G^*) \right] + \sum_{u\in V(G)} \sum_{i=1}^{p_u} \left[n_u(uu_i, G^*)n_{u_i}(uu_i, G^*) \right]$$

$$= \sum_{e=uv\in E(G)} \left[[n_u(e, G) + T_u(e, G^*)][n_v(e, G) + T_v(e, G^*)] \right]$$

$$+ \sum_{u\in V(G)} \sum_{i=1}^{p_u} \left[|V(G)| + T(G^*) - 1 \right]$$

$$= \sum_{e=uv\in E(G)} \left[n_u(e, G)n_v(e, G) + n_u(e, G)T_v(e, G^*) + n_v(e, G)T_u(e, G^*) \right.$$

$$\left. + T_u(e, G^*)T_v(e, G^*_{a,b}) \right] + \sum_{u\in V(G)} p_u \left[|V(G)| + T(G^*) - 1 \right]$$

$$= Sz(G) + \sum_{e=uv\in E(G)} \left[n_u(e, G)T_v(e, G^*_{a,b}) + n_v(e, G)T_u(e, G^*_{a,b}) \right]$$

$$+ \sum_{e=uv\in E(G)} \left[T_u(e, G^*)T_v(e, G^*_{a,b}) \right] + \left[|V(G)| + T(G^*) - 1 \right] T(G^*). \qquad \square$$

The following is a direct consequence of Proposition 8.3.2 and Lemma 8.2.2.

Corollary 8.3.4 *If G satisfies the conditions (a) and (b) of Theorem 8.1.1, then*

$$Sz(G^*_{a,b}) = (b^2 + 1)Sz(G) + (ab + a)D(G) + a^2 D^*(G) + bPI_v(G)$$
$$+ \left(2a|E(G)| + b|V(G)| \right)\left((b+1)|V(G)| + 2a|E(G)| - 1 \right).$$

Especially for a tree T, the following is proved:

$$Sz(T^*_{a,b}) = (b^2 + 1)Sz(T) + (ab + a)D(T) + a^2 D^*(T) + b\left(|V(T)|(|V(T)| - 1) \right)$$
$$+ \left((2a+b)|V(T)| - 2a \right)\left((2a+b)|V(T)| - 2a - 1 \right).$$

Corollary 8.3.5

$$Sz\left(G\{K_{1,p}\} \right) = \left(1 + p^2 \right)Sz(G) + pPI_v(G) + p|V(G)|\left((p+1)|V(G)| - 1 \right).$$

Proposition 8.3.3

$$Sz^{\star}(G^{*}) = Sz^{\star}(G) + \sum_{e=uv\in E(G)} \left(n_u(e,G) + \frac{1}{2}n_0(e,G)\right)\left(T_v(e,G) + \frac{1}{2}T_0(e,G)\right)$$

$$+ \sum_{e=uv\in E(G)} \left(n_v(e,G) + \frac{1}{2}n_0(e,G)\right)\left(T_u(e,G) + \frac{1}{2}T_0(e,G)\right)$$

$$+ T(G^{*})\Big(|V(G)| + T(G^{*}) - 1\Big).$$

Proof. It follows from Equations 8.8 through 8.11 and 8.16 that

$$Sz^{\star}(G^{*}) = \sum_{e=uv\in E(G^{*})} \left(n_u(e,G^{*}) + \frac{n_0(e,G)}{2}\right)\left(n_v(e,G^{*}) + \frac{n_0(e,G)}{2}\right)$$

$$= \sum_{e=uv\in E(G)} \left(n_u(e,G^{*}) + \frac{n_0(e,G)}{2}\right)\left(n_v(e,G^{*}) + \frac{n_0(e,G)}{2}\right)$$

$$+ \sum_{u\in V(G)} \sum_{i=1}^{p_u} \left(n_u(uu_i,G^{*}) + \frac{n_0(uu_i,G)}{2}\right)\left(n_{u_i}(uu_i,G^{*}) + \frac{n_0(uu_i,G)}{2}\right)$$

$$= \sum_{e=uv\in E(G)} \Bigg[\left((n_u(e,G) + T_u(e,G^{*}) + \frac{1}{2}\Big(n_0(e,G) + T_0(e,G^{*})\Big)\right)$$

$$\times \left((n_v(e,G) + T_v(e,G^{*}) + \frac{1}{2}\Big(n_0(e,G) + T_0(e,G^{*})\Big)\right)\Bigg]$$

$$+ \sum_{u\in V(G)} \sum_{i=1}^{p_u} \Big(|V(G)| + T(G^{*}) - 1\Big)$$

$$= Sz^{\star}(G) + \sum_{e=uv\in E(G)} \left(n_u(e,G) + \frac{1}{2}n_0(e,G)\right)\left(T_v(e,G) + \frac{1}{2}T_0(e,G)\right)$$

$$+ \sum_{e=uv\in E(G)} \left(n_v(e,G) + \frac{1}{2}n_0(e,G)\right)\left(T_u(e,G) + \frac{1}{2}T_0(e,G)\right)$$

$$+ T(G^{*})\Big(|V(G)| + T(G^{*}) - 1\Big). \qquad \square$$

Corollary 8.3.6

$$Sz^{\star}(G^{*}_{a,b}) = Sz^{\star}(G) + \sum_{e=uv\in E(G)} \left(n_u(e,G) + \frac{1}{2}n_0(e,G)\right)$$

$$\times \left(a\alpha_v(e,G) + bn_v(e,G) + \frac{a}{2}\alpha_0(e,G) + \frac{b}{2}n_0(e,G)\right)$$

$$+ \sum_{e=uv\in E(G)} \left(n_v(e,G) + \frac{1}{2}n_0(e,G)\right)$$

$$\times \left(a\alpha_u(e, G) + bn_u(e, G) + \frac{a}{2}\alpha_0(e, G) + \frac{b}{2}n_0(e, G) \right)$$
$$+ \left(2a|E(G)| + b|V(G)| \right)\left((b+1)|V(G)| + 2a|E(G)| - 1 \right).$$

A graph is called distance balanced if $n_u(e, G) = n_v(e, G)$ for each edge $e = uv \in E(G)$.

Corollary 8.3.7 *For a d-regular and distance balanced graph G, the following hold:*

$$Sz^\star(G_{a,b}^*) = Sz^\star(G) + \frac{ad+b}{2}|E(G)||V(G)|^2$$
$$+ \left(2a|E(G)| + b|V(G)| \right)\left((b+1)|V(G)| + 2a|E(G)| - 1 \right).$$

Proof. If G is a d-regular and distance balanced graph, then for all $e = uv \in E(G)$, we have

$$\left(a\alpha_v(e, G) + bn_v(e, G) + \frac{a}{2}\alpha_0(e, G) + \frac{b}{2}n_0(e, G) \right) = \frac{ad+b}{2}\left(2n_v(e, G) + n_0(e, G) \right)$$
$$= \frac{ad+b}{2}|V(G)|.$$

In the same way, we have

$$\left(a\alpha_u(e, G) + bn_u(e, G) + \frac{a}{2}\alpha_0(e, G) + \frac{b}{2}n_0(e, G) \right) = \frac{ad+b}{2}\left(2n_u(e, G) + n_0(e, G) \right)$$
$$= \frac{ad+b}{2}|V(G)|.$$

It follows from Corollary 8.3.6 that

$$Sz^\star(G_{a,b}^*) = Sz^\star(G) + \sum_{e=uv\in E(G)} \frac{ad+b}{2}|V(G)|$$
$$\times \left(n_u(e, G) + \frac{1}{2}n_0(e, G) + n_v(e, G) + \frac{1}{2}n_0(e, G) \right)$$
$$+ \left(2a|E(G)| + b|V(G)| \right)\left((b+1)|V(G)| + 2a|E(G)| - 1 \right)$$
$$= Sz^\star(G) + \frac{ad+b}{2}|E(G)||V(G)|^2$$
$$+ \left(2a|E(G)| + b|V(G)| \right)\left((b+1)|V(G)| + 2a|E(G)| - 1 \right). \qquad \square$$

Corollary 8.3.8

$$Sz^\star\left(G\{K_{1,p}\} \right) = (2p+1)Sz^\star(G) + p|V(G)|\left((p+1)|V(G)| - 1 \right).$$

Proposition 8.3.4

$$PI_e(G^*) = PI_e(G) + \sum_{e=uv \in E(G)} \left[T_u(e, G^*) + T_v(e, G^*) \right]$$

$$+ T(G^*) \left[|E(G)| + T(G^*) - 1 \right],$$

and

$$PI_e(G^*) = PI_e(G) + 2|E(G)|T(G^*) + T(G^*) \left[T(G^*) - 1 \right]$$

if and only if G is bipartite.

Proof. It follows from Equations 8.8 and 8.13 that

$$PI_e(G^*) = \sum_{e=uv \in E(G^*)} \left[m_u(e, G^*) + m_v(e, G^*) \right]$$

$$= \sum_{e=uv \in E(G)} \left[m_u(e, G^*) + m_v(e, G^*) \right]$$

$$+ \sum_{u \in V(G)} \sum_{i=1}^{p_u} \left[m_u(uu(i), G^*) + m_{u(i)}(uu(i), G^*) \right]$$

$$= \sum_{e=uv \in E(G)} \left[m_u(e, G) + T_u(e, G^*) + m_v(e, G) + T_v(e, G^*) \right]$$

$$+ \sum_{u \in V(G)} p_u \left[|E(G)| + T(G^*) - 1 \right].$$

Consequently,

$$PI_e(G^*) = PI_e(G) + \sum_{e=uv \in E(G)} \left[T_u(e, G^*) + T_v(e, G^*) \right]$$

$$+ T(G^*) \left[|E(G)| + T(G^*) - 1 \right].$$

The second part is a direct consequence of the first part and Lemma 8.2.1.   □

As a direct consequence of Lemma 8.2.2 and the previous proposition, the following is proved:

Corollary 8.3.9

$$PI_e(G^*_{a,b}) = PI_e(G) + bPI_v(G) + 2a|E(G)|^2 - a \sum_{e=uv \in E(G)} \alpha_0(e, G)$$

$$+ \left(2a|E(G)| + b|V(G)| \right) \left((2a+1)|E(G)| + b|V(G)| - 1 \right).$$

Corollary 8.3.10

$$PI_e\Big(G\{K_{1,p}\}\Big) = PI_e(G) + pPI_v(G) + \Big(|E(G)| + p|V(G)| - 1\Big)p|V(G)|.$$

Proposition 8.3.5

$$Sz_e(G^*) = Sz_e(G) + \sum_{uv\in E(G)} \Big[m_u(e,G)T_v(e,G^*) + m_v(e,G)T_u(e,G^*)\Big]$$

$$+ \sum_{uv\in E(G)} \Big[T_v(e,G^*)T_u(e,G^*)\Big].$$

Proof.

$$Sz_e(G^*) = \sum_{e=uv\in E(G^*)} m_u(e,G^*)\, m_v(e,G^*)$$

$$= \sum_{e=uv\in E(G)} m_u(e,G^*)m_v(e,G^*) + \sum_{u\in V(G)}\sum_{i=1}^{p_u} \Big[m_u(uu(i),G^*)m_{u(i)}(uu(i),G^*)\Big]$$

$$= \sum_{e=uv\in E(G)} \Big[(m_u(e,G) + T_u(e,G^*))(m_v(e,G) + T_v(e,G^*))\Big]$$

$$= \sum_{e=uv\in E(G)} \Big[(m_u(e,G)(m_v(e,G) + (m_u(e,G)T_v(e,G^*)$$

$$+ m_v(e,G)T_u(e,G^*) + \sum_{xy\in E(G)} p_x p_y\Big]$$

$$= Sz_e(G) + \sum_{uv\in E(G)} \Big[m_u(e,G)T_v(e,G^*) + m_v(e,G)T_u(e,G^*)\Big]$$

$$+ \sum_{uv\in E(G)} \Big[T_v(e,G^*) \sum_{y\in N_u(e,G)} p_y\Big]. \qquad \square$$

Corollary 8.3.11

$$Sz_e\Big(G\{K_{1,p}\}\Big) = Sz_e(G) + pPI_v(G) + \sum_{e=uv\in E(G)} \Big[m_u(e,G)n_v(e,G) + m_v(e,G)n_u(e,G)\Big].$$

8.4 Summary and Conclusion

In this chapter, we compute a variant Szeged and PI indices of thorn graphs. In fact, we establish a linear relation between these indices of a thorn graph and the corresponding parent graph and examine several special cases of this result.

References

1. A.T. Balaban. *Chemical Applications of Graph Theory*. Academic Press, London, U.K., 1976.

2. A. Behtoei, M. Jannesari, and B. Taeri. Generalizations of some topological indices. *MATCH: Communications in Mathematical and in Computer Chemistry*, 65:71–78, 2011.

3. D. Bonchev and D.J. Klein. On the Wiener number of thorn trees, stars, rings, and rods. *Croatica Chemica Acta*, 75:613–620, 2002.

4. F. Buckley and F. Harary. *Distance in Graphs*. Addison-Wesley, Redwood, CA, 1990.

5. L. Bytautas, D. Bonchev, and D.J. Klein. On the generation of mean Wiener numbers of thorny graphs. *MATCH: Communications in Mathematical and in Computer Chemistry*, 44:31–40, 2001.

6. A. Cayley. On the mathematical theory of isomers. *Philosophical Magazine*, 47:444–446, 1874.

7. N. De. Augmented eccentric connectivity index of some thorn graph. *International Journal of Applied Mathematical Research*, 1(4):671–680, 2012.

8. N. De. On eccentric connectivity index and polynomial of thorn graph. *International Journal of Applied Mathematical Research*, 3:931–934, 2012.

9. A.A. Dobrynin and I. Gutman. On a graph invariant related to the sum of all distances in a graph. *Publications de l'Institut Mathématique (Beograd)*, 56(70): 18–22, 1994.

10. A.A. Dobrynin and A.A Kochetova. Degree distance of a graph: A degree analogue of the Wiener index. *Journal of Chemical Information and Computer Sciences*, 34(5):1082–1086, 1994.

11. H. Dong, B. Zhou, and C. Trinajstić. A novel version of the edge–Szeged index. *Croatica Chemica Acta*, 84:543–545, 2011.

12. I. Gutman. A formula for the Wiener number of trees and its extension to graphs containing cycles. *Graph Theory Notes New York*, 27:9–15, 1994.

13. I. Gutman. Selected properties of the Schultz molecular topological indexdobkoch-deg-dis. *Journal of Chemical Information and Computer Sciences*, 34(5):1087–1089, 1994.

14. I. Gutman. Distance of thorny graphs. *Publications de l'Institut Mathématique (Beograd)*, 63(77):31–36, 1998.

15. I. Gutman. Some relations between distance-based polynomials of trees, *Bulletin Classe des sciences mathematiques et natturalles*, 30:1–7, 2005.

16. I. Gutman and A.R. Ashrafi. The edge version of the Szeged index. *Croatica Chemica Acta*, 81:263–266, 2008.

17. I. Gutman and A.A. Dobrynin. The Szeged index—A success story. *Graph Theory Notes New York*, 34:37–44, 1998.

18. I. Gutman, B. Furtula, J. Tošović, M. Essalih, and M. El Marraki. On-terminal Wiener indices of kenograms and plerbograms. *Iranian Journal of Mathematical Chemistry*, 4(1):77–89, 2013.

19. I. Gutman and I.O.E. Polansky. *Mathematical Concepts in Organic Chemistry*. Springer-Verlag, Berlin, Germany, 1986.

20. F. Harary. *Graph Theory*. Addison-Wesley, Reading, MA, 1969.

21. A. Heydari and I. Gutman. On the terminal Wiener index of thorn graphs. *Kragujevac Journal of Science*, 32:57–64, 2010.

22. P.V. Khadikar. On a novel structural descriptor PI. *National Academy Science Letters*, 23:113–118, 2000.

23. P.V. Khadikar. Padamakar–Ivan index in nanotechnology. *Iranian Journal of Mathematical Chemistry*, 1:7–42, 2010.

24. M.H. Khalifeh, H. Yousefi-Azari, and A.R. Ashrafi. Vertex and edge PI indices of Cartesian product graphs. *Discrete Applied Mathematics*, 156(10):1780–1789, 2008.

25. M.H. Khalifeh, H. Yousefi-Azari, A.R. Ashrafi, and I. Gutman. The edge Szeged index of product graphs. *Croatica Chemica Acta*, 81(2):277–281, 2008.

26. D.J. Klein, T. Došlič, and D. Bonchev. Vertex-weightings for distance moments and thorny graphs. *Discrete Applied Mathematics*, 155:2294–2302, 2007.

27. S. Li. Zagreb polynomials of thorn graphs. *Kragujevac Journal of Science*, 33:33–38, 2011.

28. G. Polya. Kombinatorische anzahlbestimmungen fr gruppen, graphen und chemische verbindungen. *Acta Mathematica*, 155(17):145–254, 1937.

29. G. Polya and R.C. Read. *Combinatorial Enumeration of Groups, Graphs, and Chemical Compounds*. Springer-Verlag, Berlin, Germany, 1987.

30. M. Randić. On generalization of Wiener index to cyclic structures. *Acta Chimica Slovenica*, 49:483–496, 2002.

31. H.P. Schultz. Topological organic chemistry. 1. Graph theory and topological indices of alkanes. *Journal of Chemical Information and Computer Sciences*, 29:227–228, 1989.

32. A.J. Schwenk. Computing the characteristic polynomial of a graph. In A. Dold and B. Eckmann, eds., *Graphs and Combinatorics*, Lecture Notes in Mathematics, vol. 406. Springer, Berlin, Germany, 1974.

33. R. Todeschini and V. Consonni. *Handbook of Molecular Descriptors*. Wiley-VCH, New York, 2000.

34. R. Todeschini and V. Consonni. *Molecular Descriptors for Chemoinformatics*. Wiley-VCH, Weinheim, Germany, 2009.

35. N. Trinajstić. *Chemical Graph Theory*. CRC Press, Boca Raton, FL, 1992.

36. D. Vukičević and A. Graovac. On modified Wiener indices of thorn graphs. *MATCH: Communications in Mathematical and in Computer Chemistry*, 50:93–108, 2004.

37. D. Vukičević and J. Žerovnik. Variable Wiener indices. *MATCH: Communications in Mathematical and in Computer Chemistry*, 53:385–402, 2005.

38. D. Vukičević, B. Zhou, and N. Trinajstić. On the Schultz index of thorn graphs. *Internet Electronic Journal of Molecular Design*, 4:501–514, 2005.

39. D. Vukičević, B. Zhou, and N. Trinajstić. Altered Wiener indices of thorn trees. *Croatica Chemica Acta*, 80:283–285, 2007.

40. H.B. Walikar, H.S. Ramane, L. Sindagi, and S.S. Shirakol. Hosoya polynomial of thorn trees, rods, rings, and stars. *Kragujevac Journal of Science*, 28:47–56, 2006.

41. H. Wiener. Structural determination of paraffin boiling points. *Journal of the American Chemical Society*, 69:17–20, 1974.

42. Y.-N. Yeh and I. Gutman. On the sum of all distances in composite graphs. *Discrete Mathematics*, 135:359–365, 1994.

43. B. Zhou. On modified Wiener indices of thorn trees. *Kragujevac Journal of Mathematics*, 27:5–9, 2005.

44. B. Zhou, A. Graovac, and D. Vukičević. Variable Wiener indices of thorn graphs. *MATCH: Communications in Mathematical and in Computer Chemistry*, 56:375–382, 2006.

45. B. Zhou and D. Vukičević. On Wiener-type polynomials of thorn graphs. *Journal of Chemometrics*, 23(12):600–604, 2009.

Chapter 9

Wiener Index of Line Graphs

Martin Knor and Riste Škrekovski

Contents

9.1 Introduction..279
9.2 Indices in Chemical Graph Theory..280
9.3 Wiener Index..282
9.4 Wiener Index of Graphs and Their Line Graphs................................286
9.5 Graphs with Large Girth...288
9.6 Second Line Graph Iteration...291
9.7 Higher Line Graph Iterations..293
9.8 Summary and Conclusion..296
Acknowledgment...297
References...297

9.1 Introduction

We consider the molecular descriptor Wiener index, W, of graphs and their line graphs. This index plays a crucial role in organic chemistry. It was studied by chemists decades before it attracted attention of mathematicians. In fact, it was studied long time before the branch of discrete mathematics, which is now known as graph theory, was born. Nowadays, there are many indices known used to describe the molecules.

In this chapter, we first introduce the concept of topological indices and list some of them. Next, in the third section, we focus on the Wiener index and expose some of its properties. Furthermore, we compare the values $W(G)$ and $W(L(G))$, in particular when they are equal for G being in various classes of graphs. In addition, we expose some bounds of the Wiener index of the line graph in terms of the Gutman index of the original graph. In the next section, we consider the equality $W(G) = W(L(G))$ for graphs with large girth. Finally, we consider the same equality for trees but for higher iterations of line graphs, $W(T) = W^i(L(T))$. The sixth section is dedicated to the case $i = 2$. In the seventh section, we show that a solution of

$$W(L^i(T)) = W(T) \quad (i \geq 3)$$

exists only for $i = 3$, and it is one particular class of trees, all homeomorphic to the letter H. The smallest such tree has 388 vertices.

9.2   Indices in Chemical Graph Theory

Graphs and networks can be described in quantitative terms using different measures or indices. They function as a universal language to describe the chemical structure of molecules, the chemical reaction networks, ecosystems, financial markets, the World Wide Web, and social networks. In chemical graph theory, we refer to these measures as topological indices or molecular descriptors.

Considering chemical structures as graphs is an important methodology for understanding chemical structures and reactivity. In molecular graphs, the atoms are represented by vertices and the bonds by edges. In chemistry, the degree of a vertex is called its valence. Double bonds or lone-pair electrons can be represented by multiple edges and self loops [55,57]. In this way, graph theory provides simple rules by which chemists may obtain qualitative predictions about the structure and reactivity of various chemical compounds [59].

Topological indices are numerical invariants of molecular graphs, and they may be used as numerical descriptors to derive quantitative structure–property relationships (QSPR) or quantitative structure–activity relationships (QSAR). Both QSAR and QSPR are showing the tendency to predict the properties of a compound based on its molecule structure.

When talking about topological indices (as quantitative graph measures), one can distinguish them in groups, for example,

- Distance based (Wiener index, etc.),

- Degree based (Zagreb indices, etc.),

- Graph spectra based (Estrada index, etc.),

- Information-theoretic indices based on Shannon's entropy.

The oldest topological index related to molecular branching is the Wiener index [61], which was introduced in 1947 as the path number. The same quantity has been studied and referred to in mathematics as the gross status [41], the distance of graphs [24], and the transmission [60]. The *Wiener* index of a graph G, denoted by $W(G)$, is the sum of distances between all (unordered) pairs of vertices of G

$$W(G) = \sum_{\{u,v\} \subseteq V(G)} d(u,v). \tag{9.1}$$

Though the Wiener index is the most common topological characteristic of a graph, nowadays, we know over 200 indices, and we devote this section to listing a few of them.

For an edge $e = ij$, let $n_e(i)$ be the number of vertices of G being closer to i than to j and $n_e(j)$ be the number of vertices of G lying closer to j than to i. The *Szeged* index of a graph G is defined by

$$Sz(G) = \sum_{e=ij\in E(G)} n_e(i)n_e(j).$$

This invariant was introduced by Gutman [30] during a stay at the Attila Jozsef University in Szeged, and he named it after this place.

The first Zagreb index M_1 and the second Zagreb index M_2 were defined in [39] as

$$M_1(G) = \sum_{v\in V(G)} d(v)^2 \quad \text{and} \quad M_2(G) = \sum_{uv\in E(G)} d(u)d(v),$$

where $d(u)$ and $d(v)$ denote the degree of u and v. Zagreb indices are used by various researchers in their QSPR and QSAR studies [12], as well as in molecular complexity [54]. In 1975, the *Randić* index $R(G)$ of a graph G was defined as

$$R(G) = \sum_{uv\in E(G)} (d(u)d(v))^{-1/2}.$$

It has been proved to be suitable for measuring the extent of branching of the carbon-atom skeleton of saturated hydrocarbons [56].

In 1989, led by the idea of characterizing the alkanes, Schultz defined a new index that is degree and distance based [58]. Recently, this index is known as *Schultz index (of first kind)*, and it is defined by

$$S(G) = \sum_{\{u,v\}\subseteq V(G)} (d(u) + d(v))\, d(u,v).$$

Inspired by the Schultz index, Gutman [31] back in 1994 introduced a new index,

$$Gut(G) = \sum_{\{u,v\}\subseteq V(G)} d(u)d(v)d(u,v),$$

and named it the *Schultz index of second kind*. Nowadays, this index is also known as the *Gutman* index.

All of the aforementioned topological indices are degree- and distance-based molecular descriptors, but there are also indices of a different kind. The *Hosoya* index, also known as the Z index, of a graph describes the total number of matchings within the graph. This index was introduced by Hosoya in [42] and is often used for investigations of organic compounds [44]. A high correlation exists between the Hosoya index and the boiling points of acyclic alkanes.

The *Estrada* index was introduced in 2000 as a measure of the degree of a protein folding [25]. Later, the Estrada index was used also to measure the centrality of other complex networks, such as communication, social, and metabolic networks [26,27]. This index includes the eigenvalues λ_i, $i = 1,\ldots,n$, of the adjacency matrix of a graph G and is defined as

$$EE(G) = \sum_{i=1}^{n} e^{\lambda_i}.$$

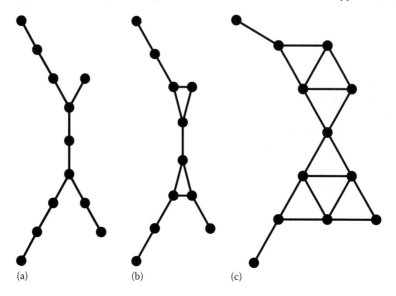

FIGURE 9.1:   (a) Graph, (b) its line graph, and (c) its second iterated line graph.

For the entropy-based indices, see the books Bonchev [6] and Dehmer [9] and the articles [10,11] as well.

Since we focus on line graphs in this chapter, let us recall the definition of (iterated) line graphs. Let G be a graph. Its *line graph*, $L(G)$, has vertex set identical with the set of edges of G and two vertices of $L(G)$ are adjacent if and only if the corresponding edges are adjacent in G (see Figure 9.1 for illustration). *Iterated* line graphs are defined inductively as follows:

$$L^i(G) = \begin{cases} G & \text{if } i = 0, \\ L(L^{i-1}(G)) & \text{if } i > 0. \end{cases}$$

We recall that although there is a characterization of line graphs by forbidden subgraphs [2], there does not exist a similar characterization for i-iterated line graphs for $i \geq 2$.

9.3   Wiener Index

At first, the Wiener index was used for predicting the boiling point of paraffin [59], but later, strong correlation between the Wiener index and the chemical properties of a compound was found. Nowadays, this index is a tool used for preliminary screening of drug molecules [1]. The Wiener index also predicts binding energy of protein–ligand complex at a preliminary stage.

Besides introducing (new) index, Wiener also stated a theorem that shows how the Wiener index of a tree can be decomposed into easily calculable edge–contributions. Denote by $N_2(F)$ the sum over all pairs of components of the product of the number of vertices of two components of a forest F, that is,

$$N_2(F) = \sum_{1 \leq i < j \leq p} n(T_i)\, n(T_j),$$

where $T_1, T_2, \ldots, T_p$ is the set of components of F. If $p = 1$, that is, if F is connected, then $N_2(F) = 0$.

Theorem 9.3.1 *[61] For a tree T, the following holds:*

$$W(T) = \sum_{e \in E(T)} N_2(T - e). \qquad (9.2)$$

Proof. Notice that in a tree any two vertices are connected by a unique path, which is the shortest one. So, an edge $e = ij$ contributes 1 to $W(T)$ for each pair of vertices for which the unique path between them contains e. And, this is a case when i and j are in distinct components of $T - e$. As the number of such paths is $N_2(F - e)$, the proof follows. □

As T is a tree, for every edge $e = ij$ of T, the forest $T - e$ is composed of two components, one of size $n_e(i)$ and the other of size $n_e(j)$, which gives $N_2(T - e) = n_e(i)n_e(j)$. Thus, one can restate (9.2) as

$$W(T) = \sum_{e=ij \in E(T)} n_e(i)n_e(j). \qquad (9.3)$$

So the Szeged index and the Wiener index coincide on trees. In fact, the Szeged index was defined from (9.3) by relaxing the condition that the graph is a tree.

In analogy to the classical Theorem 9.3.1, we have the following vertex version (see [38]):

Theorem 9.3.2 *Let T be a tree on n vertices. Then,*

$$W(T) = \sum_{v \in V(T)} N_2(T - v) + \binom{n}{2}. \qquad (9.4)$$

A theorem given by Doyle and Graver [22] is of a similar kind. In order to state it, denote by $N_3(F)$ the sum over all triplets of components of the product of the number of vertices of three components of a forest F, that is,

$$N_3(F) = \sum_{1 \leq i < j < k \leq p} n(T_i)\, n(T_j)\, n(T_k).$$

Note that if $p = 1$ or $p = 2$, then $N_3(F) = 0$. This theorem claims the following.

Theorem 9.3.3 (Doyle and Graver) *Let T be a tree on n vertices. Then,*

$$W(T) = \binom{n+1}{3} - \sum_{v \in V(T)} N_3(T - v).$$

The Wiener index is also closely related to other quantities. For example, in computer science, the average distance $\mu(G)$ is used, where

$$\mu(G) = \frac{W(G)}{\binom{|V(G)|}{2}}.$$

It is important to know the average distance traversed by a message in the network. Networks with small $\mu(G)$ are related to *small worlds*.

In the theory of social networks, the Wiener index is closely related to the betweenness centrality of a vertex that quantifies the number of times a vertex lays on a shortest path between two other vertices. More precisely, the *betweenness centrality* $B(x)$ of a vertex $x \in V(G)$ is the sum of the fraction of all-pairs shortest paths that pass through x, that is,

$$B(x) = \sum_{\substack{u,v \in V(G) \\ u \neq v \neq x}} \frac{\sigma_{u,v}(x)}{\sigma_{u,v}}, \qquad (9.5)$$

where

$\sigma_{u,v}$ denotes the total number of shortest (u, v) paths in G

$\sigma_{u,v}(x)$ represents the number of shortest (u, v) paths passing through the vertex x.

This is one of the most important centrality indices. It was introduced by Anthonisse [3] and popularized later by Freeman [28].

The following result tells that the sum of the betweenness centrality of all vertices of a graph is related to its Wiener index. Moreover, it is a generalization of (9.4) to connected graphs with cycles [38].

Theorem 9.3.4 *For any connected graph G, the following holds:*

$$W(G) = \sum_{v \in V(G)} B(v) + \binom{n}{2}.$$

Since for any pair of vertices in a tree the shortest path between them is unique, the Wiener index of a tree is much easier to compute than that of an arbitrary graph. Furthermore, it is easy to see that for trees on n vertices, the maximal Wiener index is obtained for the path P_n, and

$$W(P_n) = \binom{n+1}{3}.$$

On the other hand, the tree with minimal Wiener index is the star S_n, and

$$W(S_n) = (n-1)^2.$$

Thus, for every tree T on n vertices, we have

$$(n-1)^2 \le W(T) \le \binom{n+1}{3}.$$

As the distance between any two distinct vertices is at least one, we have that K_n has the smallest Wiener index between all graph on n vertices. So for any connected graph G on n vertices, it holds:

$$\binom{n}{2} \le W(G) \le \binom{n+1}{3}.$$

Among the 2-connected graphs on n vertices (or even more, among the graphs of minimum degree 2), the n-cycle has the largest Wiener index

$$W(C_n) = \begin{cases} \dfrac{n^3}{8} & \text{if } n \text{ is even,} \\[2ex] \dfrac{n^3 - n}{8} & \text{if } n \text{ is odd.} \end{cases}$$

The Wiener index is easy to obtain for some classes of graphs. For graphs G and H, the Wiener index of their Cartesian product $G \,\square\, H$ is

$$W(G \,\square\, H) = |n(G)|^2 \cdot W(H) + |n(H)|^2 \cdot W(G),$$

see [29]. From this result follows a simple formula for the Wiener index of hypercubes Q_n

$$W(Q_n) = n2^{2(n-1)}.$$

We conclude this section with an interesting connection between the Wiener index and Laplacian spectrum of a tree.

Theorem 9.3.5 *Let T be a tree with Laplacian eigenvalues $\lambda_1 \ge \lambda_2 \ge \cdots \ge \lambda_{n-1} > \lambda_n = 0$. Then,*

$$W(T) = n \sum_{i=1}^{n-1} \frac{1}{\lambda_i}.$$

In 1988, Hosoya [43] introduced

$$H(G,x) = \sum_{k=1}^{l} d(G,k)x^k,$$

where

G is a graph

$d(G,k)$ is the number of pairs of vertices in the graph G at distance k.

Originally, this polynomial was named Wiener polynomial, but later, it was renamed into Hosoya polynomial. The first derivative of $H(G, x)$ for $x = 1$ is equal to the Wiener index of the graph G. This property of the Hosoya polynomial gives an alternative way of calculating the Wiener index and in a way makes the Hosoya polynomial a generalization of $W(G)$. Higher derivatives of Hosoya polynomial are also used as molecular descriptors (the hyper-Wiener index, e.g., is one half of the second derivative of $H(G, x)$ for $x = 1$).

For further details and results on the Wiener index, see [15,16,23,37,40] and the references cited therein.

9.4   Wiener Index of Graphs and Their Line Graphs

The concept of line graphs and iterated line graphs in chemical graph theory is introduced in order to study the molecule complexity. The number of edges in the line graph of the molecular graph is a measure of the branching, and then the iterated line graphs are used to find complete ordering of the molecules [5]. See [33,36] for some more applications in physical chemistry.

On the other hand, mathematicians started to study the connection between $W(G)$ and $W(L(G))$. In particular, they focused on graphs G satisfying

$$W(L(G)) = W(G). \tag{9.6}$$

Although it is not clear on which graph parameters or structural properties the difference $W(L(G)) - W(G)$ depends, the problem of characterizing graphs G with $W(L(G)) = W(G)$ is interesting, and it seems to be rather difficult.

In this section, we summarize some results on this issue, for more results on the topic, see [8,18,19,33,35]. Let us remark that in the literature, one easily encounters the term *edge-Wiener index* of G, which is actually the Wiener index of the line graph, sometimes in addition shifted by $\binom{n}{2}$, see [45].

The following remark of Buckley [7] is a pioneering work in this area.

Theorem 9.4.1 (Buckley, 1981) *For every tree T, $W(L(T)) = W(T) - \binom{n}{2}$.*

Proof. Let u, v be two distinct vertices of T. On their shortest (and unique) path in T, let e_u be the edge incident with u and e_v the edge incident with v. Notice that e_u and e_v coincide when u and v are adjacent. There is an obvious one-to-one correspondence between the pairs of distinct vertices u, v and their corresponding pairs of edges e_u, e_v. As there are $\binom{n}{2}$ pairs of vertices, and for each such pair u, v, it holds $d_{L(T)}(e_u, e_v) = d_T(u, v) - 1$, we conclude the statement of the theorem.   □

In particular, the aforementioned result tells that regarding the acyclic graphs, the Wiener index of a line graph is strictly smaller than the Wiener index of the original graph. An interesting generalization of this was given by Gutman [32]:

Theorem 9.4.2 *If G is a connected graph with n vertices and m edges, then*

$$W(L(G)) \geq W(G) - n(n-1) + \frac{1}{2}m(m+1).$$

In addition, regarding Theorem 9.4.1, Gutman and Pavlović [35] showed that the Wiener index of the line graph is smaller than the Wiener index of the original graph even if we allow just one cycle in the graph.

Theorem 9.4.3 *If G is a connected unicyclic graph with n vertices, then $W(L(G)) \leq W(G)$, with equality if and only if G is a cycle of length n.*

In connected bicyclic graphs, all the three cases $W(L(G)) < W(G)$, $W(L(G)) = W(G)$, and $W(L(G)) > W(G)$ occur [35]. It is known that the smallest bicyclic graph with the property $W(L(G)) = W(G)$ has nine vertices and it is unique. There are already 26 ten-vertex bicyclic graphs with the same property [34].

The following result tells us that in most cases (9.6) does not hold for graphs of minimum degree at least 2.

Theorem 9.4.4 *Let G be a connected graph with $\delta(G) \geq 2$. Then,*

$$W(L(G)) \geq W(G).$$

Moreover, the equality holds only for cycles.

This was proved independently and simultaneously in [7,62]. In [7], a direct proof of this result is given. On the other hand, Wu [62] obtained it as a corollary from his interesting result on the bounds of the Wiener index of line graphs in terms of the Gutman index:

Theorem 9.4.5 *Let G be a connected graph of size m. Then, it holds*

$$\frac{1}{4}(\text{Gut}(G) - m) \leq W(L(G)) \leq \frac{1}{4}(\text{Gut}(G) - m) + \binom{m}{2}.$$

Moreover, the lower bound is attained if and only if G is a tree.

Let $\kappa_i(G)$ denote the number of i-cliques in a graph G. In [48], the lower bound of the aforementioned theorem is improved in the following way.

Theorem 9.4.6 *Let G be a connected graph. Then,*

$$W(L(G)) \geq \frac{1}{4}\text{Gut}(G) - \frac{1}{4}|E(G)| + \frac{3}{4}\kappa_3(G) + 3\kappa_4(G) \qquad (9.7)$$

with the equality in (9.7) if and only if G is a tree or a complete graph.

The aforementioned theorem implies the following interesting corollary.

Corollary 9.4.1 *Let G be a connected graph of minimal degree $\delta \geq 2$. Then,*

$$W(L(G)) \geq \frac{\delta^2}{4} W(G) - \frac{1}{4} |E(G)| \geq \frac{\delta^2 - 1}{4} W(G).$$

Proof. Note that

$$\text{Gut}(G) = \sum_{\{u,v\} \subseteq V(G)} d(u)d(v)d(u,v) \geq \sum_{\{u,v\} \subseteq V(G)} \delta^2 d(u,v) = \delta^2 W(G).$$

Now, since $\delta \geq 2$, the graph G is not a tree, and so Theorem 9.4.6 implies the first inequality in the corollary. The second inequality then follows, if one observes that, since every pair of adjacent vertices contributes exactly 1 to the Wiener index of the graph (while the nonadjacent ones contribute even more), we have that $|E(G)| \leq W(G)$.   □

We expect that Corollary 9.4.1 can be improved to $W(L(G)) \geq \frac{\delta^2}{4} W(G)$, with equality holding for cycles, which would correspond to the result of Wu for $\delta = 2$.

9.5   Graphs with Large Girth

A connected graph G is isomorphic to $L(G)$ if and only if G is a cycle. Thus, the cycles provide a trivial infinite family of graphs for which $W(G) = W(L(G))$. In addition, for every positive number g, there exists a graph G with girth g for which $W(G) = W(L(G))$.

Dobrynin and Mel'nikov [17] have constructed infinite family of graphs of girths 3 and 4 with the property $W(G) = W(L(G))$ and stated the following problem.

Problem 9.5.1 (Dobrynin and Mel'nikov) *Is it true that for every integer $g \geq 5$ there exists a graph $G \neq C_g$ of girth g, for which $W(G) = W(L(G))$?*

The aforementioned problem was solved by Dobrynin [14] by considering the following construction. Let $G_g(d, s, r)$ be a graph of girth g constructed from a path P_d by

1. Identifying a vertex of a distinct copy of the g-cycle C_g with each of the end vertices of P_d

2. Identifying the center of disjoint copies of the S_{s+1}-star with one end vertex and of the S_{r+1}-star with the other end vertex of P_d

Observe that G is a bicyclic graph of girth g with $2g + d + s + r - 2$ vertices (see Figure 9.2). In [14], the following result is shown.

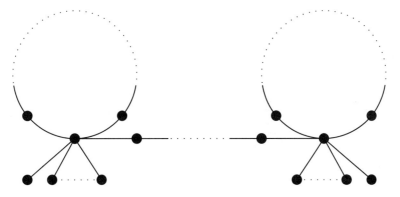

FIGURE 9.2: The graph $G_g(d, r, s)$ with $W(G_g(d, r, s)) = W(L(G_g(d, r, s)))$.

Theorem 9.5.1 (Dobrynin) *The Wiener indices of $G_g(d, s, r)$ and $L(G_g(d, s, r))$ coincide provided that the graph parameters satisfy the following relations:*

(a) *For every even $g \geq 6$,*

$$d = (g^2 - 6g + 4)/4, \quad s = (g^2 - 6g + 8)/8, \quad r = (g^2 - 6g + 16)/8.$$

(b) *For every odd $g \geq 9$,*

$$d = (g^2 - 8g + 3)/4, \quad s = (g^2 - 8g + 15)/8, \quad r = (g^2 - 8g + 23)/8.$$

Theorem 9.5.1 overlooks the values $g = 5$ and $g = 7$. For $g = 5$, see the graph G_5 on Figure 9.3. Both G_5 and its line graph $L(G_5)$ have Wiener indices 288. For $g = 7$, the graph $G_7 = G_7(6, 4, 5)$ satisfies $W(G_7) = W(L(G_7)) = 1698$.

The authors of [7] showed that for infinitely many girths, there exist infinitely many solutions of Problem 9.5.1.

Theorem 9.5.2 *For every positive integer g_0, there exists $g \geq g_0$ such that there are infinitely many graphs G of girth g satisfying $W(G) = W(L(G))$.*

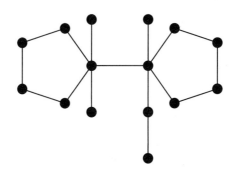
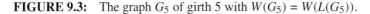

FIGURE 9.3: The graph G_5 of girth 5 with $W(G_5) = W(L(G_5))$.

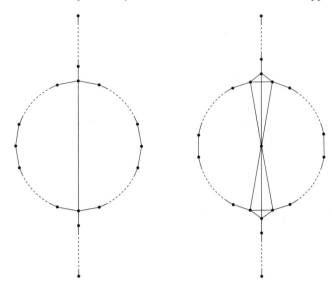

FIGURE 9.4:   Graphs $\Phi(k, p, q)$ and $L(\Phi(k, p, q))$.

The aforementioned result encourages the authors of [7] to state the following conjecture. Notice that it is true for girths 3 and 4, see [17].

Conjecture 9.5.1 *For every integer $g \geq 3$, there exist infinitely many graphs G of girth g satisfying $W(G) = W(L(G))$.*

In what follows, we give a sketch of the proof of Theorem 9.5.2. For positive integers k, p, q, we define the graph $\Phi(k, p, q)$ as follows (see Figure 9.4 for an illustration). The graph $\Phi(k, p, q)$ is simple and composed of two cycles, $C_1 = u_1 u_2 \ldots u_{2k+1}$ and $C_2 = v_1 v_2 \ldots v_{2k+1}$, and two paths $P_p = x_1 x_2 \ldots x_p$ and $P_q = y_1 y_2 \ldots y_q$ such that all the vertices are distinct except for $v_1 = u_1 = x_1$ and $y_1 = v_{2k+1} = u_{2k+1}$.

We are now interested in computing the difference $W(L(\Phi(k, p, q))) - W(\Phi(k, p, q))$, which is used in the proof of Theorem 9.5.4. The proof is straightforward and rather technical.

Theorem 9.5.3 *For integers, $k, p, q \geq 1$, let $G = \Phi(k, p, q)$ with girth $g = 2k + 1$. Then,*

$$W(L(G)) - W(G) = \frac{1}{2}(g^2 + (p - q)^2 + 5(p + q - 3) - 2g(p + q - 3)).$$

Theorem 9.5.3 implies the following result:

Theorem 9.5.4 *For every nonnegative integer h, there exist infinitely many graphs G of girth $g = h^2 + h + 9$ with $W(L(G)) = W(G)$.*

Theorem 9.5.2 is an immediate corollary of Theorem 9.5.4. For every positive integer g_0, we can choose a nonnegative integer h such that $g = h^2 + h + 9 \geq g_0$. By Theorem 9.5.4, there are infinitely many graphs G of girth g with $W(L(G)) = W(G)$.

9.6 Second Line Graph Iteration

The graph $L^2(G) = L(L(G))$ is also called the *quadratic line graph* of G. As mentioned earlier, for nontrivial tree T, we cannot have $W(L(T)) = W(T)$ although there are graphs G such that $W(L(G)) = W(G)$. For quadratic line graphs, we can have

$$W(L^2(T)) = W(T), \tag{9.8}$$

even if T is a tree (see [13,20,21]). Obviously, the simplest trees are such which have a unique vertex of degree greater than 2. Such trees are called *generalized stars*. More precisely, *generalized t-star* is a tree obtained from the star $K_{1,t}$, $t \geq 3$, by replacing all its edges by paths of positive lengths. In [17], we have the following theorem.

Theorem 9.6.1 *Let S be a generalized t-star with q edges and branches of length* $k_1, k_2, \ldots, k_t$*. Then,*

$$W(L^2(S)) = W(S) + \frac{1}{2}\binom{t-1}{2}\left(\sum_{i=1}^{t} k_i^2 + q\right) - q^2 + 6\binom{t}{4}. \tag{9.9}$$

Based on this theorem, it is proved in [17] that $W(L^2(S)) < W(S)$ if S is a generalized 3-star, and $W(L^2(S)) > W(S)$ if S is a generalized t-star where $t \geq 7$. Thus, property (9.8) can hold for generalized t-stars only when $t \in \{4, 5, 6\}$. In [17], for every $t \in \{4, 5, 6\}$, several generalized t-stars with property (9.8) are found. The smallest generalized t-stars with property (9.8) are listed in Table 9.1 (see also [17]).

From Table 9.1, one can expect that it might be easier to find generalized t-stars with property (9.8) when $t \in \{5, 6\}$ than in the case $t = 4$. Indeed, in [17], the authors

TABLE 9.1: Smallest Generalized t-Stars with the Property (9.8)

t	q	k_1	k_2	k_3	k_4	k_5	k_6	q	k_1	k_2	k_3	k_4	k_5	k_6
4	27	1	2	3	21	—	—	90	3	7	8	72	—	—
	42	1	2	6	33	—	—	102	2	3	16	81	—	—
	69	2	6	6	55	—	—	105	4	5	12	84	—	—
	72	1	3	11	57	—	—	105	2	9	10	84	—	—
	90	4	5	9	72	—	—	111	4	9	9	89	—	—
5	18	2	3	3	3	7	—	30	4	4	4	4	14	—
	24	2	3	3	6	10	—	30	3	3	3	8	13	—
	24	2	2	5	5	10	—	30	1	4	5	7	13	—
	24	1	4	4	5	10	—	36	4	4	4	7	17	—
	24	1	2	6	6	9	—	36	3	4	6	6	17	—
6	50	7	7	7	8	10	11	60	7	8	8	10	13	14
	50	6	7	8	9	9	11	60	6	8	9	11	12	14
	50	5	8	9	9	9	10	60	6	7	10	12	12	13
	60	8	8	8	9	12	15	60	5	10	10	10	11	14
	60	6	9	10	10	10	15	60	5	8	11	12	12	12

TABLE 9.2:   Infinite Families of Generalized t-Stars, $t \in \{5, 6\}$, with the Property (9.8)

t	k_1	k_2	k_3	k_4	k_5	k_6
5	1	2	$2k^2 - k + 5$	$2k^2 - k + 5$	$2k^2 + 2k + 5$	—
	1	2	$2k^2 - 2k + 5$	$2k^2 + k + 5$	$2k^2 + k + 5$	—
	1	2	$2k^2 + 6$	$2k^2 + 3k + 6$	$2k^2 + 3k + 9$	—
6	3	$4k^2 + 33$	$4k^2 - k + 36$	$4k^2 - k + 36$	$4k^2 + k + 36$	$4k^2 + k + 36$

TABLE 9.3:   Infinite Families of Generalized 4-Stars with the Property (9.8)

k_1	k_2	k_3	k_4
$a_1(k)$	$a_1(k + 1)$	$4(k_1 + k_2) - 3$	$4(k_1 + k_2 + k_3) - 3$
1	$a_2(k)$	$a_2(k + 1)$	$4(k_1 + k_2 + k_3) - 3$
2	$a_3(k)$	$a_3(k + 1)$	$4(k_1 + k_2 + k_3) - 3$
3	$a_4(k)$	$a_4(k + 1)$	$4(k_1 + k_2 + k_3) - 3$
4	$a_5(k)$	$a_5(k + 1)$	$4(k_1 + k_2 + k_3) - 3$
5	$a_6(k)$	$a_6(k + 1)$	$4(k_1 + k_2 + k_3) - 3$
6	$a_7(k)$	$a_7(k + 1)$	$4(k_1 + k_2 + k_3) - 3$
1	2	$a_8(k)$	$a_8(k + 1)$
4	5	$a_9(k)$	$a_9(k + 1)$
4	5	$a_{10}(k)$	$a_{10}(k + 1)$

found infinite families of these generalized t-stars for $t \in \{5, 6\}$. They found three infinite families for $t = 5$ and one for $t = 6$, see Table 9.2, where k is nonnegative integer. Observe that the first two infinite families of generalized 5-stars can be regarded as one provided that $k \in \mathbb{Z}$ only.

The problem of existence of an analogous infinite family of generalized 4-stars is left open in [17]. This problem is solved in [53], where several infinite families of generalized 4-stars with the property (9.8) are constructed. The constructions are grouped into three classes, and there is an infinite number of families in two of these three classes. Some of these constructions, grouped into families, are listed in Table 9.3, where $k \in \mathbb{Z}$. As regards the values of sequences $a_1, \ldots, a_{10}$, we have

$$a_1(j) = \frac{1}{4}\left(5 + \sqrt{3}\right)\left(2 - \sqrt{3}\right)^j + \frac{1}{4}\left(5 - \sqrt{3}\right)\left(2 + \sqrt{3}\right)^j + \frac{3}{2}$$

$$a_2(j) = \frac{1}{12}\left(15 + 3\sqrt{3}\right)\left(2 - \sqrt{3}\right)^j + \frac{1}{12}\left(15 - 3\sqrt{3}\right)\left(2 + \sqrt{3}\right)^j - \frac{1}{2}$$

$$a_3(j) = \frac{1}{12}\left(21 + 3\sqrt{3}\right)\left(2 - \sqrt{3}\right)^j + \frac{1}{12}\left(21 - 3\sqrt{3}\right)\left(2 + \sqrt{3}\right)^j - \frac{5}{2}$$

$$a_4(j) = \frac{1}{12}\left(33 + 9\sqrt{3}\right)\left(2 - \sqrt{3}\right)^j + \frac{1}{12}\left(33 - 9\sqrt{3}\right)\left(2 + \sqrt{3}\right)^j - \frac{9}{2}$$

$$a_5(j) = \frac{1}{12}\left(69 - 33\sqrt{3}\right)\left(2 - \sqrt{3}\right)^j + \frac{1}{12}\left(69 + 33\sqrt{3}\right)\left(2 + \sqrt{3}\right)^j - \frac{13}{2}$$

$$a_6(j) = \frac{1}{12}\left(75 - 33\sqrt{3}\right)\left(2 - \sqrt{3}\right)^j + \frac{1}{12}\left(75 + 33\sqrt{3}\right)\left(2 + \sqrt{3}\right)^j - \frac{17}{2}$$

$$a_7(j) = \frac{1}{12}\left(69 + 21\sqrt{3}\right)\left(2 - \sqrt{3}\right)^j + \frac{1}{12}\left(69 - 21\sqrt{3}\right)\left(2 + \sqrt{3}\right)^j - \frac{21}{2}$$

$$a_8(j) = \frac{1}{12}\left(45 - 21\sqrt{3}\right)\left(2 - \sqrt{3}\right)^j + \frac{1}{12}\left(45 + 21\sqrt{3}\right)\left(2 + \sqrt{3}\right)^j - \frac{9}{2}$$

$$a_9(j) = \frac{1}{12}\left(153 - 75\sqrt{3}\right)\left(2 - \sqrt{3}\right)^j + \frac{1}{12}\left(153 + 75\sqrt{3}\right)\left(2 + \sqrt{3}\right)^j - \frac{33}{2}$$

$$a_{10}(j) = \frac{1}{12}\left(297 - 165\sqrt{3}\right)\left(2 - \sqrt{3}\right)^j + \frac{1}{12}\left(297 + 165\sqrt{3}\right)\left(2 + \sqrt{3}\right)^j - \frac{33}{2}.$$

These results suggest the following conjecture:

Conjecture 9.6.1 *Let T be a nontrivial tree such that $W\left(L^2(T)\right) = W(T)$. Then, there is an infinite family of trees T' homeomorphic to T, such that $W(L^2(T')) = W(T')$.*

Of course, more interesting is the question which types of trees satisfy (9.8). Perhaps such trees do not have many vertices of degree at least 3. Let $\mathcal{T}$ be a class of trees that have no vertex of degree two, and such that $T \in \mathcal{T}$ if and only if there exists a tree T' homeomorphic to T, and such that $W\left(L^2(T')\right) = W(T')$. (Recall that graphs G_1 and G_2 are homeomorphic if and only if the graphs obtained from them by repeatedly removing a vertex of degree 2, and making its two neighbors adjacent, are isomorphic.)

Problem 9.6.1 *Characterize the trees in $\mathcal{T}$. In particular, prove that $\mathcal{T}$ is finite.*

By the aforementioned results, among the stars, only $K_{1,4}$, $K_{1,5}$, and $K_{1,6}$ are in $\mathcal{T}$. We expect that no tree in $\mathcal{T}$ has a vertex of degree exceeding 6. Based on our experience, we also expect that there is a constant c such that no tree in $\mathcal{T}$ has more than c vertices of degree at least 3. Consequently, we believe that the set $\mathcal{T}$ is finite.

9.7 Higher Line Graph Iterations

As we have seen, there is no nontrivial tree for which $W(L(T)) = W(T)$ and there are many trees T, satisfying $W(L^2(T)) = W(T)$. However, it is not easy to find a tree T and $i \geq 3$ such that $W(L^i(T)) = W(T)$. In [15], the following problem was posed.

Problem 9.7.1 *[15] Is there any tree T satisfying equality $W(L^i(T)) = W(T)$ for some $i \geq 3$?*

Observe that if T is a trivial tree, then $W(L^i(T)) = W(T)$ for every $i \geq 1$, although here the graph $L^i(T)$ is empty. The real question is, of course, if there is a nontrivial tree T and $i \geq 3$ such that $W(L^i(T)) = W(T)$. The same question appeared 4 years later in [17] as a conjecture. The authors expressed their belief that the problem has no nontrivial solution.

Conjecture 9.7.1 (Dobrynin, Entringer) *There is no tree T satisfying equality* $W(T) = W(L^i(T))$ *for any $i \geq 3$.*

In a series of papers [4,47,49–52], Conjecture 9.7.1 was disproved. In fact, all solutions of Problem 9.7.1 were found. The smallest tree disproving Conjecture 9.7.1 has 388 vertices (see the remark below Theorem 9.7.7) and this tree is unique. If we take in mind that there are approximately $7.5 \cdot 10^{175}$ nonisomorphic trees on 388 vertices while the number of atoms in the entire universe is estimated to be only within the range of 10^{78} to 10^{82}, then to find *a needle in a haystack* is a trivially easy job compared to finding a counterexample when using only the brute force of (arbitrarily many) real computers.

A function $f : \{0, 1, \ldots\} \to \mathbb{R}$ is *convex* if $f(i) + f(i + 2) \geq 2f(i + 1)$ for every $i \geq 0$. If the inequality is strict, then f is *strictly convex*. In [50], it is proved that for every connected graph G, the function $f_G(i) = W(L^i(G))$ is convex in variable i. Moreover, $f_G(i)$ is strictly convex if G is distinct from a path, a cycle, and the claw $K_{1,3}$. The following result is a straightforward consequence of this fact.

Theorem 9.7.1 *Let T be a tree such that $W(L^3(T)) > W(T)$. Then, for every $i \geq 3$, the inequality $W(L^i(T)) > W(T)$ holds.*

Let G be a graph. A *pendant path* (or a *ray* for short) R in G is a (directed) path; the first vertex of which has degree at least 3, its last vertex has degree 1, and all of its internal vertices (if any exist) have degree 2 in G. Observe that if R has length t, $t \geq 2$, then the edges of R correspond to vertices of a ray $L(R)$ in $L(G)$ of length $t - 1$. In [50], we have the following theorem.

Theorem 9.7.2 *Let T be a tree distinct from a path and the claw $K_{1,3}$ such that all of its rays have length 1. Then, $W(L^3(T)) > W(T)$.*

In [49], this statement was extended to trees with arbitrarily long rays. Denote by H a tree on six vertices, two of which have degree 3 and the remaining four have degree 1. That is, H is the graph which *looks* like the letter H. The main result of [49] is the following theorem.

Theorem 9.7.3 *Let T be a tree not homeomorphic to a path, claw $K_{1,3}$, and H. Then, $W(L^3(T)) > W(T)$.*

Combining Theorems 9.7.1 and 9.7.3, we obtain the following consequence, which proves Conjecture 9.7.1 for trees T satisfying the assumption in Theorem 9.7.3.

Theorem 9.7.4 *Let T be a tree not homeomorphic to a path, claw $K_{1,3}$, and H. Then, $W(L^i(T)) > W(T)$ for every $i \geq 3$.*

Since the case when T is a path is trivial (in this case, $W(L^i(T)) < W(T)$ whenever $i \geq 1$, and T has at least two vertices), it remains to consider graphs homeomorphic to the claw $K_{1,3}$ and those homeomorphic to H.

First, consider the case of the claw $K_{1,3}$ itself. Then, $L^i(K_{1,3})$ is a cycle of length 3 for every $i \geq 1$. Since $W(K_{1,3}) = 9$ and the Wiener index of the cycle of length 3 is 3, we have $W(L^i(K_{1,3})) < W(K_{1,3})$ for every $i \geq 1$. For other trees homeomorphic to $K_{1,3}$, the opposite inequality is proved in [52], provided that $i \geq 4$.

Theorem 9.7.5 *Let T be a tree homeomorphic to $K_{1,3}$, such that $T \neq K_{1,3}$. Then, $W(L^i(T)) > W(T)$ for every $i \geq 4$.*

Notice that it was enough to prove the aforementioned theorem for the case $i = 4$, since then the inequalities with higher powers follow from the convexity of $f_G(i) = W(L^i(G))$. Analogous statement for trees homeomorphic to H was proved in [51].

Theorem 9.7.6 *Let T be a tree homeomorphic to H. Then, $W(L^i(T)) > W(T)$ for every $i \geq 4$.*

Consequently, with the exception of paths and the claw $K_{1,3}$, for every tree T, it holds $W(L^i(T)) > W(T)$ whenever $i \geq 4$. We can summarize the results for $i \geq 4$ as follows.

Corollary 9.7.1 *Let T be a tree and $i \geq 4$. Then, the following holds:*

$$W\left(L^i(T)\right) < W(T) \quad \text{if } T \text{ is the claw } K_{1,3} \text{ or a path } P_n \text{ with } n \geq 2;$$
$$W\left(L^i(T)\right) = W(T) \quad \text{if } T \text{ is the trivial graph } P_1;$$
$$W\left(L^i(T)\right) > W(T) \quad \text{otherwise.}$$

Hence, Conjecture 9.7.1 is true for $i \geq 4$. However, for $i = 3$, the infinite class of trees described in Theorem 9.7.7 violates Conjecture 9.7.1 (see [47]).

Let $H_{a,b,c}$ be a tree on $a + b + c + 4$ vertices, out of which two have degree 3, four have degree 1, and the remaining $a+b+c-2$ vertices have degree 2. The two vertices of degree 3 are connected by a path of length 2. Finally, there are two pendant paths of lengths a and b attached to one vertex of degree 3 and two pendant paths of lengths c and 1 attached to the other vertex of degree 3 (see Figure 9.5 for $H_{3,2,4}$). We have

Theorem 9.7.7 *For every $j, k \in \mathbb{Z}$, define*

$$a = 128 + 3j^2 + 3k^2 - 3jk + j,$$
$$b = 128 + 3j^2 + 3k^2 - 3jk + k,$$
$$c = 128 + 3j^2 + 3k^2 - 3jk + j + k.$$

Then, $W(L^3(H_{a,b,c})) = W(H_{a,b,c})$.

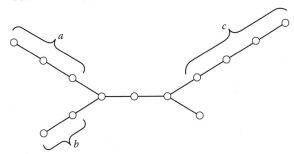

FIGURE 9.5: Graph $H_{a,b,c}$.

Let $\ell \in \{j, k, j + k\}$. Since for every integers j and k the inequality $3j^2 + 3k^2 - 3jk + \ell \geq 0$ holds, we see that $a, b, c \geq 128$ in Theorem 9.7.7. Therefore, the smallest graph satisfying the assumptions is $H_{128,128,128}$ on 388 vertices, obtained when $j = k = 0$.

The case $i = 3$ and T being homeomorphic either to the claw or H is rather interesting. In some cases (when there are not *many* long rays), we can prove $W(L^3(T)) > W(T)$, but in other ones, only the congruence arguments yield $W(L^3(T)) \neq W(T)$. These results are summarized in [46] in the following two statements.

Theorem 9.7.8 *Let T be a tree homeomorphic to $K_{1,3}$. Then, $W(L^3(T)) \neq W(T)$.*

Theorem 9.7.9 *Let G be a graph homeomorphic to H. Then, the equation $W(L^3(G)) = W(G)$ has a solution if and only if G is of type $H_{a,b,c}$ as stated in Theorem 9.7.7.*

These statements not only disprove Conjecture 9.7.1 (and give a positive answer to Problem 9.7.1) but completely characterize trees T and integers $i \geq 3$ such that $W(L^i(T)) = W(T)$. This may be surprising since for $i = 2$, we do not know answers to much weaker problems (see the previous section).

We conclude this chapter with two problems.

Problem 9.7.2 *Find all graphs (with cycles) G and powers i for which*

$$W(L^i(G)) = W(G).$$

This problem for $i = 1$ is obviously very rich with many different solutions, so it probably will not be possible to find all of them. But still, stating it as a problem could serve as a motivation for searching of various graph classes that satisfy the equation. Nevertheless, we want to emphasize the case $i \geq 2$. In this case, the problem is still rich with many solutions, particularly among the trees, but abandoning the class of trees can reduce the solutions significantly. At the moment, cycles are the only known cyclic graphs G for which $W(L^i(G)) = W(G)$ holds for some $i \geq 3$. Thus, we formulate another, much weaker problem:

Problem 9.7.3 *Let $i \geq 3$. Is there a graph G, different from a cycle and a tree, such that*

$$W(L^i(G)) = W(G)?$$

9.8 Summary and Conclusion

In this chapter, we studied the Wiener index of graphs and their (iterated) line graphs and also some related problems. Although the situation is completely solved for trees and their iterated line graphs, other than quadratic, for general graphs, the problem is still open. Of course, from the chemist's perspective, important role is played by *chemical graphs* and by graphs that do not have many cycles. We expect that this area can be very prolific, producing many nice results in the future.

Acknowledgment

The first author acknowledges partial support by Slovak research grants VEGA 1/0781/11, VEGA 1/0065/13, and APVV 0223-10. Both authors are partially supported by Slovenian ARRS Program P1-00383 and Creative Core FIŠ NM 3330-13-500033.

References

1. V. K. Agrawal, S. Bano, K. C. Mathur, and P. V. Khadikar. Novel application of Wiener vis-á-vis szeged indices: Antitubercular activities of quinolones. *Proceedings of the Indian Academy of Science (Chemical Sciences)*, 112:137–146, 2000.

2. I. Anderson. Selected topics in graph theory (Academic Press, 1979), pp. 451. *Proceedings of the Edinburgh Mathematical Society (Series 2)*, 26:398–398, 10 1983.

3. J. M. Anthonisse. The rush in a directed graph. Technical report, Stichting Mathematisch Centrum, Amsterdam, the Netherlands, October 1971.

4. S. Bekkai and M. Kouider. On mean distance and girth. *Discrete Applied Mathematics*, 158(17):1888–1893, October 2010.

5. S. H. Bertz. Branching in graphs and molecules. *Discrete Applied Mathematics*, 19:65–83, 1998.

6. D. Bonchev. *Information Theoretic Indices for Characterization of Chemical Structures*. Chemometrics research studies series. Research Studies Press, Chichester, U.K., 1983.

7. F. Buckley. Mean distance of line graphs. *Congr. Numer.*, 32:153–162, 1981.

8. P. Dankelmann, I. Gutman, S. Mukwembi, and H. C. Swart. The edge-Wiener index of a graph. *Discrete Mathematics*, 309(10):3452–3457, 2009.

9. M. Dehmer. Information-theoretic concepts for the analysis of complex networks. *Applied Artificial Intelligence*, 22(7 and 8):684–706, 2008.

10. M. Dehmer, N. Barbarini, K. Varmuza, and A. Graber. A large scale analysis of information-theoretic network complexity measures using chemical structures. *PLoS ONE*, 4(12):e8057+, December 2009.

11. M. Dehmer, A. Mowshowitz, and F. Emmert-Streib. Connections between classical and parametric network entropies. *PLoS ONE*, 6(1):e15733, January 2011.

12. J. Devillers and A. T. Balaban. *Topological Indices and Related Descriptors in QSAR and QSPR*. Gordon & Breach, Amsterdam, the Netherlands, 1999.

13. A. A. Dobrynin. Distance of iterated line graphs. *Graph Theory Notes of New York*, 37:8–9, 1999.

14. A. A. Dobrynin. The Wiener index for graphs of arbitrary girth and their edge graphs. *Sib. zhurn. industr. matem.*, 12:44–50, 2009.

15. A. A. Dobrynin, R. Entringer, and I. Gutman. Wiener index of trees: Theory and applications. *Acta Applicandae Mathematicae*, 3:211–249, 2001.

16. A. A. Dobrynin, I. Gutman, S. Klavžar, and P. Žigert. Wiener index of hexagonal systems. *Acta Applicandae Mathematicae*, 72:247–294, 2002.

17. A. A. Dobrynin and L. S. Mel'nikov. Some results on the Wiener index of iterated line graphs. *Electronic Notes in Discrete Mathematics*, 22:469–475, 2005.

18. A. A. Dobrynin and L. S. Mel'nikov. Wiener index for graphs and their line graphs with arbitrary large cyclomatic numbers. *Applied Mathematics Letters*, 18(3):307–312, 2005.

19. A. A. Dobrynin and L. S. Mel'nikov. Wiener index, line graph and the cyclomatic number. *MATCH Communications in Mathematical and in Computer Chemistry*, 53:209–214, 2005.

20. A. A. Dobrynin and L. S. Mel'nikov. Trees and their quadratic line graphs having the same Wiener index. *MATCH Communications in Mathematical and in Computer Chemistry*, 50:145–161, 2004.

21. A. A. Dobrynin and L. S. Mel'nikov. Trees, quadratic line graphs and the Wiener index. *Croatica Chemica Acta*, 77:477–480, 2004.

22. J. K. Doyle and J. E. Graver. Mean distance in a directed graph. *Environment and Planning B: Planning and Design*, 5(1):19–29, 1978.

23. R. C. Entringer. Distance in graphs: Trees. *Journal of Combinatorial Mathematics and Combinatorial Computing*, 24:65–84, 1997.

24. R. C. Entringer, D. E. Jackson, and D. A. Snyder. Distance in graphs. *Czechoslovak Mathematical Journal*, 26:283–296, 1976.

25. E. Estrada. Characterization of 3d molecular structure. *Chemical Physics Letters*, 319:713–718, 2000.

26. E. Estrada and J. A. Rodríguez-Velázquez. Spectral measures of bipartivity in complex networks. *Physical Review E*, 72(4):046105+, October 2005.

27. E. Estrada and J. A. Rodríguez-Velázquez. Subgraph centrality in complex networks. *Physical Review E*, 71:056103, 2005.

28. L.C. Freeman. A set of measures of centrality based upon betweenness. *Sociometry*, 40:35–41, 1977.

29. A. Graovac and T. Pisanski. On the Wiener index of a graph. *Journal of Mathematical Chemistry*, 8:53–62, 1991.

30. I. Gutman. A formula for the Wiener number of trees and its extension to graphs containing cycles. *Graph Theory Notes N. Y.*, 27:9–15, 1994.

31. I. Gutman. Selected properties of the Schultz molecular topological index. *Journal of Chemical Information and Computer Sciences*, 34(5):1087–1089, September 1994.

32. I. Gutman. Distance of line graphs. *Graph Theory Notes N. Y.*, 31:49–52, 1996.

33. I. Gutman and E. Estrada. Topological indices based on the line graph of the molecular graph. *Journal of Chemical Information and Computer Sciences*, 36(3):541–543, 1996.

34. I. Gutman, V. Jovašević, and A. Dobrynin. Smallest graphs for which the distance of the graph is equal to the distance of its line graph. *Journal of Chemical Information and Computer Sciences*, 33:19–19, 1997.

35. I. Gutman and L. Pavlović. More on distance of line graphs. *Graph Theory Notes N. Y.*, 33:14–18, 1997.

36. I. Gutman, L. Pavlović, B. K. Mishra, M. Kuanar, E. Estrada, and N. Guevara. Application of line graphs in physical chemistry. Predicting the surface tensions of alkanes. *Journal of the Serbian Chemical Society*, 26:1025–1029, 1997.

37. I. Gutman and O. E. Polansky. *Mathematical Concepts in Organic Chemistry*. Springer, London, U.K., 2011.

38. I. Gutman and R. Škrekovski. Vertex version of the Wiener theorem. *MATCH Communications in Mathematical and Computational Chemistry*, 72:295–300, 2014.

39. I. Gutman and N. Trinajstić. Graph theory and molecular orbitals, total π-electron energy of alternant hydrocarbons. *Chemical Physics Letters*, 17:535–538, 1972.

40. I. Gutman, Y. N. Yeh, S. L. Lee, and Y. L. Luo. Some recent results in the theory of the Wiener number. *Indian Journal of Chemistry*, 32A:651–661, 1993.

41. F. Harary. Status and contrastatus. *Sociometry*, 22:23–43, 1959.

42. H. Hosoya. Topological index. A newly proposed quantity characterizing the topological nature of structural isomers of saturated hydrocarbons. *Bulletin of the Chemical Society of Japan*, 44(9):2332–2339, 1971.

43. H. Hosoya. On some counting polynomials in chemistry. *Discrete Applied Mathematics*, 19:239–257, 1998.

44. H. Hosoya. The topological index z before and after 1971. *Internet Electronic Journal of Molecular Design*, 1:428–442, 2002.

45. M. H. Khalifeh, H. Yousefi-Azari, A. R. Ashrafi, and S. G. Wagner. Some new results on distance-based graph invariants. *European Journal of Combinatorics*, 30(5):1149–1163, 2009.

46. M. Knor, M. Mačaj, P. Potočnik, and R. Škrekovski. Complete solution of equation $w(l^3(t)) = w(t)$ for Wiener index of iterated line graphs of trees. *Discrete Applied Math.*, 171:90–103, 2014.

47. M. Knor, M. Mačaj, P. Potočnik, and R. Škrekovski. Trees t satisfying $w(l^3(t)) = w(t)$. FILOMAT (To appear.)

48. M. Knor, P. Potočnik, and R. Škrekovski. Relationship between edge-Wiener index and Gutman index of a graph. *Discrete Applied Math.*, 167:197–201, 2014.

49. M. Knor, P. Potočnik, and R. Škrekovski. On a conjecture about Wiener index in iterated line graphs of trees. *Discrete Applied Mathematics*, 312(6):1094–1105, 2012.

50. M. Knor, P. Potočnik, and R. Škrekovski. The Wiener index in iterated line graphs. *Discrete Applied Mathematics*, 160(15):2234–2245, 2012.

51. M. Knor, P. Potočnik, and R. Škrekovski. Wiener index of iterated line graphs of trees homeomorphic to h. *Discrete Mathematics*, 313(10):1104–1111, 2013.

52. M. Knor, P. Potočnik, and R. Škrekovski. Wiener index of iterated line graphs of trees homeomorphic to the claw $k_{1,3}$. *Ars Mathematica Contemporanea*, 6:211–219, 2013.

53. M. Knor and R. Škrekovski. Wiener index of generalized 4-stars and of their quadratic line graph. *Australasian J. Comb.*, 58(1):119–126, 2014.

54. S. Nikolić, I. M. Tolić, N. Trinajstić, and I. Baučić. On the Zagreb indices as complexity indices. *Croatica Chemica Acta*, 73:909–921, 2000.

55. L. Pogliani. From molecular connectivity indices to semiempirical connectivity terms: Recent trends in graph theoretical descriptors. *Chemical Reviews*, 100:3827–3858, 2000.

56. M. Randić. Characterization of molecular branching. *Journal of the American Chemical Society*, 97:6609–6615, 1975.

57. M. Randić. Aromaticity of polycyclic conjugated hydrocarbons. *Chemical Reviews*, 103:3449–3606, 2003.

58. H. P. Schultz. Topological organic chemistry. 1. Graph theory and topological indices of alkanes. *Journal of Chemical Information and Computer Sciences*, 29(3):227–228, 1989.

59. N. Trinajstić. *Chemical Graph Theory*, 2nd edn. CRC Press, Boca Raton, FL, February 1992.

60. L. Šoltés. Transmission in graphs: A bound and vertex removing. *Mathematica Slovaca*, 1:11–16, 1991.

61. H. Wiener. Structural determination of paraffin boiling points. *Journal of the American Chemical Society*, 69:17–20, 1947.

62. B. Wu. Wiener index of line graphs. *MATCH: Communications in Mathematical and in Computer Chemistry*, 64:699–706, 2010.

Chapter 10

Single-Graph Support Measures

Toon Calders, Jan Ramon, and Dries Van Dyck

Contents

10.1   Introduction..303
10.2   Preliminaries...305
     10.2.1   Graphs..305
     10.2.2   Isomorphisms..306
     10.2.3   Support Measures...307
10.3   Nonoverlap-Graph-Based Measures...307
     10.3.1   Key-Based Support Measures.......................................307
     10.3.2   Minimal Image Count Support Measure..............................308
10.4   Overlap-Graph-Based Measures..308
     10.4.1   Pairwise Overlap Graph...309
     10.4.2   Overlap-Graph-Based Support Measure..............................310
     10.4.3   Alternative Characterization for Antimonotonicity..................311
     10.4.4   Bounding Theorem..312
         10.4.4.1   MCP Measure..312
         10.4.4.2   Theorem..314
     10.4.5   Lovász and Schrijver Graph Measures as Support Measures..........314
10.5   Object-Specific Overlap Hypergraphs.......................................315
     10.5.1   Support Measure Based on Relaxed Maximal Independent Set......316
         10.5.1.1   Conditions for Antimonotonicity........................317
         10.5.1.2   Sufficient Condition...................................318
         10.5.1.3   Necessary Condition....................................318
     10.5.2   Bounding Theorem..318
     10.5.3   Relaxation of the OGSM MIS..319
10.6   Bounding the Variance of Sample Estimates Using the s Measure............319
10.7   Discussion and Conclusion...321
References..322

10.1   Introduction

Over the last several years, graph mining has emerged as a new field within contemporary data mining. One of the central tasks is the search for subgraphs, called *patterns*, that occur frequently in either a collection of graphs (e.g., databases of molecules [6], game positions [15], scene descriptions) or in a single large graph (e.g., the Internet, citation networks [16], social networks [12], protein interaction networks [10]). In the literature, the terms *frequency* and *support* have been used interchangeably to denote the measure to quantify the prevalence of a pattern. In the single-graph setting, however, the notion of frequency is not at all straightforward to define. For example, the obvious definition of taking the number of instances of a pattern as its frequency has the undesirable property that extending the pattern (i.e., making it more restrictive) may actually increase its frequency. Indeed, consider for instance, the unlabeled k-clique K_k as the single graph in which we want to find patterns. There are $\binom{k}{2}$ different embeddings under subgraph isomorphism of the unlabeled path of length 1 in K_k, whereas there are $3\binom{k}{3}$ embeddings of the path of length 2 in K_k. In fact, the number of different embeddings may increase exponentially in the size of the pattern. Hence, as pointed out by Vanetik et al. [17], a good frequency measure must be such that the frequency of a superpattern is always at most as high as that of a subpattern. This property is called the *antimonotonicity*. Also, for reasons of efficiency, antimonotonicity of the frequency measure is highly desirable, as it allows for pruning large parts of the search space in a general-to-specific exploration. The efficiency and correctness of most existing graph pattern miners relies critically on the antimonotonicity of the frequency measure being used.

In this chapter, we give an overview of measures that have been defined in the literature for assessing the frequency of graph patterns in one large graph. We divide the measures into two groups: the ones that are based on the notion of the so-called overlap graph and those that are not. An overlap graph of a pattern graph in a single data graph is itself a graph again that expresses how the different occurrences of the pattern are connected to each other in the data graph. Every node in the overlap graph denotes an occurrence of the pattern and two nodes are connected by an edge if the corresponding occurrences of the pattern graph have an overlap. In Figure 10.1, two examples of the overlap graph of a pattern in another graph have been given.

In Section 10.2, we formally define important notions such as embedding, overlap graphs, and antimonotonic support measures. Then, for reasons of completeness,

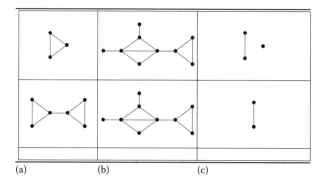

(a)   (b)   (c)

FIGURE 10.1:   Two examples of an overlap graph of H in G. (a) Pattern graph H, (b) database graph G, and (c) overlap graph of H in G.

we start our discussion of graph-support measures with nonoverlap-graph-based measures in Section 10.3, although in the rest of the chapter, we will mainly concentrate on the class of overlap-graph-based measures. This important class of graph measures is then introduced in Section 10.4. In this section, we survey important results connecting the antimonotonicity of the support-graph-based measure directly to properties of the overlap graph itself.

In Section 10.5, the results are extended to overlap *hypergraphs* that are able to express the way instances or embeddings overlap in a much more subtle and exact way. The alternative characterization and bounding theorems of the overlap-graph-based support measures are extended to this more fine-grained setting.

Section 10.6 describes an important application of the study of support measures: statistical analysis on graph datasets. Statistical theory often assumes that the objects over which summary statistics are computed are drawn independently. In networked data, however, different occurrences of subgraphs in the single-graph settings are dependent. Therefore, in Section 10.6, we relate the statistical power of a set of observations to its s-measure. In particular, if a pattern has a number of overlapping embeddings in a database graph and every embedding has some properties, one can estimate the distribution of these properties from a sample. We are interested in bounding the variance of such estimates.

Finally, Section 10.7 concludes the chapter.

10.2 Preliminaries

In this section, we introduce important graph-related notions such as graph isomorphisms and embeddings as well as antimonotonicity of graph-support measures. We assume that the reader is familiar with basic graph theoretic notions and with computational complexity. Textbooks in these areas, such as [7] and [14], supply the necessary background.

10.2.1 Graphs

A graph $G = (V, E)$ is a pair in which V is a (nonempty) set of *vertices* or *nodes* and E is either a set of *edges* $E \subseteq \{\{v, w\} \mid v, w \in V, v \neq w\}$ or a set of *arcs* $E \subseteq \{(v, w) \mid v, w \in V, v \neq w\}$. In the latter case, we call the graph *directed*. A *labeled* graph with labels from Σ is a triple $G = (V, E, \lambda)$, with (V, E) a graph, and λ a function $V \to \Sigma$ assigning labels to the vertices. We will use the notation $V(G)$, $E(G)$, and λ_G to refer to the set of vertices, the set of arcs (edges), and the labeling function of a graph G, respectively. Unless explicitly stated otherwise, we will assume to be working over undirected labeled graphs in this chapter. By $\mathcal{G}_\lambda$ and $\mathcal{G}$, we denote, respectively, the set of all labeled graphs and the set of all unlabeled graphs.

A graph $G = (V, E, \lambda)$ is said to be a subgraph of graph $H = (V_H, E_H, \lambda_H)$, denoted $G \subseteq H$, if $V \subseteq V_H$, $E \subseteq E_H$, and $\lambda = \lambda_H|_V$.

For $G \in \mathcal{G}_\lambda$,

$$\overline{G} := (V(G), \{\{v, w\} \mid v, w \in V\} \setminus E(G), \lambda_G)$$

denotes the complement graph of G. By $K_k \in \mathcal{G}$, we denote the *complete graph* on k vertices, that is,

$$K_k := (\{v_1, \ldots, v_k\}, \{\{v_i, v_j\} \mid 1 \leq i \neq j \leq k\}).$$

A subgraph $K \subseteq G$ on k vertices for which all vertices are adjacent to all other vertices is called a *k-clique*. A *cycle* of length k is a connected subgraph on k vertices each of which is incident with exactly two edges.

An undirected unlabeled *hypergraph* is a pair (V, E) where V is a set of vertices and $E \subseteq 2^V$ is a set of (hyper)edges, each of which is a subset of the set of vertices. We denote the set of all (undirected, unlabeled) hypergraphs with $\mathcal{H}$.

10.2.2 Isomorphisms

The following concepts introduced in terms of $\mathcal{G}_\lambda$ are also valid for directed and/or unlabeled graphs by adding the direction of the edges and/or dropping the labels of the vertices. For a complete set of definitions of all cases, we refer the interested reader to [4].

A *homomorphism* π from $H = (V_H, E_H, \lambda_H)$ to $G = (V, E, \lambda)$ is a mapping from $V_H \rightarrow V$, such that $\forall \{v, w\} \in E_H : \{\pi(v), \pi(w)\} \in E$. We say that H is homomorphic to G.

An *isomorphism* from H to G is a bijective homomorphism π from H to G. In that case, we say that H is isomorphic to G and write $H \cong G$. We use $H \subseteq G$ to denote that $H \cong g$, for some subgraph g of G.

By an *instance* of P in G, we refer to a subgraph g of G such that P and g are isomorphic. Any isomorphism π between P and one of its instances g is called an *embedding*. We denote the set of all instances of a pattern P in the graph G by $\mathrm{Img}(P, G)$, and the set of all embeddings by $\mathrm{Emb}(P, G)$. Notice that the number of instances does not necessarily equal the number of embeddings of P into G, as some embeddings may have the same image.

Example 10.2.1 *Consider the following datagraph G and pattern graph P. The subscripts in the labels have been added for ease of reference only. For instance, the nodes with label a have been annotated* $a_1, a_2, \ldots$

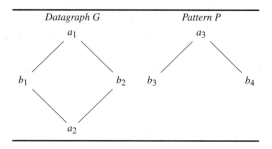

In this example, P has two instances in G (λ denotes the labeling function of G):

$$\left(\{a_1,b_1,b_2\},\{\{a_1,b_1\},\{a_1,b_2\},\{b_1,b_2\}\},\lambda|_{\{a_1,b_1,b_2\}}\right) \text{ and}$$
$$\left(\{a_3,b_1,b_2\},\{\{a_3,b_1\},\{a_3,b_2\},\{b_1,b_2\}\},\lambda|_{\{a_3,b_1,b_2\}}\right)$$

but the number of embeddings of P in G is 4:

$\begin{cases} a_3 &\mapsto a_1 \\ b_3 &\mapsto b_1 \\ b_4 &\mapsto b_2 \end{cases}$	$\begin{cases} a_3 &\mapsto a_2 \\ b_3 &\mapsto b_1 \\ b_4 &\mapsto b_2 \end{cases}$	$\begin{cases} a_3 &\mapsto a_1 \\ b_3 &\mapsto b_2 \\ b_4 &\mapsto b_1 \end{cases}$	$\begin{cases} a_3 &\mapsto a_2 \\ b_3 &\mapsto b_2 \\ b_4 &\mapsto b_1 \end{cases}$

For a more complete treatment including homeomorphisms and the extension of the notion of an instance to all morphism types, unlabeled, and directed graphs, we refer the reader to [4].

10.2.3 Support Measures

One of the key elements in a graph mining algorithm is the support measure; that is, a measure expressing the prevalence of a pattern graph in a larger database graph:

Definition 10.2.1 *A support measure on $\mathcal{G}$ is a function $f : \mathcal{G} \times \mathcal{G} \to \mathbb{N}$ that maps (P, G) to $f(P, G)$ where P is called the pattern, G is called the database graph, and $f(P, G)$ is called the* support *of P in G.*

For efficiency reasons, most graph mining algorithms use a level-wise or depth-first approach to generate frequent patterns, expanding smaller patterns to larger ones. Such an approach requires the support measure being antimonotonic in order to prune efficiently:

Definition 10.2.2 *A support measure f on is* antimonotonic *if for all patterns p and P and database graph G it holds that if $p \subseteq P$, then $f(P, G) \leq f(p, G)$. That is, the support in a graph G does not increase from a subpattern p to a superpattern P.*

In the two next sections, we will see multiple examples of graph-support measures, many of which are antimonotone.

10.3 Nonoverlap-Graph-Based Measures

The first type of single-graph-support measures we consider are those that are not based upon the overlap graph. The advantage of these measures is that they do not require the costly step of building up the overlap graph. All results have been stated in function of isomorphisms but can be extended easily to other morphism types.

10.3.1 Key-Based Support Measures

One approach that is commonly used is to select a fixed key pattern K consisting of a number of isolated vertices and to consider as pattern language the space of all superpatterns of K. This support measure is one of the first ones that was considered in relational learning and is related to the *learning from entailment* setting in the field of inductive logic programming [5]. The K-support of a pattern P in graph G is only defined if K is a subpattern of P and is defined as

$$keycount_E(P, G) = |\{\pi|_K \mid \pi \in \text{Emb}(P, G)\}|$$

where $\pi|_K$ is the restriction of the mapping to K. Clearly, *keycount* is antimonotonic. Indeed, if $p \subseteq P$ and $\pi \in \text{Emb}(P, G)$, then $\pi|_p \in \text{Emb}(p, G)$. Furthermore, if $\pi_1, \pi_2 \in \text{Emb}(P, G)$, and $\pi_1|_p = \pi_2|_p$, then also $\pi_1|_K = \pi_2|_K$. Therefore, $keycount_E(P, G) \leq keycount_E(p, G)$.

The same holds for the image-based version of this support measure:

$$keycount_I(P, G) = |\{\pi(K) \mid \pi \in \text{Emb}(P, G)\}|$$

10.3.2   Minimal Image Count Support Measure

In [3], the authors proposed an antimonotonic support measure named *min-image-based support*. For notational consistency, we give a slightly alternative definition.

$$minImage(P, G) = \min_{v \in V(P)} |\{\pi(v) \mid \pi \in \text{Emb}(P, G)\}| \qquad (10.1)$$

This support counts for every vertex, the number of nodes in the data graph to which the vertex can be mapped in an embedding. The measure is the minimum of this number over all vertices of the pattern graph. The antimonotonicity of this support is obvious, and it can be computed very efficiently. It has, however, several drawbacks as we demonstrate next.

First, from a statistical point of view, *minImage* overestimates the evidence. In particular, as Figure 10.2 shows, a vertex can be counted arbitrarily many times.

Second, *minImage* is not additive. Given a subgraph pattern P, if a database graph G has n ($n \geq 2$) connected components, that is, $G = \bigcup_{1 \leq i \leq n} G_i$, then $minImage(P, G) \geq \sum_{1 \leq i \leq n} minImage(P, G_i)$. For many realistic database graphs, strict inequality holds. In this case, it is unclear how much a connected component contributes to the whole support. Figure 10.3 shows an example.

10.4   Overlap-Graph-Based Measures

In this section, we give an overview of the most important results class of the single-graph-support measures that are the key focus of this chapter: the overlap-graph-based measures. First, we introduce the notion of an overlap graph. Then different measures based on the overlap graph and the important characterization of the monotone overlap-graph-based measures by Vanetik et al. [17] and its extension by Calders et al. [4] are discussed. After that, we extend to overlap-hypergraph measures that capture more subtle differences in how instances overlap than plain overlap graphs.

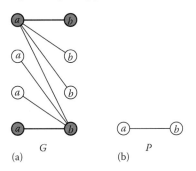

(a) G   (b) P

FIGURE 10.2: (a) Database graph G contains two independent images of (b) the subgraph pattern P. However, $minImage(P, G) = 4$ and we can make this value arbitrarily large by adding more vertices with label b (resp. a) and link them to the top-left vertex with label a (resp. bottom-right vertex with label b). As a consequence, if we remove just a single vertex (the top-left or bottom-right one) the support of the pattern in the network can suddenly drop to one.

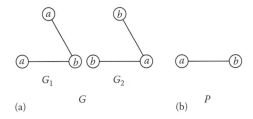

G_1 G_2
(a) G   (b) P

FIGURE 10.3: (a) A database graph G has two connected components G_1 and G_2. Consider (b) pattern P. $minImage(P, G_1) = minImage(P, G_2) = 1$, but $minImage(P, G) = 3 > minImage(P, G_1) + minImage(P, G_2)$.

10.4.1 Pairwise Overlap Graph

An important class of antimonotonic measures are the ones that are based on the notion of an *overlap* graph G_P [11,17].* An overlap graph summarizes not only the images of the pattern in the database graph but also how they overlap:

Definition 10.4.1 *Let $G \in \mathcal{G}$ be a database graph, P a pattern, and $g_1, g_2 \in Img(P, G)$ be two instances of P. g_1 and g_2 of G have a vertex overlap if $V(g_1) \cap V(g_2) \neq \emptyset$ and an edge overlap if $E(g_1) \cap E(g_2) \neq \emptyset$.*

For clarity of presentation, we will restrict ourselves to vertex overlap and isomorphic embeddings in this survey, but as shown in [4], all notions and results can be extended to other graph classes and overlap and morphism types.

---

* Vanetik et al. [17] uses the term *instance* graph instead of overlap graph. The term *instance* suggests the use of isomorphisms, and all properties and theorems remain valid for any kind of morphism. Therefore, we follow the terminology of [11] to avoid confusion.

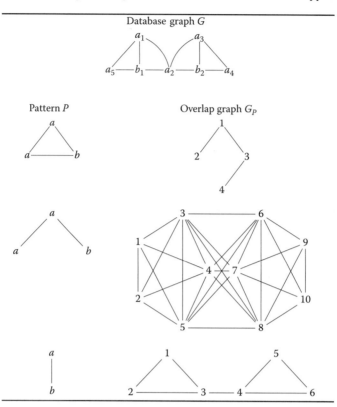

FIGURE 10.4:   Database graph G, and three patterns, with their corresponding pairwise overlap graphs G_P.

Definition 10.4.2 *The* pairwise overlap graph *(POG) G_P of a pattern P in the database graph G is an undirected, unlabeled graph in which each vertex corresponds to an instance of the pattern P. Two vertices are adjacent in G_P if the corresponding embeddings overlap.*

Figure 10.4 gives examples of POGs.

10.4.2   Overlap-Graph-Based Support Measure

We are now ready to define a POG-based support measure.

Definition 10.4.3 *A support measure f on graphs is a* pairwise overlap graph-based support measure *if there exists a graph measure $\hat{f}$ such that $\forall P, G \in \mathcal{G} : f(P, G) = \hat{f}(G_P)$.*

Informally, a POG-based support measure is a support measure that only depends on the POG. Consider, for example, the following measure based on the *maximal independent set* (MIS) of the POG. An *independent set* of a graph G is a subset I of $V(G)$ such that $\forall v, w \in I : \{v, w\} \notin E(G)$. A *MIS* of G is an independent set of maximal cardinality and its size is notated as $mis(G)$. The MIS-based POG support measure assigns to every pattern P the size of the MIS [17] of its POG G_P:

$$MIS(P, G) := mis(G_P).$$

Notice that $MIS(P, G)$ can intuitively be interpreted as the maximal number of instances that fit in G without overlap. This measure is antimonotonic.

Example 10.4.1 *Consider the example given in Figure 10.4. The POG of the triangular pattern in the data graph G consists in one path of length 3. The MISs in the POG in Figure 10.4 are $\{2, 3\}$ and $\{1, 4\}$. Hence, $MIS(P, G) = 2$.*

10.4.3 Alternative Characterization for Antimonotonicity

Vanetik et al. [17] have shown an alternative characterization of antimonotone POG-based measures. They consider three operations on the overlap graph G_P: clique contraction, edge removal, and vertex addition, as defined in the following.

Definition 10.4.4 *Let $K \subseteq G$ be a clique in $G = (V, E)$. The clique contraction $CC(G, K)$ yields a new graph $G' = (V', E')$ in which the subgraph $K \subseteq G$ is replaced by a new vertex $k \notin V$ adjacent to $\{w \mid \forall v \in V(K) : \{v, w\} \in E\}$:*

$$V' = V \setminus V(K) \cup \{k\}$$
$$E' = E \setminus \{\{v, w\} \mid \{v, w\} \cap V(K) \neq \emptyset\} \cup \{\{k, w\} \mid \forall v' \in V(K) : \{v', w\} \in E\}.$$

The edge removal $ER(G, e)$ of the edge $e = \{v, w\}$ in the graph $G = (V, E)$ yields a new graph $G' = (V, E \setminus \{\{v, w\}\})$.
    The vertex addition $VA(G, v)$ of the vertex $v \notin V$ in the graph $G = (V, E)$ yields a new graph $G' = (V \cup \{v\}, E \cup \{\{v, w\} \mid w \in V\})$.

The rationale behind these operations is that the overlap graph of a pattern P can be transformed into the overlap graph of a subpattern p of P by means of these operations.

Property 10.4.1 (Vanetik, Shimony, and Gudes). *Let G be a database graph, $p \subseteq P$ two patterns. G_P can be transformed into G_p with a sequence of CC, VA, and ER operations.*

Example 10.4.2 *In the POGs in Figure 10.1, we can transform the overlap graph of the third pattern (consisting of one edge between a node labeled a and a node labeled b) to the POG of the second pattern by a series on node additions and edge removals. These two operations together make it possible to transform a graph into any of its supergraphs. From the second overlap graph to the first, we need a series of clique contractions. We could contract subsequently $\{1, 2\}$, $\{3, 4, 5\}$, $\{6, 7, 8\}$, and $\{9, 10\}$.*

A direct result of this property is the following theorem of Vanetik et al. [17] that restates the antimonotonicity of f in function of $\hat{f}$ being nondecreasing in function of the three operations specified earlier.

Theorem 10.4.1 (Vanetik, Shimony, and Gudes). *Any overlap-graph-based support measure f is antimonotonic if and only if the associated graph measure $\hat{f}$ is nondecreasing under clique contraction, edge removal, and vertex addition.*

The proof of sufficiency, that is, that any overlap support measure f is antimonotonic if the associated graph measure $\hat{f}$ is nondecreasing under CC, VA, and ER follows immediately from the fact that G_P can be transformed into G_p by these operations.

To prove necessity, Vanetik, Shimony, and Gudes construct for every unlabeled graph H and every operation o a triple (P, p, G), where P is a superpattern, p a subpattern, and G a database graph such that $G_P \cong H$ and $G_p \cong o(H)$. Henceforth, if f would be increasing under some $o \in \{CC, ER, VA\}$, then there would be a H such that $f(H) > f(o(H))$ and one could construct a G, P, and p such that $f(G, P) > f(G, p)$, which would mean that f is not antimonotonic.

10.4.4 Bounding Theorem

The result of Vanetik et al. [17] was later extended by Calders et al. [4] to all combinations of iso-, homo-, and homeomorphisms; edge/vertex-overlap graphs; directed/undirected; and labeled/unlabeled graphs. An important consequence of the alternative characterization of the antimonotonicity of f in terms of $\hat{f}$ being nondecreasing is the bounding theorem proven by Calders et al. [4]. This theorem states that the different antimonotone measures are bounded by a natural minimal and maximal support measure; every *normalized* overlap support measure will always be between these two extremes. A normalized support measure is defined as follows.

Definition 10.4.5 *Let G be an undirected graph and $\overline{K_k}$ the graph composed of k isolated vertices.*

We call an overlap support measure f normalized if it is antimonotonic and assigns the frequency k to k nonoverlapping images, that is, $\hat{f}(\overline{K_k}) = k$.

Before we state the bounding theorem, we first introduce the minimum clique partition (MCP) measure.

10.4.4.1 MCP Measure

The first antimonotonic, normalized overlap-graph-based support measure was the MIS measure *MIS*. The *MIS* measure is defined as the size of the MIS of the overlap graph and was introduced and proven to be antimonotonic in [17]. A more compact proof of the antimonotonicity can be found in [8]. This measure was shown to be a lower bound on all normalized, antimonotonic overlap-graph-based measures. Later on, Calders et al. [4] introduced two more normalized antimonotone overlap-graph-based measures, being MCP and the Schrijver measure. We will review the Schrijver measure in Section 10.4.5.

The MCP measure is inspired by the CC-operation:

Definition 10.4.6 *A* clique partition *of an undirected graph G is a partitioning of* $V(G)$ *into* $\{V_1, \ldots, V_k\}$ *such that each* V_i *induces a complete graph in G. A MCP is a clique partition of minimum cardinality. Its cardinality is denoted* $mcp(G)$.

The MCP measure is defined by $MCP(P, G) : (P, G) \mapsto mcp(G_P)$.

Theorem 10.4.2 *[4] The MCP measure is an antimonotonic and normalized.*

It is interesting to compare *MCP* with *MIS*. Let $\chi(G)$ be the *chromatic number* of G, that is, the minimal number of colors needed to color the vertices of G such that no two vertices with the same color are adjacent, and let $\omega(G)$ be the *clique number*, the size of the largest clique in G.

First, it is known that $mcp(G) = \chi(\overline{G})$ and $mis(G) = \omega(\overline{G})$ (see, e.g., [9], Section 5.5.1). Consequently, $mcp(G) \geq mis(G)$, for all undirected graphs G, since the size of a maximum clique is a lower bound for the chromatic number.

Informally, it is easy to see why this is so: let $\{V_1, \ldots, V_k\}$ be an MCP and I a MIS for G. We know that I contains at most one vertex v_i of each V_i, $1 \leq i \leq k$. In other words, to decide whether we can include a image of V_i, *MIS* forces us to choose either no image or exactly one image v_i, which must be independent of all chosen $v_j \in V_j$. *MCP*, however, allows us to count a image in V_i as soon as there is *a* image in V_i, which does not overlap with *a* image in V_j. That is, we can make another choice for each (V_i, V_j) pair.

Example 10.4.3 *Let us look at an example: consider pattern P and the graph G as shown in Figure 10.5. The five images of P are the induced subgraphs of the database with, respectively, the nodes* $\{a, b, c, d, e\}$, $\{i, f, d, k, l\}$, $\{i, f, g, k, l\}$, $\{e, h, j, m, l\}$, *and* $\{g, h, j, m, l\}$. *The POG* G_P *of P in G is shown on the right in Figure 10.5 and is isomorphic to a pentagon. The white vertices mark the MIS* $\{2, 5\}$ *and the dashed ellipses*

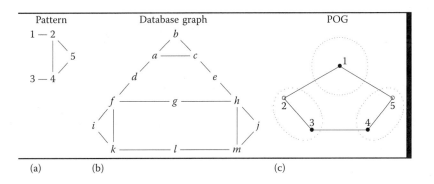

FIGURE 10.5: (a) A pattern *P* and (b) a graph *G*. (c) Overlap graph G_P with a *MCP* (dashed ellipses) and a *MIS* (white vertices).

mark a MCP consisting of the three cliques {1}, {2, 3}, and {4, 5} of G_P. Hence, if we count image 2 with MIS, we can only take image 4 or image 5 as second independent image, because 1 and 3 overlap, leading to a MIS support of 2. This is a bit unnatural, because each of the three images of the triangle can be extended to a image of P in a way that they do not overlap with each other, which would lead to a support of 3 of P.

This more natural notion of counting independent images is exactly what MCP support allows us to do: we do not count individual images, but groups of images of P sharing a image of a subpattern p (a triangle), and allow to "switch" images to decide whether a group is independent of another. In this example, the group {1} is independent of the groups {2, 3} and {4, 5}, because it does not overlap with image 3 (respectively image 4) and the group {2, 3} is independent of the group {4, 5} because, for instance, image 2 and image 5 do not overlap.

10.4.4.2   Theorem

Interestingly, *MIS* and *MCP* turn out to be the minimal and the maximal possible normalized overlap measures. The following theorem is one of our main results:

Theorem 10.4.3 *For every normalized overlap measure f, and every pattern P and database graph G, it holds that*

$$MIS(P, G) \leq f(P, G) \leq MCP(P, G).$$

This bounding theorem still leaves a lot of room to define support measures, as there can be an arbitrarily large gap between MIS and MCP [2]. The Lovász and Schrijver measures that will be discussed in Section 10.4.5 is one such measure.

Example 10.4.4 *Consider again the example given in Figure 10.5. mis(G_P) = 2 and mcp(G_P) = 3. Hence, every antimonotonic normalized overlap support measure must assign a value between 2 and 3 for P in G. Indeed, as illustrated in Figure 10.6, $\overline{K_2}$ can be transformed into G_P and G_P can on its turn be transformed into $\overline{K_3}$.*

10.4.5   Lovász and Schrijver Graph Measures as Support Measures

The first function that was shown to be a normalized antimonotonic POG-based support measure computable in polynomial time was the Lovász ϑ value of the overlap graph [4]. A similar argument can be used for the Schrijver measure [18]. Both measures are studied in depth in the graph theory literature and often also relations to the size of the maximum independent set (MIS) and other important measures are

FIGURE 10.6:   Illustration of a sequence of operations to move from a MIS to the overlap graph of Figure 10.5 to the minimal clique partition.

considered. We believe that it may be valuable for the data mining community to further explore this literature. Here, we briefly defined both measures. Interested readers can consult [4,18].

Let G be a graph. In the following, we will assume that $V(G) = \{1, \ldots, n\}$ so that we can use a vertex as an index of a vector or matrix, for example, we will denote cell i of a vector x by x_i. A *Lovász feasible matrix* A for G is a symmetric positive semidefinite matrix with (1) $A_{u,v} = 0$ for all u and v such that $\{u, v\} \in E(G)$ and (2) $Tr(A) = 1$.

Given a graph G, the *Lovász ϑ value* of G is

$$\vartheta(G) = \max \left\{ \sum_{i,j} A_{i,j} \mid A \text{ is a Lovász feasible matrix of } G \right\}.$$

The *Schrijver graph measure* of G is

$$SGM(G) = \max \left\{ \sum_{i,j} A_{i,j} \,\middle|\, \begin{array}{l} A \text{ is a Lovász feasible matrix of } G \\ \text{and } \forall i, j : A_{i,j} \geq 0 \end{array} \right\}.$$

The Schrijver graph measure has nearly the same computational complexity as the Lovász ϑ value but is closer to the MIS support measure. The latter can be an advantage for certain statistical tasks in which we want to stay as close as possible to an independent set of images.

10.5 Object-Specific Overlap Hypergraphs

The overlap-based measures we discussed up to now viewed overlap as a binary property: two instances either overlap or not. In this section, we extend this idea further with support measures taking into account *how* instances overlap. We call this object-specific overlap and will represent the overlap relationships with a hypergraph instead of a simple graph [18].

As earlier, several types of overlap exist. First, we should select the *objects* of overlap. These can be vertices, edges, or both. For a database graph G, we define

$$Obj_{vertex}(G) = V(G)$$
$$Obj_{edge}(G) = E(G)$$
$$Obj_{ev}(G) = V(G) \cup E(G).$$

These objects induce cliques in the pairwise overlap graph G_P. For instance, when considering vertex overlap of instances (as we did in the previous sections), all instances containing a vertex v of G will form a clique in G_P.

Definition 10.5.1 (Overlap hypergraph) *Let* $\gamma \in \{vertex, edge, ev\}$. *The* γ-*ins*-overlap hypergraph *of* P *in* G, *denoted by* $H_{P,\gamma}^{G,ins}$ *or more briefly* H_P^G *if the rest is* *clear from the context, is a hypergraph whose vertices are the instances $Img(G, P)$ and for each object $x \in Obj_\gamma(G)$, there is a hyperedge $e_x \in E\left(H_P^G\right)$ such that $e_x = \{g \in V\left(H_P^G\right) \mid x \in Obj_\gamma(g)\}$.*

The γ-emb-overlap hypergraph of P in G, denoted by $H_{P,\gamma}^{G,emb}$ or more briefly H_P^G if the rest is clear from the context, is a hypergraph whose vertices are the embeddings $Emb(G, P)$ and for each object $x \in Obj_\gamma(G)$ and object $y \in Obj_\gamma(P)$, there is a hyperedge $e_x \in E(H_P^G)$ such that $e_x = \{\pi \in V\left(H_P^G\right) \mid x = \pi(y)\}$.

In an overlap hypergraph $H_{P,\gamma}^{D,\delta}$, we say that a hyperedge e is *dominated* by another hyperedge e' if $e \subset e'$, and a hyperedge e is *dominating* if it is not dominated by any other hyperedge. For any D and P, we define the reduced overlap hypergraph $\tilde{H}_{P,\gamma}^{D,\delta}$ to be the hypergraph for which $V\left(\tilde{H}_{P,\gamma}^{D,\delta}\right) = V\left(H_{P,\gamma}^{D,\delta}\right)$ and $E\left(\tilde{H}_{P,\gamma}^{D,\delta}\right)$ is the set of all dominating hyperedges of $H_{P,\gamma}^{D,\delta}$. In the sequel, we only refer to $\tilde{H}_{P,\gamma}^{D,\delta}$, omitting δ and γ when they are clear from the context. We will abuse terminology and simply call $\tilde{H}_P^D$ the overlap hypergraph. See Figure 10.7 for an example.

We henceforth refer to the overlap hypergraph measures, which we denote by $f'\left(\tilde{H}_P^D\right)$, instead of referring to the induced support measure $f(D, P)$. Such induced support measures are called overlap-hypergraph-based support measures (OHSM).

10.5.1   Support Measure Based on Relaxed Maximal Independent Set

Given an overlap hypergraph $\tilde{H}_P^D$, we can derive the corresponding overlap graph G_P^D by replacing every hyperedge with a clique. Therefore, we can rephrase the definition of the MIS measure using overlap hypergraphs. Suppose D is a database graph and P is a subgraph pattern:

$$MIS(D, P) = MIS\left(\tilde{H}_P^D\right) = \max\left|\left\{I \subseteq V\left(\tilde{H}_P^D\right) \mid \forall e \in E\left(\tilde{H}_P^D\right) : |e \cap I| \leq 1\right\}\right| \tag{10.2}$$

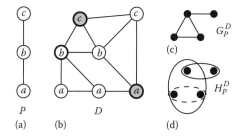

(a)   (b)   (c)   (d)

FIGURE 10.7: Overlap graph and overlap hypergraph. Given (a) a subgraph pattern P, (b) a database graph D, (c) the overlap graph G_P^D, and (d) the overlap hypergraph H_P^D are shown on the right. In the overlap hypergraph, the (dominating) hyperedges are determined by the highlighted vertices in the database graph, and a dominated hyperedge is given in a dashed ellipse.

The MIS measure requires that a vertex of an overlap (hyper)graph is either in the independent set I or not. The S measure [18] we consider in this section is a relaxation of the MIS support measure by allowing for counting vertices of an overlap hypergraph only partially.

Let $\tilde{H}_P^D$ be an overlap hypergraph. We start by assigning to each vertex v of $\tilde{H}_P^D$ a variable x_v. We then consider vectors $x \in \mathbb{R}^{V(\tilde{H}_P^D)}$ of variables where for every $v \in V\left(\tilde{H}_P^D\right)$, x_v denotes the variable (component of x) corresponding to v. x is *feasible* if and only if it satisfies

$$
\begin{array}{ll}
\text{(i)} & \forall v \in V(\tilde{H}_P^D) : 0 \leq x_v \\
\text{(ii)} & \forall e \in E(\tilde{H}_P^D) : \sum_{v \in e} x_v \leq 1.
\end{array}
\qquad (10.3)
$$

We denote the feasible region (the set of all feasible $x \in \mathbb{R}^{V(\tilde{H}_P^D)}$) by $\mathfrak{R}\left(\tilde{H}_P^D\right)$, which is a convex polytope.

Definition 10.5.2 (S support measure) *The measure S is defined by*

$$
\mathsf{S}\left(\tilde{H}_P^D\right) = \max_{x \in \mathfrak{R}\left(\tilde{H}_P^D\right)} \sum_{v \in V\left(\tilde{H}_P^D\right)} x_v. \qquad (10.4)
$$

Clearly, S is the solution to a linear program.

We will call an element $x \in \mathfrak{R}\left(\tilde{H}_P^D\right)$, which makes $\sum_{v \in V\left(\tilde{H}_P^D\right)} x_v$ maximal a *solution* to the LP of S.

There are very effective methods for solving LPs, including the simplex method, which is efficient in practice although its complexity is exponential, and the more recent interior-point methods [1]. The interior-point method solves an LP in $O(n^2 m)$ time, where n (here $\min\{|V\left(\tilde{H}_P^D\right)|, |E\left(\tilde{H}_P^D\right)|\}$) is the number of variables and m (here $|V\left(\tilde{H}_P^D\right)| + |E\left(\tilde{H}_P^D\right)|$) is the number of constraints. Usually, subgraph patterns are not large, so the LPs for computing S are sparse. Almost all LP solvers perform significantly better for sparse LPs.

10.5.1.1 Conditions for Antimonotonicity

The conditions for antimonotonicity of OHSMs are similar to the ones discussed earlier based on normal overlap graphs. In particular, we can show that an OHSM is antimonotonic if and only if it is nondecreasing under three operations on the overlap hypergraph. We begin by defining these three operations. They are similar to those on overlap graphs.

Definition 10.5.3 (Hypergraph operators) *For $H \in \mathcal{H}$, we define*

- *Vertex addition: A new vertex v is added to every existing hyperedge: $VA(H, v) = (V(H) \cup \{v\}, \{e \cup \{v\} \mid e \in E(H)\})$.*

- *Subset contraction: Let $K \subseteq V(H)$ be a set of vertices of the hypergraph such that $\exists e \in E(H) : K \subseteq e$. Then, the subset contraction operation contracts K into a single vertex k, which remains in only those hyperedges that are supersets of K. Formally, $SC(H, K, k) = (V(H) - K \cup \{k\}, E_1 \cup E_2)$ where $E_1 = \{e - K \cup \{k\} \mid e \in E(H) \text{ and } K \subseteq e\}$ and $E_2 = \{e - K \mid e \in E(H) \text{ and } K \not\subseteq e\}$).*

- *Hyperedge split: This operation splits a size k hyperedge into k hyperedges of size $(k - 1)$ each: $HS(H, e) = (V(H), E(H) - \{e\} \cup \{e - \{v\} \mid v \in e\})$, where $e \in E(H)$.*

For example, suppose H_0 is a hypergraph, $V(H_0) = \{v_1, v_2, v_3, v_4\}$, and $E(H_0)$ contains two hyperedges $\{v_1, v_2, v_3\}$ and $\{v_1, v_4\}$. Let $H_1 = VA(H_0, v_5)$, then $V(H_1) = \{v_1, v_2, v_3, v_4, v_5\}$ and $E(H_1)$ contains hyperedges $\{v_1, v_2, v_3, v_5\}$ and $\{v_1, v_4, v_5\}$. Let $H_2 = SC(H_1, \{v_1, v_3\}, v_6)$, then $V(H_2) = \{v_2, v_4, v_5, v_6\}$ and $E(H_2)$ contains hyperedges $\{v_2, v_5, v_6\}$ and $\{v_4, v_5\}$. Let $H_3 = HS(H_2, \{v_2, v_5, v_6\})$, then $V(H_3) = V(H_2)$ and $E(H_2)$ contains four hyperedges $\{v_2, v_5\}$, $\{v_2, v_6\}$, $\{v_5, v_6\}$, and $\{v_4, v_5\}$.

10.5.1.2   Sufficient Condition

We present a sufficient condition for support measure antimonotonicity in terms of the three operations on the overlap hypergraph that we have defined.

Theorem 10.5.1 *Let $f' : \mathcal{G} \times \mathcal{G} \to \mathbb{R}$ be a support measure and $f : \mathcal{H} \to \mathbb{R}$ with $f'(D, P) = f\left(\tilde{H}_P^D\right)$ be the induced OHSM. If f is nondecreasing under VA, SC, and HS, then f' is an antimonotonic support measure.*

Theorem 10.5.2 $\mathsf{s}(D, P) = \mathsf{s}(\tilde{H}_P^D)$ *is a normalized antimonotonic support measure.*

10.5.1.3   Necessary Condition

We show that the condition for antimonotonicity mentioned earlier is not only a sufficient but also a necessary condition.

Theorem 10.5.3 *Let $f' : \mathcal{G} \times \mathcal{G} \to \mathbb{R}$ be a support measure and $f : \mathcal{H} \to \mathbb{R}$ with $f'(D, P) = f\left(\tilde{H}_P^D\right)$ be the induced OHSM. If f' is antimonotonic, then f is nondecreasing under VA, SC, and HS.*

10.5.2   Bounding Theorem

In [4], the authors showed that all normalized antimonotonic OGSMs are bounded (between the MIS size and the MCP). Similarly, we prove that all normalized antimonotonic OHSMs are also bounded. We first introduce another OHSM, MSC, the size of a minimum set cover of overlap hypergraphs:

$$MSC(D, P) = MSC\left(\tilde{H}_P^D\right) = \min \left|\left\{S \subseteq E\left(\tilde{H}_P^D\right) \mid \bigcup_{e \in S} e = V\left(\tilde{H}_P^D\right)\right\}\right| \quad (10.5)$$

It is not difficult to verify that MSC is normalized and antimonotonic. Computing MSC is an NP-hard problem. The MIS size (Equation 10.2) and minimum vertex cover (Equation 10.5) are the minimally and maximally possible normalized antimonotonic OHSMs.

Theorem 10.5.4 *Given a database graph D, and a subgraph pattern P, it holds that* $MIS(D, P) \leq f(D, P) \leq MSC(D, P)$ *for every normalized antimonotonic OHSM* $f(D, P) = f'\left(\tilde{H}_P^D\right)$.

10.5.3 Relaxation of the OGSM MIS

One may ask whether the s support can be defined by relaxing the OGSM MIS instead of the OHSM MIS. In other words, is the concept of overlap hypergraphs really necessary?

Our answer is that the concept of overlap hypergraph is needed for the definition of the s support measure because it carries additional information on the overlap graph. In particular, the hyperedges show which overlaps have a common cause. If we did not have this information, we would not be able to reconstruct it. For instance, if we see a triangle in an overlap graph, we do not know whether this triangle originates from one vertex shared by the three images or from three vertices, each shared by two of the images. This additional information is needed for the definition of s, and for its mathematical properties.

10.6 Bounding the Variance of Sample Estimates Using the s Measure

An important motivation for investigating support measures is the need to perform statistical analysis on datasets. Statistical theory often assumes that data points are drawn independently. In networked data, however, where vertices are connected with edges, this is not the case anymore. In this section, we relate the statistical power of a set of observations to its s-measure. In particular, if a pattern has a number of overlapping embeddings in a database graph and every embedding has some properties, one can estimate the distribution of these properties (or its mean, variance, moment) from a sample. We are interested in bounding the variance of such estimates.

When performing statistics on a particular type of observations, we first have to define the properties that the observations of interest will need to satisfy, thus creating a subgraph pattern. For instance, suppose we want to analyze the satisfaction of clients with their first lawsuit where they are assisted by a pro-deo lawyer, that is, a lawyer paid by the government or by an association to offer legal aid services to those who cannot afford a lawyer. Then, the subgraph pattern representing the observation type of interest would consist of a client node, a lawyer node, a judge node, and a lawsuit node to which the former three are connected.

Next, let us assume that the occurrence of these observations occurs independently from the properties that are relevant for our statistical analysis. In our example, in order to ensure impartiality, the court randomly assigns judges to cases and the lawyer association randomly assigns pro-deo lawyers to cases. Hence, in order to explore the relationships between the properties of the case and its outcome, we do not need to take into account the dependency between occurrences of the subgraph pattern and its properties.

This simplifying assumption does not imply, however, that we can treat the properties of the nodes of the embeddings of the pattern as independent, since embeddings may share nodes. In our example, the same parties, lawyers, or judges may participate in different lawsuits.

Consider the simple task of estimating the expected value of a function over the properties of nodes participating in a random embedding. In particular, let $f(\cdot)$ be a function on embeddings and let μ be its expected value and σ its standard deviation. Consider also a sample, that is, a set of possibly overlapping embeddings, and the problem of estimating μ as accurately as possible. In our example, f could be the measurement of client satisfaction with the outcome of the lawsuit depending on properties of the client, the lawsuit, the lawyer, or the judge.

We will now present two approaches for deriving a relation between sample size and the variance on the estimate obtained.

In a first approach, we take a maximal independent set S_{MIS} of vertices of the overlap hypergraph H_p^D. As we assumed that the embeddings are independent from the properties of the nodes they connect, all elements in S_{MIS} are independent and the values $f(v)$ of the observations v in S_{MIS} are distributed independently with $\mathbb{E}\left[(f(v) - \mu)^2\right] = \sigma^2$. Consider now the estimator

$$\hat{\mu}_{\text{MIS}} = \frac{\sum_{u \in S_{\text{MIS}}} f(u)}{|S_{\text{MIS}}|}.$$

As the terms in the sum are independent random variables,

$$\mathbb{E}\left[(\hat{\mu}_{\text{MIS}} - \mu)^2\right] = \frac{\sigma^2}{|S_{\text{MIS}}|}. \tag{10.6}$$

We will now present a second approach based on our s measure. Suppose that we have a set $V\left(H_p^D\right)$ of observations (embeddings of the pattern p in the database graph D), whose overlaps are given by the overlap hypergraph H_p^D, and a vector x of weights x_v for the $v \in V\left(H_p^D\right)$, which is a feasible solution to the s measure related linear program (10.3). We define the estimator:

$$\hat{\mu}_s\left(f, V\left(H_p^D\right), x\right) = \frac{\sum_{v \in V\left(H_p^D\right)} x_v f(v)}{\sum_{v \in V\left(H_p^D\right)} x_v} \tag{10.7}$$

We will now prove the following:

Theorem 10.6.1 *Let p be a pattern graph with $V(p) = \{i\}_{i=1}^{k}$ and $D = \cup_{i=1}^{k} D_i$ be a database with $k = |V(p)|$ domains. Let the set of embeddings of p in D be k-partite, that is, $\text{Emb}(D, p) \subseteq D_1 \times \cdots \times D_k$. Let the overlap hypergraph H_p^D represent this set of embeddings and their overlaps, that is, two vertices $u, v \in V\left(H_p^D\right)$ overlap if and only if $u(i) = v(i)$ for some $i \in \{1 \ldots k\}$. Assume the nodes in D have properties that are independent of these embeddings. Let x be a vector of weights for the embeddings satisfying (10.3). Let f be a function on the properties of the nodes participating in an embedding. Assume that for a randomly chosen embedding u, $\mathbb{E}\left[(f(u) - \mu)^2\right] = \sigma^2$. Then,*

$$\mathbb{E}\left[\left(\hat{\mu}_s\left(f, V\left(H_p^D\right), x\right) - \mu\right)^2\right] \leq \frac{\sigma^2}{\sum_{v \in V\left(H_p^D\right)} x_v}$$

In conclusion, if we choose x such that $\sum_v x_v = s\left(H_p^D\right)$, we get

$$\mathbb{E}\left[\left(\hat{\mu}_s\left(f, V\left(H_p^D\right), x\right) - \mu\right)^2\right] \leq \frac{\sigma^2}{s\left(H_p^D\right)}.$$

Because $s \geq |MIS|$, the second approach yields a better estimate of μ. Even though we had to make a number of assumptions, this first result linking s and the statistical power of a sample suggests that closer analysis of its properties may be a valuable direction for further research. Note that the assumptions on which the first method (using a MIS) relies are not necessarily much weaker than those made for the method using s.

10.7 Discussion and Conclusion

Next to the types of overlap we considered in this paper, other types of overlap may be of interest. For example, the following notions of overlap could also be considered:

- *Two-vertex (edge) overlap*: Two images overlap if and only if they share two or more common vertices (edges).

- *Label-specific overlap*: Two images overlap if and only if they share a common vertex (edge), which has a label in a certain set.

- *Distance-based overlap*: Two images u and v overlap if u has a vertex x and v has a vertex y such that the distance between x and y is smaller than a specified constant *min_dist*.

In each of these cases, small patterns need to be treated with caution, but an anti-monotonic support measure is obtained for patterns of minimal size.

The choice of overlap notion may be inspired by several factors, one of the main ones being the statistical assumptions made and the task to be performed. For instance, in Section 10.6, we presented a derivation for the statistical power of a sample assuming that the property of interest only depends on the properties of the nodes participating in the embedding. Suppose now that this assumption does not hold. For instance, in our lawsuit example, clients belonging to the same family might share common properties or be influenced by each other and hence might not be independent. We could then add family relations to the graph and say that two embeddings (*client*1, *lawyer*1, *judge*1, *case*1) and (*client*2, *lawyer*2, *judge*2, *case*2) overlap if *client*1 and *client*2 are members of the same family (i.e., have a distance of at most 1 in the family relationship graph). In this way, we can relax our assumptions by strengthening our notion of overlap.

Frequency is not a perfect indicator of a patterns interestingness. In most cases, frequent subgraphs are trivial patterns. Therefore, Milo et al. [13] proposed methods based on statistical hypothesis testing to filter out insignificant patterns. They first assume a null model that generates networks by preserving the network degree distribution. Subsequently, every frequent subgraph is checked against the null model in a randomization test. This approach effectively filters out a lot of trivial frequent patterns. Importantly, it is computationally demanding because randomization tests require generation of random samples of the entire network and perform frequency counting on these large samples.

In this chapter, we studied support measures in the single-graph context. Most of the results in this chapter concerned the so-called overlap-graph-based support measures for which alternative characteristics, a bounding theorem, and extensions to hypergraphs were shown. We also show one example of an application, being bounding the variance of sample estimates.

References

1. S.P. Boyd and L. Vandenberghe. *Convex Optimization*. Cambridge University Press, Cambridge, U.K., 2004.

2. V.E. Brimkov. Clique, chromatic, and Lovasz numbers of certain circulant graphs. *Electronic Notes in Discrete Mathematics*, 17:63–67, 2004.

3. B. Bringmann and S. Nijssen. What is frequent in a single graph? In *Proceedings of Mining and Learning with Graphs (MLG) 2007*, Florence, Italy, 2007.

4. T. Calders, J. Ramon, and D. Van Dyck. All normalized anti-monotonic overlap graph measures are bounded. *Data Mining and Knowledge Discovery*, 23(3):503–548, 2011.

5. L. De Raedt. Logical settings for concept learning. *Artificial Intelligence*, 95:187–201, 1997.

6. L. De Raedt and S. Kramer. The levelwise version space algorithm and its application to molecular fragment finding. In B. Nebel, ed., *Proceedings of the 17th International Joint Conference on Artificial Intelligence*, pp. 853–862. Morgan Kaufmann, San Francisco, CA, 2001.

7. R. Diestel. *Graph Theory*. Springer-Verlag, New York, 2000.

8. M. Fiedler and C. Borgelt. Support computation for mining frequent subgraphs in a single graph. In *Proceedings of the Fifth Workshop on Mining and Learning with Graphs (MLG'07)*, Firenze, Italy, 2007.

9. J.L. Gross and J. Yellen. *Handbook of Graph Theory*. CRC Press, Boca Raton, FL, 2004.

10. H. He and A.K. Singh. Efficient algorithms for mining significant substructures in graphs with quality guarantees. *Data Mining, IEEE International Conference on*, 0:163–172, 2007.

11. M. Kuramochi and G. Karypis. Finding frequent patterns in a large sparse graph. *Data Mining and Knowledge Discovery*, 11(3):243–271, 2005.

12. M. McGlohon, J. Leskovec, C. Faloutsos, M. Hurst, and N. Glance. Finding patterns in blog shapes and blog evolution. In *Proceedings of the International Conference on Weblogs and Social Media*, pp. 26–28, Boulder, CO, March 2007.

13. R. Milo, S. Shen-Orr, S. Itzkovitz, N. Kashtan, D. Chklovskii, and U. Alon. Network motifs: Simple building blocks of complex networks. *Science*, 298(5594):824–827, 2002.

14. C.H. Papadimitriou. *Computational Complexity*. Addison-Wesley, Reading, MA, 1994.

15. J. Ramon, T. Francis, and H. Blockeel. Learning a Tsume-Go heuristic with Tilde. In *Proceedings of CG2000, the Second International Conference on Computers and Games, volume 2063 of Lecture Notes in Computer Science*, pp. 151–169, Hamamatsu, Japan. Springer-Verlag, Heidelberg, Germany, 2000.

16. H. Tong, C. Faloutsos, B. Gallagher, and T. Eliassi-Rad. Fast best-effort pattern matching in large attributed graphs. In *KDD'07: Proceedings of the 13th ACM SIGKDD International Conference on Knowledge Discovery and Data Mining*, pp. 737–746. ACM, New York, 2007.

17. N. Vanetik, S.E. Shimony, and E. Gudes. Support measures for graph data. *Data Mining and Knowledge Discovery*, 13(2):243–260, 2006.

18. Y. Wang and J. Ramon. An efficiently computable subgraph pattern support measure. *Knowledge Discovery and Data Mining*, 27(3):444–477, 2013.

Chapter 11

Network Sampling Algorithms and Applications

Michael Drew LaMar and Rex K. Kincaid

Contents

11.1   Introduction..325
11.2   Notations and Definitions...326
11.3   Spectral Properties and Network Dynamics.................................329
11.4   Network Sampling...333
        11.4.1   Degree Distributions..335
        11.4.2   Node-Degree Correlation...336
        11.4.3   Edge-Degree Correlation...336
11.5   Applications..337
        11.5.1   Synchronization on Networks......................................337
                11.5.1.1   Neuronal Networks.....................................338
                11.5.1.2   Pulse-Coupled Oscillators...............................338
        11.5.2   Learning from Internet Design......................................341
11.6   Optimization: Perfect b-Matching and $s(G)$...............................342
        11.6.1   Formulation and Complexity..343
        11.6.2   Solving the Minimum Randić Index Problem.......................345
        11.6.3   Heuristic for Disconnected Realizations...........................348
11.7   Conclusion...348
References...349

11.1   Introduction

Networks appear throughout the sciences, forming a common thread linking research activities in many fields, such as sociology, biology, chemistry, engineering, marketing, and mathematics. For example, they are used in ecology to represent food webs and in engineering and computer science to design high-quality Internet router connections. Depending on the application, one network structural property may be more important than another. The structural properties of networks (e.g., degree distribution, clustering coefficient, assortativity) are usually characterized in terms of invariants [8], which are functions on networks that do not depend on the labeling of the nodes. In this chapter, we focus on network invariants that are quantitative, that is, they can be characterized as network measures. Examples of network measures includes degree-based measures (Randić index, assortativity), distance-based measures (Wiener, efficiency complexity), eigenvalue-based measures (Laplacian), and entropy measures [13,14].

An increasingly important application area is how network invariants affect the dynamics of a process on the network (e.g., respiration, current, traffic) [38]. In order to study the potential effect of incremental changes in network invariants on network dynamics, one or more network invariants are held constant, thereby creating a family of networks. In this chapter, the degree distribution of a network is held constant, while other network invariants are examined. In particular, we examine the effects of assortativity, the Randić index, and eigenvalues of the Laplacian on network dynamics.

11.2   Notations and Definitions

We use the terms network and graph interchangeably. We assume the reader to have a knowledge of graph theory (see, e.g., [56]). Let a graph $G = (V, E)$ be given where V is the set of nodes and E is the set of edges. For a directed graph, we use the similar notation $G = (V, A)$, where A is the set of arcs. We specify, when necessary, whether G is directed or not. For undirected graphs, we use the notation d_i to denote the *degree* of the node i, that is, the number of edges incident to i. For directed graphs, we use the notation d_i^- and d_i^+ to denote the in- and out-degree of the node i, that is, the number of arcs with i as their head and tail, respectively. We use subscripts when there may be confusion on the graph in question, for example, $d_i^-(G)$ or $d_i^+(G)$. When we discuss subsets of graphs, we use calligraphic font, for example, $\mathcal{G}$. For undirected and directed graphs, we define $A(G)$ to be the node–node adjacency matrix and $D(G)$ to be the diagonal matrix with the degree sequence d (or d^+ in the directed case) along the main diagonal. We omit reference to G when the graph is clear from context.

For what follows, we assume that any set of graphs we use has a fixed number of nodes $n = |V|$ for all graphs in the set. We also assume that any set of graphs consists of either all directed graphs or all undirected graphs. We denote the set of undirected graphs by $\mathcal{U}$ and the set of directed graphs by $\mathcal{W}$.

For an undirected graph G, we define the *Laplacian* of G as the $n \times n$ matrix $L(G) = D(G) - A(G)$. The spectrum of $L(G)$ is well studied. We simply note that $L(G)$ is symmetric and therefore has all real eigenvalues. As stated earlier, we omit reference to G when the graph meant is clear. There are multiple definitions for the Laplacian of a *directed* graph. The Laplacian of a strongly connected directed graph G is defined as

$$L'(G) = D_\phi - \frac{D_\phi P + P^T D_\phi}{2} \tag{11.1}$$

where
$P = D(G)^{-1} A(G)$
D_ϕ is a diagonal matrix

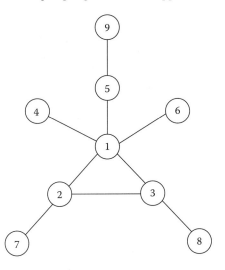

FIGURE 11.1: Undirected graph $G(V, E)$ with $|V| = 9$ and $|E| = 9$.

with ϕ solving $\phi P = \phi$ subject to $\|\phi\|_1 = 1$ [10,58]. Since L' is symmetric and has zero row sum, the eigenvalues of L' are real and nonnegative.

For an undirected graph G, the *degree sequence* of G is the nonincreasing sequence of degrees of nodes of G, such as

$$d_1 \geq d_2 \geq \cdots \geq d_n,$$

For example, the degree sequence for the undirected graph in Figure 11.1 is

$$\{5, 3, 3, 1, 2, 1, 1, 1, 1\}.$$

We denote the eigenvalues of a graph G by $\lambda_i(G)$ where

$$0 = \lambda_1(G) \leq \lambda_2(G) \leq \lambda_3(G) \leq \cdots \leq \lambda_n(G)$$

are the n eigenvalues of L or L'. We define the *algebraic connectivity* (or *spectral gap*) [18] of G as λ_2. Our choice of L' earlier is motivated by the fact that λ_2 for L' has similar properties to λ_2 for L in the undirected case [58].

For an undirected graph G, we define the *generalized Randić index* [36,48] $s(G)$ by

$$s(G) = \sum_{(i,j) \in E} d_i d_j. \qquad (11.2)$$

Section 11.4.3 discusses the extension of $s(G)$ to directed graphs. Its relation to assortativity [45,59] is discussed in the following.

Other common network invariants include assortativity, clustering coefficient, and average shortest path distance. Network assortativity is typically scaled between $[-1, 1]$ with values less than zero indicating that high-degree nodes are more likely to be adjacent to low-degree nodes (disassortativity). Assortativity, $r(G)$, can be shown to be equivalent to $s(G)$ using the equation

$$r(G) = \frac{\left[\sum_{(i,j)\in E} d_i d_j\right] - \left[\sum_{(i)\in V} \frac{1}{2}d_i^2\right]^2 \Big/ |E|}{\left[\sum_{(i)\in V} \frac{1}{2}d_i^3\right] - \left[\sum_{(i)\in V} \frac{1}{2}d_i^2\right]^2 \Big/ |E|} \tag{11.3}$$

where $|E|$ denotes the number of edges in the graph [34]. As you can see, $r(G)$ is linearly related to $s(G)$ since $s(G)$ is in the numerator and only the number of edges and degrees scale it.

The clustering coefficient of a network computes the frequency of complete subgraphs on three nodes, a triangle (in social networks, a triangle denotes that *a friend of your friend is my friend*). One possible definition [43] for local clustering is given by

$$C_i = \frac{[\text{number of triangles in which node } i \text{ is incident}]}{[\text{number of three tuples of connected nodes centered on node } i]}.$$

Then, the clustering coefficient $C = 1/n \cdot \Sigma_i C_i$. For the graph in Figure 11.1, $C = 1/9 \cdot (1/3 + 1/3 + 1/10) \approx 0.09$.

Given the shortest path distance matrix D of a graph, the average shortest path distance can be calculated by averaging the nonzero entries in D. The longest shortest path for each node i is called the *eccentricity* of i. In Figure 11.1, the eccentricity of node 1 is 2, of node 2 is 3, and of node 7 is 4. The *diameter* of a network is the length of the longest shortest path (the maximum node eccentricity).

We make use of three types of undirected graphs, Erdős-Rényi, geometric, and scale-free, whose structure depends on the parameters chosen.

1. *Erdős-Rényi G(n, p) graphs.* A number of nodes n and a probability of connection p are chosen. A uniform random number is generated for each possible edge. If the number generated is less than p, then an edge is added [16,22].

2. *Geometric graphs.* The number of nodes n is chosen and placed on a unit square (or unit circle) at random. This gives each node i coordinates x, y. A radius r is chosen. An edge is placed between nodes i and j if $(x_i - x_j)^2 + (y_i - y_j)^2 \leq r^2$ [54].

3. *Power-law graphs.* A preferential attachment algorithm is used to create graphs whose degree sequences follow a power-law distribution. Following the convention in the literature, we will refer to these graphs as scale-free. The number of nodes n is chosen. New nodes are added and connected to existing nodes, based on a probability proportional to the current degree of the nodes, until n nodes are generated, making it more likely that a new node will be connected to a higher-degree node [54]. The algorithm allows a minimum node degree to be specified.

11.3 Spectral Properties and Network Dynamics

There are a number of well-studied network invariants associated with the spectrum of a graph [8]. In particular, λ_2, λ_n, and λ_n/λ_2 have been shown to have a direct impact on a network's ability to synchronize flow activities at the nodes. In this section, we investigate how the node (e.g., airport, neuron, oscillator) connectivity influences the flow (e.g., traffic, information, current) on a network. Of importance here is the *tunability* of a given network invariant. We realize tunability via optimization. Atay et al. [2,3] provide definitions of node synchrony for networks. Instead of making direct use of these definitions, we focus on algebraic connectivity, λ_2, a graph invariant that has been shown to correlate well (see [3,4,57]) with a network's capacity to synchronize. Intuitively, networks with small λ_2 are easier to *pull apart*. In particular, if $\lambda_2 = 0$, then the network is disconnected [18] and synchronization is impossible. Without more details regarding the flow on a network, it is difficult to make definitive statements regarding synchronization. However, in general, the flow on a network is less likely to synchronize if λ_2 is small. The two upper networks in Figure 11.2 have identical degree distributions, but the network on the left is more weakly connected (e.g., the removal of a single edge can disconnect the network). Note that the two lower networks are identical to the two upper networks except for the addition of a leaf. Although both networks in (a) and (c) now have an edge connectivity of one (cutting one edge breaks the graph apart), λ_2 still reflects the higher *global* connectivity of the graphs.

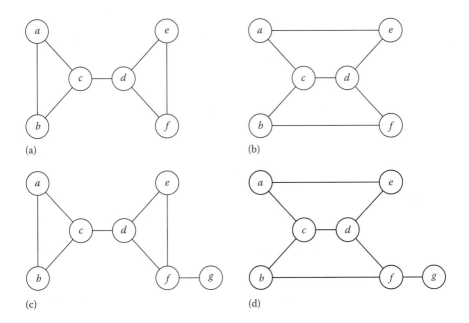

FIGURE 11.2: (a) $\lambda_2 = 0.44$, (b) $\lambda_2 = 1.00$, (c) $\lambda_2 = 0.34$, (d) $\lambda_2 = 0.60$.

For certain processes on networks (often described as a complex system), synchronization is an essential feature. For example, in mammals, a small group of neurons (roughly 200) is responsible for generating a regular rhythmic output to motor cells that initiate a breath (see Section 11.5.1.1). Without synchronization of the neuronal output, breathing would be ragged or not occur at all. Synchronization, as described here, leads to nodes (neurons) behaving in concert. In an air transportation setting, such synchronization would be undesirable. Think of the airports as the neurons in our mammalian respiratory example. Inhaling means all planes land at all airports simultaneously. Exhaling means they depart together. The result is severe congestion. Thus, for this definition of synchronization, one would like an air transport network design to minimize synchronization.

Figure 11.3 displays the result of optimizing λ_2 while holding the degree distribution fixed and $s(G)$ fixed. The graphs are generated by `socnetv`* and optimized with tabu search [29]. The position of each node in the plots is given with respect to the reciprocal of the eccentricity of each node i. The goal of the plots is to uncover qualitative differences between the graphs with small and large values of the second eigenvalue of the Laplacian.[†] Nodes with equal eccentricity values are plotted on the same (dashed line) circles. The circles with larger radii have larger eccentricity. Consequently, nodes near the center have shorter longest paths. The paired plots exhibit large qualitative differences in the eccentricity pattern.

Qualitatively, when λ_2 is small, the patterns are less organized, and the eccentricity plot in Figure 11.3a is more dispersed and consists of many rings of constant eccentricity. The eccentricity plot for the larger λ_2 in Figure 11.3b is more organized, with fewer rings of constant eccentricity. Specifically, Figure 11.3a has 11 rings, and Figure 11.3b has only 5 rings. The range of eccentricity values for the large λ_2 geometric graph, between 4 and 8, dominates the range for the small λ_2 plot which is between 26 and 42. That is, the eccentricity pattern in Figure 11.3b is non overlapping and interior to the one for Figure 11.3a.

The diameter of the graph in Figure 11.3a is 42, while the graph diameter in Figure 11.3b is 8. For graphs with a fixed degree distribution and a fixed value of $s(G)$, this result—with λ_2 inversely proportional to the eccentricity—appears to hold in general. We know of no theorem that proves this result, but numerous computational tests support this claim. Moreover, the inverse relationship between λ_2 and the eccentricity does not hold if $s(G)$ is allowed to vary. (The interested reader is referred to [29] for further examples.)

Several authors [2,3,15,41,46] have examined network dynamics as a function of network topology and have shown that different constrained topology optimization problems, such as maximizing synchronization or node proximity, lead to optimal topologies that, although not identical, share common features. Donetti et al. [15] call these optimal networks *entangled*. The result appears relevant to optimization of transport networks. Donetti et al. provide an illustrative example foreshadowing

---

[*] The source code and documentation can be found at http://socnetv.sourceforge .net/.

[†] We leave it to the reader to become acquainted to the variety of measures and display features in `socnetv`. For the purposes of this exposition, we are interested only in the qualitative differences between the plots.

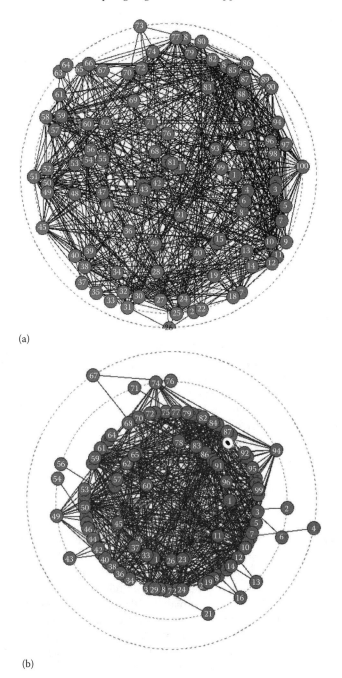

(a)

(b)

FIGURE 11.3: Geometric graphs: 100 nodes, $s(G) = 0.971$, fixed degree distribution. (a) $\lambda_2 = 0.009$, $e(V) = (26, 42)$, $c(G) = 0.426$. (b) $\lambda_2 = 0.314$, $e(V) = (4, 8)$, $c(G) = 0.297$.

the topological significance of the spectral gap for network dynamics. If a network consists of a number of disconnected subgraphs, its Laplacian is block diagonal, and the multiplicity of the trivial zero eigenvalue equals the number of disconnected subgraphs. Connecting the subgraphs weakly introduces small eigenvalues with nearly constant corresponding eigenvectors. This feature (small spectral gaps) provides a criterion for graph partitioning in well-known algorithms [44]. Intuitively, small λ_2 values imply the existence of well-defined modules that can be disconnected by cutting a small number of links, while large λ_2 values point to unstructured (entangled) graphs. Several authors [4,15,41,46] study synchronizability of diffusive processes on networks with identical nodes, considering a general dynamical process:

$$\dot{x}_i = F(x_i) + \sigma \sum_{i=1}^{N} L_{ij}H(x_j), \quad i = 1, \ldots, N, \tag{11.4}$$

where
    x_i are dynamical variables,
    F is an evolution function,
    H is a coupling function, and
    σ is a coupling constant.

Although diffusive processes are known to have synchronous states, the question is, under what conditions these states are stable? A linear stability analysis [4] reveals that synchronized states are more stable for smaller λ_N/λ_2. Since the variability of the maximum eigenvalue is bounded [40], increasing stability of synchronized states amounts to maximizing the spectral gap λ_2. Other authors [3,57] have used the spectral gap as an indicator of synchronization for discrete systems.

The normalized Laplacian, $\tilde{L} = D^{-\frac{1}{2}}LD^{-\frac{1}{2}}$, and its eigenvalues $\{\lambda_i\}$ also play an important role, especially in the study of random walks, a subject relevant to propagation of traffic through networks. Large spectral gaps increase the rate at which random walks move and disseminate. A class of graphs with large spectral gaps, known as Ramanujan graphs (see Figure 11.4), is described by Donetti et al. [15]. These graphs are regular and have a vanishing clustering coefficient, a small average shortest path distance, and a large girth (number of edges in the smallest cycle denotes the girth). Engineered systems do not typically fall into the class of Ramanujan graphs. For example, the nodes and node degree are clearly not identical for Internet router networks or air transportation networks. However, changing the coupling constant in Equation 11.4 to σ/d_i, and thus normalizing the effect of the neighboring nodes (in turn, increasing the relevance to traffic networks), results in an optimal topology when the normalized spectral gap is maximized. These graphs are not characterized as nicely as the Ramanujan graphs. In particular, the networks are no longer regular and the degree distribution is not Poisson. In [15], a plot of one instance of a graph with this degree distribution can be found. The plot is strikingly similar to one in [12] in which an optimization model was employed to construct the network. The optimization model employed in [12] minimizes a weighted graph distance that attempts to capture two conflicting objectives: avoidance of long paths (minimize diameter)

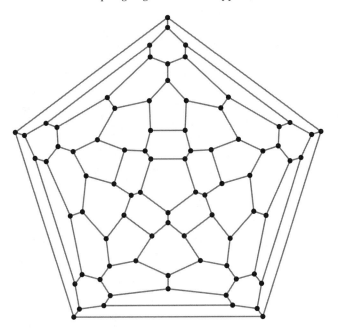

FIGURE 11.4: A Ramanujan or expander graph.

and avoidance of heavy traffic (minimize node degree). Both [15] and [12] note that the degree distribution for the network appears to decay faster than an exponential distribution and that the graph avoids construction of long paths. However, neither reference verifies the decay rate of the degree distribution.

11.4 Network Sampling

Many applications require the ability to uniformly sample networks with constrained graph invariants [11,37]. One of the most well-studied examples is the uniform sampling of networks with a fixed degree sequence. Other constraints can be added to this, including fixed assortativity or edge-degree correlation [29]. The well-known algorithm of Havel–Hakimi [23,24] constructs networks from a degree sequence, albeit in a nonuniform manner. For the directed case, an analogous algorithm to Havel–Hakimi was developed by Kleitman and Wang [31].

Another algorithm to construct directed networks from fixed degree sequences uses the Ford–Fulkerson maximum flow algorithm [19]. To see this, suppose we have a directed degree sequence $\{(d_1^+, d_1^-), \ldots, (d_n^+, d_n^-)\}$. Create a directed flow network F with $2n + 2$ nodes $V = \{v_1, \ldots, v_{2n}, s, t\}$, where s is the source node and t is the target node. The arc set is given by

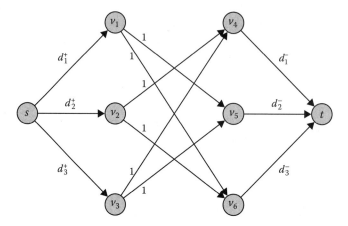

FIGURE 11.5:   Example flow network used in the Ford–Fulkerson maximal flow algorithm to construct a network with degree sequence $\{(d_1^+, d_1^-), (d_2^+, d_2^-), (d_3^+, d_3^-)\}$.

$$A = \{(s, v_i), (v_{n+i}, t)\}_{i=1}^{n} \cup \{(v_i, v_{n+j}) \mid i \neq j\}_{i,j=1}^{n}$$

with capacities

$$\begin{aligned}
c(s, v_i) &= d_i^+, & i = 1, \ldots, n, \\
c(v_i, v_{n+j}) &= 1, & i, j = 1, \ldots, n \text{ with } i \neq j, \\
c(v_{n+i}, t) &= d_i^-, & i = 1, \ldots, n.
\end{aligned}$$

Solving for a maximal flow using the Ford–Fulkerson algorithm will give a network realization of the degree sequence. An example flow network F for a degree sequence with three nodes is given in Figure 11.5.

The algorithms of Havel–Hakimi and Kleitman–Wang unfortunately do not construct networks from the set of all realizations with equal probability. In order to achieve a uniform sample from the space of realizations, more sophisticated algorithms must be used. There are two main classes of algorithms that are used in these circumstances. The first consists of importance sampling algorithms, where each network G in the sample space $\mathcal{G}$ of realizations has a positive probability $p_G > 0$ of being sampled, with the exact probability p_G known. One can then get an unbiased estimator to a population measure $f(G)$ by performing a weighted sum over a set of samples, with each sample weight $w_G = 1/p_G$. There are modifications of the Havel–Hakimi and Kleitman–Wang algorithms that have been used with importance sampling [7,21,28].

The second method to achieve an approximately uniform sample consists of using a simple algorithm like Havel–Hakimi or Kleitman–Wang to first construct a network realization and then using edge or arc-switching techniques [11,52]. For the directed case, an example of an arc switch is shown here, where we move from two arcs $\{(i, j), (m, n)\} \subset A$ to $\{(i, n), (m, j)\} \subset A$.

Note that arc switches preserve the degree sequence. In these algorithms, since any two sample networks $G_1, G_2 \in \mathcal{G}$ are connected by a series of edge or arc switches [52], we can then perform a Monte Carlo random walk on $\mathcal{G}$. Using different algorithmic modifications, such as Metropolis–Hastings, we can make the random walk's stationary distribution uniform [11]. The main drawback of these techniques is the lack of analytical measures of the mixing time, although some have been shown in certain cases [6].

11.4.1 Degree Distributions

The most studied graph invariant is the degree distribution. Some of the most important types include binomial (Erdős-Rényi networks), power law (scale-free networks), and Poisson (geometric random networks).

There are many techniques to construct random networks with a prescribed degree distribution [16,22,47]. The technique we discuss in this chapter is the use of inversion methods to randomly sample a degree sequence from a specified distribution and then sample a network uniformly from this degree sequence. To randomly sample a degree sequence, we use the *probability integral transform*, which states that if X is a continuous random variable with cumulative distribution given by F, then $U = F(X)$ is a random variable on $[0, 1]$ with a uniform distribution. In theory, to generate a random number X from the distribution F, one only needs to generate a uniform random variable U and then compute $X = F^{-1}(U)$. This is of limited use in general as it is computationally intractable to compute the inverse cumulative distribution (also known as the *quantile function*), except, for example, in the case of discrete distributions. As degree distributions are discrete, the inversion method can be translated into an algorithm as follows. Given a discrete random variable k denoting node degree with the degree distribution $p_i \equiv P(k = i)$, we can compute the cumulative distribution $F_k = \sum_{i=0}^{k} p_i$. Now draw a uniform random variable U on the interval $[0, 1]$ and choose k such that $F_{k-1} < U < F_k$, where $F_{-1} \equiv 0$.

One way to generalize the inversion method to directed networks is via the use of *copulas* [42], that is, a bivariate probability distribution with uniform marginals and specified correlation ρ between the random variables. If we denote the cumulative distribution function of the copula by $C(U_1, U_2)$ and the desired marginals' CDFs by $F_1(X_1) \equiv \text{Prob}[X \le X_1]$ and $F_2(X_2) \equiv \text{Prob}[X \le X_2]$, then we can generate a pseudorandom pair (X_1, X_2) from our desired bivariate probability distribution with marginals F_1 and F_2 by first drawing a random sample (U_1, U_2) from C and then constructing $\left(X_1 = F_1^{-1}(U_1), X_2 = F_2^{-1}(U_2)\right)$. In this way, X_1 and X_2 will have the approximate correlation between U_1 and U_2 as specified in the construction of C, as well as be representative samples from marginal distributions F_1 and F_2, respectively.

The next two subsections describe work in sampling networks with a fixed degree sequence and specified node-degree and/or edge-degree correlation.

11.4.2   Node-Degree Correlation

The *node-degree correlation* [49] of a finite network can be quantified in several ways, perhaps most intuitively using the Pearson correlation coefficient

$$\rho = \frac{1}{N} \sum_{i=1}^{N} \left(\frac{d_i^+ - \mu^+}{\sigma^+} \right) \left(\frac{d_i^- - \mu^-}{\sigma^-} \right) \equiv \frac{\text{cov}(d^+, d^-)}{\sigma^- \sigma^+},$$

where

μ^+ and σ^+ are the mean and standard deviation of the out-degrees of the nodes (similarly for μ^- and σ^-), and

$\text{cov}(d^+, d^-)$ represents the covariance between d^+ and d^-.

As described in the previous section, sampling networks from degree sequences with a prescribed node-degree correlation only require specification of the correlation coefficient between the in- and out-degrees, as well as the marginal in- and out-degree distributions.

In [33], the relationship between node-degree correlation and synchronization of pulse-coupled oscillators was explored. Examples in nature of pulse-coupled oscillators include fireflies in Southeast Asia as well as tonically firing neurons. Of particular interest in these situations is synchronization of the phases of each oscillator (see Section 11.5.1.2).

11.4.3   Edge-Degree Correlation

A natural next step is to sample networks with a fixed degree sequence and desired assortativity (11.3) or edge-degree correlation [29,43]. To define edge-degree correlation, it is easiest to start with the directed case. Thus, for a directed graph, we define the edge-degree correlation as the Pearson correlation coefficient between the in-degrees (out-degrees) at the tail and in-degrees (out-degrees) at the head of every arc. This is given by

$$\rho_e(G) = \frac{1}{M} \sum_{(i,j) \in A} \left(\frac{d_i^p - \mu_1^p}{\sigma_1^p} \right) \left(\frac{d_j^q - \mu_2^q}{\sigma_2^q} \right) = \frac{\frac{1}{M} \sum_{(i,j) \in A} d_i^p d_j^q - \mu_1^p \mu_2^q}{\sigma_1^p \sigma_2^q}, \quad (11.5)$$

where

M is the number of arcs

$p, q \in \{-, +\}$

μ_k^p, σ_k^q are the mean and standard deviation of d^p and d^q for vertices at the tail ($k = 1$) or head ($k = 2$) of all arcs.

In the undirected case, by transforming every edge into a bidirectional arc, it can be shown that the edge-degree correlation in (11.5) is equivalent to (11.3), in other words, $\rho_e(G) = r(G)$.

Due to the relationship mentioned in Section 11.2 between $s(G)$ and $\rho_e(G)$, many people choose to use the metric $s(G)$ when working with edge-degree correlations, which we do as well. One of the techniques [45,49] to increase or decrease the edge-degree correlation is to do two swaps between two edges (i, j) and $(m, n) \in E$ when the *edge-degree increment* $\Delta\big((i, j), (m, n)\big) = d_i d_j + d_m d_n - d_i d_n - d_m d_j = (d_i - d_m)(d_j - d_n)$ is positive (decrease edge-degree correlation) or negative (increase edge-degree correlation).

In the directed case, we have four different measures of edge-degree correlation [59] given by

$$s^{pq}(G) = \sum_{(i,j) \in A} d_i^p d_j^q,$$

where $p, q \in \{-, +\}$ (see (11.2)). This can be seen as a natural extension of $s(G)$ to the directed case and has a similar relationship to 11.5 as edge-degree correlation and assortativity have in the undirected case. Now we have four edge-degree increments given by

$$\Delta^{pq}\big((i, j), (m, n)\big) = \left(d_i^p - d_m^p\right)\left(d_j^q - d_n^q\right), \qquad (11.6)$$

where $p, q \in \{-, +\}$.

We will now illustrate an algorithm on directed graphs, similar to the undirected version [45,49], which attempts to sample networks with a fixed degree sequence and desired edge-degree correlations $\{s^{--}, s^{-+}, s^{+-}, s^{++}\}$. The key observation is that certain arc swaps modify only one of the four edge-degree correlations at a time. To see this, considering (11.6), if you want to vary correlation s^{-+} and leave the others fixed, for example, then you only consider swaps between an arc $(i, j) \in A$ with arcs in the set $A_{ij}^{-+} \equiv \{(m, n) \in A \mid d_m^+ = d_i^+ \text{ and } d_n^- = d_j^- \}$. Thus, for $p = +$ or $q = -$, we have $\Delta^{pq}((i, j), A_{ij}^{-+}) = 0$, and thus any two swap between (i, j) and an arc in A_{ij}^{-+} leaves s^{+-}, s^{++}, and s^{--} fixed. For general s^{pq}, we construct the set A_{ij}^{pq} as $A_{ij}^{pq} \equiv \{(m, n) \in A \mid d_m^q = d_i^q \text{ and } d_n^p = d_j^p \}$. The general algorithm then cycles through the four edge-degree correlations and performs an arc swap between a random arc (i, j) and an arc in A_{ij}^{pq} if the supposed swap moves the edge-degree correlation s^{pq} toward the desired value. Similar algorithms exist (e.g., see [59]) to sample networks with desired edge-degree correlations. Note that there is a dependency between the node-degree and edge-degree correlations [49], so that we may or may not be able to achieve a network with our desired correlation structure.

11.5 Applications

11.5.1 Synchronization on Networks

Synchronization of processes is ubiquitous in the biological sciences, for example, the synchronization of neurons in the pre-Bötzinger complex that drives the breathing rhythm [17], the synchronization of repressilator networks in gene transcription [20], and the synchronous release of oxytocin from magnocellular hypothalamic neurons in neuroendocrinology [50]. Network structure and oscillator dynamics play fundamental roles in the dynamics of the system. There are many models for the oscillator dynamics that are considered, ranging from the complexity of Hodgkin–Huxley neuron models to phase models where we track only the phase of the oscillators and ignore their positions. The three phase models that have received the most attention are the Kuramoto oscillator, the Laplacian oscillator (11.4), and pulse-coupled oscillators. In the next two subsections, we discuss the effect of network structure on the synchronous bursting of neurons in the pre-Bötzinger complex and the phase synchronization of pulse-coupled oscillators.

11.5.1.1   Neuronal Networks

In mammals, a small group of neurons in the brain stem, the pre-Bötzinger complex [17], is responsible for generating a regular rhythmic output to motor cells that initiate a breath. When disconnected, these neurons are unable to provide sufficient output to activate the motor neurons, but their interconnected network structure allows them to synchronize without any external influence and produce regular bursts. This is a clear example of an important neuronal network where robustness (and synchronization [29]) is essential.

In [51], Del Negro et al. developed a physiologically realistic mathematical model of neurons in the pre-Bötzinger complex that demonstrates the capability of the breathing rhythm to be an emergent phenomena of the network and not explicitly controlled by central pattern generators. Although there is very little known regarding the network structure of the pre-Bötzinger complex, it is possible, using the model in [51], to test various network structures. Two geometric graphs with extreme values of λ_2 were tested in [25,29]. One of the geometric networks had a value of $\lambda_2 = 0.025$, and a second had a value of $\lambda_2 = 0.974$. The rhythmic output from the network with $\lambda_2 = 0.025$ was ragged with fuzzy bursts, while outputs from the network with $\lambda_2 = 0.974$ were sharp with clear, regular bursts. The results of the two simulations, depicted in Figure 11.6, provide compelling evidence for the utility of λ_2 as a predictor of synchronization. It is easy to see that the network with higher λ_2 in Figure 11.6b synchronizes more strongly than the other network in Figure 11.6a. These experiments provide further evidence that λ_2 can be used to identify graphs (networks) that are not likely to synchronize.

11.5.1.2   Pulse-Coupled Oscillators

In [33], the effect of synchronization of homogeneous pulse-coupled oscillators on node-degree correlation (see Section 11.4.2) was studied. The dynamics of pulse-coupled oscillators is given by

$$\frac{d\phi_i}{dt} = 1 + \frac{k}{n}\sum_{j=1}^{n} A_{ji}H(\phi_j)\Delta(\phi_i), \quad \phi_i \in [0,1]$$

with $\Delta(\phi)$ as the sensitivity function (phase response curve) and $H(\phi_j) = \delta(t - t_j)$ as the pulsatile interaction function. In this notation, δ is a delta function with infinite

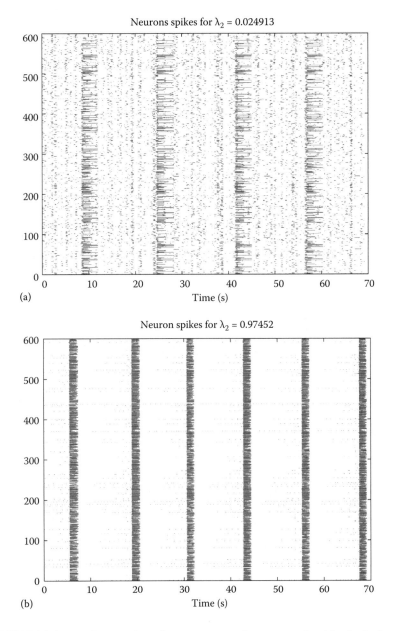

FIGURE 11.6: (a) Raster plot of neuron output for a network with algebraic connectivity $\lambda_2 = 0.024913$. A point at (x, y) indicates neuron x is spiking at time y. (b) Raster plot of neuron output for a network with a higher algebraic connectivity $\lambda_2 = 0.97452$. The higher λ_2 network in (b) displays much stronger synchronization among all nodes as predicted, as well as a quicker breath frequency.

point mass at 0 and t_j is the firing time for oscillator j. The matrix A_{ji} is the node–node adjacency matrix for the network (not the Laplacian as in (11.4)), while k/n is the coupling constant, which includes the $1/n$ so that it is well behaved in the thermodynamic limit $n \to \infty$. Mirollo and Strogatz [39] showed that in the case of all-to-all coupling, there is a globally synchronous state $\phi_1 = \cdots = \phi_n$ for almost all initial conditions, that is, the initial conditions that do not reach global synchrony are a set of measure zero.

The top panel of Figure 11.7a shows the phases $\phi_k(t)$ for 200 oscillators on a network with negative node-degree correlation $\rho \approx -1$. The global synchronization

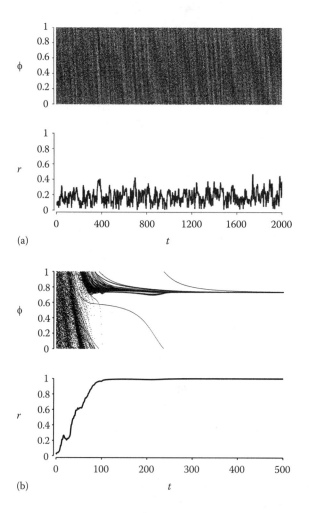

FIGURE 11.7:   (a) The top panel shows the phase ϕ of 200 oscillators as a function of time for a network with node-degree correlation $\rho \approx -1$, with the corresponding global synchronization measure $r(t)$ shown in the bottom panel. (b) Same as in (a) except with $\rho \approx 1$.

measure (or coherence measure) given by $r(t) = |\sum_{k=1}^{n} \exp(2\pi i \phi_k(t))|/n$ is displayed in the bottom panel. Note that the global synchronization measure is 1 when the oscillators are globally synchronized. Contrary to Figure 11.7a is a similar plot in Figure 11.7b for a network with positive node-degree correlation $\rho \approx 1$. In this case, complete synchronization occurs at approximately $T = 500$ when $r(500) \approx 1$.

Numerical experiments in [33] demonstrated that the proportion of initial conditions resulting in a globally synchronous state is an increasing function of node-degree correlation. For those networks observed to globally synchronize, both the mean and standard deviation of time to synchronization decrease as node-degree correlation increases. Many networks with negatively correlated node degree exhibited multiple coherent attracting states, with trajectories performing fast transitions between them. A similar phenomenon was reported in [30,55] in networks of pulse-coupled oscillators with delay.

As stated in Section 11.3, the algebraic connectivity λ_2 is known to have an effect on the rate of convergence to the globally synchronous state when identical oscillators are coupled as in the Kuramoto model or via an undirected Laplacian [1,27]. Similarly, the eigenvalue ratio λ_N/λ_2 is related to the propensity of identical oscillators coupled via an undirected Laplacian to have a stable synchronous state [4]. Figure 11.8 shows the mean Laplacian eigenvalue λ_2 (panel (a), solid line) and the ratio λ_N/λ_2 (panel (b)) are monotone increasing and decreasing functions of node-degree correlation, respectively (average over 1000 random networks for each ρ). For pulse-coupled networks with negatively correlated node degree, slow synchronization and reduced percentage of initial conditions that reached global synchronization are observed.

11.5.2 Learning from Internet Design

The problems faced by designers of air transport networks share some aspects with the design of an Internet router network. Many authors have contributed to

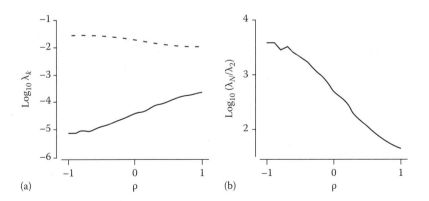

FIGURE 11.8: Laplacian eigenvalues λ_N (dashed line) and λ_2 (solid line) in (a) and ratio λ_N/λ_2 in (b). Each point on the curves shows an average over 1000 random networks.

TABLE 11.1:   Analogy between Internet Router and Air Transport Networks

	Internet	**Air Transport**
Product	Packets	Planes (loaded)
Constraint	Bandwidth	Airport capacity
Competitors	ISPs	Airlines
Links	Hardwired	FAA/Airlines
Distributors	Routers	Airports

investigations of how a router network is constructed. Two references in this field, [34] and [35], contain ideas central to our consideration of the design of air transport networks. At one level of resolution, Table 11.1 points out the analogies between these two network design problems. With regard to bandwidth, the Internet router designer must weigh the trade-offs between many low bandwidth connections and fewer high bandwidth connections. These trade-offs are akin to choosing between a few hub airports in a hub-and-spoke system and choosing lower- frequency airports that might arise in a direct route system. Of course, there are many differences as well. The variation in the size of the packets for the Internet is not nearly as great as the number of passengers on planes of different sizes. In addition, although the FAA* clearly defines the routes allowed between airports, the links are as not hardwired as they are in the Internet model. Still there is much to be learned from the research efforts on the design of effective Internet router networks.

There has been an inordinate amount of interest in networks whose complementary cumulative degree distributions have a fat tail or follow a power law. In [34], a figure is presented that plots $s(G)$ versus throughput when the degree distribution is fixed. The figure highlights the error in focusing exclusively on the form of the degree distribution. A normalized $s(G)$ value is plotted along the x-axis, and a throughput metric is plotted along the y-axis. Each data point represents the performance of a network, each of which has an identical degree distribution following a power law. An unexpectedly wide variance in the throughput performance of these networks in which the degree distribution is an invariant is observed. Moreover, low (disassortative) $s(G)$ instances lead, in general, to better throughput performance. The authors in [34] point out that when sampling from this invariant degree distribution, it was much more likely to draw an instance in which $s(G)$ is large (assortative).

11.6   Optimization: Perfect b-Matching and $s(G)$

There has been little work done with regard to classifying optimization problems associated with graph invariants. In this section, we address optimization problems associated with $s(G)$ [32]. In this context, a natural optimization problem is as follows:

---

* Federal Aviation Administration.

Minimum Randić index problem. Given a degree sequence, what is a graph realization with the minimum Randić index?

We define the *connected minimum Randić index problem* as the minimum Randić index problem with the additional constraint that the graph realization is connected. For a graph $G = (V, E)$ and a positive integer vector $b = (b_1, \cdots, b_n) \in \mathbb{Z}^n$, a *perfect b-matching* is a subset of edges $M \subseteq E$ such that for node $i \in V$, the degree of i in the graph (V, M) is b_i.

An associated optimization problem is as follows:

Minimum weight perfect b-matching problem. Given a positive integer vector b, a graph $G = (V, E)$, and a set of edge weights $w : E \to \mathbb{R}$, find a perfect b-matching with minimum weight.

The minimum Randić index problem is equivalent to the minimum weight perfect b-matching problem on a complete graph G with an appropriate choice of weights [32]. In [32], it was shown that by constraining the matchings to be connected, for an arbitrary graph G, the minimum weight perfect b-matching problem becomes NP-hard or computationally intractable. In 2008, Beichl and Cloteaux [5] investigated how well random networks generated with a chosen $s(G)$ can model the structure of real networks such as the Internet. The graphs produced optimizing the $s(G)$ resulted in better models than the ones that used simple uniform sampling.

11.6.1 Formulation and Complexity

The minimum Randić index problem can be formulated as a *minimum weight perfect b-matching problem*, which is solvable in polynomial time [53]. Note that the perfect b-matching problem does not enforce connectivity. When connectivity of solutions is desired, in [32], it is shown that even approximating the minimum weight perfect b-matching problem with connectivity is NP-hard.

Consider a graph $G = (V, E)$, a positive integer vector $b = (b_1, \ldots, b_n) \in \mathbb{Z}^n$, and $M \subseteq E$, a perfect b-matching. For a given b-matching, M, the graph induced by M is (V, M). Denote the set of perfect b-matchings of a graph G by $\mathcal{P}_b(G)$. For edge weights $w : E \to \mathbb{R}$, the *minimum weight perfect b-matching problem* requires finding the perfect b-matching with minimum weight, that is, to calculate

$$M^*(G) := \arg \min \left\{ \sum_{e \in M} w(e) : M \in \mathcal{P}_b(G) \right\}. \qquad (11.7)$$

To formulate an instance of the minimum Randić index problem as the minimum weight perfect b-matching problem, set

$$w_{ij} = b_i \cdot b_j. \qquad (11.8)$$

For these weights, solve (11.7) to obtain $M^*(G)$. $G^* = (V, M^*(G))$ is an optimal solution to the minimum Randić index problem instance. Note first that it is feasible since the degree of a node $i \in V$ is b_i by the definition of the perfect b-matching problem. Note also that any feasible graph to the minimum Randić index problem is also a perfect b-matching because the degree of any node i is equal to b_i. Moreover, (11.8) implies

$$R(G^*) = \sum_{(i,j) \in M^*(G)} b_i \cdot b_j = \sum_{(i,j) \in M^*(G)} w_{ij}.$$

Since any graph that is feasible to the minimum Randić index is also a b-matching, the optimality of $M^*(G)$ implies the optimality of G^*.

Therefore, an instance of a minimum weight perfect b-matching on a complete graph can be constructed to solve the minimum Randić index problem. Since the b-matching problem can be solved in polynomial time, finding the minimum Randić index of a graph can also be done in polynomial time. Optimal solutions, however, are not necessarily connected.

Consider the following example. Given the degree sequence $d = (3, 2, 2, 2, 2, 1)$, what is a graph realization with the minimum Randić index? Let $V = \{v_1, v_2, v_3, v_4, v_5, v_6\}$ with $b = \begin{bmatrix} 3 & 2 & 2 & 2 & 2 & 1 \end{bmatrix}$ be given. Next form the complete graph G, with weights corresponding to $b_i \cdot b_j$ for every node $v_i, v_j \in V$.

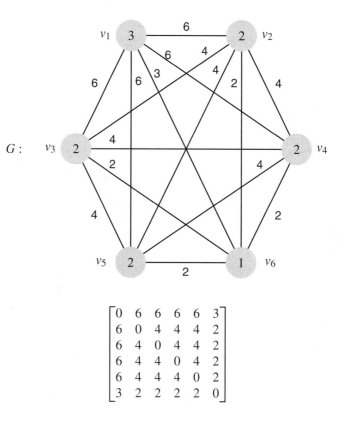

$$\begin{bmatrix} 0 & 6 & 6 & 6 & 6 & 3 \\ 6 & 0 & 4 & 4 & 4 & 2 \\ 6 & 4 & 0 & 4 & 4 & 2 \\ 6 & 4 & 4 & 0 & 4 & 2 \\ 6 & 4 & 4 & 4 & 0 & 2 \\ 3 & 2 & 2 & 2 & 2 & 0 \end{bmatrix}$$

Solve the minimum weight perfect *b*-matching for *G* and obtain *G'*:

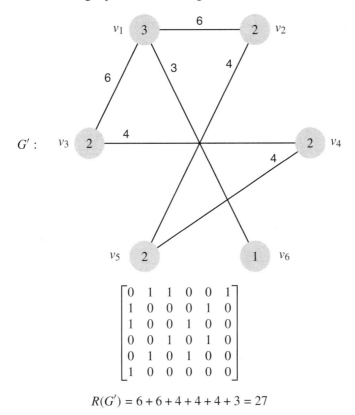

$$\begin{bmatrix} 0 & 1 & 1 & 0 & 0 & 1 \\ 1 & 0 & 0 & 0 & 1 & 0 \\ 1 & 0 & 0 & 1 & 0 & 0 \\ 0 & 0 & 1 & 0 & 1 & 0 \\ 0 & 1 & 0 & 1 & 0 & 0 \\ 1 & 0 & 0 & 0 & 0 & 0 \end{bmatrix}$$

$$R(G') = 6 + 6 + 4 + 4 + 4 + 3 = 27$$

G' is an optimal solution for the minimum weight perfect *b*-matching. The sum of the weights is the minimum Randić index and the unweighted adjacency matrix is the corresponding graph realization. Note that there are other solutions to the matching that will produce the minimum Randić index and a different realization. That is, the solution is not unique.

11.6.2 Solving the Minimum Randić Index Problem

In this section, the input graph *G* is assumed to be complete. A code written by Huang and Jebara [26] that makes use of the GOBLIN graph library [60] is designed to solve a *maximum* weight perfect b-matching problem. Given a weight matrix *H*, we transform these weights into a matrix H_2 such that the maximum matching using H_2 will yield the same solution as the minimum matching using *H*. To do this, we take a matrix *M* with ones in all positions except for the diagonal, which has zeros. We then multiply every entry by one more than the maximum entry of *H*. Then *H* is subtracted from *M* yielding H_2. An algorithm that will solve the minimum Randić index problem for a given degree sequence is given in the following.

Algorithm to solve minimum Randić index with b-matching

 Inputs: A, an adjacency matrix with degree sequence, d
 Outputs: G, the new adjacency matrix with degree sequence, d
 and minimized Randic index, r

 Create a complete graph H of degree products
 Transform H to H_2 for b-matching code
 Use b-match solver to get adjacency matrix, G of optimal solution
 Calculate $r = R(G)$
 return G and r

FIGURE 11.9: Solving minimum Randić index with b-matching.

TABLE 11.2: 100 Node Graphs

Graph Type	Connected	Disconnected	No Connected Realizations
Erdős-Rényi	16	2	82
Geometric	30	6	64
Scale-free	91	8	1

The algorithm in Figure 11.9 returns the minimum Randić index of a graph and a realization. The b-matching code runs in polynomial time [53], and it is easy to see that the transformation steps are done in polynomial time as well. Three types of randomly generated graphs to test the algorithm performance are Erdős-Rényi, geometric, and scale-free. The computational experiments are limited to graphs for which connected realizations are known to exist. The Randić index before and after the optimization is recorded. After the optimization, the graph realization with the minimum Randić index is checked to see if it is connected. In [32], computational experiments for a number of graph sizes and types are reported. Here, we include the results for 100 replications of three types of 100 node graphs in Table 11.2. Note that the number of graphs connected after the run plus the number of graphs disconnected plus the number of graphs with no connected realizations is 100 for each graph type.

The MATLAB functions used to generate the geometric and scale-free graphs are from CONTEST, a controllable test matrix toolbox for MATLAB [54]. In addition, the necessary and sufficient conditions for a set $\{a_i\}$ to be realizable (as the degrees of the nodes of a connected graph) are that $a_i \neq 0$ for all i and the sum of the integers a_i is even and not less than $2(n-1)$. This condition was used to discard graphs with a degree sequence that had no connected realizations [9].

The left box plots for each of the 100 node graphs in Figure 11.10a show the percent difference between the graph's original Randić index and the minimum Randić index. The percent difference is calculated from $\frac{original-minimum}{minimum} \times 100$. In each pair of plots, the right box plot describes the performance of the heuristic to make disconnected optimal solutions connected, which is described in the next section.

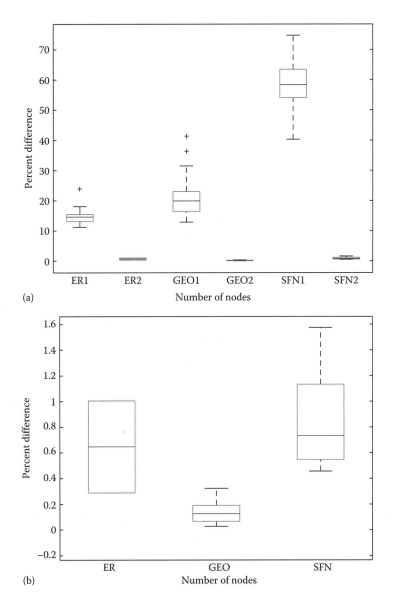

FIGURE 11.10: 100 node graph results. (a) Comparing percent differences for Erdös-Rényi (ER1), geometric (GEO1), and scale-free (SFN1) graphs. (b) Comparing percent differences between optimal and heuristic solutions for Erdös-Rényi (ER), geometric (GEO), and scale-free (SFN) graphs. Box plots ER2, GEO2, and SFN2 in (a) are the same box plots as in (b).

Two-switch Heuristic

> **Inputs:** A, an adjacency of disconnected graph
> **Outputs:** A, the new adjacency matrix of connected graph
>
> **while** the number of connected components in A is ≥ 2
>   **do** a two switch with components 1 and 2 to connect them
>   using two randomly chosen edges from each component
>
> **return** A

FIGURE 11.11:   Connecting disconnected graph with two-switch heuristic.

11.6.3   Heuristic for Disconnected Realizations

The complexity of the minimum Randić index problem subject to a connectivity constraint is known to be NP-Hard [32]. However, since some of the graph realizations with the minimum Randić index were disconnected, we developed a heuristic using two switches to connect these realizations (see Figure 11.11 for the algorithm.) The heuristic performs a two switch between every component until all the components are connected. We know that a two switch exists between any two connected components because they do not share any edges. Any edge can be used.

The heuristic was applied to all optimal solutions that were disconnected. In general, the difference in Randić index from the minimum was not significant. The Randić index changes the least after the heuristic in the Erdős-Rényi graphs. This percent difference is calculated with $\frac{after\ heuristic - minimum}{minimum} \times 100$. The number of graphs that used the heuristic depended on the number of optimal graph realizations that were disconnected. Note that this is a different number for each graph type and size (see Table 11.2 for the 100 node graph numbers).

Note that the method to connect the disconnected realizations may not produce graphs with the best structure since there is only one edge connecting one component to another. Also note that we do not need to check whether the randomly chosen edges are adjacent or not since they are in separate connected components. In addition, once we connect components 1 and 2, component 2 becomes part of component 1, and component 3 becomes the new component 2. Therefore, we can always connect components 1 and 2.

11.7   Conclusion

The importance of how network invariants affect the dynamics of a process on the network (e.g., respiration, current, traffic) has been highlighted. In studying the potential effect of incremental changes in network invariants on network dynamics, the degree distribution of a network was held constant, while other network invariants were examined. In particular, results demonstrating the effects of assortativity, $s(G)$, and eigenvalues of the Laplacian on network dynamics were presented.

In Section 11.3, the connection between λ_2 and other network dynamics was studied. Research supporting the link between λ_2 and synchronization was provided as well as an inverse relationship between λ_2 and the diameter of the network. Section 11.4 summarize algorithms that allow sampling (sometimes uniform) from the family of networks with a fixed degree sequence. The ability to sample uniformly is critical in any research attempting to discover the effects of network invariants. In Section 11.5, a number of network applications were described including neuronal networks for respiration in mammals, pulse-coupled oscillators, Internet router networks, and air transportation route networks. Section 11.6 focus on optimizing $s(G)$, a network assortativity metric. A novel connection with minimum weight perfect b-matching problem as well as computational results were given.

References

1. J. Almendral and A. Díaz-Guilera, Dynamical and spectral properties of complex networks, *New Journal of Physics*, 9 (2007), 187–187.

2. F. Atay, T. Biyikoglu, and J. Jost, Synchronization of networks with prescribed degree distributions, *Circuits and Systems I: Regular Papers, IEEE Transactions on*, 53 (2006), 92–98.

3. F. Atay, J. Jost, and A. Wende, Delays, connection topology, and synchronization of coupled chaotic maps, *Physical Review Letters*, 92 (2004), 144101.

4. M. Barahona and L. Pecora, Synchronization in small-world systems, *Physical Review Letters*, 89 (2002), 54101.

5. I. Beichl and B. Cloteaux, Measuring the effectiveness of the s-metric to produce better network models, in *Simulation Conference, 2008. WSC 2008, Miami, FL. Winter*, IEEE, Piscataway, NJ, 2008, pp. 1020–1028.

6. A. Berger and M. Müller-Hannemann, Uniform sampling of undirected and directed graphs with a fixed degree sequence, arXiv, math.DM (2009).

7. J. Blitzstein and P. Diaconis, A sequential importance sampling algorithm for generating random graphs with prescribed degrees, *Internet Mathematics*, 6 (2011), 489–522.

8. V. Chandrasekaran, P.A. Parrilo, and A.S. Willsky, Convex graph invariants, *SIAM Review*, 54 (2012), 513–541.

9. W. Chen, *Graph Theory and Its Engineering Applications*, World Scientific Pub. Co. Inc., Singapore, 1997.

10. F. Chung, Laplacians and the Cheeger inequality for directed graphs, *Annals of Combinatorics*, 9 (2005), 1–19.

11. G.W. Cobb and Y.-P. Chen, An application of Markov chain Monte Carlo to community ecology, *The American Mathematical Monthly*, 110 (2003), 265–288.

12. V. Colizza, J. Banavar, A. Maritan, and A. Rinaldo, Network structures from selection principles, *Physical Review Letters*, 92 (2004), 198701.

13. M. Dehmer, S. Borgert, and F. Emmert-Streib, Entropy bounds for molecular hierarchical networks, *PLoS ONE*, 3 (2008), e3079.

14. M. Dehmer and F. Emmert-Streib, Structural information content of networks: Graph entropy based on local vertex functionals, *Computational Biology and Chemistry*, 32 (2008), 131–138.

15. L. Donetti, F. Neri, and M. Muñoz, Optimal network topologies: Expanders, Cages, Ramanujan graphs, Entangled networks and all that, *Journal of Statistical Mechanics: Theory and Experiment*, 2006 (2006), P08007.

16. P. Erdös and A. Rényi, On the evolution of random graphs, *Publication of the Mathematical Institute of the Hungarian Academy of Sciences*, 5 (January 1960), 17–61.

17. J.L. Feldman and C.A.D. Negro, Looking for inspiration: New perspectives on respiratory rhythm, *Nature Reviews Neuroscience*, 7 (2006), 232–242.

18. M. Fiedler, Algebraic connectivity of graphs, *Czechoslovak Mathematical Journal*, 23(98) (1973), 298–305.

19. L.R. Ford and D.R. Fulkerson, Maximal flow through a network, *Canadian Journal of Mathematics*, 8 (1956), 399–404.

20. J. Garcia-Ojalvo, M.B. Elowitz, and S.H. Strogatz, Modeling a synthetic multicellular clock: Repressilators coupled by quorum sensing, *Proceedings of the National Academy of Sciences USA*, 101 (2004), 10955–10960.

21. C.I.D. Genio, H. Kim, Z. Toroczkai, and K.E. Bassler, Efficient and exact sampling of simple graphs with given arbitrary degree sequence, *PLoS ONE*, 5 (2010), 1–7.

22. E.N. Gilbert, Random graphs, *Annals of Mathematical Statistics*, 30 (1959), 1141–1144.

23. S. Hakimi, On realizability of a set of integers as degrees of the vertices of a linear graph. I, *Journal of the Society for Industrial and Applied Mathematics*, 10 (1962), 496–506.

24. V. Havel, A remark on the existence of finite graphs, *Casopis Pest. Mat.* 80 (1955), 477–480.

25. M. Holroyd and R. K. Kincaid, Synchronizability and connectivity of discrete complex systems, in *Proceedings of the International Conference on*

Complex Systems 2006, Boston, MA. New England Complex Systems Institute, Cambridge, MA, June 2006.

26. B. Huang and T. Jebara, Loopy belief propagation for bipartite maximum weight b-matching, in *Artificial Intelligence and Statistics (AISTATS)*, San Juan, Puerto Rico, 2007.

27. A. Jadbabaie, N. Motee, and M. Barahona, On the stability of the Kuramoto model of coupled nonlinear oscillators, *American Control Conference*, 5 (2004), 4296–4301.

28. H. Kim, C.I. Del Genio, K.E. Bassler, and Z. Toroczkai, Constructing and sampling directed graphs with given degree sequences, *New Journal of Physics*, 14 (2012), 023012.

29. R. Kincaid, N. Alexandrov, and M. Holroyd, An investigation of synchrony in transport networks, *Complexity*, 14 (2009), 34–43.

30. C. Kirst and M. Timme, From networks of unstable attractors to heteroclinic switching, *Physical Review E*, 78 (2008), 065201–065204.

31. D. Kleitman and D. Wang, Algorithms for constructing graphs and digraphs with given valences and factors, *Discrete Mathematics*, 6 (1973), 79–88.

32. S.J. Kunkler, M.D. LaMar, R.K. Kincaid, and D. Phillips, *Algorithm and Complexity for a Network Assortativity Measure*, arXiv, (2013).

33. M.D. LaMar and G.D. Smith, Effect of node-degree correlation on synchronization of identical pulse-coupled oscillators, *Physical Review E*, 81 (2010), 046206.

34. L. Li, D. Alderson, J. Doyle, and W. Willinger, Towards a theory of scale-free graphs: Definition, properties, and implications, *Internet Mathematics*, 2 (2005), 431–523.

35. L. Li, D. Alderson, W. Willinger, and J. Doyle, A first-principles approach to understanding the internet's router-level topology, *ACM SIGCOMM Computer Communication Review*, 34 (2004), 3–14.

36. X. Li, Y. Shi, and L. Wang, An updated survey on the Randić index, *Recent Results in the Theory of Randić Index, Mathematical Chemistry Monographs*, 6 (2008), 9–47.

37. F. Liljeros, C.R. Edling, L.A. Amaral, H.E. Stanley, and Y. Aberg, The web of human sexual contacts, *Nature*, 411 (2001), 907–908.

38. Y.-Y. Liu, J.-J. Slotine, and A.-L. Barabasi, Controllability of complex networks, *Nature*, 473 (2011), 167–173.

39. R. Mirollo and S. Strogatz, Synchronization of pulse-coupled biological oscillators, *SIAM Journal on Applied Mathematics*, 50 (1990), 1645–1662.

40. B. Mohar, Some applications of Laplace eigenvalues of graphs, *Graph Symmetry: Algebraic Methods and Applications*, 497 (1997), 227–275.

41. A. Motter, Bounding network spectra for network design, *New Journal of Physics*, 9 (2007), 182.

42. R.B. Nelsen, *An Introduction to Copulas*, Springer, New York, 1998.

43. M. Newman, J. 2003. The structure and function of complex networks, *SIAM Review*, 45 (2003), F00.

44. M. Newman and M. Girvan, Finding and evaluating community structure in networks, *Physical Review E*, 69 (2004), 26113.

45. M.E.J. Newman, Assortative mixing in networks, *Physics Review Letters*, 89 (2002), 208701.

46. T. Nishikawa and A. Motter, Maximum performance at minimum cost in network synchronization, *Physica D: Nonlinear Phenomena*, 224 (2006), 77–89.

47. M. Pósfai, Y.-Y. Liu, J.-J. Slotine, and A.-L. Barabási, Effect of correlations on network controllability, *Scientific Reports*, 3 (2013), 1067.

48. M. Randić, Characterization of molecular branching, *Journal of the American Chemical Society*, 97 (1975), 6609–6615.

49. J.G. Restrepo, E. Ott, and B.R. Hunt, Approximating the largest eigenvalue of network adjacency matrices, *Physical Review E*, 76 (2007), 056119.

50. E. Rossoni, J. Feng, B. Tirozzi, D. Brown, G. Leng, and F. Moos, Emergent synchronous bursting of oxytocin neuronal network, *PLoS Computational Biology*, 4 (2008), e1000123.

51. J.E. Rubin, J.A. Hayes, J.L. Mendenhall, and C.A.D. Negro, Calcium-activated nonspecific cation current and synaptic depression promote network-dependent burst oscillations, *Proceedings of the National Academy of Sciences USA*, 106 (2009), 2939–44.

52. H. Ryser, Combinatorial properties of matrices of zeros and ones, *Canadian Journal of Mathematics*, 9 (1957), 6.

53. A. Schrijver, *Combinatorial Optimization*, Springer, Berlin, Germany, 2003.

54. A. Taylor and D. Higham, Contest: A controllable test matrix toolbox for MATLAB, *ACM Transactions on Mathematical Software (TOMS)*, 35 (2009), 1–17.

55. M. Timme, F. Wolf, and T. Geisel, Prevalence of unstable attractors in networks of pulse-coupled oscillators, *Physics Review Letters*, 89 (2002), 154105.

56. D. West, *Introduction to Graph Theory*, 2nd edn., Prentice Hall, Upper Saddle River, NJ, 2001.

57. C. Wu, Perturbation of coupling matrices and its effect on the synchronizability in arrays of coupled chaotic systems, *Physics Letters A*, 319 (2003), 495–503. Preprint 0307052 (2006) available at http://arXiv/pdf/nlin.CD/.

58. C. Wu, On Rayleigh-Ritz ratios of a generalized Laplacian matrix of directed graphs, *Linear Algebra and Its Applications*, 402 (2005), 207–227.

59. G. Zamora-López, C. Zhou, V. Zlatić, and J. Kurths, The generation of random directed networks with prescribed 1-node and 2-node degree correlations, *Journal of Physics A: Mathematical and Theoretical*, 41 (2008), 224006.

60. C. Fremuth-Paeger, GOBLIN graph library, http://goblin2.sourceforge.net, Accessed July 19, 2013.

Chapter 12

Discrimination of Image Textures Using Graph Indices

Martin Welk

Contents

12.1 Introduction...356
    12.1.1 Quantitative Graph Theory..356
    12.1.2 Graph Models in Image Analysis......................................357
    12.1.3 Texture...358
        12.1.3.1 Texture and Shape..358
        12.1.3.2 Texture Descriptors...359
        12.1.3.3 Texture Segmentation.......................................359
12.2 Morphological Amoebas...361
    12.2.1 Amoeba Construction..361
        12.2.1.1 Amoeba Metric...362
        12.2.1.2 Differences to Lerallut et al.'s Amoeba Construction......362
        12.2.1.3 Amoeba Structuring Element.................................363
    12.2.2 Building Graphs from Amoebas..363
        12.2.2.1 Modified Patch Graphs......................................363
    12.2.3 Texture Representation by Amoeba and Patch Graphs................363
12.3 Quantitative Graph Descriptors...364
    12.3.1 Distance-Based Indices...364
        12.3.1.1 Wiener and Harary Index...................................364
        12.3.1.2 Balaban Index..365
    12.3.2 Information-Theoretic Indices...365
        12.3.2.1 Bonchev–Trinajstić Information Indices....................366
        12.3.2.2 Dehmer Entropies..367
12.4 Experimental Test of Texture Discrimination..............................369
    12.4.1 Test Images..370
    12.4.2 Evaluation Procedure for Texture Features..........................370
        12.4.2.1 Haralick Features..370
        12.4.2.2 Graph Indices..371
    12.4.3 Experimental Results on Texture Discrimination....................373
        12.4.3.1 Overlap of Features..377
12.5 Conclusion...381
References..381

12.1   Introduction

In this chapter, the applicability of quantitative graph theory methods in the context of image analysis is studied. Whereas graph models themselves have been used for image analysis for a long time (see the edited book by Lezoray and Grady [38] for an overview of methods), graph indices have to our knowledge not played a significant role in this area so far. The work presented here is to be understood as a first step in this direction. As such, it focuses on one specific problem field, texture analysis, where a statistical approach to information encoded in large collections of graphs can be expected to have great potential.

The chapter is structured as follows. The remainder of Section 12.1 surveys ideas from different fields of research that coalesce in this work, including quantitative graph theory, Section 12.1.1; graph models in image analysis, Section 12.1.2; and texture analysis, Section 12.1.3. Mathematical morphology as a further source field is addressed in Section 12.2 whose main goal is to introduce the concept of morphological amoebas, a class of image-adaptive structuring elements that provide a way to construct graph collections representing local image structure. Section 12.3 presents a selection of graph indices, which will be used for texture analysis later on, together with adaptations of these graph indices to edge-weighted graphs. In Section 12.4, a test scenario for texture analysis with quantitative graph methods is developed, along with a similar setup based on Haralick features [26] that serves for comparison. Experimental results for both sets of texture features are shown, which confirm the viability of the approach on a proof-of-concept level. A short conclusion, Section 12.5, ends the chapter.

12.1.1   Quantitative Graph Theory

As a field of research, quantitative graph theory emerged in the last decade from the study of biological, chemical, and other networks mathematically described by graphs. It aims at characterizing properties of such networks by quantitative methods (see, e.g., the edited book [14]). The approach relies on work on quantitative measures for graphs that has been performed mainly in mathematical chemistry over half a century, in which *graph indices* have been introduced in order to analyze molecular graphs (see, e.g., [4,28,33,46,64]). In quantitative graph theory, application of statistical methods to graph indices of large sets of graphs is a central principle.

Literature offers an impressive diversity of graph indices that are derived from edge connectivity, vertex degrees, distances, or information-theoretic concepts. A large-scale statistical evaluation [15] of more than 900 descriptors on different test data sets shows, however, a high redundancy within this variety: Many of the descriptors, across all of the aforementioned classes, aggregate in a few large clusters such that within each cluster, all participating descriptors yield highly correlated results.

An important thread of recent work in quantitative graph theory focuses on the capability of graph indices to distinguish and classify individual graphs. The *discrimination power* [3,16] measures the ability of an index to *uniquely* distinguish large sets of individual graphs. Discrimination power is understood as opposed to *degeneracy* of indices, that is, their property to yield equal values for different graphs.

A related direction in research are methods for *inexact graph matching*. In order to quantify the difference between graphs by suitable distance measures, such methods aim at analyzing structural differences between given graphs, particularly by identifying substructures (in the simplest case, subgraphs) that can be matched. The extent of nonmatched components then indicates the degree of nonisomorphy of graphs. A well-established concept that offers more flexibility than pure subgraph isomorphism is *graph edit distance* [21,52]. It counts elementary operations such as insertion and deletion of vertices or edges that are necessary to transform one graph into the other. In view of the complexity of subgraph isomorphism or graph edit distance algorithms, computationally cheaper alternatives are investigated (see, e.g., [10,47,60]). With the application of graph indices, this search has recently further developed into a branch of quantitative graph theory (see, e.g., [12,13,19]). In [65], such an approach to measure graph distances is integrated into a clustering framework in order to classify graphs.

Whereas discrimination of graphs demands that graph indices should distinguish different graphs as clearly as possible, quantification of graph nonisomorphy leads to the additional requirement that graph indices should change their values not too much if a graph undergoes only minor modifications. This is captured by the concept of *smoothness* as introduced in [20]. To investigate smoothness of graph indices in more detail, two numerical quantities have been associated with it in [22]: *structure sensitivity* and *abruptness*. Both quantities are derived from the relative changes of the value of a graph index under elementary modifications of graphs, corresponding to graph edit distance 1. Structure sensitivity of a graph index near a given graph G is defined as the average of these relative changes over all graphs H within graph edit distance 1, while abruptness is defined as the maximum of all these relative changes. In order for a graph index to be well suited for graph classification purposes, it should combine, as pointed out in [22], large structure sensitivity with small abruptness.

As our present contribution, too, is concerned with using graph indices to discriminate graph structures, it is natural that a concept similar to the discrimination power mentioned previously is needed and that the properties of large structure sensitivity and small abruptness are desirable for graph indices also in this context. An important difference to the aforementioned task of unique identification of graphs by indices is, however, that we will generate large sets of graphs and our goal will be to distinguish these sets from each other rather than individual graphs.

12.1.2 Graph Models in Image Analysis

The concept of graphs has been used in image processing in various ways (see [38]). A key principle for many of these approaches is the interpretation of the regular mesh of a digital image as a graph with pixels as vertices and edges connecting neighboring pixels based on either a 4- or an 8-neighborhood in 2D or similar choices in 3D (see [37, Section 1.5.1]). This construction also underlies the popular *graph cut* methods [31], which are applied, for example, for image segmentation [32,56] or correspondence problems [5,49] (see also the further references in [31]).

Other models in image analysis construct graphs only after some preprocessing steps and associate vertices to entire image regions [37, Section 1.5.2]. This is fruitful, for example, in establishing partition trees as a means to high-level semantic image interpretation. Also, existing applications of graph edit distance [21] work on this level.

On the other hand, the graph viewpoint allows to transfer image processing methods from the regular mesh to more general meshes, such as in graph-based mathematical morphology [41] or variational and partial differential equation (PDE) models on graphs [18]. Irregular meshes arise in a natural way in image processing in connection with nonlocal methods [8] or when processing scanned surface data [9].

Following the first approach mentioned, we will in this work derive graphs from the regular mesh of the image. As a difference to the aforementioned methods, we will not convert the entire image mesh into one graph but instead consider graphs within local neighborhoods. These neighborhoods can be defined either in a nonadaptive or a structure-adaptive way. Local image content will be encoded in the structure of these graphs via the edge connectivity and/or edge weights. We obtain thus a large collection of graphs attached to image locations. As a consequence, we are less interested in detail differences of individual graphs, which may be influenced by image noise and the precise cutoff parameters for the neighborhoods. Instead, the image information of interest for us is expected to manifest itself in the ensemble of the graphs within an image region, calling for a statistical treatment of the graph indices.

12.1.3   Texture

Textures in images have been investigated for decades (see, e.g., [25,26,48,57,66] for early work on the subject).

12.1.3.1   Texture and Shape

The concept of *texture* is to be understood as complementary to the notion of *shape*. In computer graphics, that is, image synthesis, a scene is typically modeled as a collection of objects that are geometrically described by shapes. Texture is then a gray value or color pattern that is mapped to the surface of a shape in the process of rendering.

In image analysis, where the goal is to extract information from given images, the same shape–texture dichotomy arises from the observation that many images are composed of regions that are in the one or other sense homogeneous within themselves. Shape is then the part of image information that describes the regions on a global level. This includes typically large-scale geometric aspects. On the other hand, texture refers to the intrinsic structure of these regions and generally involves the small-scale distribution of image intensities. Interpreting images, similarly as in image synthesis, as depicting a real-world scene consisting of distinct objects, shape again comprises the representation of the geometry and arrangements of the objects within the scene. Texture, in contrast, reflects the appearance of the object surface (in the case of reflection imaging, e.g., photography of opaque objects) or the object interior (in the case of transmission imaging, such as transmission microscopy, x-ray, or magnetic resonance imaging).

A word of caution needs to be said. The frontier between what is considered as shape and what is subsumed to texture depends to some extent on the model being used in analyzing an image. Part of the intensity variations of an object surface results from small-scale geometric details. As soon as a very detailed shape model is used, these intensity variations become part of the shape information, while with a coarser shape description, the same information becomes part of the texture. This can be illustrated by the texture samples in Figure 12.1 where part of the texture information could be interpreted as shape under a refined geometric model that would treat individual bricks, coffee beans, etc., as objects.

In the frequent case that the regions (representing objects) do not overlap (e.g., when photographing opaque objects) and are delineated by smooth contours, the shape–texture decomposition can be mathematically captured by the *cartoon–texture model* that underlies many works on image restoration and enhancement (see, e.g., [42] where it is proposed to decompose [space-continuous] images into the sum of a cartoon component from the BV space of functions and a texture component belonging to a suitable Sobolev space). A refined version of this decomposition distinguishes even three contributions—cartoon, texture, and noise—and assigns specific function spaces to all of them [1].

12.1.3.2 Texture Descriptors

Looking at discrete images, it happens only in simple cases, where homogeneity of a region manifests itself in almost constant gray values, that texture information can be captured on a single-pixel level. In general, it is localized to regions but cannot be properly detected below a minimum neighborhood size that differs from texture to texture. The structural variety of textures ranges from periodic or almost periodic patterns via quasiperiodic structures to irregular patterns whose homogeneity is revealed only in statistical properties of the intensity data. Once more, the examples in Figure 12.1 can serve to illustrate this bandwidth.

The polymorphy of textures is a lasting challenge to texture analysis. Different descriptors have been designed over the years. Focusing on periodic and quasiperiodic textures, a variety of frequency-based models have been proposed [23,34,51]. Other approaches that analyze texture differences in images use statistics on intensities [48] or on derived quantities such as gradients [26]. Also, the use of structure tensor entries (thus, local averages of derivative quantities) as texture features in [7] can be counted among the latter approaches. A generative approach to texture via grammars is found in [39].

12.1.3.3 Texture Segmentation

The task to segment an image into regions of different textures requires not only the discrimination of different textures but also a localization of the boundary between

FIGURE 12.1:   Nine textures (256 × 256 each). (a) bricks, (b) fabric, (c) flowers, (d) food, (e) leaves, (f) metal, (g) stone, (h) water, (i) wood. All images are converted to grayscale and downscaled from original 512 × 512 color images from the *VisTex* database [44]. ©1995 Massachusetts Institute of Technology. Developed by Rosalind Picard, Chris Graczyk, Steve Mann, Josh Wachman, Len Picard, and Lee Campbell at the Media Laboratory, MIT, Cambridge, Massachusetts. Under general permission for scholarly use.

them, that is, an assignment of individual pixels to regions. The latter obviously conflicts with the fact that texture cannot in general be detected on a single-pixel level. This makes the design of texture descriptors suitable for segmentation even more difficult.

Work on texture segmentation goes back more than 40 years [48]. Applications in, for example, diagnostic radiography were envisioned soon [57,66]. Since that time, numerous approaches have been proposed. Recent approaches to integrate texture descriptors into image segmentation frameworks include models based on active contours or active regions [7,43,50,51] as well as clustering-based approaches [24]. Despite the realm of existing models, texture segmentation remains a subject of ongoing research.

In this chapter, we propose the construction of texture descriptors from graph representations of local image structure. To the best of our knowledge, this is the first time that graph models are used to discriminate textures. Existing graph-based approaches to texture segmentation use, for example, graph cuts for the segmentation procedure, but the incorporation of texture information in these frameworks is done by non-graph-based models (see, e.g., [56]).

12.2 Morphological Amoebas

Mathematical morphology [27,40,53,54] is a well-established and powerful theory in image processing. Its core component is formed by image filters that are applied pixelwise to discrete images, such as dilation, erosion, or median filtering. The principle of these filters is to apply some operation like maximum (for dilation), minimum (for erosion), or median to the intensities of pixels within a *structuring element*, a predefined neighborhood of the pixel being processed.

Adaptive morphology aims at designing morphological operations more specific to local image information: instead of processing equally shaped neighborhoods of all pixels (the classic approach), structuring elements are tailored to suit the image content around each given pixel. Adaptivity allows to construct filters with superior capability to preserve important image structures [6,59].

One type of adaptive structuring elements is given by morphological amoebas. They were introduced by Lerallut et al. [35,36] to devise structure-adaptive denoising algorithms. Their construction relies on *amoeba distances*, that is, metrics on the image domain that combine spatial distance with intensity contrast information, thus preferring pixels whose intensity values are close to that of the pixel being processed. Amoebas may grow around corners or along oriented image structures. They can also be related to PDE-based image filters with structure-preserving properties [62,63] and have also been used to design an active contour method for image segmentation [61].

12.2.1 Amoeba Construction

In the following, the construction of amoeba structuring elements is described. The basic procedure goes back to [35,36]; it is stated here in a modified form as in [63].

We start with a 2D digital image f with pixels indexed by a suitable index set I. By f_i, we denote the intensity of the pixel with index $i \in I$. Its coordinates within the planar image domain are denoted as (x_i, y_i).

12.2.1.1   Amoeba Metric

Spatial distances of pixel locations are measured by the standard Euclidean metric,

$$d_E\big((x_i, y_i), (x_j, y_j)\big) := \|(x_i - x_j, y_i - y_j)\|_2 = \sqrt{(x_i - x_j)^2 + (y_i - y_j)^2}. \quad (12.1)$$

For the amoeba construction, however, we will use an *amoeba metric*. Its definition is based on the edge-weighted graph $G_w(f)$ whose vertices are the pixels of the image f and in which pixels i and j are connected by an edge if and only if j belongs to the 8-neighborhood of i. The edge weight $w(i, j)$ between two adjacent pixels i and j is obtained by combining their Euclidean spatial distance and intensity difference $|f_i - f_j|$ (also called *tonal distance*) via

$$w(i, j) = v\big(\|(x_i - x_j, y_i - y_j)\|_2, \beta\,|f_i - f_j|\big). \quad (12.2)$$

Herein, $v(s, t) = \sqrt{s^2 + t^2}$ is the Pythagorean, or l^2, sum, and the *contrast scale* $\beta > 0$ is a parameter that determines the relative weights of spatial and tonal distances.

The amoeba metric between two arbitrary pixels i and j is then defined as the minimal weighted path length in $G_w(f)$, that is,

$$d_A\big((x_i, y_i, f_i), (x_j, y_j, f_j)\big) = \min_P L(P) \quad (12.3)$$

where P runs over all paths $P = (i_0, i_1, \ldots, i_k)$ for which $i_0 \equiv i$, $i_k \equiv j$, subsequent pixels i_l, i_{l+1} are adjacent, and

$$L(P) = \sum_{l=0}^{k-1} w(i_l, i_{l+1}). \quad (12.4)$$

12.2.1.2   Differences to Lerallut et al.'s Amoeba Construction

The definition of the amoeba metric given here differs from the original setting of Lerallut et al. [35,36] in two details: First, Lerallut et al. define adjacency of pixels via a 4-neighborhood such that the spatial distance of adjacent pixels is always 1. Second, they combine spatial and tonal contributions by the l^1 sum $v(s, t) = s + t$ rather than the Pythagorean sum such that [35,36] work with the edge weights $w(i, j) = 1 + \beta\,|f_i - f_j|$.

The reason why we prefer the 8-neighborhood is that it achieves by little additional effort a better approximation of the true Euclidean distances in the image domain.

Choosing the Pythagorean instead of the l^1 sum in v is a somewhat more arbitrary decision. In fact, also other first-order homogeneous functions $v : \mathbb{R}^{+2} \to \mathbb{R}^+$ satisfying the triangle inequality could be used. Experimental evidence from [63] suggests that amoebas constructed by l^2 and l^1 amoeba distances behave similarly. By fixing v to one particular choice, we avoid one additional degree of freedom for the following investigation.

12.2.1.3 Amoeba Structuring Element

The *amoeba structuring element* $A_\varrho(i)$ for pixel $i \in I$ with *amoeba radius* ϱ is defined as the ϱ-neighborhood of i under the amoeba metric, that is, the set of all pixels $j \in I$ with $d_A(i, j) \leq \varrho$.

12.2.2 Building Graphs from Amoebas

To compute the structuring element $A_\varrho(i)$, one uses Dijkstra's shortest path algorithm [17] in the weighted pixel graph $G_w(f)$ with i as start vertex. The algorithm is terminated once path lengths exceed ϱ.

In morphological image filters with amoeba structuring elements, only the set $A_\varrho(i)$ is of interest. However, the Dijkstra tree arising as a by-product from the computation deserves attention of its own because it encodes in more detail the local image structure within $A_\varrho(i)$. With the edge weights inherited from $G_w(f)$, we call the so-obtained tree $T_w^A(f, i, \varrho)$. The same tree without the edge weights will be denoted as $T_u^A(f, i, \varrho)$.

By $G_w^A(f, i, \varrho)$, we denote a graph whose vertices are the pixels from $A_\varrho(i)$, with their distance matrix induced from $G_w(f)$. Thus, $G_w^A(f, i, \varrho)$ is essentially the induced subgraph of $G_w(f)$ with the amoeba pixels but has shortcut edges added between those vertices that have a shorter connection in $G_w(f)$ via nonamoeba pixels. This minor modification allows for a more efficient algorithmic treatment as it avoids recomputing a full distance matrix within each amoeba.

12.2.2.1 Modified Patch Graphs

By a straightforward modification of the algorithm, one can also compute counterparts of $G_w^A(f, i, \varrho)$, $T_w^A(f, i, \varrho)$, and $T_u^A(f, i, \varrho)$ with the pixels from a *Euclidean* ϱ-neighborhood of i as vertices: $G_w^E(f, i, \varrho)$ is a graph whose vertices are the pixels j with $\|(x_i - x_j, y_i - y_j)\| \leq \varrho$ and whose distance matrix is induced from that of $G_w(f)$. Running Dijkstra's algorithm with starting vertex i on the subgraph of $G_w(f)$ within the same Euclidean patch yields the weighted tree $T_w^E(f, i, \varrho)$ and its unweighted version $T_u^E(f, i, \varrho)$.

12.2.3 Texture Representation by Amoeba and Patch Graphs

The three graphs $G_w^A(f, i, \varrho)$, $T_w^A(f, i, \varrho)$, and $T_u^A(f, i, \varrho)$ represent, in different combinations, various pieces of information that can be expected to capture texture properties of the underlying image f, namely,

- The amoeba distances of pixels within the neighborhood of a pixel ($G_w^A(f, i, \varrho)$, $T_w^A(f, i, \varrho)$)

- The shape of the amoeba structuring element around that pixel (all three graphs)

- The structure of the (Dijkstra) shortest distance tree rooted at that pixel used in computing the amoeba (the two trees)

Similarly, the three graphs $G_w^E(f, i, \varrho)$, $T_w^E(f, i, \varrho)$, and $T_u^E(f, i, \varrho)$ represent analogous subsets of these information cues but without the shape of the amoeba structuring element.

We have thus identified six graphs encoding local texture properties of an image around a given pixel. In each of these scenarios, quantitative graph information can be studied.

12.3   Quantitative Graph Descriptors

We turn now to introduce selected graph descriptors from the literature that we will consider in our study. In order to cover representatives of important classes of measures established in the literature, our selection will include three distance-based descriptors and six information-theoretic measures. Some of the descriptors are applicable within all six scenarios stated earlier, while others make sense only for part of them.

Unless stated otherwise, we assume that descriptors are computed for a connected undirected graph G, either unweighted or edge-weighted, with n vertices and m edges. The vertices are assumed to be numbered $v_1, \ldots, v_n$. For simplicity of notation, we will also denote them by $1, \ldots, n$ where no ambiguity is possible. By E, we denote the edge set of G.

12.3.1   Distance-Based Indices

The first three graph indices to be discussed are representatives of the historically first class of quantitative graph measures, namely, those computed immediately from the vertex distances within the graph.

12.3.1.1   Wiener and Harary Index

12.3.1.1.1   Wiener Index   The Wiener index of an unweighted graph G is defined as

$$W(G) := \sum_{1 \leq i < j \leq n} d(i, j) \tag{12.5}$$

where $d(i, j)$ denotes the distance of vertices i and j.

Introduced by Wiener [64] for unweighted graphs in the context of physical chemistry, it has been further studied, for example, by Platt [45] and Hosoya [28].

Extending the original definition, the Wiener index can obviously also be computed from edge-weighted graphs by computing distances $d(i, j)$ from the edge weights.

We apply it therefore to the full local edge-weighted graph, the edge-weighted and unweighted Dijkstra trees within adaptive and nonadaptive structure elements.

12.3.1.1.2 Harary Index An index closely related to the Wiener index is the Harary index introduced by Plavšić et al. [46],

$$H(G) := \sum_{1 \leq i < j \leq n} \frac{1}{d(i, j)}. \tag{12.6}$$

Like the Wiener index, the Harary index can be computed in all of our six graph settings.

12.3.1.2 Balaban Index

Like the two preceding descriptors, the Balaban J index [2] is a distance-based measure. It is defined as

$$J(G) := \frac{m}{m + 1 - n} \sum_{(i,j) \in E} \frac{1}{\sqrt{S_i S_j}} \tag{12.7}$$

where S_i is the sum of distances from vertex i to all vertices in the graph, that is,

$$S_i := \sum_{j=1}^{n} d(i, j). \tag{12.8}$$

According to the results of [16], this index is among the best-performing classical graph descriptors with regard to its discriminative power for large sets of graphs and thus an interesting candidate for our investigation. Note, however, that it tends to become more degenerate for the larger graph sets studied in [16].

Although also $J(G)$ has originally been formulated for unweighted graphs, its derivation from distances makes it again straightforward to apply to edge-weighted graphs.

However, the appearance of the sums S_i in the definition of J suggests that it is not reasonable to apply it to the trees, so it will be applied to the full edge-weighted graphs within an amoeba structuring element or a nonadaptive image patch.

12.3.2 Information-Theoretic Indices

Information-theoretic graph indices are related to entropies of discrete probability measures on different components of a graph—vertices, vertex pairs, edges, etc. Recall that the entropy of a discrete probability measure p defined on a set $\{1, \ldots, k\}$ is defined as

$$H(p) := -\sum_{i=1}^{n} p(k) \operatorname{ld} p(k). \tag{12.9}$$

Since Shannon's seminal work [55], it is considered as the fundamental representation of the information content of the probability measure p.

12.3.2.1   Bonchev–Trinajstić Information Indices

One group of information-theoretic indices for graphs has been proposed by Bonchev and Trinajstić [4]. These indices are based on partitioning a graph invariant in order to measure the inhomogeneity in the distance distribution of vertices within graphs.

For an unweighted graph G with vertices $1, \ldots, n$ and diameter $D(G)$, denote by

$$k_d := \#\{(i, j) \mid 1 \leq i < j \leq n, d(i, j) = d\} \tag{12.10}$$

the counts of vertex pairs with distance d, $1 \leq d \leq D(G)$.

Following [4], one can then define for G the *total information on distances* $I_D^E(G)$, the *mean information on distances* $\bar{I}_D^E(G)$, the *total information on the realized distances* $I_D^W(G)$, and the *mean information on the realized distances* $\bar{I}_D^W(G)$:

$$I_D^E(G) := \binom{n}{2} \operatorname{ld} \binom{n}{2} - \sum_{d=1}^{D(G)} k_d \operatorname{ld} k_d, \tag{12.11}$$

$$\bar{I}_D^E(G) := - \sum_{d=1}^{D(G)} \frac{k_d}{\binom{n}{2}} \operatorname{ld} \frac{k_d}{\binom{n}{2}}, \tag{12.12}$$

$$I_D^W(G) := W(G) \operatorname{ld} W(G) - \sum_{d=1}^{D(G)} k_d \, d \operatorname{ld} d, \tag{12.13}$$

$$\bar{I}_D^W(G) := - \sum_{d=1}^{D(G)} k_d \frac{d}{W(G)} \operatorname{ld} \frac{d}{W(G)}, \tag{12.14}$$

where $W(G)$ denotes the Wiener index from the previous subsection. An easy rewrite shows that

$$I_D^E(G) = \binom{n}{2} \bar{I}_D^E(G), \tag{12.15}$$

$$I_D^W(G) = W(G) \bar{I}_D^W(G), \tag{12.16}$$

which justifies the nomenclature of $I_D^E(G)$ and $I_D^W(G)$ as total information versus $\bar{I}_D^E(G)$ and $\bar{I}_D^W(G)$ as mean information measures.

Formally, the definitions (12.11) through (12.14) can equally be applied to edge-weighted graphs, provided sums are taken over all values d occurring as pairwise vertex distances. In generic cases, however, these distances will be distinct such that all k_d will equal 1. The indices $I_D^E(G)$ and $\bar{I}_D^E(G)$ would therefore be essentially independent on the graph structure, just being functions of the number of vertices.

For the remaining two indices, it is more useful to rewrite (12.13) and (12.14) to eliminate the reference to k_d and sum over vertex pairs, which yields

$$I_D^W(G) := W(G) \operatorname{ld} W(G) - \sum_{1 \le i < j \le n} d(i, j) \operatorname{ld} d(i, j), \tag{12.17}$$

$$\bar{I}_D^W(G) := - \sum_{1 \le i < j \le n} \frac{d(i, j)}{W(G)} \operatorname{ld} \frac{d(i, j)}{W(G)}. \tag{12.18}$$

Based on this discussion we will investigate $I_D^W(G)$ and $\bar{I}_D^W(G)$ on full local edge-weighted graphs, edge-weighted and unweighted Dijkstra trees, whereas $I_D^E(G)$ and $\bar{I}_D^E(G)$ are considered only for unweighted trees.

In the light of the entropy definition (12.9), $\bar{I}_D^W(G)$ in the form (12.18) is the entropy of a discrete probability measure on the vertex pairs of a graph, where the probabilities for pairs (i, j) read as $d(i, j)/W(G)$. Similarly, $\bar{I}_D^E(G)$ as stated in (12.12) is the entropy of a discrete probability measure on the set of distances $\{1, \ldots, D(G)\}$ with $p(d) = k_d/\binom{n}{2}$.

The corresponding total information measures $I_D^E(G)$ and $I_D^W(G)$ are then just entropies rescaled by the total number of vertex pairs and the Wiener index, respectively, as can be read off (12.15) and (12.16).

12.3.2.2 Dehmer Entropies

Our selection of graph descriptors is complemented by two entropy indices based on probability measures on the vertex set of a graph G introduced by Dehmer [11]. For an arbitrary positive-valued function f (*information functional*) on the vertices of a graph G, one defines a set of vertex probabilities $p_i \equiv p(v_i)$ by

$$p_i := \frac{f(v_i)}{\sum_{j=1}^n f(v_j)} \tag{12.19}$$

such that the entropy of the so-obtained discrete probability measure is

$$I_f(G) := H(p) = - \sum_{i=1}^n p_i \operatorname{ld} p_i. \tag{12.20}$$

In [11], two candidates for f on unweighted graphs are proposed, both of which rely on the concept of spheres around vertices in G. The d-sphere $S_d(v_i)$ of v_i consists of all those vertices v_j in G for which $d(i, j) \le d$ holds.

With a basis $\alpha > 0$ and positive coefficients $c_1, \ldots, c_{D(G)}$ as parameters, the first information functional is based just on the cardinalities of spheres, $s_d(i) := \#S_d(v_i)$, and reads as

$$f^V(v_i) := \alpha^{\sum_{d=1}^{D(G)} c_d \, s_d(i)}. \tag{12.21}$$

The second information functional from [11] involves the sums of path lengths from v_i to all v_j within $S_d(v_i)$,

$$l_d(i) := \sum_{j:v_j \in S_d(v_i)} d(v_i, v_j), \tag{12.22}$$

and combines those again with parameters α and $c_1, \ldots, c_{D(G)}$ as earlier to obtain

$$f^P(v_i) := \alpha^{\sum_{d=1}^{D(G)} c_d l_d(i)}. \tag{12.23}$$

As [16] brought out, these indices as well as a third one, $I_{f^\triangle}$, achieve excellent discriminative power for large sets of graphs. The index $I_{f^\triangle}$ is based on sequences of vertex degrees. A reason to not consider it here lies in the structure of the graphs we use. As these are constructed from the regular mesh of an image, their vertices have degrees globally bounded by the neighborhood size (in our experiments, 8), leading to fairly degenerate degree sequences. Thus, it is conjectured that only an insufficient representation of image information is found in degree sequences. Forthcoming work will, however, include a more detailed investigation of this conjecture.

We turn now to state variants of the information functionals f^V and f^P that can be applied to edge-weighted graphs. We notice first that in the exponent of (12.21), each vertex j contributes to all summands s_d with $d \geq d(i, j)$. Thus, the exponent of (12.21) can be rewritten as

$$\sum_{d=1}^{D(G)} c_d s_d(i) = \sum_{j=1}^{n} \sum_{d=d(i,j)}^{D(G)} c_d = \sum_{j=1}^{n} C_{d(i,j)} \tag{12.24}$$

with

$$C_{d_0} := \sum_{d=d_0}^{D(G)} c_d. \tag{12.25}$$

Provided that the sequence $C_0, C_1, \ldots, C_{D(G)}$ is interpolated by a nonnegative function C on $[0, D(G)]$ for an edge-weighted graph G, the functional

$$f^V(v_i) = \alpha^{\sum_{j=1}^{n} C(d(i,j))} \tag{12.26}$$

is then an obvious generalization of (12.21).

Similarly, the exponent of (12.23) can be rewritten as

$$\sum_{d=1}^{D(G)} c_d l_d(i) = \sum_{j=1}^{n} \sum_{d=d(i,j)}^{D(G)} c_d d(i, j) = \sum_{j=1}^{n} C_{d(i,j)} d(i, j) \tag{12.27}$$

with $C_{d(i,j)}$ as earlier. Interpolating again the coefficient series by a function, one has the functional

$$f^P(v_i) = \alpha^{\sum_{j=1}^{n} C(d(i,j)) d(i,j)} \tag{12.28}$$

as generalization of (12.23) to edge-weighted graphs. In this form, both functionals differ only by the choice of the function C, as the substitution $\tilde{C}(d) := C(d)\,d$ turns f^P into an instance of f^V.

To specify suitable coefficient functions $C(d)$ for our investigation, we notice first that the basis α in (12.21) and (12.23) can be fixed to Euler's constant e by rescaling all c_d with the natural logarithm of α.

Sacrificing some generality in the choice of the coefficients c_d, we focus to what is called *exponential weighting scheme* in [16] and assume that the c_d is given by a geometric series, $c_d = \Lambda\,q^d$, $0 < q < 1$, for $d \leq D(G) - 1$, and $c_{D(G)} = \frac{\Lambda}{1-q}q^{D(G)} = \Lambda \sum_{d=D(G)}^{\infty} q^d$. In this case, the cumulated coefficients C_d from (12.25) become simply

$$C_{d_0} = \Lambda \left(\sum_{d=d_0}^{D(G)-1} q^d + \frac{q^{D(G)}}{1-q} \right) = \frac{\Lambda\,q^{d_0}}{1-q}, \tag{12.29}$$

which is immediately generalized to the coefficient function

$$C(d) = M\,q^d \tag{12.30}$$

where the constants have been combined into $M := \Lambda/(1-q)$. Inserting this into (12.26) and (12.28), we obtain therefore

$$f^V(v_i) = e^{M \sum_{j=1}^{n} q^{d(i,j)}}, \tag{12.31}$$

$$f^P(v_i) = e^{M \sum_{j=1}^{n} q^{d(i,j)} d(i,j)} \tag{12.32}$$

as generalized information functionals for our experimental investigation.

12.4 Experimental Test of Texture Discrimination

The test scenario developed in this section aims at exploring the potential of quantitative graph descriptors on local neighborhoods to discriminate texture patches.

Let us mention that this is a rather limited task, designed to suit the Haralick features that are chosen for comparison. By computing the *local* graph-based features and then aggregating them over entire image regions, the locality of our approach is not properly exploited. At the same time, the computational cost for this locality has to be paid, which makes the graph-based features in this setting more expensive than Haralick features that are per se region based.

This relation will change as soon as graph-based features will be applied for more advanced tasks like texture segmentation. Integration into a segmentation framework will allow to capitalize directly on the possibility to evaluate graph-based features at each single pixel location. This is a major goal for future work on the subject.

12.4.1   Test Images

The starting point in constructing the data set for our tests consisted of nine images from the *VisTex* collection [44] showing different textures.

Each original image (512 × 512 pixels RGB) was converted to grayscale and downsampled to 256 × 256 pixels, resulting in the images shown in Figure 12.1. From each of these images, the upper left and lower right quarters were clipped, leading to an effective test image size of 128 × 128 pixels.

For all comparisons between different textures, the upper left quarters were used. Additionally, the upper left and lower right patch of each texture were compared in order to assess the variability of texture measures within the same texture.

12.4.2   Evaluation Procedure for Texture Features

We turn now to describe the procedures used to evaluate the capabilities of the two groups of features for texture discrimination. To distinguish it from the concept of discrimination power of graph descriptors as mentioned in the introduction that refers to individual graphs, we will speak of *texture discrimination power* here. For the time being, this is to be understood as an intuitive concept, still not in a final mathematical formulation. In the following, we will specify quantities to be used in our experiments that try to capture this intuitive concept in the case of the Haralick texture features and our graph-based texture features, adapted to our test scenario. We hope that these measures can be developed into a more rigorous concept of texture discrimination power in further work.

12.4.2.1   Haralick Features

First introduced in [26], Haralick features are a well-established approach to quantitative description of texture within image regions. Despite the introduction of numerous further texture descriptors during the four decades since then, they continue to be a valuable tool for texture analysis [30,58], and their capability is confirmed also in comparative studies [29].

The definition of Haralick features relies on the *gray-tone spatial-dependence matrix* [26], also denoted as *co-occurrence matrix* [25]. Assume an image or image patch is sampled on a 2D regular grid $\Omega \subset \mathbb{Z}^2$ with quantized gray values in $\{0, \ldots, g-1\}$. With a fixed direction vector $d \in \mathbb{Z}^2$, one defines $\hat{p}_{d,i,j}$ for $i, j = 0, \ldots, g-1$ as the count of pairs of pixels $x, y \in \Omega$ with $x - y = \pm d$, for which x has gray value i, whereas y has gray value j. The co-occurrence matrix P_d w.r.t. direction d is then the symmetric $g \times g$ matrix obtained by normalizing these values, that is, whose entries are

$$p_{i,j} := \frac{\hat{p}_{d,i,j}}{\sum_{i,j=0}^{g-1} \hat{p}_{d,i,j}}. \tag{12.33}$$

The actual texture features are then given by statistics on the co-occurrence matrix. From the 14 features proposed in [26], we select the following six features for our experiments:

- Energy $\sum_{i,j=0}^{g-1} p_{i,j}^2$

- Entropy $\sum_{i,j=0}^{g-1} p_{i,j} \operatorname{ld} p_{i,j}$

- Maximum probability $\max_{i,j \in \{0,\dots,g-1\}} p_{i,j}$

- Correlation $\sum_{i,j=0}^{g-1} \frac{(i-\mu)(j-\mu)p_{i,j}}{\sigma^2}$, where μ and σ are the mean value and standard deviation of the gray values within Ω

- Contrast $\sum_{i,j=0}^{g-1} (i-j)^2 p_{i,j}$

- Inverse difference moment $\sum_{i,j=0}^{g-1} \frac{p_{i,j}}{1+(i-j)^2}$

Note that these are six of the seven features exposed in the later paper [25] as the most commonly used features; we drop only the last entry from this short list, a directional run-length probability for gray values.

Each of the features was applied to the complete 128×128 pixel images of homogeneous textures. Computations were done based on co-occurrence matrices in four directions in 45-degree steps, that is, $d = (1,0)^T, d = (1,1)^T, d = (0,1)^T$, and $d = (-1,1)^T$, and the values of each feature for the four directions were averaged.

For each of the Haralick features, we assessed first its variability within a single texture. To this end, the absolute differences of its values within each pair of patches of the same texture were collected, that is, brick (upper left) versus brick (lower right) and flowers (upper left) versus flowers (lower right), etc. Ordering the so-obtained nine difference values by size, $d_1 \geq d_2 \geq \cdots \geq d_9 \geq 0$, two thresholds were derived: $T_1 := 2 d_1$, the double of the largest difference, as minimal difference for the certain detection of different textures, and $T_2 := d_3$, the third largest difference, as minimal difference for the detection of possibly different textures.

Then the absolute differences of feature values for different textures were computed, that is, brick (upper left) versus flowers (upper left), etc., using consistently the upper left patches. The resulting 36 differences were classified into three groups: differences below T_2 as "$-$" (no detection of different textures), "$\circ$" (uncertain), and "$+$" (detection of different textures). Table 12.1 demonstrates this proceeding for the Haralick energy feature.

12.4.2.2 Graph Indices

To avoid extensive parameter testing in our experiments on graph indices, we fixed the amoeba contrast scale to $\beta = 0.01$ and the amoeba radius to $\varrho = 5$ throughout the test series. In the exponential weighting scheme (12.30) for the I_{fV} and I_{fP} graph indices, we fixed the parameters to $M = 1, q = \exp(-0.1)$.

Our further evaluation was carried out by a similar procedure as described earlier for the Haralick features. As graph indices can be computed for each single pixel location in an image (using information from some neighborhood of that pixel), we computed for each index statistical parameters of its value distribution within each

TABLE 12.1:   Differences of Haralick Energy Feature between Textures

	Bricks	Fabric	Flowers	Food	Leaves	Metal	Stone	Water	Wood
Bricks	92.87	948.33	904.43	948.70	973.85	893.69	659.92	1581.03	6916.25
Fabric	o	10.19	43.90	0.38	25.53	54.64	288.41	2529.36	7864.58
Flowers	o	–	1.29	44.28	69.43	10.74	244.51	2485.46	7820.68
Food	o	–	–	0.18	25.15	55.02	288.78	2529.73	7864.95
Leaves	o	–	–	–	2.29	80.17	313.94	2554.88	7890.10
Metal	o	–	–	–	–	59.56	233.77	2474.72	7809.94
Stone	o	o	o	o	o	o	98.83	2240.95	7576.17
Water	o	o	o	o	o	o	o	3325.73	5335.22
Wood	+	+	+	+	+	+	+	o	86.11

Notes: Values on the diagonal represent absolute differences between disjoint cutouts of the same texture image. Off-diagonal entries in the upper right part are absolute differences between the different textures. The lower left part shows the result of thresholding these entries with the two thresholds $T_1 = 6651.46$ (double the largest diagonal entry) and $T_2 = 92.87$ (third largest diagonal entry): Values greater than T_1 are classified as +, those between T_1 and T_2 as o, and values less or equal to T_2 as –.

image. To avoid artifacts near the image boundary, only the values within a 96×96 region of each 128×128 test image (dropping a 16-pixel wide margin) were included in the statistics. For our subsequent evaluation, we computed for each pair of patches the test quantity

$$u := \frac{|\mu_1 - \mu_2|}{\sigma} \tag{12.34}$$

borrowed from a standard parametric test for difference of mean values. Herein, μ_1 and μ_2 denote the mean values of the feature in the two patches to be compared, and σ their joint standard deviation,

$$\sigma := \sqrt{\frac{\sigma_1^2}{n_1} + \frac{\sigma_2^2}{n_2}}. \tag{12.35}$$

Similar as with Haralick features, feature values varied considerably, and large values of u were thus observed, even between different patches of the same texture sample. This clearly precluded basing texture discrimination on the standard thresholds of the statistical test. Instead, we applied to the difference measures u the same procedure as earlier for the absolute differences of Haralick feature values: first computing the u measures between each pair of patches of the same textures, selecting T_1 as twice the largest and T_2 as the third largest of these nine values, and dividing the u measures between pairs of different texture patches into the groups "–", "o", "+" by the thresholds.

To exemplify the proceeding, Table 12.2 shows u values and classification for one graph-based feature, namely, the Dehmer entropy index I_{fv} computed on full edge-weighted graphs within amoeba patches.

TABLE 12.2: Absolute Difference Measures u (12.34) of Graph Index I_{fv} Computed on Edge-Weighted Graphs G_w^A between Textures

	Bricks	Fabric	Flowers	Food	Leaves	Metal	Stone	Water	Wood
Bricks	1.460	48.679	39.191	45.334	48.134	56.343	35.657	30.868	40.360
Fabric	+	3.618	26.947	16.480	4.868	11.927	39.284	50.549	51.030
Flowers	+	o	2.124	12.869	29.616	20.386	19.656	43.195	44.209
Food	+	o	o	3.507	20.321	6.282	30.435	48.298	49.052
Leaves	+	−	+	o	5.270	16.287	40.179	49.727	50.139
Metal	+	o	o	−	o	14.031	40.109	59.486	60.277
Stone	+	+	o	+	+	+	6.747	44.409	46.568
Water	+	+	+	+	+	+	+	11.782	27.474
Wood	+	+	+	+	+	+	+	o	1.162

Notes: Values on the diagonal represent u values between disjoint cutouts of the same texture image. Off-diagonal entries in the upper right part are u values between the different textures. The lower left part shows the result of thresholding these entries with the two thresholds $T_1 = 28.062$ (double the largest diagonal entry) and $T_2 = 6.747$ (third largest diagonal entry): Values greater than T_1 are classified as +, those between T_1 and T_2 as o, and values less or equal to T_2 as −.

12.4.3 Experimental Results on Texture Discrimination

The complete results for 42 graph-based and six Haralick texture features after the thresholding procedure are collected in Table 12.3. For later use, we attach labels A1, . . . , H6 to the features.

As could be expected, some texture pairs such as *bricks* versus *fabric* were easily distinguished by almost each single measure. Others were hard to tell from each other almost regardless which measure was used; the pair *water* versus *wood* remained uncertain or undistinguishable with all features considered. The remaining texture pairs yielded mixed results depending on the choice of the feature.

Tables 12.4 and 12.5 state the numbers of certain/uncertain/failed discrimination for each texture pair across all indices, while Tables 12.6 and 12.7 state how many texture pairs were certainly or uncertainly discriminated by each method. These values can serve to assess the texture discrimination power of the different features. The labels H1, . . . in Table 12.7 correspond to Table 12.3, whereas Ax, . . . and x1, . . . in Table 12.6 refer to the first and second characters of the corresponding labels.

It is evident from these tables that the features differ greatly in their detection rates. Among the Haralick features tested, contrast (H1), correlation (H2), and entropy (H4) distinguish most of the texture pairs, while the energy measure (H3) contributes less to the overall texture discrimination power of the feature set. Inverse different moment (H5) and maximum probability (H6) do not achieve distinctions with certainty in this particular setting. Among all features from both groups, only the Haralick correlation feature (H2) tells *flowers* from *metal* on the "+" level.

TABLE 12.3: Discrimination of 42 Texture Pairs by 48 Features

Feature		Label	Water	Stone		Metal			Leaves				Food					Flowers						Fabric							Bricks							
			Wood	Wood	Water	Wood	Water	Stone	Wood	Water	Stone	Metal	Wood	Water	Stone	Metal	Leaves	Wood	Water	Stone	Metal	Leaves	Food	Wood	Water	Stone	Metal	Leaves	Food	Flow.	Wood	Water	Stone	Metal	Leaves	Food	Flow.	Fabric
G_w^E	Wiener	A1	o	+	o	+	+	+	+	+	+	o	o	o	o	−	o	o	o	o	o	o	o	+	+	+	o	o	o	o	o	o	o	+	+	o	+	+
	Harary	A2	o	+	o	+	+	+	+	+	+	o	o	o	o	−	o	o	o	o	o	o	o	+	+	+	o	o	o	o	o	o	o	+	+	o	o	+
	Balaban	A3	o	o	o	+	+	+	+	+	+	o	o	o	o	−	o	o	o	o	o	o	o	+	+	+	o	o	o	o	o	o	o	+	+	o	+	+
	I_D^W	A6	o	+	o	+	+	+	+	+	+	o	o	o	o	−	o	o	o	o	o	o	o	+	+	+	o	o	o	o	o	o	o	+	+	o	o	+
	I_D^{ZW}	A7	−	o	−	−	−	−	o	o	o	o	+	+	+	+	−	+	+	+	o	−	o	o	o	o	o	o	o	o	−	−	−	−	o	o	+	o
	I_V^J	A8	o	−	+	+	+	+	+	+	+	o	+	+	o	−	o	+	+	o	o	+	o	+	+	+	o	−	o	o	o	o	o	−	+	+	+	+
	I_P^J	A9	−	+	+	+	+	+	+	+	+	+	+	+	+	+	o	+	+	+	o	+	+	+	+	+	+	+	o	o	o	o	o	o	+	+	+	+
T_w^E	Wiener	B1	o	+	o	+	+	+	+	+	+	o	o	o	o	−	o	o	o	o	o	o	o	+	+	+	o	o	o	o	o	o	o	o	o	+	+	+
	Harary	B2	o	o	o	+	+	+	+	+	+	o	+	+	+	−	o	o	o	o	o	o	o	+	+	+	o	o	o	o	o	o	o	o	o	+	+	+
	I_D^W	B6	o	+	o	o	o	o	+	+	+	o	o	o	o	o	o	o	o	o	o	o	o	+	+	+	o	o	o	o	o	o	o	o	+	+	+	+
	I_D^{ZW}	B7	−	o	−	+	+	o	+	+	+	o	o	o	o	−	o	o	o	o	o	o	o	+	+	+	o	−	−	o	o	o	o	o	+	+	+	+
	I_V^J	B8	o	−	+	o	o	o	+	+	+	o	o	o	o	o	o	o	o	o	o	o	o	+	+	+	+	o	o	o	−	−	−	−	o	o	o	o
	I_P^J	B9	o	+	o	o	o	o	+	+	+	o	+	+	+	−	o	o	o	o	o	o	o	+	+	+	+	o	o	o	o	o	o	o	+	+	+	o
T_u^E	Wiener	C1	−	o	−	o	o	o	+	+	+	+	o	o	o	−	−	o	o	o	−	+	o	+	+	+	o	o	o	+	o	o	o	+	o	o	−	o
	Harary	C2	−	−	−	o	o	o	+	+	+	+	o	o	o	−	+	−	−	−	o	+	o	+	+	+	o	o	o	+	−	−	−	o	o	−	o	o
	I_D^E	C4	−	−	−	o	o	o	+	+	+	+	o	o	o	−	+	o	o	o	o	+	o	+	+	+	o	o	o	+	−	−	−	−	−	o	−	o
	$I_{\!E}^E$	C5	−	−	−	o	o	o	+	+	+	+	o	o	o	−	o	−	−	−	o	+	o	+	+	+	o	o	o	+	−	−	−	o	o	o	o	o
	I_D^W	C6	−	−	−	o	o	o	+	+	+	+	o	o	o	−	o	−	−	−	o	+	o	+	+	+	o	o	o	+	−	−	−	o	o	−	−	o
	I_V^J	C8	−	−	−	−	−	−	+	+	+	+	−	−	−	−	+	−	−	−	o	o	o	+	+	+	o	o	o	+	−	−	−	o	o	−	o	o
	I_D^{ZW}	C7	−	−	−	o	o	o	+	+	+	+	o	o	o	−	+	−	−	−	o	+	−	+	+	+	o	−	+	+	−	−	−	−	−	o	o	o
	I_P^J	C9	−	−	−	o	o	o	o	o	o	−	o	o	o	o	o	o	o	o	o	−	−	o	o	o	o	o	o	o	−	−	−	−	o	o	o	o

		D1	D2	D3	D6	D7	D8	D9	E1	E2	E6	E7	E8	E9	F1	F2	F4	F5	F6	F7	F8	F9	H1	H2	H3	H4	H5	H6
G_w^A	Wiener	o	o	o	o	−	o	−	−	−	−	−	−	o	−	−	−	−	−	−	−	−	−	o	o	o	o	o
	Harary	−	o	o	−	o	+	+	o	o	o	o	+	o	o	o	o	o	o	o	o	o	o	o	+	+	o	o
	Balaban	−	o	−	−	o	+	+	o	o	−	o	o	o	o	o	o	o	o	o	o	o	−	o	+	o	+	o
	t_D^W	+	+	+	+	+	+	+	+	+	+	o	+	+	+	+	+	+	+	+	+	+	+	+	+	o	o	
	t_D^{rW}	+	+	+	+	+	+	+	+	+	+	o	+	+	+	+	+	+	+	+	+	+	+	o	+	o	o	o
	l_f^V	+	+	+	+	+	+	+	+	+	+	o	+	+	+	+	+	+	+	+	+	+	+	o	o	o	−	
	l_f^P	o	o	o	o	o	o	+	o	o	o	o	o	o	o	o	o	o	o	o	o	o	+	+	−	o	−	−
T_w^A	Wiener	+	+	+	+	+	+	+	+	+	+	o	+	+	+	+	o	+	+	+	+	+	−	+	+	o	o	
	Harary	+	+	+	+	+	+	+	+	+	+	o	+	+	+	+	o	+	+	+	+	+	o	o	+	o	o	
	t_D^W	+	o	+	+	o	+	+	+	+	+	+	+	+	+	+	o	+	+	+	+	+	o	o	o	o	−	
	t_D^{rW}	−	−	−	−	−	−	−	o	−	−	−	−	o	o	−	−	−	−	−	−	−	o	−	+	−	o	−
	l_f^V	o	o	o	o	o	o	o	o	o	o	o	o	o	o	o	o	o	o	o	o	+	−	−	o	o	o	
	l_f^P	o	o	o	o	o	+	+	o	o	o	o	o	o	o	o	o	o	o	o	o	o	o	+	+	o	o	
T_u^A	Wiener	o	o	o	o	o	+	+	o	o	o	o	o	o	o	o	o	o	o	o	o	o	o	−	o	+	o	o
	Harary	o	o	o	o	o	o	+	o	o	o	o	o	o	o	o	o	o	o	o	o	o	o	+	o	o	−	−
	t_D^E	o	o	o	o	o	o	o	−	o	o	o	o	o	o	o	o	o	o	o	o	o	+	−	−	o	−	
	t_D^{rE}	+	+	+	+	+	+	+	o	+	o	+	o	+	+	+	o	+	+	+	+	o	−	o	−	o	o	−
	t_D^W	o	o	o	o	o	o	o	o	o	o	o	o	o	o	o	o	o	o	o	o	o	−	o	o	o	−	
	t_D^{rW}	+	+	+	+	+	+	+	+	+	+	+	+	+	+	+	+	+	+	+	+	+	+	+	+	o	o	
	l_f^V	+	+	+	+	+	+	+	+	+	+	+	+	+	+	+	+	+	+	+	+	+	+	o	+	o	o	o
	l_f^P	+	+	+	+	+	+	+	+	+	+	+	+	+	+	+	+	+	+	+	+	+	+	o	o	o	−	
Haralick	Contrast	o	o	o	o	−	+	o	o	o	o	−	o	o	−	−	−	−	−	−	−	−	−	o	+	+	o	o
	Corr.	o	o	o	o	o	−	+	o	o	−	o	−	−	o	o	o	−	o	−	−	o	o	+	+	−	o	−
	Energy	+	+	+	+	+	o	o	o	+	o	o	o	o	o	+	+	o	+	o	o	o	+	+	−	o	o	−
	Entr.	o	o	o	o	o	−	+	o	o	o	o	−	o	o	−	−	−	−	−	−	−	−	−	o	+	+	o
	IDM	o	o	o	o	o	−	+	o	−	o	−	o	−	−	−	−	−	−	−	−	−	−	−	o	o	o	o
	Max. pr.	−	o	o	−	o	−	o	+	+	o	+	o	−	o	−	−	−	−	−	−	−	−	−	o	+	o	o

TABLE 12.4: Numbers of Graph Index Features Yielding Certain (+) and Uncertain (o) Discrimination for Each Texture Pair

+/o	Fabric	Flowers	Food	Leaves	Metal	Stone	Water	Wood
Bricks	40/2	5/33	23/17	39/3	30/10	1/26	1/18	1/21
Fabric		16/26	1/40	1/27	2/40	38/4	40/2	40/2
Flowers			1/39	26/15	0/39	3/35	5/33	5/33
Food				5/35	2/7	21/19	24/16	24/16
Leaves					10/31	40/2	39/3	39/3
Metal						29/11	30/10	30/10
Stone							5/22	12/18
Water								0/16

Notes: In each cell, the failed discrimination ($-$) count equals 42 minus the sum of the two values given.

TABLE 12.5: Numbers of Haralick Features Yielding Certain (+) and Uncertain (o) Discrimination for Each Texture Pair

+/o	Fabric	Flowers	Food	Leaves	Metal	Stone	Water	Wood
Bricks	3/3	1/4	2/4	2/4	3/3	1/5	0/4	2/3
Fabric		2/2	1/2	2/1	0/3	1/4	3/3	4/2
Flowers			0/4	1/3	1/2	1/3	1/4	2/4
Food				1/3	1/2	1/4	2/4	3/2
Leaves					2/1	2/3	2/4	3/2
Metal						2/3	3/3	4/2
Stone							2/4	2/4
Water								0/5

Notes: In each cell, the failed discrimination ($-$) count equals 6 minus the sum of the two values given.

It can also be seen that the overall texture discrimination power of the graph-index-based texture features is well comparable to that of the Haralick features. Note that I_{fV} on G_w^A (**D8**) is the only candidate among all features tested that strongly discriminates the *bricks* and *water* samples. Within the graph index features, distinction of *bricks* versus *stones* and *wood* is also unique to this feature. Only I_{fP} on G_w^E (**A9**) is able to discriminate *fabric* from *leaves* and *flowers* from *food*.

Although the T_u^E graph setting yields rather low overall discrimination rates with practically all graph indices (**C1**, ..., **C9**), the $\bar{I}_D^W$ index on this graph (**C7**) is the only graph-based feature that tells *fabric* from *food*, thus demonstrating that also this graph setting has some merit.

We note at this point that the highest rates of certain discrimination (+) by a single feature are achieved by the I_{fP} features on G_w^E (**A9**) and the I_{fP} and I_{fV} features on G_w^A. These three features even outperform in our test setting each single one of the six Haralick features under consideration.

TABLE 12.6: Numbers of Texture Pairs with Certain (+) and Uncertain (o) Discrimination for Each Graph Index Feature

+/o		G_w^E	T_w^E	T_u^E	G_w^A	T_w^A	T_u^A
		Ax	Bx	Cx	Dx	Ex	Fx
Wiener	x1	13/22	13/22	12/17	18/14	16/16	17/14
Harary	x2	13/22	15/20	12/13	17/18	18/15	18/13
Balaban	x3	12/23	—	—	18/15	—	—
I_D^E	x4	—	—	11/18	—	—	18/13
$\bar{I}_D^E$	x5	—	—	11/14	—	—	12/20
I_D^W	x6	13/22	13/22	12/17	18/14	16/14	18/13
$\bar{I}_D^W$	x7	9/15	7/22	14/0	17/15	17/14	17/14
I_{f^V}	x8	21/13	13/21	12/13	25/9	7/28	17/14
I_{f^P}	x9	28/7	11/23	0/24	24/9	17/18	18/14

Notes: In each cell, the failed discrimination (−) count equals 36 minus the sum of the two values given.

TABLE 12.7: Numbers of Texture Pairs with Certain (+) and Uncertain (o) Discrimination for Each of the Haralick Texture Features Considered

+/o	Contrast	Correl.	Energy	Entropy	Inv. Diff. M.	Max. Prob.
	H1	H2	H3	H4	H5	H6
	21/11	16/14	7/19	19/15	0/32	0/22

Notes: In each cell, the failed discrimination (−) count equals 36 minus the sum of the two values given.

12.4.3.1 Overlap of Features

Among the large group of graph-index-based features, there are many that provide identical or almost identical discriminations, that is, they succeed and fail on the same texture pairs. This is consistent with the findings from [15] that analyzed the correlations between graph indices by clustering analysis.

Given that our present investigation involves only 36 texture pairs as test cases, which due to the multiple use of the same texture samples cannot be assumed independent, a statistical analysis similar to [15] cannot be expected to yield valid results for the time being. To give some indication on what features are most different from each other, we resort therefore to a different procedure.

To this end, we define a distance between features based on the thresholding results in Table 12.3. Denote by $D_{i,k}$ the entry (+, o, or −) of Table 12.3 in the row belonging to feature F_i ($i = $ A1, ..., H6) and texture pair P_k ($k = 1, ..., 36$). For any two given features F_i, F_j and texture pair P_k, let

TABLE 12.8: Mutual Distances (12.37) between Texture Features

	A2	A3	A6	A7	A8	A9	B1	B2	B6	B7	B8	B9	C1	C2	C4	C5	C6	C7	C8	C9	D1	D2	D3	D6	D7	D8
A1	0	1	0	47	9	19	0	2	0	16	3	13	17	21	16	20	17	42	21	28	12	6	9	12	9	13
A2		1	0	47	9	19	0	2	0	16	3	13	17	21	16	20	17	42	21	28	12	6	9	12	9	13
A3			1	44	10	20	1	3	1	13	4	12	14	18	13	17	14	39	18	25	9	5	8	9	8	14
A6				47	9	19	0	2	0	16	3	13	17	21	16	20	17	42	21	28	12	6	9	12	9	13
A7					48	44	47	45	47	23	48	24	34	46	31	43	34	59	46	29	41	45	42	41	42	56
A8						14	9	7	9	27	8	16	26	36	25	35	26	61	36	45	15	7	12	15	10	4
A9							19	17	19	31	22	20	32	44	31	43	32	75	44	55	23	17	20	23	18	16
B1								2	0	16	3	13	17	21	16	20	17	42	21	28	12	6	9	12	9	13
B2									2	18	5	11	19	23	18	22	19	48	23	34	10	4	7	10	7	11
B6										16	3	13	17	21	16	20	17	42	21	28	12	6	9	12	9	13
B7											19	13	7	11	6	10	7	26	11	14	16	18	17	16	15	37
B8												14	20	24	19	23	20	41	24	31	17	9	14	17	12	12
B9													20	24	17	21	20	43	24	33	13	11	12	13	14	18
C1														4	1	5	0	17	4	21	13	15	14	13	12	36
C2															5	1	4	13	0	25	17	19	18	17	16	46
C4																4	1	18	5	20	12	14	13	12	11	35
C5																	5	14	1	24	16	18	17	16	15	45
C6																		17	4	21	13	15	14	13	12	36
C7																			13	28	46	46	47	46	43	73
C8																				25	17	19	18	17	16	46
C9																					36	36	37	36	33	5
D1																						4	1	0	7	19
D2																							3	4	3	11
D3																								1	6	16
D6																									7	19
D7																										18
D8																										

D9	E1	E2	E6	E7	E8	E9	F1	F2	F4	F5	F6	F7	F8	F9	H1	H2	H3	H4	H5	H6
15	8	9	10	10	10	8	10	11	11	4	11	10	10	12	13	26	21	13	18	32
15	8	9	10	10	10	8	10	11	11	4	11	10	10	12	13	26	21	13	18	32
16	7	8	9	9	11	7	9	10	10	3	10	9	9	11	12	25	22	14	17	31
15	8	9	10	10	10	8	10	11	11	4	11	10	10	12	13	26	21	13	18	32
42	39	42	37	41	29	39	41	42	42	41	42	41	41	37	46	53	38	42	27	37
8	9	10	13	9	17	9	9	10	10	13	10	9	9	15	16	41	26	8	25	41
6	19	16	23	19	25	11	19	20	20	21	20	19	19	15	16	35	46	14	41	65
15	8	9	10	10	10	8	10	11	11	4	11	10	10	12	13	26	21	13	18	32
13	6	7	8	8	12	6	8	9	9	6	9	8	8	10	11	30	21	11	20	34
15	8	9	10	10	10	8	10	11	11	4	11	10	10	12	13	26	21	13	18	32
27	16	17	14	16	16	16	16	17	17	10	17	16	16	14	19	36	29	25	18	28
16	9	12	13	11	9	11	11	12	12	7	12	11	11	17	18	29	20	12	17	29
16	11	12	13	13	11	9	13	14	14	15	14	13	13	13	18	33	20	12	17	29
26	17	14	15	15	19	17	15	14	14	11	14	15	15	13	12	39	42	32	25	41
38	21	18	19	19	23	21	19	18	18	15	18	19	19	17	16	41	48	42	27	43
25	16	13	14	14	18	16	14	13	13	10	13	14	14	12	13	36	39	31	24	40
37	20	17	18	18	22	20	18	17	17	14	17	18	18	16	17	38	45	41	26	42
26	17	14	15	15	19	17	15	14	14	11	14	15	15	13	12	39	42	32	25	41
65	48	47	46	46	36	50	46	45	45	36	45	46	46	47	62	63	67	40	50	
38	21	18	19	19	23	21	19	18	18	15	18	19	19	17	16	41	48	42	27	43
49	36	37	34	36	24	36	36	37	37	22	37	36	36	36	41	36	31	45	14	20
19	6	5	6	8	20	6	8	7	7	12	7	8	8	8	9	38	35	21	26	46
13	6	3	8	6	16	4	6	5	5	8	5	6	6	8	7	30	29	13	22	40
16	5	4	5	7	17	5	7	6	6	11	6	7	7	7	8	37	32	18	25	45
19	6	5	6	8	20	6	8	7	7	12	7	8	8	8	9	38	35	21	26	46
12	7	2	7	3	19	5	3	2	2	5	2	3	3	5	4	31	34	18	25	43
12	13	14	19	15	19	11	15	16	16	21	16	15	15	21	22	45	28	10	29	47
D9	13	10	15	11	21	9	11	12	12	15	12	11	11	11	12	37	34	10	35	53
	E1	5	2	4	14	4	4	5	5	8	5	4	4	8	11	38	25	15	22	38
		E2	5	3	17	3	3	2	2	7	2	3	3	5	4	31	32	16	25	45
			E6	4	16	6	4	5	5	8	5	4	4	6	11	40	27	19	24	40
				E7	16	4	0	1	1	6	1	0	0	4	7	36	31	19	24	42
					E8	14	16	17	17	14	17	16	16	18	23	26	19	19	10	26
						E9	4	5	5	8	5	4	4	4	7	30	29	13	22	42
							F1	1	1	6	1	0	0	4	7	36	31	19	24	42
								F2	0	7	0	1	1	5	6	35	34	20	25	45
									F4	7	0	1	1	5	6	35	34	20	25	45
										F5	7	6	6	8	9	26	27	19	20	34
											F6	1	1	5	6	35	34	20	25	45
												F7	0	4	7	36	31	19	24	42
													F8	4	7	36	31	19	24	42
														F9	5	32	37	21	28	48
															H1	33	44	22	33	53
																H2	49	41	32	52
																	H3	20	15	13
																		H4	25	33
																			H5	10

$$d_{P_k}(F_i, F_j) := \begin{cases} 0, & D_{i,k} = D_{j,k}, \\ 1, & \{D_{i,k}, D_{j,k}\} \in \{\{\circ, +\}, \{\circ, -\}\}, \\ 4, & \{D_{i,k}, D_{j,k}\} = \{+, -\}. \end{cases} \tag{12.36}$$

Aggregating the contributions of all texture pairs, set

$$d(F_i, F_j) := \sum_{k=1}^{36} d_{P_k}(F_i, F_j). \tag{12.37}$$

The resulting distances are listed in Table 12.8. To visualize these distances approximately, Figure 12.2 shows a 2D map in which each feature is assigned a point in the plane such that Euclidean distances between points represent approximately the distances from Table 12.8. Of course, not all distance relations can appropriately be satisfied within the 2D plane, particularly within the lower middle region where many points form a rather dense cluster.

From that clustering, it can be seen that, as mentioned earlier, many of the graph-based features across classes yield very similar results. The six Haralick features are

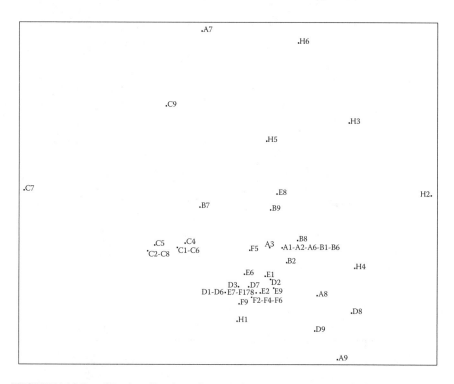

FIGURE 12.2: 2D visualization of approximate distances (12.37) between texture features. Labels such as D1-D6 collect features that are not separated (distance 0); E7-F178 abbreviates E7-F1-F7-F8.

fairly well separated in this map, indicating that they complement each other comparably well.

In order to establish a subset of the graph-based features whose members complement each other to give a high overall sensitivity of texture discrimination, one would choose only few features from the central clusters and include some of the features that lie particularly far away from these clusters. Once more, it is evident that most of these deviating and therefore distinctive candidates are based on the $\bar{I}_D^W$ (**x7**), I_{fV} (**x8**), and I_{fP} (**x9**) graph indices. Together with the findings about detection rates, this is a strong indication that these graph descriptors are particularly well suited for our texture analysis task.

12.5   Conclusion

The work presented in this chapter is the first exploration of an application of quantitative graph theory methods in image analysis. Focusing on the limited task of discriminating homogeneous texture patches, it led to promising results that are a starting point for further investigation. An important goal of this future work is to integrate the graph-based texture descriptors into segmentation frameworks based on active contour/active region models or clustering approaches. The intrinsic locality of the patch and amoeba graphs will be exploited in such a context much better.

A limitation to the evaluation consists in the fact that no large database but just a rather small collection of test images was considered so far, such that no statistical analysis could be performed. It is therefore a desiderate to analyze a suitable selection of graph-based features on larger test sets. We mention also that the exclusion of some further graph indices in this work, like $I_{f\triangle}$, was based on rather heuristic arguments that call for additional validation. Also, the values of several parameters in the computation of features were fixed in the experiments. This leaves the optimization of these parameters as a further source of improvement for further studies.

Notwithstanding these limitations, we believe that we have demonstrated the usefulness of graph descriptors in texture analysis. We hope that further work bringing together ideas from quantitative graph theory and image analysis will be beneficial for the development of both research areas.

References

1. J.-F. Aujol and A. Chambolle. Dual norms and image decomposition models. *International Journal of Computer Vision*, 63(1):85–104, 2005.

2. A.T. Balaban. Highly discriminating distance-based topological index. *Chemical Physics Letters*, 89:399–404, 1982.

3. D. Bonchev, O. Mekenyan, and N. Trinajstić. Isomer discrimination by topological information approach. *Journal of Computational Chemistry*, 2(2):127–148, 1981.

4. D. Bonchev and N. Trinajstić. Information theory, distance matrix, and molecular branching. *Journal of Chemical Physics*, 67(10):4517–4533, 1977.

5. Y. Boykov, O. Veksler, and R. Zabih. Markov random fields with efficient approximation. In *Proc. 1998 IEEE Computer Society Conference on Computer Vision and Pattern Recognition*, pp. 648–655, Santa Barbara, CA, June 1998.

6. U.M. Braga-Neto. Alternating sequential filters by adaptive neighborhood structuring functions. In P. Maragos, R.W. Schafer, and M.A. Butt, eds., *Mathematical Morphology and Its Applications to Image and Signal Processing*, volume 5 of Computational Imaging and Vision, pp. 139–146. Kluwer, Dordrecht, the Netherlands, 1996.

7. T. Brox, M. Rousson, R. Deriche, and J. Weickert. Colour, texture, and motion in level set based segmentation and tracking. *Image and Vision Computing*, 28:376–390, 2010.

8. A. Buades, B. Coll, and J.-M. Morel. A non-local algorithm for image denoising. In *Proc. 2005 IEEE Computer Society Conference on Computer Vision and Pattern Recognition*, vol. 2, pp. 60–65, San Diego, CA. IEEE Computer Society Press, Los Alamitos, CA, June 2005.

9. U. Clarenz, M. Rumpf, and A. Telea. Surface processing methods for point sets using finite elements. *Computers and Graphics*, 28:851–868, 2004.

10. A.D.J. Cross, R.C. Wilson, and E.R. Hancock. Inexact graph matching using genetic search. *Pattern Recognition*, 30(6):953–970, 1997.

11. M. Dehmer. Information processing in complex networks: Graph entropy and information functionals. *Applied Mathematics and Computation*, 201:82–94, 2008.

12. M. Dehmer and F. Emmert-Streib. Comparing large graphs efficiently by margins of feature vectors. *Applied Mathematics and Computation*, 188:1699–1710, 2007.

13. M. Dehmer, F. Emmert-Streib, and J. Kilian. A similarity measure for graphs with low computational complexity. *Applied Mathematics and Computation*, 182:447–459, 2006.

14. M. Dehmer, F. Emmert-Streib, and A. Mehler, eds. *Towards an Information Theory of Complex Networks: Statistical Methods and Applications*. Birkhäuser Publishing, Basel, Switzerland, 2012.

15. M. Dehmer, F. Emmert-Streib, and S. Tripathi. Large-scale evaluation of molecular descriptors by means of clustering. *PloS ONE*, 8(12):e83956, 2013.

16. M. Dehmer, M. Grabner, and K. Varmuza. Information indices with high discriminative power for graphs. *PLoS ONE*, 7(2):e31214, 2012.

17. E.W. Dijkstra. A note on two problems in connexion with graphs. *Numerische Mathematik*, 1:269–271, 1959.

18. A. Elmoataz, O. Lezoray, V.-T. Ta, and S. Bougleux. Partial difference equations on graphs for local and nonlocal image processing. In O. Lezoray and L. Grady, eds., *Image Processing and Analysis with Graphs: Theory and Practice*, chapter 7, pp. 174–206. CRC Press, Boca Raton, FL, 2012.

19. F. Emmert-Streib and M. Dehmer. Topological mappings between graphs, trees and generalized trees. *Applied Mathematics and Computation*, 186:1326–1333, 2007.

20. F. Emmert-Streib and M. Dehmer. Exploring statistical and population aspects of network complexity. *PLoS ONE*, 7(5):e34523, 2012.

21. M. Ferrer and H. Bunke. Graph edit distance—Theory, algorithms, and applications. In O. Lezoray and L. Grady, eds., *Image Processing and Analysis with Graphs: Theory and Practice*, chapter 13, pp. 383–422. CRC Press, Boca Raton, FL, 2012.

22. B. Furtula, I. Gutman, and M. Dehmer. On structure-sensitivity of degree-based topological indices. *Applied Mathematics and Computation*, 219:8973–8978, 2013.

23. D. Gabor. Theory of communication. *Journal of the Institution of Electrical Engineers*, 93:429–457, 1946.

24. B. Georgescu, I. Shimshoni, and P. Meer. Mean shift based clustering in high dimensions: A texture classification example. In *Proc. 2003 IEEE International Conference on Computer Vision*, vol. 1, pp. 456–463, Nice, France, October 2003.

25. R.M. Haralick. Statistical and structural approaches to texture. *Proceedings of the IEEE*, 67(5):786–804, May 1979.

26. R.M. Haralick, K. Shanmugam, and I. Dinstein. Textural features for image classification. *IEEE Transactions on Systems, Man, and Cybernetics*, 3(6):610–621, November 1973.

27. H.J.A.M. Heijmans. *Morphological Image Operators*. Academic Press, Boston, MA, 1994.

28. H. Hosoya. Topological index: A newly proposed quantity characterizing the topological nature of structural isomers of saturated hydrocarbons. *Bulletin of the Chemical Society of Japan*, 44(9):2332–2339, 1971.

29. P. Howarth and S. Rüger. Evaluation of texture features for content-based image retrieval. In P. Enser, Y. Kompatsiaris, N.E. O'Connor, A.F. Smeaton, and A.W.M. Smeulders, eds., *Image and Video Retrieval*, volume 3115 of Lecture Notes in Computer Science, pp. 326–334. Springer, Berlin, Germany, 2004.

30. K. Huang and R.F. Murphy. Automated classification of subcellular patterns in multicell images without segmentation into single cells. In *Proc. 2004 IEEE International Symposium on Biomedical Imaging*, vol. 2, pp. 1139–1142, Arlington, VA, April 2004.

31. H. Ishikawa. Graph cuts—Combinatorial optimization in vision. In O. Lezoray and L. Grady, eds., *Image Processing and Analysis with Graphs: Theory and Practice*, chapter 2, pp. 25–63. CRC Press, Boca Raton, FL, 2012.

32. H. Ishikawa and D. Geiger. Segmentation by grouping junctions. In *Proc. 1998 IEEE Computer Society Conference on Computer Vision and Pattern Recognition*, pp. 125–131, Santa Barbara, CA, June 1998.

33. O. Ivanciuc, T.-S. Balaban, and A.T. Balaban. Design of topological indices. Part 4. Reciprocal distance matrix, related local vertex invariants and topological indices. *Journal of Mathematical Chemistry*, 12(1):309–318, 1993.

34. G. Lendaris and G. Stanley. Diffraction pattern sampling for automatic pattern recognition. *Proceedings of the IEEE*, 58(2):198–216, 1970.

35. R. Lerallut, E. Decencière, and F. Meyer. Image processing using morphological amoebas. In C. Ronse, L. Najman, and E. Decencière, eds., *Mathematical Morphology: 40 Years On*, volume 30 of Computational Imaging and Vision, pp. 13–22. Springer, Dordrecht, the Netherlands, 2005.

36. R. Lerallut, E. Decencière, and F. Meyer. Image filtering using morphological amoebas. *Image and Vision Computing*, 25(4):395–404, 2007.

37. O. Lezoray and L. Grady. Graph theory concepts and definitions used in image processing and analysis. In O. Lezoray and L. Grady, eds., *Image Processing and Analysis with Graphs: Theory and Practice*, chapter 1, pp. 1–24. CRC Press, Boca Raton, FL, 2012.

38. O. Lezoray and L. Grady, eds. *Image Processing and Analysis with Graphs: Theory and Practice*. CRC Press, Boca Raton, FL, 2012.

39. S.Y. Lu and K.S. Fu. A syntactic approach to texture analysis. *Computer Graphics and Image Processing*, 7:303–330, 1978.

40. G. Matheron. *Eléments pour une théorie des milieux poreux*. Masson, Paris, France, 1967.

41. L. Najman and F. Meyer. A short tour of mathematical morphology on edge and vertex weighted graphs. In O. Lezoray and L. Grady, eds., *Image Processing and Analysis with Graphs: Theory and Practice*, chapter 6, pp. 141–173. CRC Press, Boca Raton, FL, 2012.

42. S. Osher, A. Solé, and L. Vese. Image decomposition and restoration using total variation minimization and the H^{-1} norm. *Multiscale Modeling and Simulation*, 1(3):349–370, 2003.

43. N. Paragios and R. Deriche. Geodesic active regions: A new paradigm to deal with frame partition problems in computer vision. *Journal of Visual Communication and Image Representation*, 13(1/2):249–268, 2002.

44. R. Picard, C. Graczyk, S. Mann, J. Wachman, L. Picard, and L. Campbell. Vistex database. Online resource, http://vismod.media.mit.edu/vismod/imagery/VisionTexture/vistex.html, 1995. Accessed on November 20, 2013.

45. J.R. Platt. Prediction of isomeric differences in paraffin properties. *Journal of Physical Chemistry*, 56(3):328–336, 1952.

46. D. Plavšić, S. Nikolić, and N. Trinajstić. On the Harary index for the characterization of chemical graphs. *Journal of Mathematical Chemistry*, 12(1):235–250, 1993.

47. K. Riesen and H. Bunke. Approximate graph edit distance computation by means of bipartite graph matching. *Image and Vision Computing*, 27:950–959, 2009.

48. A. Rosenfeld and M. Thurston. Edge and curve detection for visual scene analysis. *IEEE Transactions on Computers*, 20(5):562–569, 1971.

49. S. Roy and I. Cox. Maximum-flow formulation of the *n*-camera stereo correspondence problem. In *Proc. 1998 IEEE International Conference on Computer Vision*, pp. 492–499, Mumbai, India, January 1998.

50. C. Sagiv, N.A. Sochen, and Y.Y. Zeevi. Integrated active contours for texture segmentation. *IEEE Transactions on Image Processing*, 15(6):1633–1646, June 2006.

51. B. Sandberg, T. Chan, and L. Vese. A level-set and Gabor-based active contour algorithm for segmenting textured images. Technical Report CAM-02-39, Department of Mathematics, University of California at Los Angeles, CA, July 2002.

52. A. Sanfeliu and K.-S. Fu. A distance measure between attributed relational graphs for pattern recognition. *IEEE Transactions on Systems, Man, and Cybernetics*, 13(3):353–362, 1983.

53. J. Serra. *Image Analysis and Mathematical Morphology*, vol. 1. Academic Press, London, U.K., 1982.

54. J. Serra. *Image Analysis and Mathematical Morphology*, vol. 2. Academic Press, London, U.K., 1988.

55. C.E. Shannon. A mathematical theory of communication. *Bell System Technical Journal*, 27:379–423, 623–656, 1948.

56. J. Shi and J. Malik. Normalized cuts and image segmentation. *IEEE Transactions on Pattern Analysis and Machine Intelligence*, 22(8):888–906, 2000.

57. R.N. Sutton and E.L. Hall. Texture measures for automatic classification of pulmonary disease. *IEEE Transactions on Computers*, 21(7):667–676, 1972.

58. L. Tesař, A. Shimizu, D. Smutek, H. Kobatake, and S. Nawano. Medical image analysis of 3D CT images based on extension of Haralick texture features. *Computerized Medical Imaging and Graphics*, 32(6):513–520, 2008.

59. J.G. Verly and R.L. Delanoy. Adaptive mathematical morphology for range imagery. *IEEE Transactions on Image Processing*, 2(2):272–275, 1993.

60. J.T.L. Wang, K. Zhang, and G.-W. Chen. Algorithms for approximate graph matching. *Information Sciences*, 82:45–74, 1995.

61. M. Welk. Analysis of amoeba active contours. Technical Report cs:1310.0097, arXiv.org, 2013.

62. M. Welk and M. Breuß. Morphological amoebas and partial differential equations. In *Advances in Imaging and Electron Physics*, P. Hawkes, (ed.), Elsevier, vol. 185, in press.

63. M. Welk, M. Breuß, and O. Vogel. Morphological amoebas are self-snakes. *Journal of Mathematical Imaging and Vision*, 39:87–99, 2011.

64. H. Wiener. Structural determination of paraffin boiling points. *Journal of the American Chemical Society*, 69(1):17–20, 1947.

65. L. Zhu, W.K. Ng, and S. Han. Classifying graphs using theoretical metrics: A study of feasibility. In J. Xu, G. Yu, S. Zhou, and R. Unland, eds., *Database Systems for Advanced Applications*, volume 6637 of Lecture Notes in Computer Science, pp. 53–64. Springer, Berlin, Germany, 2011.

66. S. Zucker. Toward a model of texture. *Computer Graphics and Image Processing*, 5:190–202, 1976.

Chapter 13

Network Analysis Applied to the Political Networks of Mexico

Philip A. Sinclair

Contents

13.1   Introduction...387
13.2   Definitions for Networks...389
13.3   Centrality and Centralization...390
13.4   Institutional Revolutionary Party and the Presidential Succession............390
13.5   Network Cohesion and Centralization 393
13.6   Sensitivity and Reliability of Scores 397
13.7   Networks and Decompositions..399
13.8   Conclusion...403
References..404

13.1   Introduction

The network paradigm has been found to be useful and indeed necessary to the study of power in society [13]. The aim of the present chapter is to present and discuss the application of node centrality and network centralization to the analysis of the political networks of Mexico, as described in publications by Jorge Gil and Samuel Schmidt and their collaborators, [14–18] (see also [12]). Brass and Krackhardt, in [4], provide an introduction to methodological issues and analytical techniques used in power research. They sum up the network philosophy:

> While personal attributes and strategies may have an important effect on power acquisition,...structure imposes the ultimate constraints on the individual.

While not an advocate of the network approach, Camp [8] makes it clear that it is not enough to just study the formal structures to understand important political processes such as recruitment and career progression in the Mexican political system. The Mexican political culture has unwritten rules and informal network structures through which politicians get together to form working groups.

At the National Autonomous University of Mexico (UNAM), the group led by Jorge Gil Mendieta* developed the REDMEX databases [12]; these contain information on all the politicians active in Mexico from the end of the Mexican Revolution to the end of the twentieth century. The following analyses about node centrality and network centralization are based around the political networks of the elite as described in the book, "Estudios sobre la Red Política de México" (ERPM) [18]. The 11 networks, each of which contains approximately 20 nodes, were generated from the REDMEX database cross sectionally at 5 yearly intervals from 1940 to 1990. The nodes represent the most powerful and influential politicians in Mexico and the edges a variety of relationships including political, family, friendship, and military subordination or some combination of these—one edge can indicate one or more relationships existing between the two politicians that are represented by the end nodes of the edge. An example, the 1965 network, is shown in Figure 13.1. In [17], Gil and Schmidt describe the *Mexican network of power* (MNP), which can be constructed by combining all 11 networks to form one connected network that shows the line of the presidents and how the presidents and other politicians exerted influence over time. The MNP is available as a Pajek data file [2]; there are differences between the set of eleven of networks and the whole connected network, some of which will be discussed later. Gil and Schmidt [18] make use of these networks and others to help to explain several

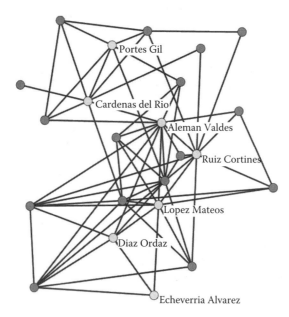

FIGURE 13.1: The 1965 Mexican political network of the elite, with presidents labeled.

* Professor Jorge Gil Mendieta (1937–2010).

interrelated aspects of the Mexican political system: its relative stability, the changing composition of the elite, and the transmission of power.

13.2 Definitions for Networks

In their analyses of the Mexican political elite, Gil and Schmidt calculate several different network parameters. In addition to the degree distributions of the networks, they also calculate the distributions for geodesic distances and cliques. Since politicians need to work in cliques to achieve and maintain power, much of the analyses of the political networks by Gil and Schmidt has been to do with the existence of cliques and their continuity and development over time. The networks have diameter greater than 2, perhaps reflecting the competition between different groups in the networks. Table 13.1 gives some parameter values for the 11 political networks of the elite.

Notation and basic network terms are as follows. The number of nodes and edges in a network $G = G(V, E)$ will be denoted $p = |V|$ and $q = |E|$, respectively. The networks are simple and undirected, and the *density*, δ, of the network is equal to the proportion of edges in the network. Let $d(u)$ denote the *degree* of a node u, and let Δ denote the *maximum degree* in the network. A *subnetwork* $H \subseteq G$ is a network $H = H(V', E')$ such that $V' \subseteq V$, $E' \subseteq E$ and each $e \in E'$ has end nodes in V'. A *clique* is a maximal subnetwork of order at least three in which every pair of nodes is adjacent. Let K denote the order of the largest clique. Let $d(u, v)$ denote the *distance* or length of a shortest path between the nodes u and v. The maximum distance over all pairs of nodes in a connected network is called the *diameter* of the network and is denoted Γ. For a node u, let $p_j(u)$ be the number of nodes at distance j from u in G, and let $e(u) = \max \{d(u, v) \text{ for } v \in V(G)\}$ be the *eccentricity* of u. The term *random network* refers to an instance of a network chosen at random from a set of networks such that all networks have an equal probability of being selected. The *set of networks* is usually all simple connected networks with given order and density (p and q fixed).

TABLE 13.1: Some Parameter Values for the Political Networks

Year	p	q	δ	Δ	K	Γ
1940	19	53	0.310	14	6	3
1945	19	49	0.287	13	5	3
1950	24	67	0.243	13	5	3
1955	24	69	0.250	13	5	3
1960	25	75	0.250	13	5	4
1965	23	65	0.257	12	5	4
1970	21	50	0.238	11	5	5
1975	18	40	0.261	10	5	5
1980	14	30	0.330	9	4	4
1985	13	29	0.372	9	4	3
1990	11	22	0.400	8	4	4

13.3   Centrality and Centralization

The basic idea for a centrality index is a function to assign a value to each node in a network that in some sense reflects the *position* of the node in the network. The meaning of the term position depends on the context, and often, there is a role or status that might be inferred. One rational for using a centrality index is that the connections that politicians have to others allow for the flow of resources and information. Thus, an influential politician will, during his career, build up many relationships that enable him to go about his business, fight battles, gain and give favors, and so forth. For Gil and Schmidt [18], a plausible use for a centrality index was to develop a heuristic that would help to describe the presidential succession in Mexico; the candidate with the most connections within the political elite would be the one most likely to be nominated to be the next president.

The simplest of the centrality indices is degree centrality, which assigns to each node a count of the number of edges incident to a node. However, degree centrality does not really reflect the basic tenet of network analysis, namely, that what is being measured is not just a local phenomenon but a network one. Influence is not understood to stop with those directly connected to a politician, but spreads further throughout the network. The power index, introduced by Gil and Schmidt [14], generalizes the degree centrality index by including a weighted count of indirect contacts. The contribution made to the index by nodes connected indirectly is weighted according to their distance from the node being indexed—closer contacts being more valuable.

Let G be a connected network with $p > 1$. For $u_j \in V(G)$, define

$$I(u_j) = \frac{1}{p-1} \sum_{u_k \in V(G)} \frac{1}{d(u_j, u_k)}$$

$I(u_j)$ is then the *harmonized* sum of the shortest distances from u_j to each other node in G. The term $1/(p-1)$ normalizes the index so that $0 < I(u_j) \leq 1$. In the case of a disconnected network, let G be the component that u_j belongs to; nodes in other components do not affect the value of $I(u_j)$. For $p = 1$, put $I(u_j) = 0$.

In general, care must be taken when comparing the centrality of nodes in different networks or in different components of the same network. To take an extreme example, since the index is normalized, any politician that belongs to a component that is a clique will have centrality value one, and this is regardless of the size of the clique. Typically, the larger the network, the greater the potential for power that members of the network have.

13.4   Institutional Revolutionary Party and the Presidential Succession

The party that was in power during the period 1940–1990 (and through to 2000) was called the Institutional Revolution Party (PRI). During the Mexican Revolution

TABLE 13.2:   Presidents: Their Centrality Ranks in Network and Office prior to Presidency

Year	President (Sexenio)	Rank	Office
1940	Lázaro Cárdenas del Rio (1934–1940)	1	National defense
1945	Manuel Ávila Camacho (1940–1946)	2	National defense
1950	Miguel Alemán Valdés (1946–1952)	2	Government
1955	Adolfo Ruiz Cortines (1952–1958)	3	Government
1960	Adolfo López Mateos (1958–1964)	5	Employment
1965	Gustavo Diaz Ordaz (1964–1970)	10	Employment
1970	Luis Echeverria Álvarez (1970–1976)	19	Government
1975	Luis Echeverria Álvarez (1970–1976)	14	
1980	José López Portillo (1976–1982)	9	Treasury
1985	Miguel de la Madrid Hurtado (1982–1988)	6	Planning
1990	Carlos Salinas de Gortari (1988–1994)	5	Planning

(1910–1921), armies from disparate regions and under different generals fought against the regime. The party was created in 1929 after many years of regional conflicts, its purpose being to resolve conflicts, to distribute wealth, and to reach compromises through a centralized institution [9]. The president is the most powerful actor in the system; however, the 1917 Constitution limits any president to just one 6-year term or *sexenio*. This means that the presidential succession is a key political process. Commentators make it clear that the incumbent president designates his successor, or equivalently the PRI nominee [8]. However, there is a competition within the PRI between presidential candidates, and there is often thought to be some compromise between the competing groups and ideologies. In [15], Gil and Schmidt propose that the presidential succession can be best understood from a network perspective, through the identification of significant relationships that individual politicians have built up during their career.

Table 13.2 gives the centrality rank of each president who was incumbent during the year of each of the 11 networks. It should be understood that the 11 networks are part of the *MNP* and that they are created in the context of previous and future developments. For example, in [18], Gil and Schmidt consider the transition from Alemán to Cortines in some depth, and although Cortines has a high centrality score in the 1950 network, he was not at the time the most likely successor to Alemán. Earlier presidents remain in the networks after the end of their sexenio and continue to exert influence. In Figure 13.2, the centrality scores of the presidents are plotted over time—the stars indicate when a particular president was in office. It can be seen that Alemán, for example, achieved high centrality scores to 1980 long after his sexenio. However, even allowing for the relatively high centrality of early presidents in the networks, it is clear that around the 1970s, the centrality score of a politician is no longer a good indicator of whether he was likely to be the next president. Echeverría (president 1970–1976) has rank 19 out of 21 in the 1970 network that is rather low, even allowing for the four ex-presidents in the network. The idea that something might be amiss is reflected in the presidential succession at this time:

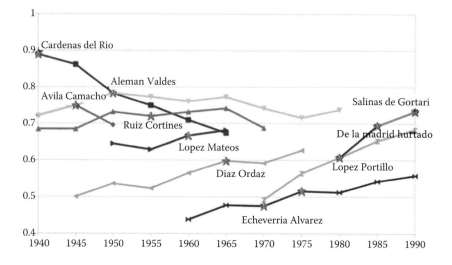

FIGURE 13.2:   The centrality of the Presidents in network by year.

> The choice of Lopez Portillo surprised almost all observers, as he was the
> darkest horse in the race for the presidency. His opposition included some
> formidable politicians, each with a solid base of support in one of the wings
> of the party. ([9], pp. 153–154)

In [18], it is suggested that the lower centrality scores might indicate some kind of
weakening of the function of the president. An idea as to what was happening can be
had by considering the competition within the network at that time. In [15], Gil and
Schmidt use a structural block modelling algorithm to identify two groups within the
MNP, which they name the *politicos* and *financiers*. The 1970s represent a transition
period between these two groups in that up until 1976 the presidents were from the
politico block and from 1982 onwards the presidents were from the financier block.
Other authors have also suggested the existence of two main camps within the PRI;
in [9], Centeno argues that prior to the 1970s, two ministries, the Ministry of Government and the Ministry of Treasury, acted like vice presidencies and limited the
president's power somewhat. An important figure was Ortiz Mena who was never
elected president, but who was the secretary of the Ministry of Treasury from 1958
to 1970 and a member of seven of the eleven networks. Between 1946 and 1976 prior
to ascending to the presidency each president was either a secretary of the Ministry
of Government or of the Ministry of Employment (see Table 13.2). From the 1982
election, all presidents have a financial background either in the Ministry of Treasury or in the Ministry of Planning. The Ministry of Planning came into being in the
1970s when structural changes were made to the government with the intention of
strengthening the position of the president [9].

The first of the political elite were the generals of the Revolution, and even if
an early president such as Aléman did not have a military background, he still had
close connections to the Revolution, for example, via his mentor, General Aguilar.

Over time, political recruitment tended to move away from the military to the universities, and important connections and networks often originated at the UNAM and later foreign universities such as Harvard. However, through the network of power, the political elite kept close to the original generals of the Revolution. Later, presidents are still part of a cohesive and connected network of the elite. In a detailed longitudinal analysis of the elite network around Salinas (president 1988–1994), Gil and Schmidt [18] conclude that the reason Salinas became president was not because he was a financier, but because he belonged to the network of the power elite.

13.5   Network Cohesion and Centralization

A large body of research in social networks has focused on how macrostructures, such as hierarchy, clustering, or centralization, might have arisen in a network through individual actors following simple rules of behavior, for example, in friendship formation. On the other hand, the structure of a network can act to either allow or prohibit processes that take place within the network. For example, in their study [21] on the rise of the Medici in Renaissance Florence, Padgett and Ansell appeal, at least in part, to the centralized structure of the network built around the Medici family group to explain the effectiveness of the Medici party action. For Mexico, centralization was important to the development of the political system after the Revolution, and the tendency over time has been for the formal government structure to become increasingly hierarchical. In [9], Centeno explains that the creation of the political party and centralization

> …freed the population from the arbitrary control of local bosses, but it also led to an increasing monopolization of power in the hands of a relatively small political elite.

In comparison to other Latin American countries in the twentieth century, Mexico is seen as being remarkably stable; the PRI held power for 71 years in total. In [18], Gil and Schmidt argue that this stability is best explained from a network perspective. The political elite formed a strong and cohesive network, but a network that was also developing and changing over time. One way in which to describe the differences between the 11 networks is to calculate and compare parameters for the networks, in particular the centralization index.

A *network centralization index* can be defined as follows. For a node centrality index, F, and a connected network of order p,

$$F(G) = \frac{\sum_j \{F(u_1) - F(u_j)\}}{\max \sum_j \{F(u_1) - F(u_j)\}} \tag{13.1}$$

where j runs through $1 \ldots p$ and u_1 is a node that achieves maximum centrality value. The denominator acts to normalize $F(G)$ so that values lie between 0 and 1; the maximum is usually taken with reference to all connected networks on p nodes. In [24], the following result is proved.

TABLE 13.3:   Maximum Centrality and Centralization
Values for the Networks

Year	p	δ	$I(u_1)$	$I(G)$
1940	19	0.310	0.889	0.560
1945	19	0.287	0.861	0.535
1950	24	0.243	0.783	0.412
1955	24	0.250	0.783	0.410
1960	25	0.250	0.771	0.394
1965	23	0.257	0.773	0.392
1970	21	0.238	0.767	0.445
1975	18	0.261	0.784	0.465
1980	14	0.330	0.846	0.502
1985	13	0.372	0.875	0.503
1990	11	0.400	0.883	0.511

Theorem 13.5.1 *For a connected network on p nodes,*

$$\sum_k \{I(u_1) - I(u_k)\} \leq \frac{p-2}{2}$$

The maximum is obtained uniquely by the star network on p nodes.

The numerator of Equation 13.1 can be rewritten as

$$\sum_k \{F(u_1) - F(u_k)\} = n \times \left(F(u_1) - \overline{F}\right)$$

where $\overline{F}$ is the mean centrality score. In [15], the *maximum centrality* score is used as a network index that measures the level of network cohesion. Table 13.3 gives the maximum centrality and centralization scores for the 11 networks, as presented in [18].

Methodologists have argued that when centralization and other global network indices are normalized, the network density should be taken into account [6]. A value that can typically be considered a high maximum centrality or network centralization score varies with both the order and the density of the networks. The plots in Figure 13.3 show how the distribution of the maximum centrality and the network centralization scores vary with density for random networks. Each point represents the index value of a random network sampled from the set of all simple connected networks on 20 nodes with density as indicated. The impression is that the range of values realized by density is restricted for both indices and therefore that interpretation should be made in the context of both the order and density of the network. In [6], Butts proves that network degree centralization values are restricted to about half of the density by centralization region for networks on 20 nodes.

In order to decide whether the values of the maximum node centrality and the network centralization indices calculated for the Mexican political networks are high or

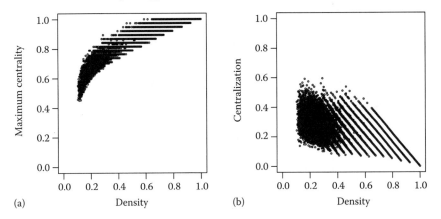

(a)   Density   (b)   Density

FIGURE 13.3: Samples of simple connected random networks on 20 nodes.

not, the distribution of these scores was simulated for random networks. The box plots of Figure 13.4 show the distributions of the values for the indices for random networks generated from the set of simple connected networks with order and density equal to that of the empirical network for the year given. In the first box plot, the maximum node centrality scores for the Mexican political networks are indicated by a diamond, $\diamond$. There is some interesting variation from year to year; note, for instance, that the fairly high centrality score for the maximum centrality for the political network in 1960 would be an average score in 1980. The conclusion would appear to be that for all networks, there is a high maximum node centrality score, which suggests that the networks are cohesive, and this could be a reason for the robustness of the elite. In the second plot, the $\diamond$s indicate the value of the network centralization index for each of the Mexican political networks. All the networks have a high level of centralization that suggests that they are hierarchical.

As already discussed, the 1960s and 1970s might represent some kind of transition period between competing groups of politicos and financiers within the network. From the late 1970s, the PRI faced severe problems, left wing factions split away from the party to form opposition parties, and local elections were lost to another opposition party, the National Action Party (PAN), which went on to win the 2000 Presidential election. Thus, although the indices increase after the 1980s, there is an underlying movement toward a multiparty system.

A criticism of using random network distributions for the previous type of analysis is that political networks may have values for certain parameters that may not be common in the population of random networks sampled from; for example, in the set of simple connected networks with a given order and density, there may be a high proportion of networks with diameter two, but in a more restricted population of political networks, networks with diameter two may not be plausible. However, unless there is a good theory that says why this is the case for each such parameter, further modelling could lead to spurious conclusions.

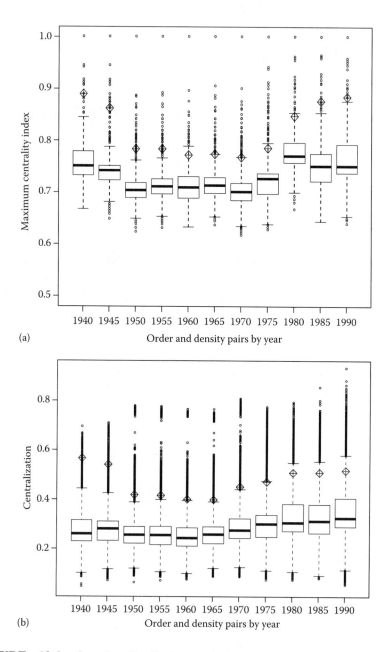

(a)

(b)

FIGURE 13.4: Sample distributions of index values for order–density combinations.

13.6   Sensitivity and Reliability of Scores

It is clear that a substantial amount of effort went into obtaining accurate information about the networks discussed in the published works of Gil and Schmidt and their collaborators. Nevertheless, there are well-known problems in generating political networks: defining boundaries, measuring relationships, and so forth. In an interesting article [22], Schmidt discusses at some length the practical problems of trying to obtain accurate information from which the political networks were created. The main source of data for the REDMEX databases is the *Diccionario Biográfico del Gobierno Mexicano* [18], and while some ties are unambiguous, for example, family relationships, other ties are implied by joint membership of a group. In the latter case, it is not always clear what the nature of such ties is. Camp [7] explains that a problem with studies that just make use of formal networks is that the important ties are often informal ones. For example, with company director board interlocks, individuals are assumed to network because they sit on the same company board; however, the important ties are often friendships that existed before a member joined the board. Further sources of information for the networks were interviews with politicians and political bibliographies. When a network is not too large, the detailed information obtained from such sources can be used to check the accuracy of the network. With the longitudinal nature of the data, another issue is the length of time a tie can be thought to be in existence for, and this can depend, in part, on the type of tie. A tie arising through an administrative position, for example, may only be relevant while the politician is holding that position; in contrast, a family tie might be expected to be more permanent. Even with a degree of certainty as to the existence of a relationship at a particular time, social network research suggests that the utility of different relationships is dependent on the purpose that they are used for and the context in which they arise within the network [5,20].

Given some uncertainty about the networks, it is useful to be aware as to how sensitive or not an index is to changes in the network. One opportunity to observe such changes occurs when there is uncertainty as to the composition of the final network. As mentioned earlier, in the 1980s, a substantial number of politicians broke away from the PRI to form opposition parties, and one of these was Cuauhtémoc, the son of Cárdenas del Rio (president 1934–1940). In the MNP, Cuauhtémoc is part of a connected network of power with the politician Velasco also included in the network, whereas in the analysis of the 1989 and 1990 networks in ERPM, Cuauhtémoc is not part of a connected network of power, and Velasco is not included in those networks at all. The effect on the index values can be seen in Table 13.4.

Another technique that has been used to compare the sensitivity of node and network indices is to take networks, make random changes to them, and then gauge the effect. For example, in [3], Bolland investigates the effect of random changes in networks on centrality scores, and in [10], Costenbader and Valente calculate the degree to which centrality indices for empirical networks and samples from the networks are correlated. The interpretation of results from such simulation studies should be made with caution. The empirical study [27] found that the type of relationship and the density of the network can have an effect on the sensitivity of different centrality indices

TABLE 13.4:   Comparing Index Scores for Different Networks

Year	ERPM		MNP	
	$I(u_1)$	$I(G)$	$I(u_1)$	$I(G)$
1985	0.875	0.503	0.810	0.499
1990	0.883	0.511	0.806	0.494

to changes in the network. The findings in [26] further suggest that the effect of random changes on different networks, and for a variety of centrality indices, is very variable, depending on the characteristics of the network such as the level of clustering and the skewness of the degree distribution. A random change in a network might be completely unreasonable in real terms and destroy some characteristic of the network that exists because it is a political network.

An informal exploration using an igraph routine called rewire [11,25] was made. Rewire makes changes to the edge set by swapping pairs of edges, keeping the network simple. This approach is of interest because it allows the degree sequence to be fixed.* As might be expected, the maximum centrality does not show much variation, since the degree remains constant for the node with maximum centrality. In Figure 13.5, the variation in centralization score is shown for when just one rewire is made from the original network. There is a considerable range of values for some

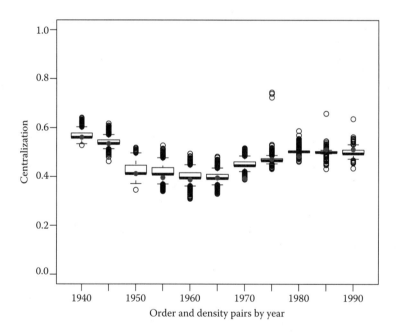

FIGURE 13.5:   Results from single rewire of political networks.

---

* Note that this routine does not enable us to take a random sample from the set of all networks with a given degree sequence.

networks, although the values are still high when compared to the sample distributions of Figure 13.4.

13.7   Networks and Decompositions

Often, network analysis involves the transformation of a network to create a new network that is easier to interpret, for example, in [21], a block modelling algorithm is used to reduce networks of 92 families of the Florentine political elite to ones with just 33 family blocks. For the Padgett and Ansell study, the transformation meant that only relevant information was included in the networks and the smaller networks are drawn so as to emphasize pertinent qualities. A further advantage to using small networks for political network analysis is that a detailed examination of the data can be made to try to ensure its accuracy.

The REDMEX III political database maintained at the Network Laboratory at the UNAM contains information on over 7000 members of the governments of Mexico [18]. Many networks can be derived from the database, including cross-sectional ones for given years that typically contain several hundred nodes. In the case of the *MNP* [15], a network has been extracted from an implicit larger network based on the perceived importance of the actors and their relationships. This technique of extracting a subnetwork can make use of different criteria, for example, subnetworks can be extracted according to the geographical region that politicians were active in, or their professional background [19]. A general description of the network approach is given in [1]. Following [1], a *cut* is defined as a subnetwork with values of a given property above a fixed threshold. For example, a node centrality index can be used to assign values to nodes, and then, the subnetwork induced by nodes with centrality value above a fixed cutoff point k can be extracted. When degree centrality is used to define a cut, the resulting network is often referred to as the k-core in the social network literature. There are many functions in use for assigning values to nodes, and several for assigning values to the edges of the network so that the extraction of a subnetwork can be made based on edge weight.

A fundamental relationship in Mexican society is that of the mentor and disciple [7]. Mentors are key to providing contacts in both formal and informal networks, and the majority of politicians start and progress in their career with the help of a patron or *godfather*. In Mexican politics, there exists a special type of informal network called a *camarilla*. In [8], Camp states that

> Mexican politics, from the postrevolutionary generation onwards has been built on the interrelationships between the camarillas.

Camarillas are hierarchical, based on loyalty and trust, and their members are often named after the political head of the camarilla, for example, alemanistas or cardenistas. As the career of a politician progresses, he or she can belong to several camarillas, and the potential for power and influence arises from belonging to several groups. Gil and Schmidt [18] assume that camarillas are fundamental to the presidential succession and that the ties in a camarilla can last for decades, for example, in the analysis of Alemán's camarilla, they find connections linking Alemán (presidency, 1946–1952) to Salinas (presidency, 1988–1994). The 1988 competition for the presidency is seen as a struggle between the alemanistas and cardenistas.

In [1], some decomposition procedures are described that make a contraction of one or more clusters of nodes in a network. First, a partition is made of all the nodes in the network into *color classes*. The partition might be obtained through some network algorithm designed to identify cohesive subgroups within the network or by using some node attribute. Second, a contraction of a color class can be made replacing all nodes of a particular color by a single node. The new node is then joined to each node outside the chosen color class that was joined to a node in the class. Define a *color class* C as a nonempty subset of nodes from a network, and define $E_C = \{e = uv \in E(G) | u, v \in C\}$ and $A_C = \{v \in V(G) | v \notin C, uv \in E(G), u \in C\}$. The *contraction* G/C is the network that is formed by replacing the nodes of C by a single node c and defining the new network G/C, as $V(G/C) = (V(G) - C) \cup \{c\}$ and $E(G/C) = (E(G) - E_C) \cup \{cv | v \in A_C\}$. A second color class can be contracted in the new network and so on. Depending on the number of color classes that are contracted, the resulting network will show more or less of the global structure of the network.

An example of a similar kind of decomposition that shows the clique/co-clique structure of the network was described in [23]. If we consider certain cross-sectional networks that can be created from the REDMEX database, then it is often found to be the case that there are large cliques of politicians. In the decomposition, these cliques are contracted; politicians that belong to two or more cliques are identified and form part of the reduced network. There are two ideas behind the decomposition: first, to create a smaller network so that the global structure of the network can be viewed using a network package such as Pajek, and, second, to reveal the brokerage type role that certain politicians might play. An example of such a decomposition is given in Figure 13.6, from [23]. The nodes x, y, and z are a separating set, and thus potentially, these politicians have a broker role, though of course they might not fully utilize such a role.

Let $\mathcal{C}(G) = \{C_1, \dots, C_s\}$ be the set of all cliques of G. Let $\mathcal{A}(G) = \{a_1, \dots, a_t\}$ be the set of all broker nodes, those that belong to more than one clique of $\mathcal{C}(G)$. Then, the *clique broker network* of G, J_A, is the bipartite network with $V(J_A) = \mathcal{C}(G) \cup \mathcal{A}(G)$ and $E(J_A) = \{a_k c_i$ iff $a_k \in V(C_i) \cap V(C_j), C_i, C_j \in \mathcal{C}(G)\}$, where $c_i \in V(J_A)$ is the node in $V(J_A)$ corresponding to clique C_i of G. In Theorem 2 and Theorem 3, proved in [23], G is a connected network with $V(G) = \cup_i V(C_i)$. The calculation of centrality for nodes from G using the network J_A is straightforward and allows for the comparison of centrality between different groups for such networks. For $u \in V(J_A)$, put $I'(u) = I(u) \times (p - 1)$.

Theorem 13.7.1 *Suppose that there exists $u \in V(C_1) \backslash \mathcal{A}(G)$, for $C_1 \in \mathcal{C}(G)$. If $e(c_1)$ is odd, then*

$$I'(u) = |V(C_1)| - 1 + \sum_{i=1}^{m} \frac{\sum_{d(c_1,c_j)=2i} \{|V(C_j)| - d(c_j)\} + p_{2i+1}(c_1)}{i+1}$$

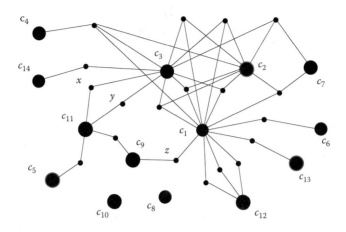

FIGURE 13.6: The clique broker network of the political network of Mexico for 1987.

where m = $\dfrac{e(c_1) - 1}{2}$.

If $e(c_1)$ is even, then

$$
I'(u) = |V(C_1)| - 1 + \sum_{i=1}^{m} \frac{\sum_{d(c_1,c_j)=2i}\left\{\left|V(C_j)\right| - d\left(c_j\right)\right\} + p_{2i+1}\left(c_1\right)}{i+1}
$$
$$
+ \frac{2\sum_{d(c_1,c_j)=e(c_1)}\left\{\left|V(C_j)\right| - d\left(c_j\right)\right\}}{e(c_1)+2}
\tag{13.2}
$$

where m = $\dfrac{e(c_1)}{2} - 1$.

Centrality scores for a nonbroker politician in each clique are given in Table 13.5; note that all nodes in clique C_2 belong to more than one clique and thus Theorem 2 is not applicable. An example calculation for $u \in V(C_1)$ is as follows: $e(c_1) = 6$ and therefore Equation 13.2 is used, with $m = 2$. From the network, $d(c_1, c_j) = 2$ for $j = 2, 3, 4, 6, 7, 9, 12, 13$, $d(c_1, c_j) = 4$ for $j = 11, 14$ and $d(c_1, c_j) = 6$ for $j = 5$.

$$
I'(u) = 65 + \frac{198}{2} + \frac{35}{3} + \frac{2 \times 16}{8}
$$
$$
= 179.6667
$$

Using Theorem 1, from [23],

$$
|V(G)| = \sum_{i} |V(C_i)| - \sum d(c_i) + |\mathcal{A}(G)| = 363 - 47 + 19 = 335
$$

which means that the order of the component that u belongs to is 315. Thus,

$$
I(u) = \frac{179.6667}{314} = 0.572.
$$

TABLE 13.5:   14 Cliques, Their Respective Orders,
Node Degree, and Centrality Scores for a Nonbroker Node

Clique	Order (n)	$d(c_i)$	$I(u)$
C_1	66	13	0.572
C_2	8	8	—
C_3	62	10	0.531
C_4	8	1	0.408
C_5	17	1	0.307
C_6	43	1	0.441
C_7	27	2	0.448
C_8	5	0	1.000
C_9	25	2	0.419
C_{10}	15	0	1.000
C_{11}	14	4	0.391
C_{12}	17	3	0.385
C_{13}	31	1	0.415
C_{14}	25	1	0.383

Similarly, the centrality scores can be calculated for individual broker nodes. Typically, broker nodes have higher centrality scores than nonbroker nodes within cliques since they have a higher number of direct ties.

Theorem 13.7.2 *Let $v \in \mathcal{A}(G)$.*
If $e(v)$ is even, put $m = \dfrac{e(v)}{2}$, then

$$I'(v) = \sum_{i=1}^{m} \frac{\sum_{d(v,c_j)=2i-1} \left\{ |V(C_j)| - d(c_j) \right\} + p_{2i}(v)}{i}$$

If $e(v)$ is odd, put $m = \dfrac{e(v) - 1}{2}$, then

$$I'(v) = \sum_{i=1}^{m} \frac{\sum_{d(v,c_j)=2i-1} \left\{ |V(C_j)| - d(c_j) \right\} + p_{2i}(v)}{i}$$

$$+ \frac{2\sum_{d(v,c_j)=e(v)} \left\{ |V(C_j)| - d(c_j) \right\}}{e(v) + 1} \tag{13.3}$$

An example calculation for the node, z, shown in the network of Figure 13.6 is as follows. Clearly, $d(z, c_j) = 1$ for $j = 1, 9$, $d(z, c_j) = 3$ for $j = 2, 3, 4, 6, 7, 11, 12, 13$, and $d(z, c_j) = 5$ for $j = 5, 14$. Then, the sum in Equation 13.3 of Theorem 3 is

$$I'(z) = 89 + \frac{185}{2} + \frac{40}{3} = 194\frac{5}{6}$$

so that

$$I(z) = \frac{194.833}{314} = 0.620.$$

TABLE 13.6: Centrality Values for Nodes in J_A

	c_1	c_2	c_3	c_4	c_5	c_6	c_7
$I(c_i)$	0.651	0.502	0.575	0.338	0.229	0.295	0.361
	c_8	c_9	c_{10}	c_{11}	c_{12}	c_{13}	c_{14}
$I(c_i)$	—	0.352	—	0.382	0.339	0.295	0.279

The network J_A can be thought of as an affiliation network with clique nodes forming one side of the node bipartition and broker nodes the other. The edges then indicate which cliques the broker nodes belong to. The information about the structure of the network can be incorporated into a centralization index for cliques, calculated from J_A as follows. Let $C(G)$ be labeled such that c_1 is a most central node of $C(G)$. The clique centralization index is calculated using

$$\sum_{c_i \in V(H_A) \cap C(G)} \{I(c_1) - I(c_i)\} \tag{13.4}$$

where H_A is the main component of Figure 13.6. The $I(c_i)$ values, calculated using (13.1), are given in Table 13.6.

For the network of Figure 13.6,

$$\sum_{c_i \in V(H_A) \cap C(G)} \{I(c_1) - I(c_i)\} = 3.214$$

In [24], a formula is given for the maximum value that the sum of (13.4) can obtain for connected bipartite networks with a given order in each partition. In terms of the network H_A, the theorem can be written as follows:

Theorem 13.7.3 *Let* $|C| = s$, $|A| = t$ *and* $s - 1 = (b-1)t + r$ *where* $b > 0$ *and* $r \geq 0$ *are both integers. Then,*

$$\sum_{c_i \in V(H_A)} \{I(c_1) - I(c_i)\} \leq \frac{1}{p-1} \left\{ (s-1)t + \frac{1}{2}(s-1)^2 \right.$$

$$-\frac{1}{12}(b-1)(t-r)(3b+3s+4t+2)$$

$$\left. -\frac{1}{12}rb(3b+3s+4t+5) \right\}$$

Thus, the maximum value for (13.4) is 5.317, and

$$I(H_A) = \frac{3.214}{5.317} = 0.605$$

13.8 Conclusion

Gil, Schmidt, and their collaborators have successfully developed and applied the power centrality index to several problems. In making calculations from the

REDMEX database, the centrality index is useful in describing the distribution of power among politicians. Through the extraction of the core of the political elite, they have been able to use the centrality index to throw light on the stability of the political system over time and the dynamics of the elite. The maximum centrality value, interpreted as an index of network cohesion and controlling for order and density, shows that the political elite is cohesive, even through what appears to be a major transition between competing groups in the 1970s.

The in-depth analysis made in [18] of the presidential successions has shown this to be a fascinating area of study, but that the centrality function can only give an indication of the state of affairs in a network. This partly reflects the fact that the incumbent president's rationale for his choice of successor was in some cases idiosyncratic. In [22], Schmidt has put forward proposals for a more sophisticated index that includes weights on both nodes and ties. The weights depend on individual factors, such as administrative position held and type of relationship that exists between politicians. Such weightings assume an underlying scale and can appear arbitrary, but perhaps, the utility of such modelling is the insight gained by attempting to identify and quantify the relevant factors. An alternative approach is to treat different types of tie as quantitatively different and then to incorporate them in a multiplex model.

Network decompositions are a useful technique in network analysis that make large networks more amenable to visualization and analysis. Cliques are fundamental to the structure of political networks, and being able to view both the clique and broker structure at the same time can provide insight into important groups and individual politicians.

References

1. V. Batagelj. Social network analysis, large-scale. In B. Meyers and J. Scott, eds., *Encyclopedia of Complexity and System Science*, pp. 355–375. Springer Verlag, New York, 2009.

2. V. Batagelj and A. Mrvar. Pajek data sets: `http://pajek.imfm.si/doku.php?id=data:index`, 2003. Accessed June 13, 2014.

3. J.M. Bolland. Sorting out centrality: An analysis of the performance of four centrality models in real and simulated networks. *Social Networks*, 10:233–253, 1988.

4. D.J. Brass and D.M. Krackhardt. Power, politics and social networks in organizations. In G.R. Ferris and D.C. Treadway, eds., *Politics in Organizations Theory and Research Considerations*, pp. 355–375. Routledge, New York, 2012.

5. R.S. Burt. *Structural Holes: The Social Structure of Competition*. Harvard University Press, Cambridge, MA, 1992.

6. C.T. Butts. Exact bounds for degree centralization. *Social Networks*, 28:283–296, 2006.

7. R.A. Camp. *Mexico's Mandarins. Crafting a Power Elite for the Twenty-First Century.* University of California Press, Berkeley, CA, 2002.

8. R.A. Camp. *Politics in Mexico. The Democratic Consolidation*, 5th edn. Oxford University Press, New York, 2007.

9. M.A. Centeno. *Democracy Within Reason. Technocratic Revolution in Mexico*, 2nd edn. The Pennsylvania State University Press, University Park, PA, 1997.

10. E. Costenbader and T.W. Valente. The stability of centrality measures when networks are sampled. *Social Networks*, 25:283–307, 2003.

11. G. Csárdi and T. Nepusz. igraph library: `http://igraph.sourceforge.net`, 2003. Accessed June 13, 2014.

12. Universidad Nacional Autónoma de México. Laboratorio de redes. `http://harary.iimas.unam.mx/publicaciones.php`, 2005–2013. Accessed June 13, 2014.

13. G.W. Domhoff. WhoRulesAmerica.net Power, politics and social change. `http://www2.ucsc.edu/whorulesamerica`, 2012. Accessed June 13, 2014.

14. J. Gil and S. Schmidt. The origin of the Mexican network of power. *International Social Network Conference*, pp. 22–25, San Diego, CA, 1996.

15. J. Gil and S. Schmidt. The political network in Mexico. *Social Networks*, (18):355–381, 1996.

16. J. Gil and S. Schmidt. *La Red Política en México: modelación y análisis por medio de la teoria de gráficas.* IIMAS, UNAM, México City, México, 1999.

17. J. Gil, S. Schmidt, J. Castro, and A. Ruiz. A dynamic analysis of the mexican network of power. *Connections*, 20(2):34–55.

18. J. Gil, S. Schmidt, and in collaboration with A. Ruiz. *Estudios sobre la red política de México.* IIMAS, UNAM, México City, México, 2005.

19. J. Gil Mendieta and L. Lomnitz Adler. Networks of lawyers and economists and the introduction of the new liberal policies in Mexico. *Conferecia International y V Europea de Análisis de Redes Sociales*, 17(7), 1999.

20. M. Granovetter. The strength of weak ties. *American Journal of Sociology*, 78(6):1360–1380, 1973.

21. J.F. Padgett and C.K. Ansell. Robust action and rise of the Medici, 1400–1434. *American Journal of Sociology*, 98(6):1259–1319, 1993.

22. S. Schmidt. La dificultad de medir. *REDES—Revista hispana para el analisis de redes sociales*, 17(7):163–183, 2006.

23. P.A. Sinclair. A representation for the Mexican political networks. *Social Networks*, 29:81–92, 2007.

24. P.A. Sinclair. Network centralization with the Gil Schmidt power centrality index. *Social Networks*, 31(3):214–219, 2009.

25. R Development Core Team. R: A language and environment for statistical computing. R foundation for statistical computing, Vienna, Austria. version 2.10.1. `http://www.r-project.org/`, 2009. Accessed June 13, 2014.

26. D.J. Wang, X. Shi, D.A. McFarland, and J. Leskovec. Measurement error in network data: A re-classification. *Social Networks*, 34:396–409, 2012.

27. B. Zemlič and V. Hlebec. Reliability of measures of centrality and prominence. *Social Networks*, 27:73–88, 2005.

Chapter 14

Social Network Centrality, Movement Identification, and the Participation of Individuals in a Social Movement: The Case of the Canadian Environmental Movement

David B. Tindall, Joanna L. Robinson, and Mark C.J. Stoddart

Contents

14.1   Introduction..407
14.2   Literature...408
      14.2.1   Social Network Analysis...408
      14.2.2   Collective Action and Social Movements..........................411
      14.2.3   Identification...412
14.3   Research Questions, Hypotheses, and Theoretical Model......................412
14.4   Methods..415
14.5   Measures...416
14.6   Results..417
14.7   Discussion...418
Acknowledgment...420
References...420

14.1   Introduction

The objective of this chapter is to illustrate some ways in which social network analysis is employed in sociology by examining the role that personal networks play in a social movement. With regard to the latter, we examine survey data collected from a nationwide random sample of members of environmental organizations in Canada. We examine the empirical relationships between social network centrality, level of identification in the environmental movement, and level of participation in the environmental movement (or level of activism). We employ a quantitative analysis. In this regard, we calculate several measures of social network centrality based on affiliations and ties with environmental organizations and then use these measures as independent variables in a series of ordinary least squares (OLS) multiple regression analyses to statistically explain identification with, and participation in the environmental movement.

The environmental movement has been described as one of the most important in the twentieth century (Castells 2004). Further, various commentators have described the environmental movement as one of the most important social movements in Canada. Greenpeace was formed in Vancouver, Canada, in 1971, and has gone on to become one of the world's most high-profile environmental organizations. Initially, Greenpeace was concerned with nuclear weapons testing and then whaling. Within Canada, environmental movements have campaigned around a variety of issues. From the 1980s through the 2000s, there were high-profile campaigns over-protecting old-growth forests. A number of campaigns have focused on protecting wilderness areas and protected areas, endangered species, and so on (Magnusson and Shaw 2002, Tindall 2002, Wilson 1998). In recent years, there has been much focus on addressing global warming and climate change issues (Davidson and Gismondi 2011, Saxifrage et al. 2012). Part of this involves resisting the expansion of bitumen extraction in the Alberta's oil sands and expansions of pipelines for transporting bitumen. These are but a few of the diverse range of issues that environmental movements in Canada have responded to.

14.2   Literature

14.2.1   Social Network Analysis

The term social network refers to the pattern of ties (e.g., relationships or social connections) among social units termed nodes (e.g., individuals, groups, organizations, and nation-states) within a bounded social grouping (Scott 2000, Tindall and Wellman 2001). Social network analysis (Wasserman and Faust 1994) can examine the relationships amongst all of the nodes in a bounded network (whole networks) or examine the ties to alters from a particular node (personal or ego networks). (The latter sometimes include analyses of interrelationships among the alters.)

Social networks can be depicted in matrices or in graphs (Knoke and Yang 2008). Analyses can focus on visualizations of network data or on network statistics describing the properties of various aspects of the network (Borgatti et al. 2013, de Nooy et al. 2011) and the nodes contained therein (Brandes et al. 2001, Brandes and Pich 2011, Krempel 2011). Visualizations of social network data (Freeman 2000) can be very useful for analyzing relatively small networks and for illustrating theoretical concepts (e.g., Burt's (1992) notion of structural holes). Visualizations can also be particularly useful when longitudinal network data exist, which can be used to illustrate particular processes (such as diffusion). However, in larger social networks, it is often difficult to detect meaningful patterns. For example, Figure 14.1 provides the entire two-mode social network of individuals and organizations in our study, where N for individuals is >1000.

In such cases, the visualizations are usually reduced in some fashion, often by (1) only using a sample of data, (2) using mock data to illustrate a theoretical point, (3) focusing on the ego networks of specific nodes, (4) reducing ties to those between egos and categories of alters (rather than specific alters), or (5) using a statistical

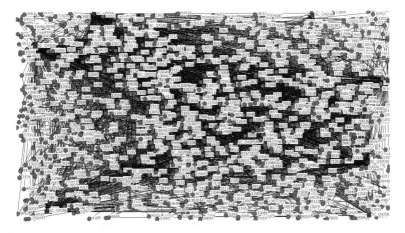

FIGURE 14.1: A complete two-mode graph.

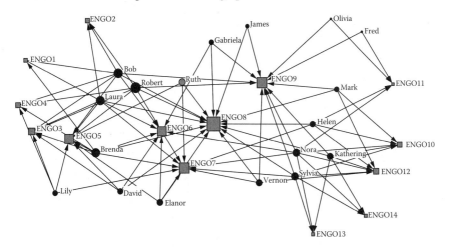

FIGURE 14.2: Example of a two-mode network: A sample of cases.

technique to reduce the model of ties and nodes to relations among certain types of actors (e.g., through the use of block modeling).

In terms of graphs, we use some mock data in Figure 14.2 in order to illustrate several key concepts (described in the following), and in Figure 14.3, we reduce a two-mode network of individuals by organizations to a one-mode network of organizations by organizations.

In one-mode networks, the rows and the columns of the matrix are the same actors (or nodes). In two-mode networks, the rows and columns are different actors (Borgatti and Everett 1997, Faust 1997). For example, the rows might be individuals, while the columns might be organizations or events. With regard to the individuals and organizations example, Breiger (1974) has introduced the concept of the duality of persons and groups where he notes that individuals are linked to one another through

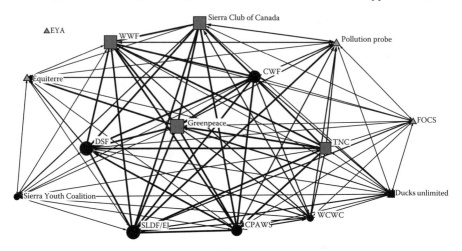

FIGURE 14.3:   Example of a one-mode network of ENGOs: Two-mode to one-mode transformation. *Notes*: Node size indicates level of eigenvector centrality. Shape indicates scale of activity. Square indicates international, circle indicates national, and triangle indicates regional. Line thickness indicates the ties between organizations in terms of overlapping memberships of individuals.

their joint membership in organizations and organizations are linked together through the joint membership of individuals.

One of the most important concepts of social network analysis is the notion of centrality (Freeman 1979). This concept was originally developed by researchers who were trying to understand the effects of social structure in human communication. The basic idea is that the more central an actor, the greater the degree of his or her involvement with others in a social network. A variety of centrality measures have been developed that correspond to different theoretical and analytical aims (Freeman 1979). Measures that focus upon a single unit or node (e.g., a single individual within a group or a single group within an intergroup network) are referred to as indicators of point centrality. Measures that focus on the overall pattern of an entire social network are referred to as measures of centralization.

A number of different centrality measures exist (Wasserman and Faust 1994) that have been specifically developed for describing different conceptualizations of centrality. Borgatti and Everett (1997) have developed four measures of centrality for two-mode networks: degree, betweenness, closeness, and eigenvector centrality. These different measures are useful for measuring different types of social processes. Degree is the most basic measure of centrality and refers to the number of different alters that a social unit is tied to. Our analysis here will focus on a particular version of degree, but for illustrative purposes, we will describe some of the other measures as well. A geodesic refers to the shortest path between two nodes. Betweenness measures the number of geodesics that a node falls on, between pairs of other actors in a social network. (This measure is sometimes used to measure processes like brokerage.) Closeness refers to the relative distance a node is from the alters he or she is tied to in geodesic terms. (This measure is sometimes used to examine diffusion processes.) Eigenvector centrality considers nodes to be central to the extent that they are tied to relatively central alters. (This measure is related to the procedures that Google uses to rank web pages.)

In Figure 14.2, we provide an example of two-mode eigenvector network centrality that illustrates a hypothetical example of the relations between individual members of environmental non-governmental organizations (ENGOs) and the organizations themselves. An individual has higher eigenvector centrality to the extent that he or she is tied to ENGOs with higher degree centrality. ENGOs have higher eigenvector centrality, to the extent that they tend to be tied to individuals with higher degree centrality. In Figure 14.2, individuals are portrayed as circles, and groups are portrayed as squares. (In Figure 14.2, node size indicates level of eigenvector centrality.)

In two-mode networks, the analyst can manipulate the data matrices to convert the two-mode network into a one-mode network of row nodes by row nodes or a one-mode network of column nodes by column nodes. The latter is illustrated by Figure 14.3, where we display a graph of organizations by organizations. (Here, the size of the nodes represents the eigenvector centrality of the organizations in the original two-mode matrix of individuals by organizations.)

A final concept we will consider here is the notion of network range. This term has been used in several different ways, but here, we use the term range to refer to the diversity of social groups that a node is linked to (Burt 1980). Differences between groups can potentially be based on various things—such as ethnicity/race, class, and religion (Côté and Erickson 2009, Erickson 1996, 2003). In this instance, we are referring to range as ties to different environmental organizations (see Tindall 2002). As will be seen in the methods section, with the data we analyze here, in practical terms, range is the same as degree. Conceptually, however, they are distinct. Range is important for a variety of reasons, but in particular, it implies diversity. People who have ties to diverse alters will be exposed to more diverse information (Erickson 1996, 2003), and this can have a variety of consequences (Erickson 2003, Tindall et al. 2012).

Social network researchers also distinguish between weak ties (those to whom we are not particularly close—such as acquaintances) and strong ties (people whom we are close to—such as close friends and immediate family members). In some contexts, weak ties are thought to be more important (such as for the acquisition of novel information) because we have more weak ties than strong ties and also because our weak ties tend to stretch further into social space (Granovetter 1973, Tindall and Malinick 2007). In the context of social movements, strong ties are thought to be more important when activities are more risky and costly (McAdam 1986). In the present study, we restrict our analysis to weak ties.

14.2.2 Collective Action and Social Movements

Collective action refers to collectivities of individuals working together to realize a common goal usually through relatively noninstitutionalized means. When the goal involves social change, we refer to such phenomena as social movements.

Common sense explanations for the participation of individuals in collective action and social movements include discontent and beliefs. It is commonly argued that people get involved in events like protests because they are very discontented about a particular situation, or they have strongly held ideological beliefs about something (Klandermans 2004, McAdam 1986).

Social movement researchers have observed that there is often a gap between feelings of discontent or beliefs and people's actions. For example, past research has shown that in many instances, the majority of the general public may share values associated with a social movement's goals, but only a small minority actually participate in any social movement activities designed to realize those goals (Klandermans and Oegema 1987). Social network scholars argue that what is crucial is that those who hold pro-movement values have ties to actors who are involved in the movement and are targeted for recruitment appeals (Diani 1995, Klandermans 2004).

14.2.3   Identification

Identification refers to the cognitive association one has with a particular group (della Porta and Diani 2006, Dunlap and McCright 2008, Tarrow 2011). This is a key concept in the social movement literature and plays several different roles in social movement accounts (Castells 2004, Hunt and Benford 2004, Melucci 1996, Stryker et al. 2000). Some explanations consider movement identification to be a key explanatory variable for understanding participation (Tindall 2002). Identities tend to be associated with bundles of values, beliefs, normative expectations, and responsibilities (Friedman and McAdam 1992). People who identify with particular social movements are more likely to participate in these movements because of their identities and feeling that by acting, they are validating these identities (Snow and McAdam 2000, Tarrow 2011).

On the other hand, some scholars have examined identity as an *outcome* variable (Poletta and Jasper 2001). For example, at certain stages at least, key goals of some movements have been to create new collective identities, as in the case of feminist and LGBT movements (Taylor and Whittier 1992). Some social movement scholars have been interested in how identification arises out of social processes, such as interactions in the context of social networks (Tindall 2004). Some social network scholars have postulated that movement identification is an intervening variable that mediates the relationship between network ties and social movement participation (Tindall 2002, 2007).

14.3   Research Questions, Hypotheses, and Theoretical Model

Our central questions are as follows: Is social network range associated with level of identification in the environmental movement? Is level of environmental identification associated with level of activism by individuals in the environmental movement?

Is social network range associated with level of activism by individuals in the environmental movement?

> H1a. Range of environmental organization memberships is positively associated with level of identification.

> H1b. Range of ties to people in different environmental organizations is positively associated with level of identification.

We argue that having ties to multiple, diverse organizations provides individuals with more diverse information (Erickson 1996, 2003; Tindall 2002, Tindall et al. 2012). Such information may be more useful in terms of becoming more informed about movement events and issues and thus lead to increased participation (Tindall 2002). Also having ties to a range of organizations affects the perception that an individual is part of a larger movement—as opposed to just being affiliated with a single organization.

Hypotheses H1a and H1b distinguish between belonging to formal environmental organizations and actually having ties to people in them. One can technically belong to an organization (e.g., by completing a membership form and paying a fee) without having any social ties to individuals in the organization. Alternatively, one can have social ties to individuals within an organization without belonging to the organization. Both of these types of connections are important. Further, it may be that different types of information are conveyed through these different types of ties.

> H2a. Range of environmental organization memberships is positively associated with level of activism.

> H2b. Range of ties to people in different environmental organizations is positively associated with level of activism.

These are the central hypotheses of this study. We argue that network centrality in terms of range of ties (as measured by two-mode degree centrality) is positively associated with activism. Greater centrality means that individuals will have greater opportunities for communication about movement events and issues. Such communication implies increased opportunities for social influence, information exchange, as well as other processes, which, we argue, results in greater levels of individual activism. Again, we argue that there are distinct effects for range of organization memberships and range of ties. For example, effects for range of organization memberships may reflect the information campaigns of these organizations, while the effects of range of organization ties may reflect somewhat more idiosyncratic processes related to interpersonal interactions.

> H3. Level of identification with the environmental movement is positively associated with level of activism.

In H3, we argue that the stronger one's level of identification with the environmental movement, the more active the individual will be. This is based on the rationale that group identities provide values, norms, and responsibilities that partly guide action.

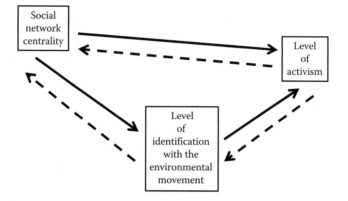

FIGURE 14.4:   Theoretical model.

Figure 14.4 illustrates the main theorized relationships that we examine in H1a through H3.

The solid lines indicate the main direction of theorized causal influence, as articulated previously. Range of environmental organization memberships and range of ties to individuals in different environmental organizations are the two indicators of social network centrality we employ in the analyses. As illustrated, our argument is that network centrality has a direct positive effect on level of activism. Also social network centrality has an indirect effect, via level of movement identification, on level of activism.

We also acknowledge that some reciprocal influences likely exist. For example, while we think that those with a greater number of ties to different organizations will be more active, it is also the case that those who are more active will form more ties to individuals and organizations that are part of the movement (Snijders et al. 2010).

Similarly, while we argue that those with higher network centrality will identify more strongly with the environmental movement, it is also the case that those who strongly identify with the movement will therefore be motivated to form more ties with other individuals and organizations that constitute the movement.

Finally, while we argue that those with higher levels of movement identification will participate more in movement activities, it is also the case that individuals will sometimes make self-attributions about their behavior, and conclude that because they participate in environmental movement activities, they must be part of the movement, and thus identify more highly (Bem 1972).

The dashed lines in Figure 14.4 indicate these reciprocal relationships. Because our study is a cross-sectional study, we cannot empirically adjudicate the relative causal direction of these relationships. However, this issue is somewhat moot, as we think there are likely reciprocal influences. Nevertheless, we can empirically examine whether or not net associations exist between the variables in the hypothesized relationships.

14.4 Methods

The data for this study were collected (with assistance from the environmental organizations listed below) using a mailed self-administered questionnaire distributed to a stratified systematic sample (with a random start) of environmental organization members across Canada. The data were collected between June 2007 and November 2007. The main national and large regional organizations in Canada participated in this survey. The participating environmental groups included the Canadian Wildlife Federation, the David Suzuki Foundation, Iquiterre, Greenpeace Canada, the Sierra Legal Defense Fund (now known as Équiterre Canada), the Sierra Club of Canada, the Sierra Youth Coalition, the Wilderness Committee (formerly known as the Western Canada Wilderness Committee), and the World Wildlife Fund of Canada.

The sample N for this survey was 1227, which represents a 32.3% response rate. (The response rate is based on 1227 returned questionnaires divided by 3799, which is the number of questionnaires sent to valid addresses.) The response rates for the surveys were relatively low in absolute terms, though not necessarily in comparative terms (see Muller and Opp 1986, Opp 1986, Tindall 2002). Indeed, they are typical to those found in similar studies.

The questionnaires were provided in English or in French depending upon database records maintained by the ENGOs regarding their members (the questionnaires were provided in French for organizations where there were substantial numbers of known Francophone members; otherwise the questionnaires were provided in English). There were 1227 completed questionnaires. The margin of error for the survey is ±2.8 at the 95% confidence level.

The sample was stratified by the different organizations and, in some instances (where there were substantial numbers of both Francophone and Anglophone members), by language group. The general procedures involved the researchers sending questionnaire packages and reminder postcards to the ENGOs. The ENGOs were given instructions on how to draw a systematic sample with a random start from their membership/supporter lists in order to print labels for the questionnaire packages and reminder postcards. In general, the ENGOs affixed address labels to the prestamped questionnaire packages and reminder postcards. They then distributed the questionnaire packages (on behalf of the researchers) and subsequently (about 2 weeks later) mailed the reminder postcards. Through this procedure, the ENGOs were able to protect the confidentiality of their membership lists.

There is no evidence of substantial sampling bias related to nonresponse. Based on typical patterns of nonresponse bias (Dillman et al. 2002), we might expect that people with more available time or those who feel more strongly about the issues covered in the questionnaire to be more likely to respond. On the other hand, an analysis of key variables such as level of participation in movement activities reveals a roughly normal distribution. Most people participated in activities to a relatively moderate extent with fewer people either being relatively highly active or having relatively low levels of activity. It is not the case the sample is made up of respondents who are either mostly highly active or mostly nonactive.

The cases were weighted in proportion to the relative size of the ENGO for the total population of ENGOs (based on these nine organizations). Values for ENGO size were based on the number of members/supporters of each organization provided in annual reports from the organizations at the time of the study and from information provided in interviews with representatives of each organization.

14.5   Measures

Level of activism: Respondents were asked about their frequency of participation in 19 different types of activities related to the environmental movement where 1 = never, 2 = occasionally, and 3 = often. These items included the following: (1) been a member of an environmental organization; (2) donated money to an environmental organization, other than purchasing a membership; (3) volunteered for an environmental organization; (4) written and sent a letter, fax, or e-mail to a government official regarding an environmental issue; (5) written and sent a letter, fax, or e-mail to a newspaper about an environmental issue; (6) written and sent a letter, fax, or e-mail to a company/corporation about an environmental issue; (7) written and sent a letter, fax, or e-mail to another organization regarding an environmental issue; (8) telephoned a government, company, or other organizational official regarding an environmental issue; (9) signed a petition about an environmental issue; (10) attended a community meeting about an environmental issue; (11) attended a rally or protest demonstration about an environmental issue; (12) participated in an information campaign for the general public about an environmental issue; (13) made a presentation to a public body about an environmental issue; (14) given a lecture on an environmental issue to a school group or voluntary organization; (15) served as a representative on an advisory board formed around an environmental issue; (16) purchased a book, t-shirt, poster, mug, or other merchandise from an environmental organization; (17) boycotted a product or company because of environmental concern; (18) worked to elect someone because of their views on the environment; (19) voted for someone because of their environmental views; and (20) any other activities that you have done to help conserve the environment.

Responses to the 19 indicators were then summed and divided by 19, in order to create an index. Cronbach's alpha coefficient is $\alpha = 0.83$. This indicates a relatively high level of inter-item reliability for the items in the scale and gives us confidence in the utility of these items for constructing an index.

Gender: Gender is a dummy variable where women = 1 and men = 0.

Age: Chronological age in years.

Income: Personal income in the past year in dollars.

Range of environmental organization memberships: Respondents were asked about whether or not they were a member in each of 16 environmental organizations. Range of environmental organization memberships is based on two-mode network degree, which is the proportion of the organizations that individuals indicated they belonged to.

Range of weak ties to environmental organizations: Respondents were asked about whether or not they knew an acquaintance in each of 16 environmental organizations. Range of environmental organization ties is based on two-mode network degree, which is the proportion of the organizations that individuals indicated they had an acquaintanceship tie to.

Level of movement identification: Respondents were asked about the extent to which they agree with the following statement: You identify yourself as a member of the environmental movement. This measure is based on a five-point scale, where response categories varied from completely disagree = 1 to complete agree = 5.

14.6   Results

We utilize OLS multiple regression to examine our hypotheses. Our strategy is to add blocks of conceptually related variables to the model in stages. In Table 14.1, level of identification with the environmental movement is the dependent variable. In model 1, a set of sociodemographic and socioeconomic control variables are entered. Age and education are significantly and positively associated with identification, while income is significantly and negatively associated with identification. There is no effect for gender. In model 1, the explained variation is quite small: 5%.

In model 2, range of environmental organization memberships and range of weak ties to people in different environmental organizations both have significant positive effects. The effect for range of memberships is stronger than the effect for range of weak ties. In model 2, inclusion of the network variables increases the explained variation by a small amount: 2%. Overall, the explained variation is quite modest: 7%.

In Table 14.2, level of activism is the dependent variable. In model 1, a set of sociodemographic and socioeconomic control variables are entered. Here, we

TABLE 14.1:   Multiple Regression Analyses Explaining Identification with the Environmental Movement (Standardized Regression Coefficients)

	Model 1	Model 2
Gender (female = 1)	.03	.02
Age	.12****	.10****
Years of education	.10****	.06
Income (personal)	−.17****	−.18****
Range of memberships	—	.16****
Range of weak ties	—	.06*
R^2	.05****	.07****
Adjust R^2	.04****	.07****
N	1005	1005

Notes: *p ≤ .05, **p ≤ .01, ***p ≤ .005, ****p ≤ .001.

TABLE 14.2: Multiple Regression Analyses Explaining Level of Activism (Standardized Regression Coefficients)

	Model 1	Model 2	Model 3
Gender (female = 1)	−.07	−.07*	−.08**
Age	−.09***	−.10****	−.14****
Years of education	.31****	.24****	.21****
Income (personal)	−.15****	−.14****	−.07*
Range of memberships	—	.23****	.16****
Range of weak ties	—	.26****	.25****
Level of movement identification	—	—	.41****
R^2	.10****	.23****	.38****
Adjust R^2	.10****	.22****	.38****
N	892	892	892

Notes: *$p \leq .05$, **$p \leq .01$, ***$p \leq .005$, ****$p \leq .001$.

find that younger respondents are more active, people with higher levels of formal education are more active, and those with lower incomes are more active. There is no significant effect for gender. Overall, these variables explain 10% of the variation in level of activism. In model 2, we add our two social network explanatory variables: range of organization memberships and range of weak ties to different environmental organizations. Both of these variables are highly significant and positively associated with level of activism. In this model, the previously entered control variables are all significant. Gender becomes significant. Education remains significant, though its coefficient diminishes somewhat. Overall, the addition of the network variables increases the explained variation by 13%.

In model 3, we add level of movement identification. The level of movement identification is positively and highly significantly associated with activism (controlling for the other variables in the model). In fact, it has the strongest beta coefficient. In model 3, the sociodemographic variables all remain significant, though notably the effect for income has diminished somewhat. The effects for the network variables remain significant, though the effect for range of memberships has decreased. Overall, the model explains 38% of the variation in activism.

This is quite a powerful model by social science standards. (In social science, R^2s are often in the 0.10–0.20 range.) The results reported here support the hypotheses we introduced earlier (H1A, H1B, H2A, H2B, and H3).

14.7 Discussion

The empirical findings are consistent with the theoretical hypotheses we have offered. Based on two measures of network range, we found that social network centrality is significantly positively associated with level of activism.

More specifically, we found those who belonged to a greater range of environmental organizations and those who had weak ties to individuals in a greater range of environmental organizations participated more frequently in a greater range of environmental movement activities. Several different processes likely underlie these effects, including communication through networks and, relatedly, information diffusion through network ties, social influence, and greater opportunities to be targeted and recruited for involvement in events (among others).

Also, our hypotheses concerning movement identification were supported. The two network range measures were positively associated with level of identification, and level of identification was positively associated with activism. Having a greater array of network ties to people in different movement organizations facilitates opportunities for social comparison (Gartrell 1987) and thus fosters identification with the movement. Identification with a movement is associated with a particular set of values and norms that provides a motivational basis for acting—in part, in order to validate a valued identity (Hunt and Benford 2004).

This is a rare example of a network study of a nationwide social movement. The method of data collection on weak ties is relatively novel in that we used a modified form of an organization roster network question to obtain data about the range of weak ties in people's personal networks.

The results here reinforce previous findings. Most notably, the present findings reinforce the importance of network range for explaining movement identification and the relative importance of level of movement identification for explaining activism (see Tindall 2002). The findings concerning the effects of network range on activism also support previous findings.

One challenge for research on this topic, as is the case for much social network research, is disentangling cause and effect. The present study used a cross-sectional design. Thus, to focus on one aspect of the analysis, it is difficult to assess the extent that individual activism increases as a function of the number of ties an individual has, versus the extent to which people who are more active develop a greater number of ties. It seems likely that both processes are at work. We provide strong evidence that network centrality and activism are empirically correlated. Panel data are required to explore the nuances of these processes (see McAdam 1989). This problem has been discussed by Snijders et al. (2010) and is often described as the problem of social influence versus social selection. The SIENA methodology is one approach to examine this problem (see Snijders et al. 2010); though for a variety of methodological and practical reasons, it is difficult to obtain the type of real-world panel data that are necessary to explore these processes more fully in the context of social movements.

The present study also points to several avenues for future research. In this study, we focus on weak ties. Future research could explore the importance of strong ties to the processes examined here (see Granovetter 1973, Tindall 2002, Tindall and Malinick 2007). Also, future research could focus more on specifying the mechanisms that underlie the correlations between network centrality and activism (see Tindall 2007), such as trust, social influence, social support, communication, among others.

Acknowledgment

This research project was supported by several research grants from the Social Sciences and Humanities Research Council of Canada. We thank Todd Malinick, Andrea Streilein, Noelani Dubeta, Andrea Rivers, and Devon Deckant for their contributions to this research. The social network diagrams in this chapter were produced using Netdraw (Borgatti 2002).

References

Bem, D.J. 1972. Self-perception theory. In: L. Berkowitz (ed.), *Advances in Experimental Social Psychology*, Vol. 6. New York: Academic Press, pp. 1–62.

Borgatti, S.P. 2002. *NetDraw: Graph Visualization Software*. Harvard, MA: Analytic Technologies.

Borgatti, S.P. and M.G. Everett. 1997. Network analysis of 2-mode data. *Social Networks* 19:243–269.

Borgatti, S.P., M.G. Everett, and J.C. Johnson. 2013. *Analyzing Social Networks*. Thousand Oaks, CA: Sage Publications.

Brandes, U. and C. Pich. 2011. Explorative visualization of citation patterns in social network research. *Journal of Social Structure* 12(8).

Brandes, U., J. Raab., and D. Wagner. 2001. Exploratory network visualization: Simultaneous display of actor status and connections. *Journal of Social Structure* 2(4).

Breiger, R.L. 1974. The duality of persons and groups. *Social Forces* 53:181–90.

Burt, R. 1980. Models of network structure. *Annual Review of Sociology* 6:79–141.

Burt, R.S. 1992. *Structural Holes: The Social Structure of Competition*. Cambridge, MA: Harvard University Press.

Castells, M. 2004. *The Power of Identity*, 2nd edn., Vol. 2. Oxford, U.K.: Blackwell Publishing.

Côté, R.R. and B.H. Erickson. 2009. Untangling the roots of tolerance: How forms of social capital shape attitudes toward ethnic minorities and immigrants. *American Behavioral Scientist* 52(12):1664–1689.

Davidson, D.J. and M. Gismondi. 2011. *Challenging Legitimacy at the Precipice of Energy Calamity*. New York: Springer.

de Nooy, W., A. Mrvar, and V. Batagelj. 2011. *Exploratory Social Network Analysis with Pajek*, revised and expanded 2nd edn. Cambridge, U.K.: Cambridge University Press.

della Porta, D. and M. Diani. 2006. *Social Movements: An Introduction*, 2nd edn. Oxford, U.K.: Blackwell Publishing.

Diani, M. 1995. *Green Networks: A Structural Analysis of the Italian Environmental Movement*. Edinburgh, U.K.: Edinburgh University Press.

Dillman, D.A., J.L. Eltinge, R.M. Groves, and R.J.A. Little. 2002. Survey nonresponse in design, data collection, and analysis. In: R.M. Groves, D.A. Dillman, J.L. Eltinge, and R.J.A. Little (eds.), *Survey Nonresponse*. New York: John Wiley & Sons, Inc., pp. 3–26.

Dunlap, R.E. and A.M. McCright. 2008. Social movement identity: Validating a measure of identification with the environmental movement. *Social Science Quarterly* 89(5):1045–1065.

Erickson, B. 1996. Culture, class and connections. *American Journal of Sociology* 102:217–251.

Erickson, B. 2003. Social networks: The value of variety. *Contexts* 2(1): 25–31.

Faust, K. 1997. Centrality in affiliation networks. *Social Networks* 19:157–191.

Freeman, L.C. 1979. Centrality in social networks: Conceptual clarification. *Social Networks* 1:215–239.

Freeman, L.C. 2000. Visualizing social networks. *Journal of Social Structure* 1(1):4.

Friedman, D. and D. McAdam. 1992. Collective identity and activism: Networks, choices, and the life of a social movement. In: A.D. Morris and C.M. Mueller (eds.), *Frontiers in Social Movement Theory*. New Haven, CT: Yale University Press, pp. 156–173.

Gartrell, C.D. 1987. Network approaches to social evaluation. *Annual Review of Sociology* 13:49–66.

Granovetter, M. 1973. The strength of weak ties. *American Journal of Sociology* 78:1360–1380.

Hunt, S.A. and R.D. Benford. 2004. Collective identity, solidarity, and commitment. In: D.A. Snow, S.A. Soule, and H. Kriesi (eds.), *The Blackwell Companion to Social Movements*. Oxford, U.K.: Blackwell Publishing, pp. 433–457.

Klandermans, B. 2004. The demand and supply of participation: Social-psychological correlates of participation in social movements. In: D.A. Snow, S.A. Soule, and H. Kriesi (eds.), *The Blackwell Companion to Social Movements*. Malden, MA: Blackwell Publishing, pp. 360–379.

Klandermans, B. and D. Oegema. 1987. Potentials, networks, motivations, and barriers: Steps towards participation in social movements. *American Sociological Review* 52:519–531.

Knoke, D. and S. Yang. 2008. *Social Network Analysis*, 2nd edn. Thousand Oaks, CA: Sage Publications.

Krempel, L. 2011. Network visualization. In: J. Scott and P.J. Carrington (eds.), *The Sage Handbook of Social Network Analysis*. Thousand Oaks, CA: Sage Publications, pp. 558–577.

Magnusson, W. and K. Shaw. 2002. *A Political Space: Reading the Global through Clayoquot Sound*. Montreal, Quebec, Canada: McGill-Queen's University Press.

McAdam, D. 1986. Recruitment to high-risk activism: The case of freedom summer. *American Journal of Sociology* 92:64–90.

McAdam, D. 1989. The biographical consequences of activism. *American Sociological Review* 54:744–760.

Melucci, A. 1996. *Challenging Codes: Collective Action in the Information Age*. Cambridge, U.K.: Cambridge University Press.

Muller, E.N. and K. Opp. 1986. Rational choice and rebellious collective action. *American Political Science Review* 80:471–487.

Opp, K. 1986. Soft incentives and collective action: Participation in the anti-nuclear movement. *British Journal of Political Science* 16:87–112.

Poletta, F. and J.M. Jasper. 2001. Collective identity and social movements. *Annual Review of Sociology* 27:283–305.

Saxifrage, C., V. Observer, and L. Solomon (eds.). 2012. *Extract: The Pipeline Wars, Vol. 1 Enbridge*. Vancouver, British Columbia, Canada: Vancouver Observer.

Scott, J. 2000. *Social Network Analysis: A Handbook*, 2nd edn. London, U.K.: Sage Publications.

Snijders, T.A.B., G.G. van de Bunt, and C.E.G. Steglich. 2010. Introduction to stochastic actor-based models for network dynamics. *Social Networks* 32:44–60.

Snow, D.A. and D. McAdam. 2000. Identity work processes in the context of social movements: Clarifying the identity/movement nexus. In: T.J.O. Stryker and R.W. White (eds.), *Self, Identity, and Social Movements*. Minneapolis, MN: University of Minnesota Press, pp. 41–67.

Stryker, S., T.J. Owens, and R.W. White (eds.). 2000. *Self, Identity, and Social Movements*. Minneapolis, MN: University of Minnesota Press.

Tarrow, S. 2011. *Power in Movement: Social Movements and Contentious Politics*. New York: Cambridge University Press.

Taylor, V. and N.E. Whittier. 1992. Collective identity in social movement communities: Lesbian feminist mobilization. In: A. Morris and C.M. Mueller (eds.), *Frontiers in Social Movement Theory*. New Haven, CT: Yale University Press, pp. 104–129.

Tindall, D.B. 2002. Social networks, identification, and participation in an environmental movement: Low-medium cost activism within the British Columbia wilderness preservation movement. *Canadian Review of Sociology and Anthropology* 39(4):413–452.

Tindall, D.B. 2004. Social movement participation over time: An ego-network approach to micro-mobilization. *Sociological Focus* 37(2):163–184.

Tindall, D.B. 2007. From metaphors to mechanisms: Some critical issues in networks and social movements research. *Social Networks* 29:160–168.

Tindall, D.B., J. Cormier, and M. Diani. 2012. Network social capital as an outcome of social movement mobilization. *Social Networks* 34:387–395.

Tindall, D.B. and T. Malinick. 2007. Weak ties (strength of). In: G. Ritzer (ed.), *The Blackwell Encyclopedia of Sociology*, Vol. 10. Oxford, U.K.: Blackwell Publishing, pp. 5222–5225.

Tindall, D.B. and B. Wellman. 2001. Canada as social structure: Social networks and Canadian sociology. *Canadian Journal of Sociology* 26:265–308.

Wasserman, S. and K. Faust. 1994. *Social Network Analysis: Methods and Applications*. Cambridge, U.K.: Cambridge University Press.

Wilson, J. 1998. *Talk and Log: Wilderness Politics in British Columbia.* Vancouver, British Columbia, Canada: University of British Columbia Press.

Chapter 15

Graph Kernels in Chemoinformatics

Benoît Gaüzère, Luc Brun, and Didier Villemin

Contents

15.1 Introduction..426
    15.1.1 General Definitions..427
15.2 Graph Similarity Measures...431
    15.2.1 Maximum Common Subgraphs.......................................431
    15.2.2 Frequent Graphs..432
    15.2.3 Graph Edit Distance..432
        15.2.3.1 Graph Edit Distance Approximation 433
        15.2.3.2 Graph Edit Distance and Machine Learning Methods.....434
15.3 Graph Embedding Methods..436
    15.3.1 Empirical Kernel Map...436
    15.3.2 Isometric Embedding...436
    15.3.3 Local Approaches..437
    15.3.4 Global Approaches...437
    15.3.5 Conclusion...438
15.4 Kernel Theory..438
    15.4.1 Definitions...439
    15.4.2 Kernel Trick..440
    15.4.3 Kernels and Similarity Measures....................................441
    15.4.4 Kernel Methods..441
        15.4.4.1 Support Vector Machines...............................441
        15.4.4.2 Support Vector Machines for Regression.................443
        15.4.4.3 Kernel Ridge Regression...............................444
    15.4.5 Graph Kernels and Chemoinformatics.............................445
15.5 Graph Kernels Based on Bags of Patterns................................445
    15.5.1 Complete Graph Kernel..446
    15.5.2 Linear Pattern Kernels...446
        15.5.2.1 Random Walks.......................................446
        15.5.2.2 Tottering...447
        15.5.2.3 Paths..447
    15.5.3 Nonlinear Pattern Kernels...448
        15.5.3.1 Graphlets...448
        15.5.3.2 Tree Patterns.......................................450
        15.5.3.3 Weisfeiler–Lehman..................................451
        15.5.3.4 Graph Fragments....................................451
        15.5.3.5 Treelets..452
    15.5.4 3D Pattern Kernel..453

  15.5.5 Cyclic Pattern Kernels...454
     15.5.5.1 Cyclic Patterns..454
     15.5.5.2 Relevant Cycle Graph....................................456
     15.5.5.3 Relevant Cycle Hypergraph.............................457
  15.5.6 Pattern Weighting..458
     15.5.6.1 A Priori Methods..459
     15.5.6.2 Branching Cardinality...................................459
     15.5.6.3 Ratio Depth/Size..459
     15.5.6.4 Multiple Kernel Learning...............................460
     15.5.6.5 Simple MKL..460
     15.5.6.6 Generalized MKL..460
     15.5.6.7 Infinite MKL...461
15.6 Experiments...461
  15.6.1 Boiling Point Prediction..462
  15.6.2 Classification...463
     15.6.2.1 AIDS Dataset...463
     15.6.2.2 PTC Dataset..464
15.7 Conclusion..465
References..466

15.1   Introduction

Graphs provide a generic data structure widely used in chemo- and bioinformatics to represent complex structures such as chemical compounds or complex interactions between proteins. However, the high flexibility of this data structure does not allow to readily combine it with usual machine learning algorithms based on a vectorial representation of input data.

Indeed, algorithms restricted to the graph domain are essentially restricted to k-nearest neighbors or k-median algorithms. These algorithms use different measures of similarity based on the set of frequent subgraphs extracted from graphs or from the size of the maximum common subgraph of two graphs. A widely used graph similarity measure is based on a measure of the distortion required to transform one graph into another. This measure of dissimilarity, called graph edit distance, is NP (non-polynomial)-hard to compute. However, this complexity can be reduced by computing a suboptimal, but usually effective, graph edit distance. Nevertheless, graph edit distance does not usually fit all the requirements of a Euclidean distance and hence cannot be readily applied in conjunction with many machine learning methods.

Graph embedding methods aim to tackle this limitation by embedding graphs into explicit vectorial representations that allow the use of any machine learning algorithm defined on vectorial representations. However, encoding graphs as explicit vectors having a limited size induces a loss of information that may reduce prediction accuracy.

Graph kernels are defined as similarity measures between graphs. Under mild conditions, graph kernels correspond to scalar products between possibly implicit graph embeddings into a Hilbert space. Thanks to this graph embedding, machine learning methods that may be rewritten so as to use only scalar products between input data, such as support vector machine (SVM), can be applied on graphs. Graph kernels thus provide a natural connection between graph space and machine learning.

A large family of kernel methods is based on a decomposition of graphs into bags of patterns. The prototypical example of this last family is the complete graph kernel that is based on a decomposition of a graph into all its substructures (subgraphs). Such a kernel being NP-hard to compute, kernels based on bags of patterns must be restricted to the extraction of a different family of substructures. Given a particular type of bag of patterns, each substructure of a bag encodes a different structural information. This information may be weighted according to the property to predict. Such a weighting scheme allows to highlight relevant patterns having a high influence on a particular property.

Graph kernels provide thus an interesting approach in order to address chemoinformatics problems since they correspond to a similarity measure between molecular graphs that can be used in conjunction with powerful machine learning methods.

15.1.1 General Definitions

This first section aims to introduce some basics of graph theory required to define graph similarity measures used in chemoinformatics.

Definition 15.1.1 (Graph) *An unlabeled graph is a pair $G = (V, E)$ such that V corresponds to a set of nodes and $E \subset V \times V$ corresponds to a set of edges connecting nodes. The size of the graph is defined by $|V|$. If $(u, v) \in E$, u is said to be adjacent to v.*

Definition 15.1.2 (Nonoriented graph) *A graph is nonoriented if for any pair of nodes $(u, v) \in E$, $(v, u) \in E$. In this case, (u, v) denotes indifferently the oriented edges (u, v) and (v, u).*

In the following, we will only use nonoriented graphs. Hence, unless otherwise stated, graphs are supposed to be nonoriented.

Definition 15.1.3 (Neighborhood) *Neighborhood relationship is encoded by the function $\Gamma : V \to \mathcal{P}(V)$ with $\Gamma(v) = \{u \in V \mid (u, v) \in E\}$ where $\mathcal{P}(V)$ denotes all subsets of V.*

Definition 15.1.4 (Degree) *The degree of a node $v \in V$ is defined as $|\Gamma(v)|$.*

Definition 15.1.5 (Subgraph) *A graph $G' = (V', E')$ is a subgraph of $G = (V, E)$, denoted by $G' \sqsubseteq G$, if $V' \subseteq V$ and $E' \subseteq E$. If $E' = E \cap (V' \times V')$, G' is called a vertex-induced subgraph of G.*

Definition 15.1.6 (Walks, trails, paths) *A walk of a graph $G = (V, E)$ is a sequence of nodes $W = (v_1 \ldots v_n)$ connected by edges: $(v_i, v_{i+1}) \in E$ for any $i \in \{1, \ldots, n-1\}$. If each edge (v_i, v_{i+1}) appears only once in W, W is called a trail. If each vertex (and thus each edge) appears only once, W is called a path. The length of a walk is defined by its number of nodes.*

Definition 15.1.7 (Distance between nodes) *The distance $d_G(u, v)$ between two nodes u and v of a graph $G = (V, E)$ is defined as the length of the shortest path between u and v in G.*

Definition 15.1.8 (Cycle) *A cycle is a path whose first node is equal to the last one. This node is the only one appearing twice in the sequence.*

Definition 15.1.9 (Connected graph) *Given a graph $G = (V, E)$, a set $U \subseteq V$ is said to be connected if there exists a path in U between any pair of distinct nodes in U. The set U is called a connected component of G if it is not included into a larger connected set. The graph G is said to be connected if all its connected components are equal to V.*

Definition 15.1.10 (Tree) *A tree is a connected graph without cycles.*

Definition 15.1.11 (Bridge) *A bridge is an edge whose removal increases the number of connected components of the graph. The set of bridges of a graph G is denoted $\mathcal{B}(G)$.*

Definition 15.1.12 (Labeled graph) *A nonoriented labeled graph $G = (V, E, \mu, \nu)$ is a nonoriented unlabeled graph $G = (V, E)$ associated to a node labeling function $\mu : V \to L_V$ and an edge labeling function $\nu : E \to L_E$, where L_V and L_E denote, respectively, the sets of node and edge labels.*

Two graphs are considered as equal if there exists a bijection between the nodes of both graphs that respect adjacency relationships. In this case, both graphs are said to be isomorphic.

Definition 15.1.13 (Graph isomorphism) *Two graphs $G = (V, E)$ and $G' = (V', E')$ are structurally isomorphic, denoted by $G \simeq_s G'$, if and only if there exists a bijection*

$$f : V \to V'$$

such that

$$(u, v) \in E \Leftrightarrow (f(u), f(v)) \in E'.$$

If G and G' correspond to labeled graphs, that is if $G = (V, E, \mu, \nu)$ and $G' = (V', E', \mu', \nu')$, a structural isomorphism between G and G' is called a graph isomorphism, denoted by $G \simeq G'$, if

$$\forall v \in V, \qquad \mu(v) = \mu'(f(v)) \ and$$
$$\forall(u, v) \in E, \quad \nu(u, v) = \nu'(f(u), f(v)).$$

Definition 15.1.14 (Partial subgraph isomorphism) *Let $G = (V, E, \mu, \nu)$ and $G' = (V', E', \mu', \nu')$ denote two graphs such that $|V| \leq |V'|$. There exists a partial structural subgraph isomorphism between G and G' if and only if there exists an injection:*

$$f : V \to V'$$

such that

$$\forall (u,v) \in V^2, (u,v) \in E \Rightarrow (f(u), f(v)) \in E'.$$

A partial structural subgraph isomorphism is a partial subgraph isomorphism, denoted $G \subseteq_p G'$, if

$$\forall v \in V, \quad \mu(v) = \mu'(f(v)) \text{ and}$$
$$\forall (u,v) \in E, \quad \nu(u,v) = \nu'(f(u), f(v)).$$

Note that the partial subgraph isomorphism relationship may also be denoted as homomorphism relationship in other references.

Definition 15.1.15 (Subgraph isomorphism) *Let $G = (V, E, \mu, \nu)$ and $G' = (V', E', \mu', \nu')$ denote two graphs such that $|V| \leq |V'|$. There exists a structural subgraph isomorphism if there exists a partial structural subgraph isomorphism between G and G' and*

$$\forall (u,v) \in V'^2, (u,v) \in E' \Leftrightarrow (u,v) \in E$$

A structural subgraph isomorphism is a subgraph isomorphism denoted $G \subseteq G'$, if

$$\forall v \in V, \quad \mu(v) = \mu'(f(v)) \text{ and}$$
$$\forall (u,v) \in E, \quad \nu(u,v) = \nu'(f(u), f(v)).$$

Hypergraphs correspond to an extension of graphs that allows to define an adjacency relationship between more than two nodes. Hypergraphs have been introduced by Claude Berge [3].

Definition 15.1.16 (Hypergraph) *A hypergraph $H = (V, E)$ is a pair of sets V, encoding hypergraph's nodes, and $E = (e_i)_{i \in I} \subseteq \mathcal{P}(V)$ encoding hyperedges:*

- $\forall i \in \{1, \ldots, |E|\}, e_i \neq \emptyset$,

- $\cup_{i \in I} e_i = V$

As for graphs, the size of a hypergraph $H = (V, E)$ is defined as $|V|$, and two nodes u and v are adjacent if there exists $e \in E$ such that $\{u, v\} \subseteq e$.

An oriented hypergraph [13] is a hypergraph where hyperedges connect two sets of nodes (Figure 15.1):

Definition 15.1.17 (Oriented hypergraph) *An oriented hypergraph $H = (V, \vec{E})$ is a pair of sets V, encoding hypergraph's nodes, and $\vec{E} = (e_i)_{i \in I} \subseteq \mathcal{P}(V) \times \mathcal{P}(V)$ encoding oriented hyperedges. An oriented hyperedge $e = (s_u, s_v) \in \vec{E}$ with $s_u = \{u_1, \ldots, u_i\} \subset \mathcal{P}(V)$ and $s_v = \{v_1, \ldots, v_j\} \subset \mathcal{P}(V)$ encodes an adjacency relationship between the two sets of node s_u and s_v (Figure 15.1).*

In the remaining part of this chapter, we consider that a hyperedge (s_v, s_u) exists in $\vec{E}$ for each $(s_u, s_v) \in \vec{E}$, and we thus do not differentiate (s_u, s_v) from (s_v, s_u). Note anyway that an oriented hypergraph remains different from a usual hypergraph since

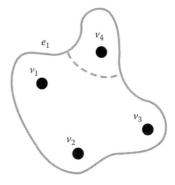

FIGURE 15.1: Oriented hypergraph with oriented hyperedge $e_1 = (\{v_1, v_2\}, \{v_3\})$.

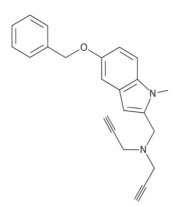

FIGURE 15.2: A molecular graph.

within an oriented hypergraph, an edge connects two sets of vertices, while within the usual hypergraph framework, an edge is composed of a single set.

A usual molecular representation is defined by the molecular graph [16,51] (Figure 15.2). A molecular graph is a nonoriented labeled graph encoding adjacency relationships between atoms of a molecule. The set of nodes encodes the set of atoms of a molecule while the set of edges encodes atomic bonds connecting these atoms. Each vertex is labeled by the chemical element of the corresponding atom, and each edge is labeled by the type of the atomic bond (simple, double, triple, or aromatic) connecting two atoms. According to the octet rule, an atom may share pairs of valence electrons with a maximum of 7 atoms. Therefore, a molecule is encoded by a molecular graph having a degree bounded by 7. A usual convention consists in not encoding hydrogen atoms within a molecular graph since they are implicitly encoded by the valency of other atoms. Another convention is to not explicitly represent carbon atoms within the graphical representation of a molecular graph (Figure 15.2).

15.2 Graph Similarity Measures

Most of the existing methods used in chemoinformatics are based on the similarity principle that states that similar compounds have similar properties [26]. If we consider the molecular representation given by the molecular graph, comparing molecular compounds thus relies on comparing molecular graphs. Therefore, in order to be able to predict physical or biological properties of molecular compounds, we have to define molecular graph similarity measures.

15.2.1 Maximum Common Subgraphs

A first similarity measure between graphs based on graph theory is defined from the maximum common subgraph. Intuitively, two graphs are considered as similar if they share a large common structure and dissimilar otherwise. In order to define the notion of maximum common subgraph, let us first define the notion of common subgraph:

Definition 15.2.1 (Common subgraph) *A graph G is a common subgraph of two graphs G_1 and G_2 if there exist two subgraphs $\hat{G}_1$ and $\hat{G}_2$ of G_1 and G_2 such that*

$$G \simeq \hat{G}_1 \simeq \hat{G}_2$$

In order to formally define maximum common subgraphs, let us additionally define the notion of maximal common subgraphs:

Definition 15.2.2 (Maximal common subgraph) *A common subgraph of two graphs G_1 and G_2 is maximal if it is a common subgraph of G_1 and G_2 and if it is not itself a subgraph of another common subgraph of G_1 and G_2.*

The notion of maximal common subgraph corresponds to a set of nodes and edges common to two graphs that cannot be enlarged. Using maximal common subgraphs, we can now define maximum common subgraph:

Definition 15.2.3 (Maximum common subgraph) *A common subgraph G is maximum if it is a maximal common subgraph of G_1 and G_2 and there does not exist a larger common subgraph of G_1 and G_2.*

A maximum common subgraph corresponds thus to the largest maximal common subgraph.

The size of a maximum common subgraph of two graphs may be seen as a similarity measure between these two graphs. Methods based on maximum common subgraph consider two graphs as similar if they share a large common structure. Conversely, graphs sharing small common structures are not considered as similar. This particularity may alter the accuracy of methods based on maximum common subgraphs on chemical prediction problems where molecular activity is due to a number of small structures.

15.2.2   Frequent Graphs

A widely used approach in chemoinformatics consists in finding substructures that are responsible for a particular activity. This family of methods is thus more dedicated to activity prediction problems since from a chemical point of view, a molecular compound may have a particular property if it contains a substructure, called a pharmacophore. This chemical consideration has sustained the emergence of a family of methods based on the discovery of frequent subgraphs [11,40,58]. These methods are based on the following assumptions:

- A substructure is considered as frequent if its number of occurrences is greater than a frequency threshold σ within a set of positive molecular compounds and is insignificant within a set of negative molecular compounds.

- A substructure of size k may be frequent if it is composed of at least two frequent substructures of size $k - 1$.

Among the set of methods using such an approach, methods described in [11,58] are based on an iterative algorithm (Algorithm 15.1) that aims to find frequent subgraphs of size k from the set of frequent subgraphs of size $k - 1$.

These approaches obtain a good prediction accuracy on many datasets within a reasonable computational time. However, the main drawback of these approaches is the frequent subgraph hypothesis. Indeed, the hypothesis that one pharmacophore is responsible of an activity is no longer valid when the activity of molecular compounds is induced by a set of different pharmacophores. In this case, each pharmacophore is associated with few molecular compounds, and each pharmacophore does not reach the minimal frequency threshold in order to be considered as a frequent substructure. Such configurations induce thus a poor prediction model.

15.2.3   Graph Edit Distance

Graph edit distance aims to define a distance between graphs. This distance is based on a measure of distortion induced by a transformation of one graph into another. Graph edit distance between two graphs is defined by the minimal cost associated with an edit path that transforms the first graph into the second one. This distance is defined as a sequence of edit operations, each of these elementary operations being defined as

---

Algorithm 15.1: Generic algorithm used by frequent subgraph methods

---

1 Find a set of subgraphs S_2 having a size $k = 2$ and a frequency greater than σ;
2 **for** $k = 3 \rightarrow k\text{-}max$ **do**
3 $\quad$ Build a set of candidates C_k of size k from S_{k-1};
4 $\quad$ $S_k \leftarrow$ frequent subgraphs in C_k;

- A node/edge insertion: adding one node/edge to the graph

- A node/edge deletion: removing one node/edge to the graph

- A node/edge substitution: replacing one node/edge label

Each of these edit operations is associated with a cost $c(.) \in \mathbb{R}_+$. Considering these edit costs, graph edit distance is defined as

$$d_{edit}(G, G') = \min_{(e_1,\ldots,e_k) \in \mathcal{C}(G,G')} \sum_{i=1}^{k} c(e_i) \qquad (15.1)$$

where $\mathcal{C}(G, G')$ encodes all edit paths transforming G into G'. Each edit path encodes a sequence of edit operations $e_1, \ldots, e_k$, each operation e_i being associated to an edit cost $c(e_i)$.

Graph edit distance may be computed using A^* algorithm [22]. This algorithm consists in building a rooted tree where each path from the root to a leaf encodes a possible edit path. Then, the root to leaf path corresponding to the lowest sum of edit operation costs is defined as the optimal edit path. However, the computational complexity induced by this method is exponential according to the number of graphs' nodes that limits its application to very small graphs.

15.2.3.1 Graph Edit Distance Approximation

In order to reduce the complexity required by the computation of graph edit distance, Riesen proposed a polynomial algorithm [46] that reduces graph edit distance complexity. Given two graphs $G = (V, E)$ of size $n = |V|$ and $G' = (V', E')$ of size $m = |V'|$, this algorithm is based on a complete bipartite graph $G_a = (V_a, E_a)$ where $V_a = V_G \cup V_{G'}$, $V_G = V \cup \{\varepsilon_1, \ldots, \varepsilon_m\}$, $V_{G'} = V' \cup \{\varepsilon'_1, \ldots, \varepsilon'_n\}$, and $E_a = \{(u, v) \mid u \in V_G \text{ and } v \in V_{G'}\}$. Bipartite graph G_a encodes a node to node matching between the two sets V_G and $V_{G'}$. A matching $u \in V \rightarrow v \in V'$ encodes a substitution, whereas a matching $u \in V \rightarrow \varepsilon'_i$ (resp. $\varepsilon_i \rightarrow v \in V'$) encodes a node deletion (resp. a node insertion). Each edge $(u, v) \in V_G \times V_{G'}$ is weighted by the cost associated to the edit operation encoded by the matching $u \rightarrow v$. Therefore, each matching $\varepsilon_i \rightarrow \varepsilon'_j$ is associated to a zero cost since they do not encode any edit operation.

Optimal matching, that is, matching from V to V' having the lowest sum of edge cost, is computed by means of Munkres' algorithm (or Hungarian algorithm) [38]. This optimal matching is then associated to an edit path, itself being associated to a cost. However, the computed cost does not correspond necessarily to graph edit distance since there is no guarantee that the computed edit path is the optimal one. Therefore, on one hand, the main drawback of this algorithm is that the computed dissimilarity measure does not correspond to the exact graph edit distance. On the other hand, this dissimilarity measure can be computed in polynomial time that allows the use of this approach on chemoinformatics problems.

15.2.3.2   Graph Edit Distance and Machine Learning Methods

If we consider that costs associated to elementary edit operations define a distance (Definition 15.2.4), graph edit distance fulfills the four properties of a metric and thus defines a distance [39].

Definition 15.2.4 (Distance) *A function* $d : \mathcal{X}^2 \to \mathbb{R}_+$ *is a distance over* $\mathcal{X}$ *if, for any* $(x_i, x_k) \in \mathcal{X}^2$, *it fulfills the four following properties:*

1. *Nonnegativity:* $d(x_i, x_j) \geq 0$.

2. *Self-distance:* $d(x_i, x_j) = 0 \Leftrightarrow x_i \simeq x_j$.

3. *Symmetry:* $d(x_i, x_j) = d(x_j, x_i)$.

4. *Triangle inequality: for any* $x_k \in \mathcal{X}$, $d(x_i, x_j) \leq d(x_i, x_k) + d(x_k, x_j)$.

However, graph edit distance does not fulfill the fifth property of a Euclidean distance (Definition 15.2.5) [10] and hence does not correspond to a distance in a Euclidean space. From a practical point of view, this limitation induces that the metric space defined by the edit distance on the set of graphs cannot be isometrically embedded (neither explicitly nor implicitly) into a Hilbert space. This last point restricts drastically the set of machine learning algorithm that may be used in conjunction with the graph edit distance.

The need of a fifth property to define Euclidean distances can be illustrated using a distance matrix completion problem. Let us consider the example defined by [10] and based on four points x_1, x_2, x_3, and x_4 forming a polyhedron (Figure 15.3). Distance

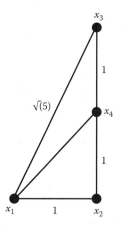

FIGURE 15.3: Four-point polyhedron and point-to-point distances. Distance between x_1 and x_4 cannot be retrieved from the first four properties of a distance. (Example taken from Dattorro, J., *Convex Optimization & Euclidean Distance Geometry*, Meboo Publishing, Palo Alto, CA, 2005.)

matrix $D \in \mathbb{R}^{4,4}$ encodes squared distances between the four points of this polyhedron. Let us consider the following incomplete distance matrix D:

$$D = \begin{bmatrix} 0 & 1 & 5 & d_{14} \\ 1 & 0 & 4 & 1 \\ 5 & 4 & 0 & 1 \\ d_{41} & 1 & 1 & 0 \end{bmatrix} \tag{15.2}$$

The first property of a distance induces that $d_{14} \geq 0$ and $d_{41} \geq 0$. The third property induces that $d_{14} = d_{41}$, and finally, the fourth property of a distance allows us to bind d_{14} by $\sqrt{5} - 1 \leq d_{14} \leq 2$. However, only one value corresponds to a Euclidean embedding since x_4 is at a distance of 1 from x_2 and x_3. Moreover, x_2 and x_3 are separated by a distance of 2, which induces that x_2, x_3, and x_4 must be collinear. All other distance values fulfill the first four distance properties, but they do not correspond to an embedding regardless of the number of dimensions [10]. Therefore, the four properties of a distance are not sufficient to ensure that a distance corresponds to a distance in a Euclidean space.

Definition 15.2.5 (Relative angle inequality [10]) *For any* $i, j, l \neq k$, $k = \{1, \ldots, N\}$, $i < j < l$. *For* $N \geq 4$, *let us consider a set of distinct points* $\mathcal{X}_k = \{x_1, \ldots, x_N\}$. *Relative angle inequality property is fulfilled if all points* $x_k \in \mathcal{X}_k$ *fulfill the following inequalities:*

$$cos(\theta_{ikl} + \theta_{lkj}) \leq cos(\theta_{ikj}) \leq cos(\theta_{ikl} - \theta_{lkj})$$
$$0 \leq \theta_{ikl}, \theta_{lkj}, \theta_{ikj} \leq \pi \tag{15.3}$$

where $\theta_{ikj} = \theta_{jki}$ *encodes angle between vectors* $\overrightarrow{x_k x_i}$ *and* $\overrightarrow{x_k x_j}$.

Proposition 15.2.1 (Euclidean distance) *A function* $d : \mathcal{X}^2 \to \mathbb{R}_+$ *corresponds to a distance in a Euclidean space* $\mathbb{R}^N$ *if d is a distance (Proposition 15.2.4) and d fulfills the relative angle inequality (Definition 15.2.5).*

Proof: see [10].

An alternative characterization of a Euclidean distance function is provided by the following proposition [10]:

Proposition 15.2.2 (Euclidean distance matrix) *The matrix* D *corresponds to a Euclidean distance if*

$$D \in \mathbb{S}_h^N \text{ and } \forall c \in \mathbb{R}^N, \quad s.t \begin{cases} e^t c = 0 \\ \|c\| = 1 \end{cases} \quad c^t D c \leq 0 \tag{15.4}$$

where $e = (1, \ldots, 1)^t$ *and* $\mathbb{S}_h^N$ *corresponds to the set of* $N \times N$ *symmetric matrices having 0 main diagonal.*

Since graph edit distance does not define a distance in a Euclidean space, it cannot be directly used in machine learning methods where an implicit or explicit vectorial embedding is mandatory. Despite this, graph edit distance encodes a dissimilarity measure between two molecular graphs that may be used in conjunction with some algorithms such as the k-nearest neighbors or k-median algorithms.

15.3   Graph Embedding Methods

Methods presented in Section 15.1 are defined within graph space. Despite the fact that this space allows to encode complex structures, it suffers from a lack of mathematical properties. Therefore, the use of many well-known machine learning methods is not possible within such a space. In order to use conjointly graphs and usual machine learning methods, one possibility consists in defining an explicit vectorial space where each graph is embedded. Using such a scheme, each graph is encoded by a vector that can be used conjointly with any machine learning method defined on vectorial representations. Such vectorial representations of graphs are called graph embeddings.

15.3.1   Empirical Kernel Map

In order to combine graph edit distance and machine learning algorithms, some methods aim to define a graph embedding based on graph edit distance. Riesen has proposed to encode a graph by an explicit vector that encodes distances between an input graph and a set of graph prototypes [44]. Given a set of prototype graphs denoted by $\mathcal{G}_{ref} = \{G_1, \ldots, G_N\}$, each graph is embedded into an explicit Euclidean space by the following embedding function $\Phi_{EKM}^{\mathcal{G}} : \mathcal{G} \to \mathbb{R}^N$:

$$\Phi_{EKM}^{\mathcal{G}_{ref}}(G) = (d(G, G_1), \ldots, d(G, G_N)) \tag{15.5}$$

This method thus encodes each graph by a vector that may be used in any machine learning method defined on Euclidean spaces. Although this method uses a widely used graph dissimilarity measure, the choice of prototype graphs strongly impacts the embedding quality. Riesen [44] addressed this last problem through different heuristics providing *good* sets of prototype graphs.

15.3.2   Isometric Embedding

Let us consider a set of graphs $\mathcal{G} = \{G_1, \ldots, G_n\}$ and a function encoding a graph dissimilarity $d : \mathcal{G} \times \mathcal{G} \to \mathbb{R}$. We define a dissimilarity matrix D by $D_{i,j} = d(G_i, G_j)^2 \in \mathbb{R}^{n \times n}$. The isometric embedding method described in [27] consists in computing n vectors x_i having p-dimensions such that the distance between x_i and x_j is as close as possible from the dissimilarity measure between G_i and G_j encoded by $D_{i,j}$.

In order to define vectorial representations $(x_i)_{i \in \{1, \ldots, n\}}$, a method consists in defining a matrix S that encodes scalar products between vectorial representations of graphs. Considering a distance matrix D, matrix S is defined by $S = (I - ee^t)D(I - ee^t)$ with $e = (1, \ldots, 1)^t$ and satisfies: $D_{i,j} = S_{i,i} + S_{j,j} - 2S_{i,j}$. This last equation corresponds to the well-known relationship between Euclidean distance and scalar products: $\|x_i - x_j\|^2 = \langle x_i, x_i \rangle + \langle x_j, x_j \rangle - 2\langle x_i, x_j \rangle$. If S is a semidefinite positive matrix, then its spectral decomposition is given by $S = V \Lambda V^t$ where Λ is a diagonal matrix containing the (positive) eigenvalues of S. Matrix S can then be written as XX^t with $X = V\left(\Lambda^{\frac{1}{2}}\right)$. Each entry $S_{i,j}$ of S is then defined as a scalar product between two rows of matrix X. Taking x_i as an embedding of graph G_i, we indeed obtain $D_{i,j} = d^2(G_i, G_j) = \|x_i - x_j\|^2$.

However, if the dissimilarity matrix D does not correspond to a distance in a Euclidean space (Definition 15.2.5), matrix S is not necessarily semidefinite positive and we have to regularize it. One possibility is to use the constant shift embedding method, which consists in removing negative eigenvalues of S by subtracting its lowest (negative) eigenvalue. Regularized matrix S' is then defined by $S' = S - \lambda_n(S)I$ where $\lambda_n(S)$ denotes the lowest negative eigenvalue of S. This regularization step alters the initial distances according to the following equation: $D' = D - 2\lambda_n(S)(ee^t - I)$. The magnitude of this alteration depends linearly on the lowest eigenvalue of S. Such a regularization step should be performed only for small values of $|\lambda_n(S)|$.

15.3.3 Local Approaches

Other graph embedding methods that are not based on graph edit distance can be defined using different approaches. A first family consists in encoding each graph by a set of information describing local adjacency relationships and labeling information. Attribute statistics-based embedding [20] defines a graph embedding where each coordinate of the vectorial representation encodes either the number of occurrences of a node label or the number of edge incident to two node labels. For $|L_V|$ node labels and $|L_E|$ edge labels, the size of a vector encoding a nonoriented graph is thus equal to $\frac{1}{2}|L_E||L_V|(|L_V| + 1)$. This type of embedding limits the size of encoded paths and thus the amount of encoded information. Indeed, including paths of size 3 induces a huge increase in the size of vectors in order to encode all possible labeled paths of size 3.

In order to encode more structural information, topological embedding [50] encodes the number of occurrences of a set of unlabeled structures corresponding to the set of isomorphic graphs having at most N nodes. Labeling information is included within a matrix representation by means of histograms: for each structure, a histogram encodes the distribution of node and edge labels among all labeled subgraphs corresponding to the unlabeled structure. This method allows to encode larger structures that encode more structural information than paths of size 2. However, the multiple configurations of labels within all subgraphs isomorphic to a given unlabeled structure are reduced to a simple histogram of node and edge labels, hence losing an important part of the label information. This loss of information is induced by the limit on the size of vectors used to encode graphs.

15.3.4 Global Approaches

Instead of using local approaches, some methods aim to define an embedding that globally encodes a graph rather than using a concatenation of local characteristics. Fuzzy multilevel graph embedding [31] defines a vectorial representation encoding both local characteristics, such as node degrees or node and edge labels, and simple global information such as the number of nodes and edges of graphs. Vectorial representation provided by a fuzzy multilevel graph embedding includes thus different levels of analysis of the graph.

Spectral embedding methods [7,29,30] base the vectorial representation of a graph on different characteristics of the spectrum of the graph Laplacian matrix. The basic idea of this family of methods is that the spectrum of the graph Laplacian matrix is insensitive to any rotation of the adjacency matrix and may thus be considered as a characteristic of the graph rather than a characteristic of its adjacency matrix. The resulting graph embedding is based on global characteristics of the graphs that are however often difficult to interpret in terms of graph properties.

15.3.5   Conclusion

Graph embedding methods define graph vectorial representations. Conversely to methods defined within graph space, these vectorial representations can be used in any machine learning methods defined on vectors. However, the encoding of a graph by an explicit vector of limited size induces necessarily a loss of information. This loss may alter prediction models since some relevant information may not be encoded by a vector of limited size.

15.4   Kernel Theory

Methods presented in Sections 15.2 and 15.3 are based on two different approaches. The first one consists in defining similarity measures between graphs without reference to any embedding. Such similarity measures may be combined with few machine learning algorithms such as the k-nearest neighbors or the k-median algorithms. The second one consists in defining vectorial representations of graphs. These methods may be combined with any machine learning algorithm, but the transformation from graphs to vectors may induce important loss of information due to the limited vector size.

In order to combine a graph similarity measure with many machine learning algorithms, we have thus to transform our graphs into vectors. In the same time, the dimension of the embedding space must be sufficiently large (even infinite) in order to avoid the loss of most of the graph information by this transformation. From this point of view, graph kernels provide a good mathematical framework that allows to define a similarity measure between graphs that corresponds to a scalar product in some Hilbert space. The key point, known as the kernel trick, is that many machine learning algorithms may be rewritten solely in terms of scalar products between input data. The substitution of scalar products by graph kernels within these methods allows to avoid the explicit transformation of graphs into vectors, hence allowing to work in Hilbert spaces of arbitrary dimension.

15.4.1 Definitions

Intuitively, a kernel $k : \mathcal{X}^2 \to \mathbb{R}$ between two objects x and x' corresponds to a similarity measure defined as a scalar product between two projections $\Phi_{\mathcal{H}}(x)$ and $\Phi_{\mathcal{H}}(x')$ of x and x' into a Hilbert space $\mathcal{H}$:

$$\forall (x, x') \in \mathcal{X}^2, \ k(x, x') = \langle \Phi_{\mathcal{H}}(x), \Phi_{\mathcal{H}}(x') \rangle. \tag{15.6}$$

In order to define a valid kernel k, it is not mandatory to explicitly define the embedding function $\Phi_{\mathcal{H}} : \mathcal{X} \to \mathcal{H}$. However, to ensure that such an embedding does exist, the kernel k has to fulfill some properties.

Definition 15.4.1 (Positive definite kernel) *A positive definite kernel on $\mathcal{X}^2$ is a function $k : \mathcal{X}^2 \to \mathbb{R}$ symmetric*

$$\forall (x, x') \in \mathcal{X}^2, k(x, x') = k(x', x) \tag{15.7}$$

and semidefinite positive (Mercer's condition [36])

$$\forall \{x_1, \ \ldots, \ x_n\} \in \mathcal{X}^n, \forall c \in \mathbb{R}^n, \sum_i^n \sum_j^n c_i k(x_i, x_j) c_j \geq 0. \tag{15.8}$$

Some usual kernels defined on vectorial representations are listed in Table 15.1.

Definition 15.4.2 (Gram matrix) *A Gram matrix K associated to a kernel k on a finite set $X = \{x_1, \ldots, x_N\}$ is a $N \times N$ matrix defined by*

$$\forall (i, j) \in \{1, \ldots, N\}^2, \ K_{i,j} = k(x_i, x_j). \tag{15.9}$$

For any finite set of objects $X = \{x_1, \ldots, x_N\}$, the Gram matrix associated to a positive definite kernel k is semidefinite positive. Conversely, if, for any set $X = \{x_1, \ldots, x_N\}$, the Gram matrix K associated to kernel k is semidefinite positive, then k is a positive definite kernel.

A positive definite kernel k can be built from a combination of positive definite kernels $k_1, \ldots, k_n$:

Proposition 15.4.1 (Kernel combination [2]) *Let k_1 and k_2 denote two definite positive kernels defined on $\mathcal{X}^2$, $\mathcal{X}$ corresponding to a nonempty space. We have*

TABLE 15.1: Usual Definite Positive Kernels between Vectors

Linear	$k(x, y) = x^T y$
Gaussian	$k(x, y) = exp(-\frac{\|x-y\|^2}{2\sigma^2})$
Polynomial	$k(x, y) = (x^T y)^d + c, c \in \mathbb{R}, d \in \mathbb{N}$
Cosine	$k(x, y) = \frac{(x^T y)}{\|x\| \|y\|}$
Intersection	$k(x, y) = \sum_{i=1}^{N} min(x_i, y_i)$

1. *The set of positive definite kernels is a closed convex cone. Therefore,*

   - *Let $w_1, w_2 \geq 0$; kernel $k_3 := w_1 k_1 + w_2 k_2$ is positive definite.*

   - *Let k_n be a sequence of positive definite kernels and $k(x, x') := \lim\limits_{n \to \infty} k_n(x, x')$; then k is a positive definite kernel.*

2. *The product of two definite positive kernels is a positive definite kernel.*

3. *Let us consider that for $i = 1, 2$, k_i is a positive definite kernel on $\mathcal{X}_i^2$, $\mathcal{X}_i$ being defined as a nonempty space, and let us consider $(x_1, y_1) \in \mathcal{X}_1^2$ and $(x_2, y_2) \in \mathcal{X}_2^2$. Then, the tensor product $k_1 \otimes k_2((x_1, x_2), (y_1, y_2)) = k_1(x_1, y_1) k_2(x_2, y_2)$ and the direct sum $k_1 \oplus k_2((x_1, x_2), (y_1, y_2)) = k_1(x_1, y_1) + k_2(x_2, y_2)$ correspond to positive definite kernels on $(\mathcal{X}_1 \times \mathcal{X}_2)^2$.*

Following [1], if k is a positive definite kernel on $\mathcal{X}$, then there exists a Hilbert space $\mathcal{H}$, having scalar product $\langle \cdot, \cdot \rangle_{\mathcal{H}}$, and an embedding $\Phi : \mathcal{X} \to \mathcal{H}$ such that

$$\forall (x, x') \in \mathcal{X}^2, \ k(x, x') = \langle \Phi(x), \Phi(x') \rangle_{\mathcal{H}} \tag{15.10}$$

The Hilbert space $\mathcal{H}$ is called the reproducing kernel Hilbert space (RKHS) of k or more usually the feature space.

15.4.2 Kernel Trick

Let us consider a polynomial kernel $k(x, y) = \langle x, y \rangle^2$ with $x = (x_1, x_2)$ and $y = (y_1, y_2) \in \mathbb{R}^2$. Kernel value $k(x, y)$ is thus equal to

$$k(x, y) = x_1^2 y_1^2 + x_2^2 y_2^2 + \sqrt{2}(x_1 x_2)\sqrt{2}(y_1 y_2) \tag{15.11}$$

Despite the fact that this last equation does not correspond to the usual equation of a scalar product, it indeed corresponds to a scalar product between a mapping of x and y through the following function:

$$\begin{pmatrix} x_1 \\ x_2 \end{pmatrix} \xrightarrow{\Phi} \begin{pmatrix} x_1^2 \\ x_2^2 \\ \sqrt{2} x_1 x_2 \end{pmatrix} \tag{15.12}$$

Note that, using a polynomial kernel of degree 2, we implicitly work in a space of dimension 3, while our original data are of dimension 2. Let us now suppose that we wish to combine our kernel with a k-nearest neighbors algorithm. We have thus to compute distances between vectors in the feature space of dimension 3. Such distances may be computed as follows:

$$\begin{aligned} d_k^2(x, y) = \| \Phi(x) - \Phi(y) \|^2 &= \langle \Phi(x), \Phi(x) \rangle + \langle \Phi(y), \Phi(y) \rangle \\ &\quad - 2\langle \Phi(x), \Phi(y) \rangle \\ &= k(x, x) + k(y, y) - 2k(x, y) \end{aligned} \tag{15.13}$$

Therefore, the k-nearest neighbors algorithm may be applied in the feature space of dimension 3 without computing the transformation Φ but by using solely our kernel between data of dimension 2. This trick, known as the kernel trick, allows us to avoid the explicit transformation of our original input data in many machine learning algorithms. This last point allows us to avoid to compute and to store vectors of very large or even infinite dimension, for example, for the Gaussian kernel (Table 15.1).

Applied to graphs, the kernel trick allows us to avoid the explicit transformation of graphs into vectors and to focus our attention on the definition of a similarity measure. The fact that the implicit Hilbert space may be of very large dimension allows to reduce the loss of information induced by the transformation from graphs to vectors.

15.4.3 Kernels and Similarity Measures

From a mathematical point of view, a kernel is defined as a scalar product between objects embedded into a Hilbert space. However, kernels are generally considered as similarity measures. This relationship between scalar product and similarity measure may be explained by the relationship between scalar product and Euclidean distance (Equation 15.13):

$$\|\Phi(x) - \Phi(y)\|^2 = k(x,x) + k(y,y) - 2k(x,y)$$

$$\Rightarrow k(x,y) = \frac{1}{2}(\|\Phi(x)\|^2 + \|\Phi(y)\|^2 - \|\Phi(x) - \Phi(y)\|^2) \qquad (15.14)$$

If we consider normalized vectorial representations, that is, $\|\Phi(x)\| = 1 \ \forall x \in \mathcal{X}$, we obtain

$$k(x,y) = 1 - \frac{1}{2}d_k^2(x,y)$$

where $d_k^2(x,y) = \|\Phi(x) - \Phi(y)\|^2$.

In this case, kernels are defined as the opposite of the Euclidean distance between vectors in the feature space. Intuitively, Euclidean distance encodes a dissimilarity between objects. Therefore, a kernel function defined as a decreasing function of the distance encodes a similarity measure between objects. A high value corresponds to a high similarity, whereas a low value, close to 0, encodes a high dissimilarity between objects.

15.4.4 Kernel Methods

15.4.4.1 Support Vector Machines

Problem definition [5]: SVM corresponds to a machine learning algorithm. The classification problem addressed by SVM is the following: given a set of objects labeled by a class $\{x_i, y_i\}_{i=1}^{n}$, $x_i \in \mathbb{R}^d$ and $y_i \in \mathcal{Y} = \{-1, +1\}$, learn a function $f : \mathbb{R}^d \to \mathcal{Y}$ such that $f(x_i) = y_i$. SVM algorithm consists in finding a hyperplane having $d - 1$ dimensions that separates data points according to their classes (Figure 15.4).

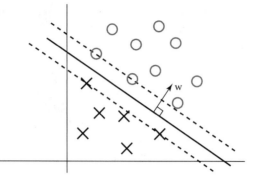

FIGURE 15.4:   Hyperplane splitting data according to their classes (circles and crosses). Dashed lines encode margins.

The optimal hyperplane is the one that maximizes the distance between the hyperplane and the closest data points of each class. Such a distance is called the margin. Considering the hyperplane defined by $\langle w, x \rangle + b = 0$, this distance is inversely proportional to the norm of vector w. Finding the best hyperplane thus relies to solve

$$\underset{w}{\text{minimize}} \ \frac{1}{2} \|w\|^2$$

subject to:

$$y_i(\langle w, x \rangle + b) \geq 1, \forall i \in \{1, \ldots, n\} \tag{15.15}$$

Considering a nonlinear separable case, that is, where a hyperplane having $d - 1$ dimensions that separates data does not exist, Cortes and Vapnik proposed to include slack variables $\xi_i \in \mathbb{R}^+$ [9]. These variables allow to take into account classification errors during training. The addressed problem is then defined by

$$\underset{w}{\text{minimize}} \ \frac{1}{2} \|w\|^2 + C \sum_{i=1}^{n} \xi_i$$

subject to:

$$y_i(\langle w, x \rangle + b) \geq 1 - \xi_i, \forall i \in \{1, \ldots, n\}$$
$$\xi_i \geq 0, \forall i \in \{1, \ldots, n\} \tag{15.16}$$

where $C \in \mathbb{R}_+$ is a regularization parameter that allows to weight the influence of errors made during training. A high C value favors a learning without errors and thus potentially an overlearning. Conversely, a low C value allows more errors during training and thus a greater generalization.

Using a kernel k together with SVM allows to find a linear hyperplane that solves Equation 15.16 in the feature space associated to kernel k. Thanks to the kernel trick, this hyperplane may be a nonlinear separator in the original data space. For example, let us consider the kernel defined by Equation 15.12. The hyperplane equation

$\langle w, x \rangle + b = 0$ is computed within the kernel feature space and corresponds to a nonlinear equation in $\mathbb{R}^2$:

$$w_1 x_1^2 + w_2 x_2^2 + w_3 \sqrt{2} x_1 x_2 + b = 0 \tag{15.17}$$

Note that such an equation corresponds to a quadric.

15.4.4.2 Support Vector Machines for Regression

Regression problems consist in predicting a continuous value, conversely to classification problems that aim to predict a discrete value encoding a class. More formally, a regression problem is defined as follows: given a learning set $\{x_i, y_i\}_{i=1}^n$, composed of a set of n objects $\mathcal{X} = \{x_1, \ldots, x_n\}$ with $x_i \in \mathbb{R}^d$, each object being associated to a value $y_i \in \mathbb{R}$, learn a prediction function $f : \mathbb{R}^d \to \mathbb{R}$ such that $\hat{y}_i \simeq f(x_i)$.

SVMs are initially defined as a solver for classification problems. In order to handle regression problems, SVM has been adapted [12] by including an ε-insensitive cost function such as defined in [53]. Instead of computing a hyperplane splitting data according to their classes, SVM for regression consists in computing a hyperplane w associated to an ε-tube that includes data points to predict (Figure 15.5).

Formally, the minimization problem addressed by SVM for regression is defined by

$$\underset{w}{\text{minimize}} \; \frac{1}{2} \|w\|^2$$

subject to:

$$\begin{cases} y_i - \langle w, x_i \rangle - b \leq \varepsilon, \forall i \in \{1, \ldots, n\} \\ \langle w, x_i \rangle + b - y_i \geq \varepsilon, \forall i \in \{1, \ldots, n\} \end{cases} \tag{15.18}$$

Minimizing Equation 15.18 relies to compute a linear function that approximates y values with an accuracy ε. In order to allow errors during training, slack variables for regression have been introduced into minimization problem. Similar to SVM classification, these slack variables allow to weight, according to C, prediction errors made during the learning step. The minimization problem is then defined as

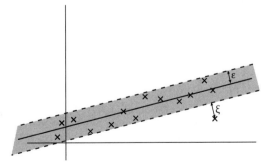

FIGURE 15.5: Hyperplane associated to an ε-tube (shaded).

$$\underset{w}{\text{minimize}} \frac{1}{2}\|w\|^2 + C\sum_{i=1}^{m}(\xi_i + \xi_i^*)$$

subject to:

$$\forall i \in \{1, \dots, n\} \begin{cases} y_i - \langle w, x_i \rangle - b \leq \varepsilon + \xi_i \\ \langle w, x_i \rangle + b - y_i \geq \varepsilon + \xi_i^* \\ \xi_i, \xi_i^* \geq 0 \end{cases}$$

(15.19)

This regression algorithm allows to compute a prediction function up to an accuracy defined by ε. This parameter is particularly interesting in chemoinformatics since it may encode inaccuracies of physical properties induced by experiments that are not predictable using molecular graphs.

15.4.4.3 Kernel Ridge Regression

Considering the ridge regression framework, a regression problem can be addressed by computing a linear function that encodes relationships between data x and responses y. Considering the $d \times n$ matrix X encoding each data $x_i \in \mathcal{X}$, this function may be computed by minimizing the following objective function:

$$\underset{w}{\text{minimize}} J(w)$$

with

$$J(w) = \|y - X^t w\|^2 + \lambda\|w\|^2$$

(15.20)

Minimizing the first term of Equation 15.20 relies to solve a least square problem corresponding to the minimization of prediction errors. The second term $(\lambda\|w\|^2)$ corresponds to a regularization term that aims to penalize vectors w having a high norm. The factor λ allows to weight the influence of the regularization term into the minimization problem. Therefore, a high λ value may prevent the prediction model from overlearning, which may induce a better generalization of the prediction model when applied on test sets.

The objective function to minimize, defined by Equation 15.20, corresponds to a sum of L_2 norms that thus defines a convex function. Global minimum is reached for a value w satisfying $\frac{\partial J}{\partial w} = 0$. This optimal w is obtained by the following analytic form:

$$w^\star = X(X^t X + \lambda I)^{-1} y$$

(15.21)

Predicting property value of a new data $x' \in \mathbb{R}^d$ relies on computing $\hat{y} = w^{\star t} x' = y(X^t X + \lambda I)^{-1} X^t x' = y(K + \lambda I)^{-1} \kappa(x')$ where K denotes the Gram matrix associated to train set, that is, $K_{i,j} = \langle x_i, x_j \rangle$ and $\kappa(x')_i = \langle x_i, x' \rangle$. We can notice that the access to data X is only performed through K and $\kappa(x')$. Therefore, the minimization of objective function and value prediction can be performed without accessing directly to raw data $\{x_1, \dots, x_n\}$ but only through the scalar product defined on these data. Within the kernel framework, XX^t corresponds to the Gram matrix K associated to the kernel

k such as $K_{i,j} = k(x_i, x_j) = \langle x_i, x_j \rangle$. In the same way, $\langle x_i, x' \rangle$ is equal to $k(x_i, x')$. Given a nonlinear kernel $k_{\mathcal{H}}$, associated to a feature space $\mathcal{H}$ and an embedding function $\Phi_{\mathcal{H}} : \mathcal{X} \rightarrow \mathcal{H}$, ridge regression using $k_{\mathcal{H}}$ relies to compute a vector w lying on feature space $\mathcal{H}$, instead of original data space $\mathcal{X}$.

15.4.5 Graph Kernels and Chemoinformatics

Considering conclusions of Sections 15.2 and 15.3, graph kernels provide a solution to both limitations involved by methods based on graph embeddings and methods defined in graph space. On one hand, graph kernels allow to define a similarity measure that is not limited by vector sizes since graph vectorial representations are not required to be explicitly and exhaustively defined. On the other hand, graph kernels guarantee that a vectorial representation exists hence allowing the conjoint use of kernels with many machine learning methods. Therefore, graph kernels define a natural connection between molecular graphs and machine learning algorithms. Many graph kernels have thus been defined to solve Quantitative structure-activity relationship/Quantitative structure property relationship (QSAR/QSPR) problems in chemoinformatics.

15.5 Graph Kernels Based on Bags of Patterns

A graph kernel family widely used in chemoinformatics consists in defining similarity between graphs from bags of patterns extracted from graphs to be compared. Formally, given a set of substructures $\mathcal{P}$, a bag of patterns extracted from a graph G is defined as $\mathcal{B}(G) = \{(p, f_p(G)), p \in \mathcal{P}, p \subseteq G\}$ where p encodes a pattern and $f_p(G)$ encodes some properties (e.g., the cardinal) of the set of p in G. Given a subkernel $k : (\mathcal{P} \times \mathbb{R})^2 \rightarrow \mathbb{R}$ defined as a tensor product of two kernels $k_1 : \mathcal{P} \times \mathcal{P} \rightarrow \mathbb{R}$ and $k_2 : \mathbb{R} \times \mathbb{R} \rightarrow \mathbb{R}$, a convolution kernel [23] $K : \mathcal{G} \times \mathcal{G} \rightarrow \mathbb{R}$ is defined as a sum of subkernel values:

$$K(G, G') = \sum_{x \in \mathcal{B}(G)} \sum_{y \in \mathcal{B}(G)} k(x, y) \tag{15.22}$$

$$K(G, G') = \sum_{x \in \mathcal{B}(G)} \sum_{y \in \mathcal{B}(G)} (k_1 \otimes k_2)(x_1, x_2, y_1, y_2) \tag{15.23}$$

$$K(G, G') = \sum_{x \in \mathcal{B}(G)} \sum_{y \in \mathcal{B}(G)} \prod_{i=1}^{2} k_i(x_i, y_i) \tag{15.24}$$

For each pair of substructures $(p, p') \in \mathcal{P}^2$, subkernel k encodes the similarity between two molecular graphs according to patterns p and p' and their set of occurrences in G and G'. Kernels presented in this section are defined as convolution kernels. Although these kernels use a common approach, they mainly differ on the set of substructures used to define them.

15.5.1   Complete Graph Kernel

A first trivial approach consists in defining the set of substructures by all possible subgraphs.

Definition 15.5.1 (Complete graph kernel) *Let G and G' be two graphs. Complete graph kernel is defined by*

$$k_{complete}(G, G') = \sum_{p \sqsubseteq G} \sum_{p' \sqsubseteq G'} k_{iso}(p, p') \tag{15.25}$$

with

$$k_{iso}(p, p') = \begin{cases} 1 \text{ iff } p \simeq p' \\ 0 \text{ otherwise} . \end{cases} \tag{15.26}$$

Complete graph kernel [15] relies on embedding graphs into a Hilbert space where each dimension encodes the number of occurrences of a subgraph of the considered graph. However, computing this kernel is NP-hard since it relies on determining if two graphs are isomorphic. Therefore, this kernel cannot be efficiently computed, which avoids using such an approach to chemoinformatics problems.

15.5.2   Linear Pattern Kernels

15.5.2.1   Random Walks

A less complex approach consists in deducing graph similarity from the number of random walks common to both graphs to be compared. Two methods [15,28] have been proposed both based on the bag of random walks $\mathcal{W}(G)$ of a graph G. Kernels defined using random walks are based on the following formulation:

Definition 15.5.2 (Marginalized kernel) *Given two graphs G and G', $\mathcal{W}(G)$ and $\mathcal{W}(G')$ encode the sets of random walks in G and G', and $k_{\mathcal{W}} : \mathcal{W} \times \mathcal{W} \to \mathbb{R}$ is defined as a kernel between walks. A marginalized kernel is then defined as*

$$k_{rw}(G, G') = \sum_{w \in \mathcal{W}(G)} \sum_{w' \in \mathcal{W}(G')} p_G(w) p_{G'}(w') k_{\mathcal{W}}(w, w') \tag{15.27}$$

where $p_G(w)$ corresponds to the probability of traversing w in G.

Kernel $k_{\mathcal{W}}$ corresponds to a kernel encoding a similarity between walks according to their labeling. Usually, this kernel is binary, that is, $k_{\mathcal{W}}$ is equal to 1 if label sequences are similar and 0 otherwise. Method defined in [15] uses a different approach based on a direct product graph but is based on the same feature space and thus encodes the same information. Different implementations have been proposed [15,28,33,55], and this kernel may be computed in polynomial time with the number of nodes of both graphs.

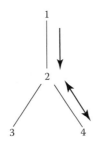

FIGURE 15.6: A path traversal altered by tottering corresponds to a label sequence equals to $1, 2, 4, 2, 4, 2, \ldots$.

15.5.2.2 Tottering

A random walk corresponds to a node sequence that may oscillate between two connected neighbors (Figure 15.6). This phenomenon, called tottering, induces random walks that are not representative of the corresponding molecular graph structure since the same information may be repeated indefinitely. In order to limit the influence of these nonrepresentative structures, Mahé et al. [33] proposed to transform a molecular graph into an oriented graph in $\mathcal{O}(|V|^2)$. Then, the kernel is computed on this new molecular representation that allows to avoid oscillations between two connected nodes and thus to reduce the tottering phenomenon.

15.5.2.3 Paths

Instead of using random walks that may suffer from tottering, other kernels base their similarity measure on paths extracted from graphs to be compared. Although enumerating all paths of a graph remains an NP-hard problem [4], a kernel can be defined using all paths composed of at most n nodes [42]. This path enumeration is performed using a depth first traversal from each node of the graph. Computing the kernel on two graphs $G = (V, E)$ and $G' = (V', E')$ relies on extracting paths from G and G', which induces $\mathcal{O}(n(|V||E| + |V'||E'|))$ operations. Enumerated paths are encoded by a vector that can be seen as a molecular descriptor where each descriptor encodes the number of occurrences of a path. The similarity between these vectors may be computed using usual kernels such as Tanimoto kernel or Min–Max kernel [42].

Another approach consists in computing similarity between graphs from the lengths of shortest paths between any pair of nodes. This method is based on Floyd's transformation [14] that consists in transforming a connected graph $G = (V, E, \mu, \nu)$ into a graph $G_F = (V, E_F, \mu, \nu_F)$ where each edge $e_f = (v_i, v_j) \in E_F$ is labeled by the length of the shortest path between v_i and v_j in G if such a path exists. Considering the two graph transformations $G_F = (V, E_F, \mu, \nu_F)$ and $G'_F = (V', E'_F, \mu', \nu'_F)$ of two graphs $G = (V, E, \mu, \nu)$ and $G' = (V', E', \mu', \nu)'$, the shortest path kernel is defined by

$$k_{sp}(G, G') = \sum_{e \subseteq E_F} \sum_{e' \subseteq E'_F} k_{path}^1(e, e') \tag{15.28}$$

where $k^1_{path}(e, e')$ is a kernel defined between edges. This kernel may be defined as a Dirac function that considers two edges $e = (u, v)$ and $e' = (u', v')$ as similar if and only if $\mu(u) = \mu'(u')$, $\mu(v) = \mu'(v')$, and $v_F(e) = v'_F(e')$. Intuitively, two edges are similar if they connect two pairs of nodes having same labels and separated by the same distance in G and G'. This kernel can be computed in polynomial time $(\mathcal{O}(|V|^4))$, most of the time being dedicated to the computation of Floyd's transformation and to the comparison of the $|V|^2$ edges of transformed graphs. However, this approach only encodes the length of a shortest path connecting two nodes, but not the sequences of node and edge labels composing these shortest paths. This loss of information leads to a less accurate kernel.

15.5.3 Nonlinear Pattern Kernels

Although linear patterns allow to define a kernel that can be computed in polynomial time, most of the structural information included within molecular graphs cannot be encoded using only linear patterns (Figure 15.7).

15.5.3.1 Graphlets

In order to increase the structural information encoded within bags of patterns, graphlet kernel [48] has been introduced. This kernel is based on bags of patterns defined as all unlabeled graphs having k nodes, $k \in \{3, 4, 5\}$ (Figure 15.8). This kernel may be efficiently computed for graphs having a bounded maximum degree (say by d). Note that a molecular graph (Section 15.1.1) has a maximal degree bounded by 7.

Graphlet enumeration is divided into two steps. The first one corresponds to the enumeration of connected graphlets and the second one to the enumeration of

FIGURE 15.7: Two different oriented graphs having the same representation in a feature space based on linear patterns.

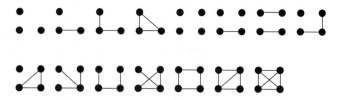

FIGURE 15.8:   All graphlets of size 3 and 4. (From Shervashidze, N. et al., Efficient graphlet kernels for large graph comparison, in *International Conference on Artificial Intelligence and Statistics*, pp. 488–495, 2009.)

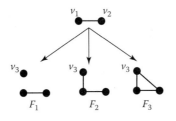

FIGURE 15.9: Different possible topological configurations involving a pair of nodes and a third node.

disconnected graphlets. Connected graphlets are divided into two classes. The first class corresponds to graphlets of size k containing at least a path of size $k-1$ and the second class to other graphlets. The first class of graphlets is enumerated from paths of length $k-1$, themselves enumerated using a depth first traversal. The second class of graphlets is enumerated from *3-star* graphlets that correspond to graphlets having 4 nodes with a central node having a degree equal to 3. Other graphlets are then enumerated by looking at the neighborhood of *3-stars*.

Concerning disconnected graphlets of size 3, their enumeration is based on pairs of nodes $(v_1, v_2) \in E$. If we consider a third node v_3 distinct from v_1 and v_2, three different configurations are possible (Figure 15.9). These different configurations may be distinguished using the neighborhood of (v_1, v_2). We have thus

- $v_3 \notin \Gamma(v_1) \cup \Gamma(v_2) \rightarrow F_1$ configuration

- $v_3 \in \Gamma(v_1) \setminus \Gamma(v_2) \vee v_3 \in \Gamma(v_2) \setminus \Gamma(v_1) \rightarrow F_2$ configuration

- $v_3 \in \Gamma(v_1) \cap \Gamma(v_2) \rightarrow F_3$ configuration

The method defined to enumerate graphlets of size 4 and 5 is based on the same scheme than the one used for graphlets of size $k = 3$: the set of 11 graphlets having 4 nodes is enumerated from the set of graphlets having 3 nodes. Similarly, the 34 graphlets having 5 nodes is enumerated from the ones composed of 4 nodes. This method allows to compute the distribution of graphlets of size k in $\mathcal{O}(nd^{k-1})$ operations that correspond to a linear complexity with the number of nodes of graphs if the maximum degree is bounded. Graphlet kernel is then defined as a scalar product between normalized vectors encoding the number of occurrences of each graphlet:

Definition 15.5.3 (Graphlet kernel) *Let $G = (V, E) \in \mathcal{G}$ and $G' = (V', E') \in \mathcal{G}$ be two graphs and D_G a vector encoding the number of occurrences of graphlets in G:*

$$D_G(i) = \frac{\#graphlet(i) \subseteq G}{\sum_j \#graphlet(j) \subseteq G} \qquad (15.29)$$

Graphlet kernel is then defined by

$$k_{graphlets} = D_G^\top D_{G'} \qquad (15.30)$$

Although this kernel allows to encode more structural information than kernels based on linear patterns, it does not encode the similarity involved by labeling. Therefore, this kernel is limited to unlabeled graphs that correspond to a limited application domain in chemoinformatics. Indeed, atom's chemical elements are an important information and must be taken into account in order to define an accurate similarity measure.

15.5.3.2   Tree Patterns

In order to include molecular graph labeling into kernel computation, tree pattern kernel [34,43] is defined as a kernel based on nonlinear and labeled patterns. This method deduces the similarity between two graphs from the number of their common tree patterns.

Definition 15.5.4 (Tree Pattern) *Let $G(V, E) \in \mathcal{G}$. If it exists a node $r \in V$, then r is a tree pattern of G rooted in r and having a height equal to 1. Let $t_1, t_2, \ldots, t_n$ be n tree patterns respectively rooted in $r_1, r_2, \ldots, r_n$, with $r_i \neq r_j, \forall i \neq j$. If $(r, r_1), (r, r_2), \ldots, (r, r_n) \in E$, then $r(t_1, t_2, \ldots, t_n)$ is a tree pattern rooted in r. r is defined as the parent of each r_i.*

Each tree pattern is associated to a dimension in the feature space. Each dimension encodes the number of tree patterns included within the two graphs to be compared.

Definition 15.5.5 (Neighborhood matching set) *Let $G = (V, E, \mu, \nu)$ and $G' = (V', E', \mu', \nu')$ denote two graphs and let us consider two nodes $r \in V$ and $s \in V'$. Neighborhood matching set $M_{r,s}$ is defined by*

$$M_{r,s} = \{R \subseteq \Gamma(r) \times \Gamma(s) \mid \quad (\forall (a,b), (c,d) \in R : a \neq c \wedge b \neq d) \wedge$$
$$(\forall (a,b) \in R : \mu(a) = \mu'(b) \wedge \nu(r,a) = \nu'(s,b))\}$$
$$(15.31)$$

Definition 15.5.6 (Kernel between tree patterns) *Let $G = (V, E, \mu, \nu)$ and $G' = (V', E', \mu', \nu')$ denote two graphs; the kernel value between two tree patterns of height h and rooted in r and s is defined by*

$$k(r, s, h) = \lambda_r \lambda_s \sum_{R \in M_{r,s}} \prod_{(r',s') \in R} k(r', s', h - 1). \qquad (15.32)$$

For $h = 1$, $k(r, s, 1)$ is equal to 1 if $\mu(r) = \mu'(s)$, 0 otherwise.

Kernel value is strictly greater than 0 if the two tree patterns are isomorphic. Parameters λ_r and λ_ν are defined as positive real numbers lower than 1 such that large tree patterns have a low contribution. Tree pattern kernel is then defined as follows:

Definition 15.5.7 (Tree pattern kernel) *Let $G = (V, E, \mu, \nu) \in \mathcal{G}$ and $G' = (V', E', \mu', \nu') \in \mathcal{G}$. Tree pattern kernel for a height h is defined by*

$$k(G, G', h) = \sum_{r \in V} \sum_{s \in V'} k(r, s, h) \qquad (15.33)$$

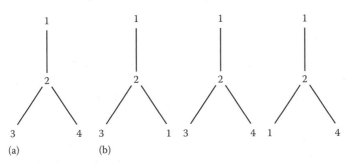

FIGURE 15.10: Difference between tree patterns and trees: graph (a) includes only one subtree of size 4 rooted in 1 (itself) but three different tree patterns (b), all rooted in 1.

15.5.3.3 Weisfeiler–Lehman

Weisfeiler–Lehman kernels [47,49] correspond to graph kernels based on a subset of tree patterns corresponding to tree structures encoding all nodes within a given radius. This algorithm, based on the Weisfeiler–Lehman isomorphism test, allows to compute a kernel based on a subfamily of tree patterns with a linear complexity according to the number of nodes of both graphs. It thus provides faster computational times than the tree pattern kernel (Section 15.5.3.2). Moreover, Weisfeiler–Lehman kernel computes an explicit representation of the feature space associated to kernels. This explicit enumeration allows to perform pattern enumeration only N times where N is the size of the training set. Using an implicit enumeration, pattern enumeration has to be performed for each pair of graph, that is, N^2 times.

One has to notice that tree patterns differ from trees (Figure 15.10). Similarly to the difference between paths and walks, a subtree pattern may include the same node twice. Therefore, information encoded by tree patterns may also be altered by tottering (Section 15.5.2.2). However, graph transformation proposed by Mahé et al. to prevent tottering in random walks kernel can also be applied to kernels based on tree patterns.

15.5.3.4 Graph Fragments

Graph fragment kernel [57] corresponds to a kernel based on a set of patterns defined as all connected subgraphs having at most l edges. This set of subgraphs, denoted GF, can be divided into different subsets:

- *PF (Path fragment)* encodes all linear patterns.

- *TF (tree fragment)* encodes all subtrees having at least one node v such that $d(v) \geq 2$.

- *AF (acyclic fragment)* is defined as $AF = TF \cup PF$, which thus corresponds to all subtrees.

- The fourth subset corresponds to the set difference $GF \setminus AF$. This subset encodes all subgraphs containing at least one cycle.

This set of patterns defines the feature space associated to this kernel where each coordinate $\Phi_i(G)$ encodes either the absence/presence or the number of occurrences of a given subgraph in G. A molecular graph G is thus encoded by a vector $\Phi(G) = (\Phi_1(G), \ldots, \Phi_d(G))$, where d encodes the number of different patterns enumerated for a given length l. The kernel between two graphs G and G' is then defined as a Min–Max kernel [42] between the two vectors $\Phi(G)$ and $\Phi(G')$.

This method, based on an exhaustive enumeration of all subgraphs, relies to compute the complete graph kernel for a l big enough (Section 15.5.1). Since this problem is an NP-hard problem and thus is not feasible for large graphs, graph fragment is generally defined for $l = 7$ in order to obtain reasonable computational times. Moreover, no significant accuracy gain is observed for length $l \geq 5$.

15.5.3.5   Treelets

Conversely to methods based on tree patterns, treelet kernel (TK) [17] uses a bag of patterns defined as a subset of strict subtrees, called treelets. Treelets are defined as the set of all labeled trees having at most 6 nodes (Figure 15.11). Conversely to methods based on a nonlimited set of substructures, such as the tree pattern kernel, considering a set of predefined structures allows to define an efficient *ad hoc* linear enumeration. This enumeration is performed in two steps:

The first step consists to identify treelet's structure in a graph. This step is based on the method used to enumerate connected graphlets [48] (Section 15.5.3.1). Graphlet enumeration method is extended to enumerate tree structures up to 6 nodes while keeping a linear complexity when considering degree bounded graphs. This first step allows to associate a structure index $(G_0, \ldots, G_{13})$ (Figure 15.11) to each treelet.

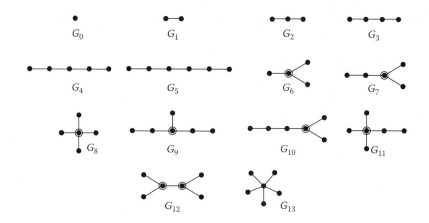

FIGURE 15.11:   Set of treelet's structures.

The second step consists to encode the label of each instance of treelet found within a graph. This step is based on a tree traversal of each instance of treelet based on Morgan algorithm [37]. This traversal provides a string that, as shown in [17], identifies the label of each node and each edge of a treelet given the index of its structure. Note that, since the size of treelet's structure is bounded, the computation of the string encoding treelet's labels is performed in constant time. This last point is an important advantage over alternative encodings used, for example, by frequent subgraphs or graph fragment methods [11].

The concatenation of treelet's index and treelet's key defines a unique code for each treelet that allows to perform an explicit enumeration of all treelets composing a graph. Based on this enumeration, we define a function f that associates to each graph G a vector $f(G)$ whose components encode the number of occurrences of each treelet t found in G:

$$f(G) = (f_t(G))_{t \in \mathcal{T}(G)} \text{ with } f_t(G) = |(t \subseteq G)| \tag{15.34}$$

where $\mathcal{T}(G)$ denotes the set of treelets extracted from G and $\subseteq$ the subgraph isomorphism relationship. Then, similarity between treelet distributions is computed using the sum of subkernels between treelet's frequencies:

$$k_{\mathcal{T}}(G, G') = \sum_{t \in \mathcal{T}(G) \cap \mathcal{T}(G')} k(f_t(G), f_t(G')) \tag{15.35}$$

where $k(.,.)$ defines any positive definite kernel between real numbers such as linear kernel, Gaussian kernel, or intersection kernel. Note that, conversely to tree pattern kernel (Section 15.5.3.2), this kernel explicitly enumerates subtrees by computing the number of occurrences of each pattern.

15.5.4 3D Pattern Kernel

Three-dimensional molecular information may be an important and useful information for some chemoinformatics problems such as docking or optical angle rotation. Among the few existing kernels including such information, 3D pharmacophore kernel [32] encodes a set of patterns corresponding to triplets of distinct atoms. Each pattern can be understood as a potential pharmacophore and encodes Euclidean distances between each pair of nodes included within a corresponding triplet. These distances are measured on the most stable 3D conformation of a molecule. Pharmacophore kernel is defined as a convolution kernel based on a kernel $k_{\mathcal{P}}$ defined on atom triplets. This last kernel is defined as a combination of two kernels:

1. A first kernel that compares node labels of the two triplets to be compared.

2. A second kernel that encodes spatial information by comparing distances between atoms by means of Gaussian kernels. More formally, if $p = (x_i)_{i=\{1,...,3\}}$ and $p' = (x'_i)_{i=\{1,...,3\}}$ are two atom triplets, this kernel is defined as:

$$k_{spatial}(p, p') = \prod_{i=1}^{3} k_{dist}(\|x_i - x_{i+1}\|, \|x'_i - x'_{i+1}\|) \qquad (15.36)$$

Kernel k_{dist} may be defined as a Gaussian kernel and aims to compare two distances.

This kernel encodes spatial information of molecular compounds and allows to define a relevant similarity measure for chemoinformatics problems including 3D information. However, distances used in this kernel are relative to a priori 3D conformations of both molecules while some alternative conformations may better explain the property to predict.

Three-dimensional molecular information may also be encoded by the chirality of molecules that allows to distinguish two stereoisomers. Tree pattern kernel has been adapted to include stereoisomerism [6] when this information can be locally characterized. Similarly, TK has been adapted to include stereoisomerism [21]. However, conversely to [6], this last adaptation allows to also characterize a nonlocal stereoisomerism.

15.5.5   Cyclic Pattern Kernels

Similarly to 3D information, cyclic information may have a particular influence on molecular properties since cycles reduce the atom's degrees of freedom.

15.5.5.1   Cyclic Patterns

A first approach consists in defining a kernel based on the set of simple cycles of a molecular graph (Figure 15.12). Considering a graph G, cyclic pattern kernel [25] is based on a decomposition of G into two subsets:

- A first subset $C(G)$ corresponding to the set of simple cycles included within G (Figure 15.12).

- A second subset $T(G)$ defined as all connected subtrees extracted from the set of bridges $B(G)$. This set of subtrees corresponds thus to atoms and edges that are not included in $C(G)$.

Considering these two subsets, cyclic pattern kernel is defined as the sum of two intersection kernels applied on $T(G)$ and $C(G)$. Considering two graphs G and G', cyclic pattern kernel is thus defined as

$$k_C(G, G') = k_{inter}(T(G), T(G')) + k_{inter}(C(G), C(G')) \qquad (15.37)$$

where $k_{inter}(T(G), T(G'))$, resp. $k_{inter}(C(G), C(G'))$, is defined as an intersection kernel that computes the number of common subtrees in $T(G)$ and $T(G')$, resp. the number of common cycles in $C(G)$ and $C(G')$. Similarity between molecular graphs is thus deduced from a sum of two kernels encoding, respectively, similarities between their cyclic and acyclic parts.

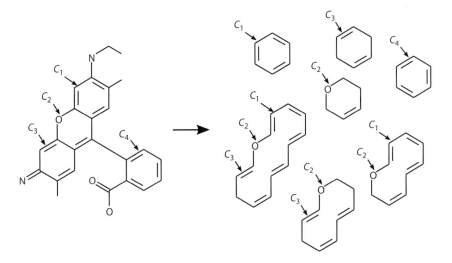

FIGURE 15.12: Decomposition of a molecular graph into its set of simple cycles.

However, computing the set of simple cycles of a graph is an NP-hard problem that may however be addressed in chemoinformatics problems since molecular graphs generally include a low number of simple cycles. In order to tackle this complexity, cyclic pattern kernel has been refined using the set of relevant cycles (denoted $\mathcal{C}_{\mathcal{R}}$) [24,56] instead of the set of simple cycles. The enumeration of relevant cycles can be performed in polynomial time that reduces kernel computational complexity. Relevant cycles can be understood as a set of cycles that defines a basis for encoding all cycles included in a molecular graph (Figure 15.13).

Cyclic pattern kernel allows to explicitly encode cyclic similarity into kernel computation that may provide an accurate similarity measure for chemoinformatics

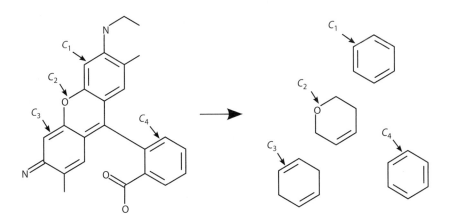

FIGURE 15.13: Decomposition of a molecular graph into its set of relevant cycles.

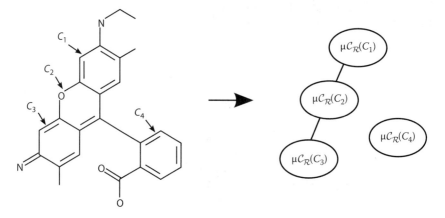

FIGURE 15.14:   Relevant cycle graph computed from a molecular graph.

problems. However, although this approach allows to catch some cyclic information, it does not encode relationships between molecular cycles.

15.5.5.2   Relevant Cycle Graph

In order to encode more cyclic information, relevant cycle relationships may be encoded using the relevant cycle graph [19]. This graph representation aims to encode adjacency relationships between relevant cycles and thus to encode the cyclic system of a molecule. This graph is defined as $G_C = (\mathcal{C_R}, E_{\mathcal{C_R}}, \mu_{\mathcal{C_R}}, \nu_{\mathcal{C_R}})$ where each vertex encodes a relevant cycle and two vertices are connected by an edge if their corresponding cycles share at least one vertex of the initial graph (Figure 15.14). Relevant cycle graph labeling functions are defined as follows:

- $\mu_{\mathcal{C_R}}(C)$: Each cycle C is defined by a sequence of edge and vertex labels encountered during the traversal of C. In order to obtain a sequence invariant to cyclic permutations, $\mu_{\mathcal{C_R}}(C)$ is defined as the sequence having the lowest lexicographic order.

- $\nu_{\mathcal{C_R}}(e)$: An edge e in G_C encodes a path shared by two cycles and corresponds to a sequence of edge and node labels. Since such a path may be traversed from its two extremities, we define $\nu_{\mathcal{C_R}}(e)$ as the sequence of lowest lexicographic order.

In order to compute a cyclic similarity using relevant cycle graphs, TK (Section 15.5.3.5) may be applied on the set of treelets extracted from relevant cycle graphs. This kernel is thus defined as follows:

$$k_C(G, G') = \sum_{t_C \in \mathcal{T}(G_C) \cap \mathcal{T}(G'_C)} k(f_{G_C}(t_C), f_{G'_C}(t_C)) \qquad (15.38)$$

where t_C encodes a configuration of adjacent cycles present in both G and G'.

Then, using cyclic pattern kernel, a global kernel based on the relevant cycle graph may be defined as a sum of a kernel encoding acyclic similarity, such as the TK k_T applied on the original molecular graphs, and a kernel k_C applied on their relevant cycle graphs:

$$k(G, G') = k_T(G, G') + \lambda k_C(G, G') \tag{15.39}$$

where $\lambda \in \mathbb{R}_+$ allows to weight the contribution of the cyclic kernel.

This kernel allows to take into account more cyclic information since, conversely to cyclic pattern kernel, relationships between relevant cycles are encoded in the relevant cycle graph. However, such as cyclic pattern kernel, global similarity measure is based on a sum that splits cyclic information and acyclic information. This last point constitutes a drawback since within the chemoinformatics framework, relationships between a cycle and its substituents may have an influence on molecular properties and must be encoded in kernel definition.

15.5.5.3 Relevant Cycle Hypergraph

Relevant cycle hypergraph [18] (Figure 15.15) corresponds to an oriented hypergraph representation [3,13] that allows to encode both cyclic and acyclic information into a single representation. This molecular representation is based on the relevant cycle graph that encodes molecular cyclic systems. Relevant cycle graph may be augmented by adding acyclic parts in order to obtain the relevant cycle hypergraph. Given a graph $G = (V, E, \mu, \nu)$, its corresponding relevant cycle hypergraph $H_{RC}(G) = (V_{RC}, E_{RC})$ is defined as follows:

1. The set of nodes V_{RC} is defined as a union of two subsets:

   a. A first subset corresponding to the set of relevant cycles

   b. A second subset corresponding to the set of atoms not included within a cycle of G

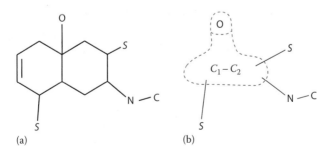

(a)　　　　　　　　　(b)

FIGURE 15.15: (a) Relevant cycle hypergraph representation of a molecule. (b) We can note that hyperedge encodes an adjacency between oxygen atom in one hand and two cycles in the other hand.

All atoms of the original molecular graph are thus encoded in the relevant cycle hypergraph by either a vertex encoding a cycle or a vertex encoding the atom itself. In the same way as for vertices,

2. The set of hyperedges E_{RC} is composed of two subsets:

   a. A set of edges composed of

      i. Edges between relevant cycle vertices, corresponding to the set of edges of the relevant cycle graph

      ii. Edges of G that connect two acyclic atoms or connect a single relevant cycle to another single relevant cycle (C_1 and C_2 in Figure 15.15(b)) or an acyclic part of G (C_1 and S in Figure 15.15(b))

   b. A set of hyperedges that allows to encode special cases where an edge connects at least two distinct relevant cycles to another part of the molecule. For example, a hyperedge is required to encode adjacency relationship between C_1 and C_2 in one hand and an oxygen atom on the other hand in Figure 15.15(b).

   Similarly to vertices, each atomic bond of a molecular graph is encoded within its relevant cycle hypergraph. Therefore, a relevant cycle hypergraph encodes all the information of a molecular graph.

Considering such a new hypergraph representation, similarity must be computed using a hypergraph kernel. In order to encode local relationships between cycles and substituents, TK has been adapted [18] to enumerate treelets included in relevant cycle hypergraphs. This kernel defines a similarity based both on an enumeration of relevant cycles of both graphs (pattern G_0 in Figure 15.11) and on an enumeration of more complex treelets encoding relationships between cycles and relationships between cyclic and acyclic parts of a molecule.

15.5.6   Pattern Weighting

Considering graph kernels based on bags on patterns, it may be interesting to weight each pattern according to its influence on a given property. Similarly to the notion of pharmacophore, some patterns may have a particular influence, whereas some do not encode any information about the property to predict. Considering a kernel defined as a sum of subkernels, where each subkernel encodes a similarity according to a given pattern $p \in \mathcal{P}$, pattern weighting relies on computing a weight for each term of the sum. The weighted kernel is defined as

$$k_{\text{weight}}(x, x') = \sum_{p \in \mathcal{P}} w(p) * k_p(x, x') \tag{15.40}$$

where $w : \mathcal{P} \rightarrow \mathbb{R}_+$ encodes the influence of each pattern: a high value $w(p)$ encodes a high influence of pattern p and thus corresponds to a high contribution to the similarity measure. Conversely, a $w(p)$ close to or equal to 0 consists in removing the contribution of pattern p to the similarity measure.

15.5.6.1 A Priori Methods

Tree pattern kernel as defined by Mahé [35] (Section 15.5.3.2) includes a tree pattern weighting. This kernel is defined on the same feature space than the original kernel but includes a weighting of each pattern:

Definition 15.5.8 (Weighted tree pattern kernel) *Let us consider two graphs $G = (V, E) \in \mathcal{G}$ and $G' = (V', E') \in \mathcal{G}$ together with the set of tree patterns of size $h : \mathcal{T}_h$ a counting function of tree patterns: $\phi_t : \mathcal{G} \to \mathbb{N}$ and a weighing function: $w : \mathcal{T}_h \to \mathbb{R}^+$. The weighted tree pattern kernel is defined by*

$$k^h(G, G') = \sum_{t \in \mathcal{T}_h} w(t) \phi_t(G) \phi_t(G'). \qquad (15.41)$$

This tree pattern kernel definition corresponds to an adaptation of Equation 15.40 to tree pattern kernel. Two weighting functions based on structural information included in tree patterns have been proposed [35].

15.5.6.2 Branching Cardinality

The first one is based on branching cardinality. Branching cardinality is defined as the number of leaves minus one. Given a tree pattern $t = (V, E)$, branching cardinality is defined by

$$branch(t) = \sum_{v \in V} \mathbf{1}(\Gamma(v) = 1) - 1 \qquad (15.42)$$

where $\mathbf{1}(x)$ is equal to 1 if x is true. Weighting function is then defined by

$$w_{branch}(t) = \lambda^{branch(t)} \qquad (15.43)$$

where $\lambda \in \mathbb{R}_+^*$. This weighting function aims to favor linear patterns if $\lambda < 1$. Conversely, structurally complex patterns will be favored if $\lambda > 1$.

15.5.6.3 Ratio Depth/Size

The second weighting function is based on a ratio between the height and the size of a tree pattern. For a given tree pattern $t = (V, E)$ of height h and rooted in $r \in V$, the size-based weighting function is defined by

$$w_{size}(t) = \lambda^{|V|-h} \qquad (15.44)$$

Similarly to Section 15.5.6.2, $\lambda < 1$ favors tree patterns having similar sizes and heights, which correspond to linear patterns. Conversely, $\lambda > 1$ favor more structurally complex tree patterns.

These two weighting functions allow to weight tree pattern influence according to their structural information. Parameter λ must be tuned using a priori chemical knowledge or cross validation. However, this weighting framework does not allow to weight specifically each tree pattern according to a property to predict.

15.5.6.4   Multiple Kernel Learning

Considering a pattern weighting problem as defined in Equation 15.40, multiple kernel learning (MKL) aims to compute an optimal weighting function w according to a given dataset. Considering a finite set of m patterns, weighting function w is encoded by a vector $d \in \mathbb{R}_+^m$ where each coordinate d_i encodes a weight associated to a pattern. The weighted kernel is then defined as

$$k_{\mathrm{MKL}}(x, x') = \sum_{i=1}^{m} d_i * k_i(x, x') \tag{15.45}$$

MKL methods consist in computing an optimal vector d according to a prediction task.

15.5.6.5   Simple MKL

Simple MKL [41] consists in optimizing problems addressed by SVM (Section 15.4.4.1) by computing an optimal weighting. Considering a kernel as defined in Equation 15.45, simple MKL consists in solving

$$\underset{d}{\text{minimize}}\ J(d) \text{ such that } \begin{cases} \sum_{i=1}^{m} d_i = 1 \\ d_i > 0,\ \forall i \in \{1, \ldots, m\} \end{cases} \tag{15.46}$$

with

$$J(d) = \begin{cases} \underset{w,b,\xi}{\min} & \frac{1}{2}\|w\|^2 + C\sum_{i=1}^{n} \xi_i \\ \text{subject to:} & y_i(\langle w, x \rangle + b) \geq 1 - \xi_i, \forall i \in \{1, \ldots, n\} \\ & \xi_i \geq 0,\ \forall i \in \{1, \ldots, n\} \end{cases} \tag{15.47}$$

This minimization problem includes a constraint on the L_1 norm of vector d, which induces sparsity on vector d. This sparsity allows to keep only most relevant patterns into kernel computation and to remove irrelevant ones. Simple MKL problem is resolved by alternating a classical SVM resolution together with a projected gradient descent according to vector d. This projected gradient step allows to minimize objective function while ensuring that constraints on vector d are fulfilled. Since problem defined in Equation 15.46 is convex, iterating these two steps leads to the optimal vector d.

15.5.6.6   Generalized MKL

In order to allow more flexibility on sparsity constraint, generalized MKL [54] defines an objective function that includes a Tikhonov regularization on vector d

instead of an equality constraint in simple MKL. Corresponding minimization problem is defined as

$$\underset{d}{\text{minimize}}\ J(d) \text{ such that } d_i > 0,\ \forall\, i \in \{1,\dots,m\} \tag{15.48}$$

with

$$J(d) = \begin{cases} \underset{w,b,\xi}{\min} & \frac{1}{2}\|w\|^2 + C\sum_{i=1}^{n}\xi_i + \sigma\|d\|_1 \\ \text{subject to :} & y_i(\langle w,x \rangle + b) \geq 1 - \xi_i, \forall i \in \{1,\dots,n\} \\ & \xi_i \geq 0,\ \forall i \in \{1,\dots,n\} \end{cases} \tag{15.49}$$

Parameter σ allows to weight the influence of the L_1 norm regularization into the minimization problem. A high σ favors vector d having a low L_1 norm and thus sparse vectors. Conversely, a low σ relaxes sparsity constraint. The method used to resolve this problem is closely similar to the one proposed to solve simple MKL method. The main difference is that the gradient is no longer projected on the simplex since equality constraint no longer exists.

15.5.6.7 Infinite MKL

In order to be able to deal with thousands of patterns, one has to either compute each Gram matrix for each subkernel at each iteration or store each Gram matrix. When considering thousands of patterns and graphs, these two options induce too much computational time or memory space to be applicable. In order to be able to handle such datasets, infinite MKL [59] defines an MKL method based on simple MKL that only considers a subset of subkernels at each iteration.

This method only performs optimization using a subset of active kernels, that is to say kernels having a weight d_i strictly greater than 0. At the end of each simple MKL optimization, an active kernel set is updated by removing kernels having a null weight and potentially active kernels are added to the set of active kernels. Potentially active kernels may be determined using an oracle based on Karush-Kuhn-Tucker (KKT) conditions and gradients associated to each subkernel [59].

This approach allows to consider a high, possibly infinite, number of subkernels and allows to select the most relevant ones according to a particular property to predict.

15.6 Experiments

Different kernels presented in Section 15.5 have been tested on several chemoinformatics datasets. These datasets are divided into regression and classification problems.*

---

* All these datasets are available on the IAPR TC15 web page: https://iapr-tc15.greyc.fr/links.html.

TABLE 15.2:   Boiling Point Prediction on Acyclic Molecule Dataset Using 90% of the Dataset as Train Set and Remaining 10% as Test Set

Method	RMSE (°C)	Gram Matrix (s)	
		Learning	**Prediction**
1. Edit distance	10.27	1.35	0.05
2. Graph embedding	10.19	2.74	0.01
3. Path kernel	12.24	7.83	0.18
4. Random walks kernel	18.72	19.10	0.57
5. Tree pattern kernel	11.02	4.98	0.03
6. TK	8.10	0.07	0.01
7. TK with MKL	5.24	70	0.01

Note: Execution times are displayed in seconds.

15.6.1   Boiling Point Prediction

The first prediction problem used to test different methods is based on a dataset composed of 185 acyclic molecules [8]. This prediction problem consists in predicting boiling point of molecules. Table 15.2 shows prediction ability of different methods together with execution times. The first line of Table 15.2 shows results obtained by a Gaussian kernel applied on an approximate edit distance (Section 15.2.3.1). Note that since graph edit distance does not define a Euclidean distance, Gram matrix is regularized in order to be semidefinite positive. The second line corresponds to the embedding method described in Section 15.3.1. We can first note that methods based on edit distance obtain intermediate results on this dataset. These results may be due either to the approximation of edit distance or the loss of information induced by regularization or embedding. Next lines correspond to graph kernels defined in Section 15.5.3. Line 3 corresponds to a kernel based on paths (Section 15.5.2.3), and line 4 corresponds to a random walks kernel (Section 15.5.2.1). These two kernels suffer from the low expressiveness of linear patterns, and we can note that the tottering phenomenon degrades prediction accuracy. Finally, the third last lines correspond to kernels based on nonlinear patterns. As we can see in line 7, best results are obtained thanks to the combination of multiple kernel learning (Section 15.5.6.4) that allows to only consider relevant patterns. Execution times show that explicit enumeration of treelets allows to compute Gram matrix in less time than tree pattern kernel or random walks kernels. Conversely, pattern weighting allows to define a more accurate similarity measure while requiring the highest computational time. Note, however, that prediction time is not altered since weighting is only computed during learning step.

This first experiment highlights the gain obtained by using nonlinear patterns instead of linear ones. Moreover, pattern weighting step allows to greatly increase prediction accuracy by removing irrelevant patterns from similarity measure computation. This weighting step allows to get best results among tested methods.

15.6.2 Classification

The two next experiments correspond to binary classification problems that consist in predicting if a molecule has a particular activity or not.

15.6.2.1 AIDS Dataset

First classification experiment has been performed on a graph database provided by [45]. This dataset, defined from the AIDS Antiviral Screen Database of Active Compounds, is composed of 2000 chemical compounds some of them being disconnected. These chemical compounds have been screened as active or inactive against HIV, and they are split into three subsets:

1. A train set composed of 250 compounds used to train SVM

2. A validation set composed of 250 compounds used to find parameter set giving the best accuracy result

3. A test set composed of the remaining 1500 compounds used to test the classification model

This dataset is composed of a large set of different chemical compounds including both cyclic and acyclic molecules and composed of several heteroatoms.

Table 15.3 shows results obtained by different methods on this dataset. First line of Table 15.3 corresponds to a classifier defined as a k-nearest neighbors algorithm using approximate graph edit distance. Line 2 corresponds to a graph embedding as described in Section 15.3.1, and line 3 corresponds to a Gaussian kernel applied on graph edit distance and regularized. We can first note that the use of k-nearest neighbors or graph embedding approach does not lead to good results on this dataset. Conversely, Gaussian kernel on edit distance (Table 15.3, line 4) combined with SVM obtains the best results on this dataset. Note that the regularization added to Gram matrix does not alter prediction accuracy. This may be explained by the fact that classification results may not be altered by a reasonable distortion of the graph edit distance induced by the regularization step. Conversely, regression problems that consist in predicting a real value instead of a binary class may be altered by any

TABLE 15.3: Results on AIDS Dataset

Method	Classification Accuracy(%)
1. KNN using graph edit distance	97.3
2. Graph embedding	98.2
3. Edit distance	99.7
4. Path kernel	98.5
5. Random walks kernel	98.5
6. TK	99.1
7. TK with MKL	99.7

modification of the initial distance. Next lines correspond to kernels using different bags of patterns. Kernels corresponding to line 4 and line 5 are based on linear patterns (Section 15.5.2). As seen in regression experiment, the low expressiveness of kernels based on linear patterns leads to poor results on this dataset. Conversely, methods based on nonlinear patterns (lines 6 and 7) obtain better results since they encode more structural information than kernels based on linear patterns. Moreover, as observed in regression experiment, multiple kernel learning step allows to reach the best results on this classification problem. Note that structures having biggest weights can be assimilated to pharmacophores. These relevant patterns according to MKL may be analyzed by chemical experts.

This experiment confirms the conclusions drawn in Section 15.6.1 on a regression problem with only acyclic molecules.

15.6.2.2 PTC Dataset

The second classification experiment is a classification problem taken from the Predictive Toxicity Challenge (PTC) [52], which aims to predict carcinogenicity of chemical compounds applied to female (F) and male (M) rats (R) and mice (M). This experiment is based on 10 different datasets, each of them being composed of 1 train set and 1 test set. Table 15.4 shows the number of correctly classified molecules over the 10 test sets for each method and for each class of animal.

Table 15.4 shows results obtained by different kernels encoding cyclic information in different ways. The first line of this table corresponds to TK that does not encode any cyclic information since it is only based on acyclic patterns. Lines 2 to 4 correspond to methods encoding different levels of cyclic information. Line 2 corresponds to cyclic pattern kernel that simply compares common cycles (Section 15.5.5.1). Line 3 corresponds to a TK applied on relevant cycle graph (Section 15.5.5.2) that encodes cycle relationships. Line 4 corresponds to results obtained by treelet kernel adapted to relevant cycle hypergraph (Section 15.5.5.3) comparison [18]. First, we can note that best results are obtained by a kernel encoding both cyclic and acyclic relationships that validates the relevance of including cyclic information and more particularly adjacency relationships between cyclic and acyclic parts of a molecule.

TABLE 15.4:   Classification Accuracy on PTC Dataset

Method	# Correct Predictions			
	MM	FM	MR	FR
1. TK	208	205	209	212
2. Cyclic pattern kernel	209	207	202	228
3. TK on relevant cycle graph	211	210	203	232
4. TK on relevant cycle hypergraph (TCH)	217	224	207	233
5. TK + MKL	217	224	223	250
6. TC + MKL	216	213	212	237
7. TCH + MKL	225	229	215	239
8. TK + λTCH	225	230	224	252

Then, lines 5 to 7 correspond to different kernels combined with multiple kernel learning. This weighting step shows that kernel based on cyclic information obtains best results on two datasets over four (line 7, datasets MM and FM) and kernel only based on acyclic patterns obtains best results over the two other datasets (line 5, datasets MR and FR). Note that pattern weighting step allows us to reduce the number of patterns included within kernel computation from about 3500 to 150, depending on dataset. Finally, since different kernels obtain best results over the four datasets, weighted combination of kernel encoding cyclic information and a kernel encoding acyclic information leads to the best results on this dataset (line 8). This combination shows the flexibility of kernel approaches by means of multiple kernel learning and linear combinations of kernels.

15.7 Conclusion

Graph kernel framework allows to define scalar products between implicit or explicit vectorial representations of graphs in a given feature space. On one hand, conversely to methods based on graph theory, graph kernels can be used in well-known and widely used machine learning methods such as SVM. On the other hand, exemption of an explicit vectorial representation allows to encode more information than methods based on an explicit and fixed size vectorial representation. This characteristic allows to define accurate graph similarity measures that encode most of the structural and labeling information encoded by molecular graphs. Therefore, graph kernels allow to combine efficient machine learning methods with accurate and expressive similarity measures.

Defining a graph kernel consists in defining a graph similarity measure that encodes a maximum of useful information and that fulfills all properties required to define a positive definite kernel. Kernels based on bags of patterns deduce molecular graph similarities from similarities between bags of patterns extracted from these graphs. Such kernels aim to encode a maximum of information while keeping an efficient computational time in order to be applicable to datasets encountered in chemoinformatics. Kernels based on nonlinear patterns encode more structural information than kernels based on linear patterns and can be computed in linear time when applied on molecular graphs. Moreover, some bags of patterns are defined such as they explicitly encode cyclic information into similarity measure. This information is particularly useful in chemoinformatics since molecular cycles have a great influence on molecular properties. Among kernels based on bags of patterns, relevant cycle hypergraph encodes both acyclic and cyclic parts and their relationships into a single representation that allows to explicitly encode adjacency relationships between a cycle and its substituents.

The kernel theory allows to define a kernel from a linear combination of subkernels. Considering kernels based on bags of patterns, each sub kernel may be defined as a kernel encoding a molecular similarity according to a particular pattern. From this point of view, multiple kernel learning methods allow to compute an optimal weight for each subkernel according to a property to predict. This weight corresponds to the contribution of each pattern to kernel computation and may be understood as a measure of the influence of each pattern. On one hand, this weighting step allows to increase the accuracy of prediction models by removing irrelevant patterns from the kernel computation. On the other hand, patterns corresponding to high weights may be seen as pharmacophores. These pharmacophores obtained without a priori chemical knowledge may be analyzed by chemical experts to understand some chemical or biological mechanisms involved in a given property.

In conclusion, graph kernels provide a useful framework that may obtain accurate prediction results by combining expressive similarity measures and powerful machine learning methods.

References

1. N. Aronszajn. Theory of reproducing kernels. *Transactions of the American Mathematical Society*, 68(3):337–404, 1950.

2. C. Berg, J.P.R. Christensen, and P. Ressel. *Harmonic Analysis on Semigroups: Theory of Positive Definite and Related Functions*. Applied Mathematical Sciences. Springer-Verlag, New York, 1984.

3. C. Berge. *Graphs and Hypergraphs*, vol. 6. Elsevier, Amsterdam, the Netherlands, 1976.

4. K.M. Borgwardt and H. Kriegel. Shortest-path kernels on graphs. In *Fifth IEEE International Conference on Data Mining (ICDM'05)*, Houston, TX, pp. 74–81, 2005.

5. B.E. Boser, I.M. Guyon, and V.N. Vapnik. A training algorithm for optimal margin classifiers. In *Proceedings of the Fifth Annual Workshop on Computational Learning Theory*, Pittsburg, PA, pp. 144–152. ACM Press, 1992.

6. J.B. Brown, T. Urata, T. Tamura, M.A. Arai, T. Kawabata, and T. Akutsu. Compound analysis via graph kernels incorporating chirality. *Journal of Bioinformatics and Computational Biology*, 8(1):63–81, 2010.

7. T. Caelli and S. Kosinov. An eigenspace projection clustering method for inexact graph matching. *IEEE Transactions on Pattern Analysis and Machine Intelligence*, 26:515–519, 2004.

8. D. Cherqaoui, D. Villemin, A. Mesbah, J.M. Cense, and V. Kvasnicka. Use of a neural network to determine the normal boiling points of acyclic ethers, peroxides, acetals and their sulfur analogues. *Journal of the Chemical Society, Faraday Transactions*, 90:2015–2019, 1994.

9. C. Cortes and V. Vapnik. Support-vector networks. *Machine Learning*, 20(3): 273–297, 1995.

10. J. Dattorro. *Convex Optimization & Euclidean Distance Geometry.* Meboo Publishing, Palo Alto, CA, 2005.

11. M. Deshpande, M. Kuramochi, and G. Karypis. Frequent sub-structure-based approaches for classifying chemical compounds. In *Proceedings of the Third IEEE International Conference on Data Mining (ICDM '03)*, pp. 35–42, Washington, DC. IEEE Computer Society, 2003.

12. H. Drucker, C.J.C. Burges, L. Kaufman, A. Smola, and V. Vapnik. Support vector regression machines. In *Advances in Neural Information Processing Systems*, Denver, CO, pp. 155–161, 1997.

13. A. Ducournau. Hypergraphes: Clustering, réduction et marches aléatoires orientées pour la segmentation d'images et de vidéo. PhD thesis, École Nationale d'Ingénieurs de Saint-Étienne., Saint-Étienne, France, 2012.

14. R.W. Floyd. Algorithm 97: Shortest path. *Communications of the ACM*, 5(6):345, 1962.

15. T. Gärtner, P.A. Flach, and S. Wrobel. On graph kernels: Hardness results and efficient alternatives. In *Proceedings of the 16th Annual Conference on Computational Learning Theory and the 7th Kernel Workshop*, pp. 129–143, Washington, DC, 2003.

16. J. Gasteiger, ed. *Handbook of Chemoinformatics*, 1st edn. Wiley-VCH, 2003.

17. B. Gaüzère, L. Brun, and D. Villemin. Two new graphs kernels in chemoinformatics. *Pattern Recognition Letters*, 33(15):2038–2047, 2012.

18. B. Gaüzère, L. Brun, and D. Villemin. Relevant cycle hypergraph representation for molecules. In W.G. Kropatsch, N.M. Artner, Y. Haxhimusa, and X. Jiang, eds., *Graph-Based Representations in Pattern Recognition*, vol. 7877 of *Lecture Notes in Computer Science*, pp. 111–120. Springer, Berlin, Germany, 2013.

19. B. Gaüzère, L. Brun, D. Villemin, and M. Brun. Graph kernels based on relevant patterns and cycle information for chemoinformatics. In *Proceedings of ICPR 2012*, Tsukuba, Japan, pp. 1775–1778. IAPR, IEEE, November 2012.

20. J. Gibert, E. Valveny, and H. Bunke. Graph embedding in vector spaces by node attribute statistics. *Pattern Recognition*, 45(9):3072–3083, 2012.

21. P.-A. Grenier, L. Brun, and D. Villemin. Treelet kernel incorporating chiral information. In *Graph-Based Representations in Pattern Recognition*, Vienna, Austria, pp. 132–141. Springer, 2013.

22. P.E. Hart, N.J. Nilsson, and B. Raphael. A formal basis for the heuristic determination of minimum cost paths. *IEEE Transactions on Systems Science and Cybernetics*, 4(2):100–107, 1968.

23. D. Haussler. Convolution kernels on discrete structures. Technical report, University of California at Santa Cruz, Santa Cruz, CA, 1999.

24. T. Horváth. Cyclic pattern kernels revisited. In *Proceedings of the Ninth Pacific-Asia Conference on Knowledge Discovery and Data Mining*, Hanoi, Vietnam, vol. 3518, pp. 791–801, 2005. Springer-Verlag.

25. T. Horváth, T. Gärtner, and S. Wrobel. Cyclic pattern kernels for predictive graph mining. In *Proceedings of the 2004 ACM SIGKDD International Conference on Knowledge Discovery and Data Mining (KDD '04)*, Seattle, WA, p. 158. ACM Press, 2004.

26. M.A. Johnson and G.M. Maggiora, eds. *Concepts and Applications of Molecular Similarity*. Wiley, 1990.

27. S. Jouili and S. Tabbone. Graph embedding using constant shift embedding. In *Proceedings of the 20th International Conference on Recognizing Patterns in Signals, Speech, Images, and Videos (ICPR'10)*, pp. 83–92, Berlin, Germany, 2010. Springer-Verlag.

28. H. Kashima, K. Tsuda, and A. Inokuchi. Marginalized kernels between labeled graphs. In *International Conference on Machine Learning*, Washington, DC, vol. 3, pp. 321–328, 2003.

29. B. Luo, R.C. Wilson, and E. Hancock. A spectral approach to learning structural variations in graphs. *Pattern Recognition*, 39(6):1188–1198, 2006.

30. B. Luo, R.C. Wilson, and E.R. Hancock. Spectral embedding of graphs. *Pattern Recognition*, 36(10):2213–2230, 2003.

31. M.M. Luqman, J.-Y. Ramel, J. Lladós, and T. Brouard. Fuzzy multilevel graph embedding. *Pattern Recognition*, 46(2):551–565, 2013.

32. P. Mahé, L. Ralaivola, V. Stoven, and J.-P. Vert. The pharmacophore kernel for virtual screening with support vector machines. *Journal of Chemical Information and Modeling*, 46(5):2003–14, 2006.

33. P. Mahé, N. Ueda, T. Akutsu, J.-L. Perret, and J.-P. Vert. Extensions of marginalized graph kernels. In *Twenty-First International Conference on Machine Learning (ICML '04)*, Banff, Alberta, Canada, p. 70. ACM Press, 2004.

34. P. Mahé and J.-P. Vert. Graph kernels based on tree patterns for molecules. *Machine Learning*, 75(1):3–35, 2008.

35. P. Mahé and J.-P. Vert. Graph kernels based on tree patterns for molecules. *Machine Learning*, 75(1)(September 2008):3–35, 2009.

36. J. Mercer. Functions of positive and negative type, and their connection with the theory of integral equations. *Proceedings of the Royal Society A: Mathematical, Physical and Engineering Sciences*, 83(559):69–70, November 1909.

37. H.L. Morgan. The generation of a unique machine description for chemical structures: A technique developed at chemical abstracts service. *Journal of Chemical Documentation*, 5(2):107–113, 1965.

38. J. Munkres. Algorithms for the assignment and transportation problems. *Journal of the Society for Industrial and Applied Mathematics*, 5(1):32–38, 1957.

39. M. Neuhaus and H. Bunke. *Bridging the Gap Between Graph Edit Distance and Kernel Machines*. World Scientific Publishing Co., Inc., River Edge, NJ, 2007.

40. G. Poezevara, B. Cuissart, and B. Crémilleux. Discovering emerging graph patterns from chemicals. In *Proceedings of the 18th International Symposium on Methodologies for Intelligent Systems (ISMIS 2009)*, pp. 45–55, Prague, Czech Republic, 2009. LNCS.

41. A. Rakotomamonjy, F. Bach, S. Canu, and Y. Grandvalet. SimpleMKL. *Journal of Machine Learning Research*, 9:2491–2521, 2008.

42. L. Ralaivola, S.J Swamidass, H. Saigo, and P. Baldi. Graph kernels for chemical informatics. *Neural Networks : The Official Journal of the International Neural Network Society*, 18(8):1093–1110, 2005.

43. J. Ramon and T. Gärtner. Expressivity versus efficiency of graph kernels. In *First International Workshop on Mining Graphs, Trees and Sequences*, Cavtat-Dubrovnik, Croatia, pp. 65–74, 2003.

44. K. Riesen. Classification and clustering of vector space embedded graphs. PhD thesis, Institut für Informatik und angewandte Mathematik Universität Bern, Switzerland, 2009.

45. K. Riesen and H. Bunke. IAM graph database repository for graph based pattern recognition and machine learning. In *Proceedings of the 2008 Joint IAPR International Workshop on Structural, Syntactic, and Statistical Pattern Recognition (SSPR & SPR '08)*, pp. 287–297, Berlin, Germany, 2008. Springer-Verlag.

46. K. Riesen and H. Bunke. Approximate graph edit distance computation by means of bipartite graph matching. *Image and Vision Computing*, 27(7):950–959, 2009.

47. N. Shervashidze and K.M. Borgwardt. Fast subtree kernels on graphs. In *Advances in Neural Information Processing Systems*, pp. 1660–1668, 2009.

48. N. Shervashidze, T. Petri, K. Mehlhorn, K.M. Borgwardt, and S. Viswanathan. Efficient graphlet kernels for large graph comparison. In *International Conference on Artificial Intelligence and Statistics*, Clearwater Beach, FL, pp. 488–495, 2009.

49. N. Shervaszide. Scalable graph kernels. PhD thesis, Universität Tübingen, Tübingen, Germany, 2012.

50. N. Sidere, P. Héroux, and J.-Y. Ramel. Vector representation of graphs: Application to the classification of symbols and letters. In *ICDAR*, pp. 681–685. IEEE Computer Society, 2009.

51. R. Todeschini and V. Consonni. *Molecular Descriptors for Chemoinformatics*, vol. 41. Wiley-VCH, Weinheim, Germany, 2009.

52. H. Toivonen, A. Srinivasan, R.D. King, S. Kramer, and C. Helma. Statistical evaluation of the predictive toxicology challenge 2000–2001. *Bioinformatics*, 19(10):1183–1193, 2003.

53. V. Vapnik. *The Nature of Statistical Learning Theory*. Springer, 1995.

54. M. Varma and D. Ray. Learning the discriminative power-invariance trade-off. In *IEEE 11th International Conference on Computer Vision, 2007 (ICCV 2007)*, Rio de Janeiro, Brazil, pp. 1–8. IEEE, 2007.

55. S.V.N. Vishwanathan, K.M. Borgwardt, and N.N. Schraudolph. Fast computation of graph kernels. *Advances in Neural Information Processing Systems*, 19:1449, 2007.

56. P. Vismara. Union of all the minimum cycle bases of a graph. *The Electronic Journal of Combinatorics*, 4(1):73–87, 1997.

57. N. Wale, I.A. Watson, and G. Karypis. Comparison of descriptor spaces for chemical compound retrieval and classification. *Knowledge and Information Systems*, 14(3):347–375, 2008.

58. X. Yan and J. Han. gspan: Graph-based substructure pattern mining. In *Proceedings of the 2002 IEEE International Conference on Data Mining*, ICDM '02, pp. 721–724, Washington, DC, 2002. IEEE Computer Society.

59. F. Yger and A. Rakotomamonjy. Wavelet kernel learning. *Pattern Recognition*, 44(10):2614–2629, 2011.

Chapter 16

Chemical Compound Complexity in Biological Pathways

Atsuko Yamaguchi and Kiyoko F. Aoki-Kinoshita

Contents

16.1 Introduction..471
16.2 Biological Pathways...472
16.3 Molecular Graphs...473
16.4 Molecular Graph Representations...475
    16.4.1 Tabular Representations..475
    16.4.2 Linear Representations...477
16.5 Treewidth and Local Treewidth...477
16.6 KEGG COMPOUND...479
16.7 Molecular Graph Complexity...481
    16.7.1 Data set...481
    16.7.2 Algorithm...482
    16.7.3 Results..484
        16.7.3.1 Treewidths Analysis....................................484
        16.7.3.2 Range Analysis...484
    16.7.4 Discussion...487
16.8 Conclusion..488
References..491

16.1 Introduction

In order to extract useful knowledge from big biological data, it is essential that the data be analyzed using computers, which requires that the data be mathematically modeled. In the areas of chemistry and biology, graph representations are frequently used for computational analyses. In particular, a graph is very suitable to represent the structure of a chemical compound. The structural formula of a chemical compound can be represented by a graph, where an atom is regarded as a vertex and a chemical bond is regarded as an edge. A graph representation of a structural formula can be applied to the computation of the similarity of compounds, the combinatorial enumeration of structural isomers, synthetic design, etc.

However, in the field of cheminformatics, methods using graph algorithms have not entered the mainstream [1] because graph problems are often intractable.

For example, the problem of finding the maximum common subgraph of two graphs is known to be NP-hard [2] although the problem is very useful to find similar chemical compounds from a chemical database.

Currently, to calculate the similarity of chemical compounds, the similarity of binary vectors, which represent whether or not small molecular fragments exist as subgraphs in the chemical compound, is often used [3,4]. For example, referring to Figure 16.2 representing shikimic acid, small molecular fragments may be defined as carbons, oxygens, hydroxyl groups, carboxyl groups, and rings, in this order in the representative vector. Since there are 10 carbons, 5 oxygens, 4 hydroxyl groups, no carboxyl groups, and one ring in shikimic acid, its representative vector would be $(1, 1, 1, 0, 1)$. However, many different chemical compounds can also be described by this same vector. In other words, such vector descriptions using small molecular fragments cannot precisely represent a chemical compound because the vector only indicates the existence of each particular small molecular fragment. Apart from chemical compounds, there have been studies on graph similarity measures for limited types of graphs. For example, [5] presents a similarity measure for directed unlabeled hierarchical graphs using graph alignment computed by a dynamic programming approach.

On the other hand, the precise representation of chemical compounds as graphs greatly increases computation time. Therefore, it is important to reduce the space of instances for graph problems to allow for polynomial-time algorithms. That is, graph theory algorithms need to be developed specifically for the domain under investigation. Rigorous mathematical descriptions of biological phenomena are difficult due to their complexity, but a quantitative method for investigating the characteristics of chemical compounds would aid in the analysis of biological pathways. Thus, the main contribution of this chapter is the analysis of the characteristics of chemical compounds in biological pathways. Moreover, we claim that the complexity of chemical compounds in biological pathways is generally low in terms of treewidth, which is introduced in Section 16.5.

16.2   Biological Pathways

In this chapter, we focus on chemical compounds in biological pathways. A biological pathway is a network of chemical reactions with molecular interactions, such as protein–ligand interactions, protein–protein interactions, and genetic interactions, in cells. Many biological pathways including metabolic pathways, genetic regulatory networks, signaling pathways, etc., have also been studied in biology. Because chemical reactions in biological pathways affect all cells, analyzing chemical compounds in these pathways is very important to understand biological systems in organisms.

Biological pathways are often represented as graphs especially for metabolic pathways, which are representative biological pathways. Metabolic pathways consist of enzymes and chemical compounds on which the enzymes react to form new chemical compounds. An enzyme is a protein that accelerates a chemical reaction by recognizing a specific chemical compound. By undergoing the reaction, the recognized compound is transformed into another different compound, which is then further recognized by other enzymes. In the metabolic pathway, a graph is formed where the vertices represent the compounds while the edges represent the enzymes. Thus, a directed graph is formed representing the entire metabolic pathway. Figure 16.1 is an example of a metabolic pathway that releases free energy by converting α-D-glucose into pyruvate. Squares represent enzymes and circles represent compounds. In Figure 16.1 in particular, the top circle is the chemical compound α-D-glucose-1P, which is catalyzed by the enzyme called pgm, a phosphoglucomutase (numbered 5.4.2.2 in the figure), thus producing the compound called α-D-glucose-6P, which is further catalyzed by an enzyme called Glucose-6-phosphate isomerase (GPI) (numbered 5.3.1.9), producing the compound called β-D-fructose-6P. In this way, various chemical compounds are produced within the cell, in a sort of chain reaction, which can be represented as directed graphs indicating the direction of the reactions.

16.3 Molecular Graphs

In this section, we describe the main data structure used to represent a chemical compound, the molecular graph. A molecular graph for a chemical compound is a simple undirected graph whose vertices correspond to the atoms of the compound and edges correspond to chemical bonds. The vertices are labeled with atom-type information and the edges are labeled with the bond type connecting the atoms. For example, Figure 16.2 shows the molecular graph for shikimic acid. Vertices labeled with "C" represent carbon atoms, vertices labeled with "O" represent oxygen atoms, and vertices labeled with "H" represent hydrogen atoms. Edges labeled with "1" represent single bonds, and edges labeled with "2" represent double bonds. Using this representation of a molecular graph, the maximum possible degree of a vertex represents the valency of the corresponding atom in the chemical compound. Because each atom has its own valence, the maximum degree of a molecular graph is generally bounded. A vertex representing hydrogen is often omitted from a molecular graph to simplify it because a hydrogen can only have a single bond, and the omission of the hydrogen atoms connecting each atom with another type can be trivially inferred from the information of the types of remaining edges connecting the atom. A molecular graph omitting the vertices of hydrogens is called a hydrogen-suppressed molecular graph. Figure 16.3 shows a hydrogen-suppressed molecular graph depicting shikimic acid. Hydrogen-suppressed molecular graphs are used in general.

From the viewpoint of graph theory, enumeration, topological analysis using invariants, and other such properties of molecular graphs have been well studied [6,7]. By the results of these studies, chemists may obtain many qualitative predictions about the structures of chemical compounds, because a molecular graph is a direct mathematical representation of the structural formulae of chemical compounds.

To quantitatively compute molecular graphs, they must be expressed in some data format. Thus, we describe the major molecular graph representations in the next section.

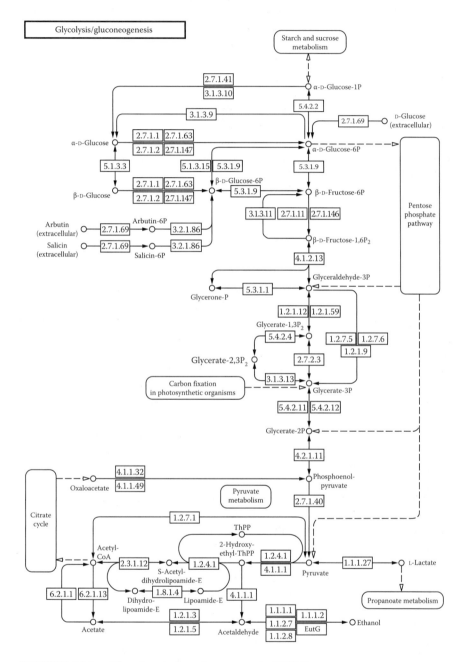

FIGURE 16.1:   The *glycolysis* pathway map in KEGG.

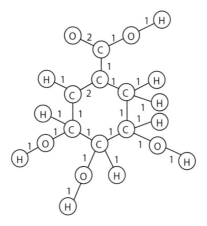

FIGURE 16.2: The molecular graph for shikimic acid. It includes 7 carbons, 5 oxygens, and 10 hydrogens.

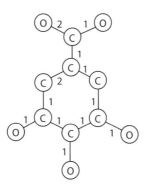

FIGURE 16.3: The hydrogen-suppressed molecular graph for shikimic acid.

16.4 Molecular Graph Representations

In the area of cheminformatics, there are two principal techniques for representing chemical structures as molecular graphs: a table representation using a connection table and adjacency matrix and a linear notation describing chemical structures as a single text string. We describe each of these representations here.

16.4.1 Tabular Representations

We first describe the table representation of chemical compounds, which includes a connection table and an adjacency matrix of a molecular graph as often used in computer science. The Molecular Design Limited (MDL) Molfile format is one of

```
Shikimic acid                                                    ⎫
generated by *** Sep 13 2013                                     ⎬— Header block
                                                                 ⎭

12 12  0  0  1  0  0  0  0  0999 V2000                           ⎬— Counts line
     22.1200 –15.1127   0.0000 C  0 0 0 0 0 0 0 0 0 0 0 0  ⎤
     22.3375 –15.8112   0.0000 C  0 0 0 0 0 0 0 0 0 0 0 0  ⎥
     20.9090 –15.8112   0.0000 C  0 0 0 0 0 0 0 0 0 0 0 0  ⎥
     22.1137 –13.7223   0.0000 C  0 0 0 0 0 0 0 0 0 0 0 0  ⎥
     23.3375 –17.2208   0.0000 C  0 0 2 0 0 0 0 0 0 0 0 0  ⎥
     20.9090 –17.2208   0.0000 C  0 0 2 0 0 0 0 0 0 0 0 0  ⎥
     23.3182 –13.0238   0.0000 O  0 0 0 0 0 0 0 0 0 0 0 0  ⎬— Atom block
     20.9025 –13.0301   0.0000 O  0 0 0 0 0 0 0 0 0 0 0 0  ⎥
     22.1200 –17.9322   0.0000 C  0 0 1 0 0 0 0 0 0 0 0 0  ⎥
     24.5485 –17.9128   0.0000 O  0 0 0 0 0 0 0 0 0 0 0 0  ⎥
     19.6978 –17.9128   0.0000 O  0 0 0 0 0 0 0 0 0 0 0 0  ⎥
     22.1137 –19.3290   0.0000 O  0 0 0 0 0 0 0 0 0 0 0 0  ⎦
  1  2  1 0    0 0                                ⎤
  1  3  2 0    0 0                                ⎥
  1  4  1 0    0 0                                ⎥
  2  5  1 0    0 0                                ⎥
  3  6  1 0    0 0                                ⎥
  4  7  1 0    0 0                                ⎥
  4  8  2 0    0 0                                ⎬— Bond block
  5  9  1 0    0 0                                ⎥
  5 10  1 1    0 0                                ⎥
  6 11  1 6    0 0                                ⎥
  9 12  1 6    0 0                                ⎥
  6  9  1 0    0 0                                ⎦
M  END
```

FIGURE 16.4:   The molecular graph in the MDL Molfile format for shikimic acid.

the major file formats that can represent chemical structures using this technique [8]. For example, using the MDL Molfile format, shikimic acid can be represented as shown in Figure 16.4. The first three lines of the file consist of the *headerblock* for describing any metadata regarding the chemical compound in the file. Metadata refers to any additional supporting information describing the file contents. The fourth line is called the *countsline*, which lists the number of atoms (=12) and bonds (=12), which corresponds with the number of lines in the following sections. The following sections in this example only contain atoms and bonds, so the *atomblock* is described from the 5th line to the 16th line, and the *bondblock* is described from the 17th line to the 28th line.

Each line in the atom block corresponds to each atom of the chemical compound. From the left, each line represents the x-, y-, and z-coordinates, the type of atom, and other information for each atom. The atom block of Figure 16.4 indicates that shikimic acid consists of seven carbon and five oxygen atoms.

Each line in the bond block corresponds to a bond. From the left, it describes the two vertices connected by the bond, the bond type, etc. In Figure 16.4, the first two lines of the bond block indicate that vertex 1 is connected to vertex 2 with a single bond and that vertex 1 is connected to vertex 3 with a double bond.

16.4.2   Linear Representations

The linear notation of chemical structures is often based on the depth-/breadth-first traversal of a molecular graph. Major formats using this representation include SMILES™ (Simplified Molecular Input Line Entry System) [9] and InChI™ (International Union of Pure and Applied Chemistry [IUPAC] International Chemical Identifier) [10]. SMILES is a linear notation proposed by Daylight Chemical Information Systems, Inc., and InChI is proposed by IUPAC. As an example of these two formats, shikimic acid can be represented using the SMILES notation as follows:

```
C1[C@H]([C@@H]([C@@H](C=C1C(=O)O)O)O)O
```

and using InChI as follows:

```
InChI=1S/C7H10O5/c8-4-1-3(7(11)12)2-5(9)6(4)10/h1,
4-6,8-10H,2H2,(H,11,12)/t4-,5-,6-/m1/s.
```

As evident from these examples, compared to the table representation like the MDL Molfile format, the linear notation is more compact, and locally connected atom structures can be read from the strings. On the other hand, it is not necessarily easy to get a grasp of the structure of complex chemical compounds using this linear notation, especially since subgraphs cannot be easily extracted using it. That is, the subgraph of a structure represented using the linear notation may not necessarily be a substring of the whole string, thus requiring the string to be first converted to a graph.

Because there are many tools that can convert data among these chemical structural formats, chemical structures can be converted to the appropriate format depending on the situation or problem at hand. In this chapter, we use the MDL Molfile format because our goal is to analyze the structures of chemical compounds as molecular graphs.

16.5   Treewidth and Local Treewidth

We use the concept of *treewidth* as a measure to characterize chemical compounds as molecular graphs in biological pathways. Here, we provide some definitions to describe this concept. A tree is a connected undirected simple graph without a cycle. Many problems proven to be NP-hard can be computable in polynomial time if input graphs are restricted to be a tree. For example, vertex cover, independent set, and subgraph isomorphism are known to be computable in polynomial time for trees [11–13]. Roughly speaking, the treewidth of a graph is a measure of tree likeness. A graph with a treewidth of 1 is a tree, and a graph with a treewidth of $n - 1$ is a complete graph, where n is the number of vertices. Figure 16.5 illustrates the same graph with varying treewidths. The leftmost graph has a treewidth of 1, the next has a treewidth of 2, and the complete graph is shown on the very right. Note that many NP-hard problems including independent set, dominating set, chromatic number, Hamiltonian

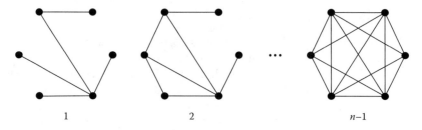

FIGURE 16.5:   Examples of graphs having treewidths 1, 2, and $n - 1$.

circuit, etc., are also known to be solvable in polynomial time if the treewidths of the input graphs are bounded [14].

The treewidth of a graph can be defined using the tree decomposition of the graph. The tree decomposition of a graph $G = (V, E)$ is a pair (T, X), where T is a tree and $X : V(T) \rightarrow 2^{V(G)}$ that satisfies the following three conditions:

1. $\cup_{t \in V(T)} X(t) = V(G)$.

2. For every edge $(u, v) \in V(G)$, there exists a vertex $t \in V(T)$ such that $u, v \in X(t)$.

3. For any three vertices $r, s, t \in V(T)$, if s is on the path from r to t, then $X(r) \cap X(t) \subseteq X(s)$.

The *width* of a tree decomposition (T, X) is $\max_{t \in V(T)} |X(t) - 1|$. The *treewidth* of a graph G is then the minimum width of all possible tree decompositions of G.

A k-tree is a graph that is defined recursively as follows: The complete graph with k vertices is a k-tree. A k-tree with $n+1$ vertices can be constructed from a k-tree with n vertices by adding a vertex adjacent to all the vertices of one of its k-vertex cliques and only to these vertices. A graph with treewidth at most k is called a partial k-tree because it is known that k-trees are the maximal graphs with treewidth k and a graph with treewidth at most k to be a subgraph of a k-tree.

The problem to determine whether or not the treewidth of a graph is at most k has been shown to be NP-complete [15]. For a fixed constant k, however, linear time algorithms to determine whether or not the treewidth of a graph is at most k and to construct the tree decomposition if the treewidth of the graph is at most k have been presented by [16,17]. These algorithms are practical only when k is small because the constant factor of the computation time for the algorithm becomes exponentially large about k. To compute the treewidth of a given graph, some practical exponential time algorithms have been presented in [18–20].

Next, to precisely analyze the characteristics of molecular graphs, we describe the concept of *local treewidth* that [21] introduced as a generalization of treewidth. For a graph $G = (V, E)$, we define the r-neighborhood $N_r(v)$ of a vertex $v \in V$ to be the set of all vertices $w \in V$ of distance at most r from v. We denote the vertex induced subgraph on $N_r(v)$ of G by $G(N_r(v))$ and the treewidth of a graph $G(N_r(v))$ by $tw(G(N_r(v)))$. Then, the local treewidth of a graph G is the function

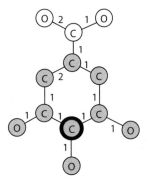

FIGURE 16.6: The vertex v and $N_3(v)$ for $G(N_3(v))$ with treewidth 2.

$$ltw^G(r) = \max\{tw(G(N_r(v))) \mid v \in V\}$$

Trivially, for a graph G with treewidth k, the lower bound of ltw^G is 1, the upper bound of ltw^G is k because $ltw^G(1) = 1$ and $ltw^G(n) = k$, and ltw^G is a monotonically decreasing function where n is the size of V.

Using the concept of local treewidth, we now define the *range* of a graph, as introduced in [22]. For a graph G with treewidth k, the range r_G can be defined as

$$r_G = \min\{r \mid ltw^G(r) = k\}$$

Using the molecular graph of shikimic acid introduced earlier in Figure 16.3, we illustrate its range in Figure 16.6. The following can be said about this graph:

- The treewidth of the graph is 2.

- $ltw^G(3) = 2$ because the treewidth of $G(N_3(v))$ is 2 when the bold vertex is v and $N_3(v)$ is the shaded section.

- $G(N_2(v))$ for any v is a tree and thus $ltw^G(3) = 1$ because the length of the only cycle in the graph is 6.

Thus, we can say that the range of shikimic acid is 3.

16.6 KEGG COMPOUND

In this work, we use the data in the (Kyoto Encyclopedia of Genes and Genomes) KEGG COMPOUND database of chemical compounds, so we describe this database here. KEGG is a collection of databases for understanding high-level functions and utilities of the biological system, such as the cell, the organism, and even the ecosystem, from molecular-level information to especially large-scale molecular data sets [23]. KEGG COMPOUND is one of the KEGG databases, and it is a collection of small molecules, biopolymers, and other chemical substances that are relevant to biological pathways [24]. KEGG also contains the KEGG PATHWAY database of various biological pathways, including metabolic and signaling pathways.

Each entry of KEGG COMPOUND contains the chemical structure and associated information. Figure 16.7 shows the entry for shikimic acid in KEGG COMPOUND. This entry includes such information as the name of the chemical compound, chemical formula, and links to other databases, among others. Chemical structures appearing in KEGG COMPOUND are provided in MDL Molfile format,

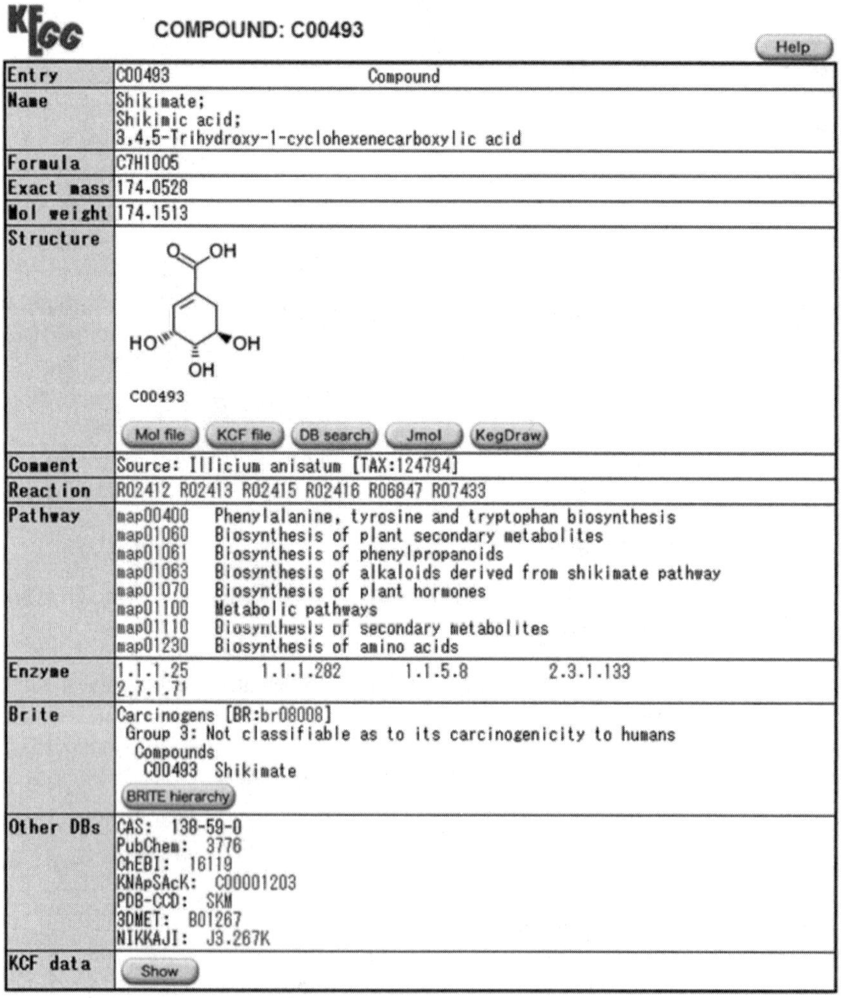

KEGG   COMPOUND: C00493   [Help]

Entry	C00493 Compound
Name	Shikimate; Shikimic acid; 3,4,5-Trihydroxy-1-cyclohexenecarboxylic acid
Formula	C7H10O5
Exact mass	174.0528
Mol weight	174.1513
Structure	 C00493 (Mol file) (KCF file) (DB search) (Jmol) (KegDraw)
Comment	Source: Illicium anisatum [TAX:124794]
Reaction	R02412 R02413 R02415 R02416 R06847 R07433
Pathway	map00400 Phenylalanine, tyrosine and tryptophan biosynthesis map01060 Biosynthesis of plant secondary metabolites map01061 Biosynthesis of phenylpropanoids map01063 Biosynthesis of alkaloids derived from shikimate pathway map01070 Biosynthesis of plant hormones map01100 Metabolic pathways map01110 Biosynthesis of secondary metabolites map01230 Biosynthesis of amino acids
Enzyme	1.1.1.25 1.1.1.282 1.1.5.8 2.3.1.133 2.7.1.71

Brite	Carcinogens [BR:br08008]
	  Group 3: Not classifiable as to its carcinogenicity to humans
	    Compounds
	      C00493 Shikimate
	(BRITE hierarchy)

Other DBs	CAS: 138-59-0 PubChem: 3776 ChEBI: 16119 KNApSAcK: C00001203 PDB-CCD: SKM 3DMET: B01267 NIKKAJI: J3.267K
KCF data	(Show)

FIGURE 16.7:   The entry of KEGG COMPOUND for shikimic acid. A variety of related information in KEGG such as pathways in which this compound is used and chemical reactions in which it is involved are provided.

as well as in KEGG Chemical Function (KCF) format defined by KEGG. The KCF format is quite similar to the MDL Molfile format.

KEGG also provides a method for accessing their data using computer programs. The KEGG API (application programming information) is a REST (Representational State Transfer)-style web API for the KEGG databases (http://www.kegg.jp/kegg/docs/keggapi.html). Thus, the KEGG data can be accessed by using just a uniform resource identifier (URI) to the corresponding data. For example, the following URI can be used to obtain the entry data for two chemical compounds having IDs C08712 and C16854:

http://rest.kegg.jp/get/C08712+C16854

The atomic information for the entries are returned in ASCII text, so that it can be easily manipulated using a computer program.

16.7   Molecular Graph Complexity

The issue of the *complexity* of molecular graphs has been discussed for many years and remains an open problem in mathematical chemistry. At the time of this writing, complexity measures can be categorized into two main types. One type is based on information content such as the Shannon entropy of a molecular graph [25,26]. The complexity measures of this type have been improved many times since the first work introduced in [27]. The other type is based on structural features of molecular graphs such as the number of paths [28], the number of walks [29], overall connectivity [30], and the number of edge biclique covers [31]. The concept of treewidth, which we introduced in Section 16.5, can be categorized into the latter type of complexity measure.

As a first step towards capturing the properties of chemical compounds in biological pathways, we examined the treewidth of molecular graphs in KEGG COMPOUND. After that, we analyzed the molecular graphs whose treewidths were relatively high by using the concept of local treewidth. These computational experiments were originally tested and reported in 2003 [22], and we retested the experiments using the latest versions of the available data set at the time of this writing.

16.7.1   Data set

We obtained 16,430 files in MDL Molfile format from KEGG COMPOUND Release 67.0 (Sep. 2013) through the KEGG API. The number of vertices in the molecular graphs in this data set ranged from 1 to 285. Figure 16.8 shows the molecular graph with the largest number of vertices in this data set (C16854). The average sizes of vertices and edges of all the molecular graphs in this data set are 25.9646 and 27.4167, respectively. Thus, just by looking at the average (number of vertices)/(number of edges) ratio of 1.05592, we can guess that the molecular graphs in

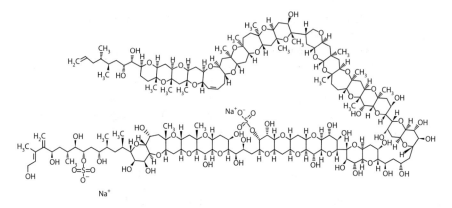

FIGURE 16.8:   The chemical structure of the entry C16854 with the largest number of vertices in KEGG COMPOUND.

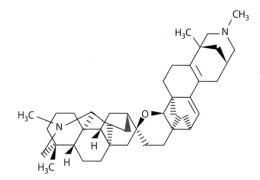

FIGURE 16.9:   The chemical structure of entry C08712, which has the largest ratio of vertices to edges (1.25532 = 59/47) in KEGG COMPOUND.

KEGG COMPOUND may be rather sparse. In fact, even the maximum (number of vertices)/(number of edges) ratio among all the molecular graphs reaches just 1.25532 (Figure 16.9). The numbers of vertices and edges for C08712 are 47 and 59, respectively.

We note here that some of the molecular graphs in KEGG COMPOUND are not connected. In these cases, we selected the connected component with the largest treewidth from these entries in our data set.

16.7.2   Algorithm

Using the files in MDL Molfile format, we examined the treewidths of the molecular graphs for each compound in KEGG COMPOUND. To compute the treewidths, we employed the algorithm presented in [15] and described in Algorithm 16.1.

Algorithm 16.1: The treewidth recognition of a graph G for k.

Data: A graph $G = (V, E)$
Result: *true* or *false*

1   Initialize $L := \emptyset$;
2   **foreach** *subset $S \subseteq V$ of size k* **do**
3 | **if** *S is a separator of G* **then**
4 | | $L := L \cup \{S\}$;
5 | | Compute the set C_S of connected components in G_S;
6 | | **foreach** $H = (V_H, E_H) \in C_S$ **do**
7 | | | **if** $|V_H| \leq k + 1$ **then**
8 | | | | $D(H) := true$;
9 | | | **end**
10 | | **end**
11 | **end**
12 **end**
13 **foreach** *graph $H = (V_H, E_H)$ in $\cup_{S \in L}\{C \mid C \in C_S\}$ in increasing order* **do**
14 | $S_0 :=$ a separator in L corresponding to H;
15 | **foreach** $v \in V_H$ **do**
16 | | Initialize $U := \emptyset$;
17 | | **foreach** $S \in L$ **do**
18 | | | **if** $S \subseteq S_0 \cup \{v\}$ **then**
19 | | | | $U := U \cup S$;
20 | | | | **foreach** $H' = (V'_H, E'_H) \in C_S$ with $D(H') = true$ **do**
21 | | | | | $U := U \cup V'_H$;
22 | | | | **end**
23 | | | **end**
24 | | **end**
25 | | **if** $V_H \subseteq U$ **then**
26 | | | $D(V_H) = true$;
27 | | | break;
28 | | **end**
29 | **end**
30 **end**
31 **foreach** $S \in L$ **do**
32 | **if** *$D(H) = true$ for any H in C_S* **then**
33 | | **return** *true*;
34 | **end**
35 **end**
36 **return** *false*;

For a graph $G = (V, E)$ and a fixed k, this algorithm determines whether or not the treewidth of G is at most k in $O(n^{k+2})$ time. As described in the previous section, the number of vertices in molecular graphs from KEGG COMPOUND is relatively small. Therefore, this algorithm is fast enough for the graphs in our experiments.

This algorithm uses what is called a *separator*. A separator S of a graph $G = (V, E)$ is a subset of vertices such that the subgraph G_S of G induced by the vertex set $\{V\} - \{S\}$ is not connected. A sufficiently large graph with treewidth k is known to have a separator with size k such that each of the connected components generated by the separator augmented by the completely connected separator vertices is a graph with treewidth k.

This algorithm investigates all the connected components generated by all the separators of size k and decides whether or not the treewidth is at most k (lines 2–12). All of the connected components satisfying this are labeled as *true*. From line 13, using a dynamic programming approach, each connected component H is evaluated in the order of their increasing sizes and given a value of *true* or *false* by determining whether or not the vertex set of H is a subset of the union of vertex sets of connected components such that H and the connected components with value *true* share the same separator. In the loop from line 31, if there exists a separator such that all the connected components for the separator have the value *true*, the algorithm returns *true*, otherwise *false*.

16.7.3   Results

16.7.3.1   Treewidths Analysis

The distribution of the treewidths of all the molecular graphs in KEGG COMPOUND and the percentage of each treewidth among all of the molecular graphs are shown in Table 16.1. In all but two of the 16,430 cases, the treewidth was between 1 and 3, and the treewidth of the two exceptions was 4. The high complexity of the structures of these two exceptions, as illustrated in Figure 16.10, may allow us to ignore them. Thus, we are able to claim that the structures of chemical compounds in biological pathways are generally simple in terms of treewidths.

Among the 16,428 molecular graphs having treewidths between 1 and 3, the molecular graphs having treewidth 1 or 2 account for 95% of them. Therefore, because molecular graphs with treewidth 3 account for just less than 5% of these molecular graphs, we can conclude that they are very complex graphs in terms of the chemical compounds generally appearing in KEGG COMPOUND (Figure 16.11).

16.7.3.2   Range Analysis

To more precisely analyze these chemical compounds having treewidth 3, we examined their ranges r_G. By definition of local treewidth and range, the size of the range may indicate the size of the most complex part of a molecular graph. In other words, if r_G is small, the complex portion of the molecular graph G is small. On the other hand, if r_G is large, the molecule does not have a small complex portion, implying that the atoms may be globally connected. Figure 16.12 shows the distribution of

TABLE 16.1: The Distribution of Treewidths and Their Percentages among All Molecular Graphs in KEGG COMPOUND

Treewidth	Number of Graphs	Percentage
1	2,972	0.180888618
2	12,665	0.770846013
3	791	0.04814364
4	2	0.000121729

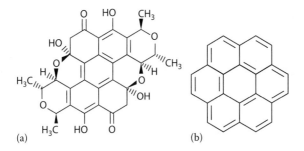

(a)   (b)

FIGURE 16.10: The chemical structures of the entries (a) C01863 and (b) C19375 having treewidth 4.

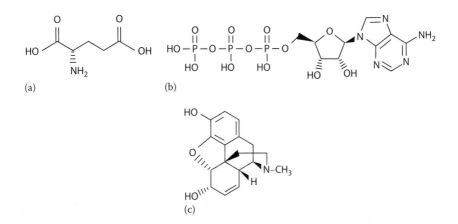

FIGURE 16.11: Examples of chemical structures having treewidth: (a) 1 (C00025), (b) 2 (C00005), and (c) 3 (C01516).

r_G for each molecular graph with treewidth 3. The number of molecular graphs with $r_G = 3$ is largest among every r_G and makes up 73.0% of the total of the molecular graphs with treewidth 3. Furthermore, there are peaks in the number of molecular graphs with ranges 5 and 8, as well as small peaks for ranges 6, 9, and 12. Figure 16.13 shows representative chemical compounds with ranges 5 (C03184) and 8 (C11956)

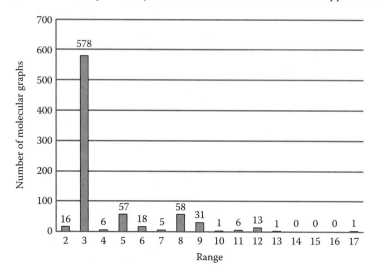

FIGURE 16.12:   A bar graph of the ranges of the local treewidths for the molecular graphs having treewidth 3.

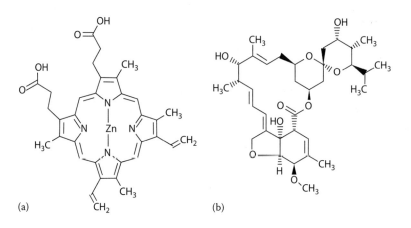

FIGURE 16.13:   Representative chemical compounds having ranges (a) 5 (C03184) and (b) 8 (C11956).

and Figure 16.14 shows representative chemical compounds with ranges 6 (C06176), 9 (C09369), and 12 (C01849).

C03184 having range 6 is a zinc protoporphyrin (ZPP), which is a compound found in red blood cells when heme production is inhibited by lead and/or by lack of iron [32]. ZPP is measured to screen for lead poisoning and for iron deficiency, since this compound incorporates a zinc ion instead of iron (which forms heme) in these situations. Note that the zinc ion is incorporated into the center of this compound,

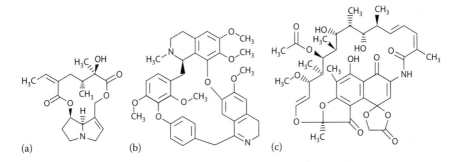

FIGURE 16.14: Representative chemical compounds having ranges (a) 6 (C06176), (b) 9 (C09369), and (c) 12 (C01849).

so there may be a need to have a reasonable range that is not too small but also not too large to accommodate this ion. C11956 is avermectin A2b without any attached sugar molecules and is a part of the macrolide synthesis pathway. Macrolides are a group of drugs of which antibiotics take a large part. Avermectins often have potent insecticidal properties [33,34] and are naturally occurring compounds that are generated as fermentation products by bacteria.

Among the representatives of the smaller peaks, it is interesting that both C06176 and C09369 are alkaloids, which are naturally occurring chemical compounds and are known as secondary metabolites. Alkaloids are produced by a large variety of organisms including bacteria, fungi, plants, and animals and often have pharmacological effects for use as medications. Both avermectins and alkaloids are naturally occurring compounds generated by bacteria and also used to fight against infection, indicating that such compounds with treewidth 3 may be important for such uses. Similar tendencies could be seen for other compounds having treewidth 3, which may be considered to be reasonably sized for biologically related chemical compounds.

16.7.4 Discussion

As we have shown in the previous section, the treewidths of the majority of molecular graphs in KEGG COMPOUND are at most 2. Moreover, the treewidths of only less than 5% of the molecular graphs are 3 or 4. These results support the results reported previously in [22]. In the previous work, there were 9712 chemical compounds in KEGG COMPOUND. The numbers of molecular graphs having treewidths 1, 2, and 3 were 1881 (19.4%), 7336 (75.7%), and 477 (4.9%), respectively. Only one molecular graph (C01863 shown in Figure 16.10) had a treewidth of 4. Therefore, from both of these results, we can conclude that molecular graphs in biological pathways have small treewidth.

Because the degree of each vertex in a molecular graph is bounded due to its valency, all the molecular graphs are sparse graphs. Therefore, it may seem obvious that the treewidths of these molecular graphs are very small. However, if the area under investigation is not limited to biological pathways, there exist well-known molecular graphs with relatively high treewidth. For example, fullerenes [35] including buckyballs and carbon nanotubes have large treewidths. Buckminsterfullerene, also known as C60, is a representative of fullerenes, and their molecular graphs have been well studied in graph theory [36,37]. In contrast, because chemical reactions in cells need relatively small energy and are basically reversible, complex compounds in terms of treewidth may not be able to be constructed by chemical reactions in biological pathways.

Based on these results of local treewidth, we more precisely analyzed the structures of the molecular graphs with the same ranges. Figure 16.15 is a plot of our results, where we found that the molecular graphs with the same ranges tended to include similar substructures in their most complex parts. For range $r_G = 11$, all six molecular graphs have the same substructures with treewidth 3.

In addition, we found that molecular graphs including the same substructures having treewidth 3 tend to be mapped into the same biological pathway. For example, as shown in Figure 16.16, seven entries (corresponding to the seven bold circles located at the bottom left of the figure) of eight entries have links to pathways with molecular graphs including the same substructures with treewidth 3 for $r_G = 12$. These can be mapped to chemical compounds in a series of consecutive chemical reactions in the same pathway (biosynthesis of ansamycins).

16.8   Conclusion

In this chapter, we have illustrated the applicability of the concept of treewidth to cheminformatics problems. In particular, we showed that it can be used to classify chemical compounds from the viewpoint of biological pathways, despite the fact that it was purely derived from the treewidth and the local treewidth of molecular graphs defined in graph theory and that biological knowledge was not used in the classification.

For future work, we can consider the application and comparison of this work with other chemical compound databases, such as PubChem Compound [38], ChEMBL [39], and ChemSpider [40], which are all well known in biochemistry. These databases contain very large amounts of data. For example, PubChem Compound includes 48 million entries as of December 2013. Because faster algorithms have also been recently presented [19,20], it may be possible to compute the treewidths for the chemical compounds in such large databases. Thus, it can be surmised that such theoretical approaches can be applied to a variety of fields beyond the scope of cheminformatics.

Moreover, recently, the metabolic pathways of KEGG were analyzed based on structural patterns of small chemical units [41]. These patterns suggested that there was a correspondence between the genomic units of the enzymes and these chemical units. Thus, the combination of graph–theoretical approaches such as treewidth analysis with genomic and proteomic data may also be of interest to biochemists.

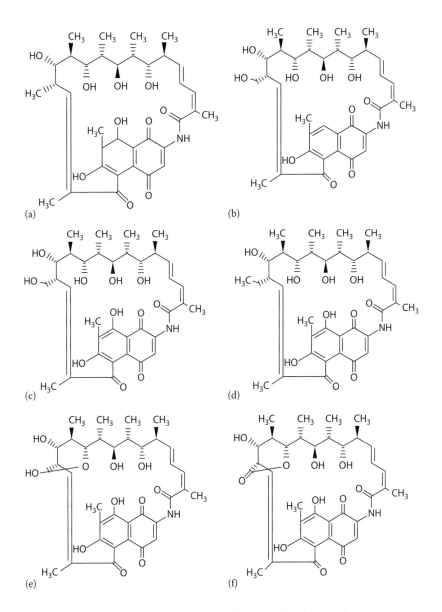

FIGURE 16.15: Chemical compounds whose molecular graphs have the same substructures with treewidth 3, for $r_G = 11$. (a) Proansamycin X (C12176), (b) Protorifamycin I (C12246), (c) Rifamycin W (C12247), (d) 34a-Deoxy-rifamycin W (C14721), (e) Rifamycin W-hemiacetal (C14722) and (f) Rifamycin Z (C14723).

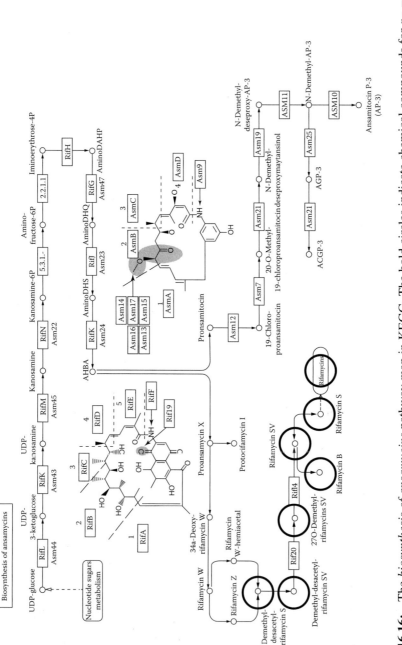

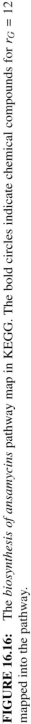

FIGURE 16.16:   The *biosynthesis of ansamycins* pathway map in KEGG. The bold circles indicate chemical compounds for $r_G = 12$ mapped into the pathway.

References

1. G. M. Maggiora and V. Shanmugasundaram. Molecular similarity measures. *Methods in Molecular Biology*, 275: 1–50, 2004.

2. M. R. Garey and D. S. Johnson. *Computers and Intractability: A Guide to the Theory of NP-Completeness*. W.H. Freeman & Co., New York, NY, 1979.

3. M. J. McGregor and P. V. Pallai. Clustering of large databases of compounds: Using the MDL Keys as structural descriptors. *Journal of Chemical Information and Computer Sciences*, 37(3): 443–448, 1997.

4. Daylight Chemical Information Systems, Inc. *Daylight Theory Manual*, Laguna Niguel, CA. http://www.daylight.com/dayhtml/doc/theory/ (release January 11, 2008).

5. M. Dehmer and A. Mehler. A new method of measuring similarity for a special class of directed graphs. *Tatra Mountains Mathematical Publications*, 36: 39–59, 2007.

6. D. Bonchev and D.H. Rouvray. *Chemical Graph Theory: Introduction and Fundamentals*. Taylor & Francis, London, U.K., 1991.

7. N. Trinajstic. *Chemical Graph Theory*. CRC Press, Boca Raton, FL, 1992.

8. A. Dalby, J. G. Nourse, W. D. Hounshell, A. K. I. Gushurst, D. L. Grier, B. A. Leland, and J. Laufer. Description of several chemical structure file formats used by computer programs developed at Molecular Design Limited. *Journal of Chemical Information and Computer Sciences*, 32(3): 244–255, 1992.

9. D. Weininger. SMILES 1. Introduction and encoding rules. *Journal of Chemical Information and Computer Sciences*, 28(1): 31–36, 1988.

10. S. E. Stein, S. R. Heller, and D. Tchekhovskoi. An open standard for chemical structure representation: The IUPAC chemical identifier. In *Proceedings of the 2003 International Chemical Information Conference (Nimes)*, Infonortics, Malmesbury, U.K., pp. 131–143, 2003.

11. R. Rizzi. A short proof of Knig's matching theorem. *Journal of Graph Theory*, 33(3): 138–139, 2000.

12. S. Pawagi. Maximum weight independent set in trees. *BIT Numerical Mathematics*, 27(2): 170–180, 1987.

13. D. W. Matula. Subtree isomorphism in $O(n^{5/2})$. *Annals of Discrete Mathematics*, 2: 91–106, 1978.

14. S. Arnborg and A. Proskurowski. Linear time algorithms for NP-hard problems restricted to partial k-trees. *Discrete Applied Mathematics*, 23(1): 11–24, 1989.

15. S. Arnborg, D. Corneil, and A. Proskurowski. Complexity of finding embeddings in a k-tree. *SIAM Journal on Matrix Analysis and Applications*, 8(2): 277–284, 1987.

16. H. L. Bodlaender. A linear time algorithm for finding tree-decompositions of small treewidth. *SIAM Journal on Computing*, 25(6): 1305–1317, 1996.

17. L. Perković and B. Reed. An improved algorithm for finding tree decompositions of small width. *International Journal of Foundations of Computer Science*, 11(3): 365–371, 2000.

18. V. Gogate and R. Dechter. A complete anytime algorithm for treewidth. In *Proceedings of the 20th Conference on Uncertainty in Artificial Intelligence (AUAI '04)*, AUAI Press, Arlington, TX, pp. 201–208, 2004.

19. F. V. Fomin, D. Kratsch, I. Todinca, and Y. Villanger. Exact algorithms for treewidth and minimum fill-in. *SIAM Journal on Computing*, 38(3): 1058–1079, 2008.

20. H. L. Bodlaender, F. V. Fomin, A. M. C. A. Koster, D. Kratsch, and D. M. Thilikos. On exact algorithms for treewidth. *ACM Transactions on Algorithms*, 9(1): Article No. 12, 2012.

21. D. Eppstein. Diameter and treewidth in minor-closed graph families. *Algorithmica*, 27(3–4): 275–291, 2000.

22. A. Yamaguchi, K. F. Aoki, and H. Mamitsuka. Graph complexity of chemical compounds in biological pathways. *Genome Informatics*, 14: 376–377, 2003.

23. M. Kanehisa, S. Goto, Y. Sato, M. Furumichi, and M. Tanabe. KEGG for integration and interpretation of large-scale molecular data sets. *Nucleic Acids Research*, 40(Database issue): D109–D114, 2012.

24. S. Goto, Y. Okuno, M. Hattori, T. Nishioka, and M. Kanehisa. LIGAND: Database of chemical compounds and reactions in biological pathways. *Nucleic Acids Research*, 30(1): 402–404, 2002.

25. S. H. Bertz. On the complexity of graphs and molecules. *Bulletin of Mathematical Biology*, 45(5): 849–855, 1983.

26. M. Dehmer, K. Varmuza, S. Borgert, and F. Emmert-Streib. On entropy-based molecular descriptors: Statistical analysis of real and synthetic chemical structures. *Journal of Chemical Information and Modeling*, 49(7): 1655–1663, 2009.

27. N. Rashevsky. Life, information theory, and topology. *Bulletin of Mathematical Biophysics*, 17(3): 229–235, 1955.

28. M. Randić, G. M. Brissey, R. B. Spencer, C. L. Wilkins. Search for all self-avoiding paths for molecular graphs. *Computers & Chemistry*, 3(1): 5–13, 1979.

29. I. Lukovits, A. Miličević, S. Nikolić, N. Trinajstić. On walk counts and complexity of general graphs. *Internet Electronic Journal of Molecular Design*, 1(8): 388–400, 2002.

30. D. Bonchev. Overall connectivities/topological complexities: A new powerful tool for QSPR/QSAR. *Journal of Chemical Information and Computer Sciences*, 40(4): 934–941, 1999.

31. S. H. Bertz and C. M. Zamfirescu. New complexity indices based on edge covers. *Communications in Mathematical and in Computer Chemistry*, 42: 39–70, 2000.

32. R.F. Labbe, H.J. Vreman, and D.K. Stevenson. Zinc protoporphyrin: A metabolite with a mission. *Clinical Chemistry*, 45(12): 2060–2072, 1999.

33. S. Omura and K. Shiomi. Discovery, chemistry, and chemical biology of microbial products. *Pure and Applied Chemistry*, 79(4): 581–591, 2007.

34. T. Pitterna, J. Cassayre, O. F. Huter, P. M. Jung, P. Maienfisch, F. M. Kessabi, L. Quaranta, and H. Tobler. New ventures in the chemistry of avermectins. *Bioorganic and Medicinal Chemistry*, 17(12): 4085–4095, 2009.

35. H. Zhang and F. Zhang. New lower bound on the number of perfect matchings in fullerene graphs. *Journal of Mathematical Chemistry*, 30(3): 343–347, 2001.

36. S. Jendrol' and P. J. Owens. Longest cycles in generalized Buckminsterfullerene graphs. *Journal of Mathematical Chemistry*, 18(1): 83–90, 1995.

37. N. Trinajstić, Z. Mihalić, and F. E. Harris. A note on the number of spanning trees in buckminsterfullerene. *International Journal of Quantum Chemistry*, 52(S28): 525–528, 1994.

38. E. E. Bolton, Y. Wang, P. A. Thiessen, and S. H. Bryant. PubChem: Integrated platform of small molecules and biological activities. *Annual Reports in Computational Chemistry*, 4: 217–241, 2008.

39. A. Gaulton, L. J. Bellis1, A. P. Bento, J. Chambers, M. Davies, A. Hersey, Y. Light, S. McGlinchey, D. Michalovich, B. Al-Lazikani and J. P. Overington. ChEMBL: A large-scale bioactivity database for drug discovery. *Nucleic Acids Research*, 40(Database issue): D1100–D1107, 2012.

40. H. E. Pence and A. Williams. ChemSpider: An online chemical information resource. *Journal of Chemical Education*, 87(11): 1123–1124, 2010.

41. M. Kanehisa. Chemical and genomic evolution of enzyme-catalyzed reaction networks. *FEBS Letters*, 587(17): 2731–2737, 2013.

Index

A

Abelian groups, 144, 146, 157–158
Activism, 416–418
Acyclic digraphs, 215
Additive-degree-Kirchhoff index, 64–68
Additive Kirchhoff index
    d-regular graph, 72–73
    full binary tree, 73–74
    semiregular graph, 73
Adjacency matrix, 113
Adjusted betweenness centrality, 236
AIDS dataset, 463–464
Alkanes, 261
Amoebas
    building graphs from, 363
    construction, 361
    distances, 361
    Lerallut, 362
    metric, 362
    modified patch graphs, 363
    structuring element, 363
    texture representation by, 363–364
Ansamycins pathway, 488, 490
Antimonotonicity
    definition, 304
    object-specific overlap hypergraphs, 317–318
    overlap-graph-based measures, 311–312
    pairwise overlap graph, 311–312
Arithmetic-geometric-harmonic mean inequalities, 36
Atom–bond connectivity (ABC) index, 12
Attribute statistics-based embedding, 437

B

Balaban index, 11, 13–14, 17, 365
Benzenoid hydrocarbons, 82
Betweenness centrality
    analogues, 235–237
    definition, 235
    graph invariants, 235–237
    with graph operations, 242–243
    with graph parameters, 237–242
    possible values of, 237–242
    properties, 234–235
    spectral relations, 251–254
    Wiener index, 284
Betweenness-uniform graphs, 244–251
Bicyclic graphs, metric-extremal, 126–128
Biological pathways, chemical compound in, 472–473
Bipartite graph, 238–239, 263
Bi-rank width, 212–215
Boiling point prediction, 462
Bonchev–Trinajstić information indices, 366
Boolean networks, 184, 225
Bounding theorem
    object-specific overlap hypergraphs, 318–319
    overlap-graph-based measures, 312–314

Average

Average distance, 116
Average edge betweenness centrality, 236, 242
Average vertex betweenness centrality, 235, 241

Brambles, in directed graphs, 198–201
Branching cardinality, 459
Branching time logics, 224
Bridge, 428

C

Cactus, 121
Camarilla, 399–400
Cartoon–texture model, 359
Catacondensed hexagonal system, 83
Cauchy–Schwarz inequality, 36
c-cyclic graphs, 49
Centrality
    and centralization, 390
    eigenvector, 411
    index, 233–234
    measurement, 410–411
    social network analysis, 407
        activism level, 416–418
        concepts of, 410–411
        environmental movement, 408
        environmental organization,
          416–417
        Greenpeace, 408
        hypotheses, 413–414
        literature, 408–411
        network range measures,
          418–419
        OLS multiple regression, 417
        research questions, 412–413
        self-administered
          questionnaire, 415
        theoretical model, 414
        two-mode network, 409–411
        visualizations, 408–409
Chemical compounds
    biological pathways, 472–473
    k-tree, 478
    local treewidth, 478–479
    Molecular Design Limited Molfile
      format, 475–476
    molecular graph, 473–475
      complexity measures, 481

KEGG COMPOUND (*see* KEGG
    COMPOUND)
    linear representations, 477
    range analysis, 484–487
    tabular representations, 475–476
    structural formula of, 471
    treewidth, 477–478
    zinc protoporphyrin, 486
Chemical graph theory, 280–282
Chemoinformatics, 445
Circuit, 144
Clique width, 212–215
Clustering, 380–381
Collective action, and social movements,
    411–412
Color classes, 400
Combined hexagonal chains, 94
Common subgraph, 431
Comparative graph analysis
    graph edit distance, 6–7
    graph kernels, 8–9
    isomorphism-based measures, 5–6
    iterative methods, 8
    molecular similarity, 9–10
    similarity of document structures, 9
    statistical graph matching, 10–11
    string-based measures, 8
    tree similarity, 9
Compatible graph, 151
Complete graph kernel, 446
Complexity theory, 186–187
Computations, 166–168
Computing DAG decompositions, 206
Computing directed tree decompositions,
    193–197
Conjecture engines, 19–20
Connected graph, 428
Connectivity index, 12
Constant shift embedding method, 437
Co-occurrence matrix, 370
Cop monotone, 189
Cops and robber games, 187
    directed, 189–191
    monotonicity in, 189–191
    on undirected graphs, 188

Counting principles
    first reduction principle, 152–153
    flows on wheels, 156–158
    numbers of partitions, 155–156
    planar case, 154–155
    for simple networks, 150–152
Cubic graphs, edge colorings of,
        174–178
Cycle
    definition, 428
    rank, 215
Cycle-containing graphs, 116
Cyclical edge connectivity, 159–161
Cyclically k-edge connected graph, 147
Cyclic graph, 144, 147
Cyclic pattern kernels
    cyclic patterns, 454–456
    relevant cycle graph, 456–457
    relevant cycle hypergraph, 457–458

D

DAG, *see* Directed acyclic graph (DAG)
d-cube graph, 72
Decompositions, networks and, 399–403
Degeneracy
    of graph measures, 13–15
    of Wiener index, 102–106
Degree-based indices
    over edges
        generalized sum-connectivity
            index, 52
        general Randić index, 50–52
    over vertices
        first general Zagreb index, 48–49
        first multiplicative Zagreb
            index, 49
Degree-based measures, 12
Degree distributions, 335
Degree of node, 427
Degree of vertex, 234
Degree sequence, for undirected
        graph, 327
Dehmer entropies, 367–369
Descriptive approaches, 2

Descriptive graph theory, 3–4
Digraphs
    of bounded directed tree width, 183
    decomposition of, 184
    havens in, 197–198
    linkedness of, 201
    nowhere crownful classes of,
        215–218
    structural properties, 184–185
    undirected graphs to, 183
    vertex elimination, 209
    well-linkedness, 201
Dihedral groups, reductions by, 165–166
Dirac function, 448
Directed acyclic graph (DAG), 186
    decompositions, 206
    width, 204–206
Directed circuit, 144
Directed clique width, 183–184
Directed cops and robber games,
        189–191
Directed decompositions, 215
Directed dominating sets, 216,
        225–226
Directed elimination ordering, 209
Directed graph, 183
    brambles in, 198–200
    Laplacian of, 326–327
    network sampling, 326
Directed path width, 183, 211–212
Directed tree, 186
    decompositions, 191–193
        computing, 193–197
    width, 183
        brambles in directed graphs,
            198–200
        havens in digraphs, 197–198
        well-linked sets, 200–201
Directed width-measures, 190–191, 218
Discrimination power, 356–357
Disjoint paths, 219–223
Distance, 434–435
    matrix, 113
    between nodes, 428
Distance balanced graph, 273

Distance-based indices
  Balaban index, 365
  Wiener and Harary index, 364–365
Distance-based measures, 11
Distance-regular graph, 245, 251
Document structures similarity, 9
Dominating set problem, 187
Double bonds, 280
d-regular graphs, 259, 273
  additive Kirchhoff index, 72–73
  lower bounds for, 71–72
  upper bounds, 72
d-thorn graphs, 265
Dual graphs, 148–150
D-width, 202–204

E

EAID numbers, 14
Eccentricity, 328
Edge colorings, of cubic graphs, 174–178
Edge-degree correlation, 336–337
Edge expansion index, 253
Edge-forwarding index, 253–254
Edge-Szeged index, 263
Edge-vertex-Szeged index, 264
Edge-Wiener index, 116, 286
Eigenvalue-based indices
  energy index, 53–54
  Laplacian indices, 54–57
  normalized Laplacian indices,
    54–57
Eigenvalue-based measures, 12, 14
Eigenvector, centrality, 411
Empirical kernel map, 436
Energy index, 53–54
Environmental non-governmental
    organizations (ENGOs)
  one-mode network of, 410
  social movement, 411, 415–416
Erdos-Rényi undirected graphs, 328
Estrada index, 281
Euclidean distance, 441
  definition, 435
  matrix, 435

Euler graph, 3
Euler's theorem, 177
Expander graph, 333
Explicit information inequalities, 16–17
Exponential weighting scheme, 369

F

𝔽-bi-rank width, 213, 215
Fibonacenes, 84–85
  2-linked family of, 88
  3-linked family of, 89
First general Zagreb index, 48–49
First multiplicative Zagreb index, 49
First reduction principle, 152–153
Fixed-parameter algorithms, 226
Fixed-parameter tractable (fpt), 187
Flows, 143–146, 156–158
Forbidden network, 162–163
𝔽-or bi-rank width, 183–184
Ford–Fulkerson maximum flow
    algorithm, 333–334
Four-color theorem, 174, 177
𝔽-rank width, 213, 215
Frequent graphs, 432
Full binary tree, 73–74
Fuzzy multilevel graph embedding, 437

G

GED, *see* Graph edit distance (GED)
Generalized MKL, 460–461
Generalized Randić index, 327
Generalized stars, 291
Generalized sum-connectivity index
  degree-based indices, 52
  numerical results, 70
Generalized trees, 8
Generalized t-star, 291–292
General Randić index
  degree-based indices, 50–52
  numerical results, 68–70
Geometric undirected graphs, 328, 331
Γ-equivalence classes, 152
Γ-expression, 213

Girths
    graphs with, 288–290
    restrictions of, 163–164
Global synchronization, 340–341
Glycolysis pathway, 474
Gram matrix, 439, 444–445
Graph-based document structures, 9
Graph edit distance (GED), 6–7,
        357, 426, 432–433
    approximation, 433
    and machine learning methods,
        434–435
Graph embedding methods, 426
    empirical kernel map, 436
    global approaches, 437–438
    isometric embedding, 436–437
    local approaches, 437
Graph energy, 14
Graph fragments, 451–452
Graph indices, 356, 371–373
Graph invariants, 20, 235–237
Graph isomorphism, 428
Graph kernels, 8–9, 426–427, 445
    and chemoinformatics, 445
    complete, 446
    cyclic pattern kernels
        cyclic patterns, 454–456
        relevant cycle graph, 456–457
        relevant cycle hypergraph,
            457–458
    3D pattern kernel, 453–454
    linear pattern kernels
        paths, 447–448
        random walks, 446–447
        tottering, 447
    nonlinear pattern kernels
        graph fragments, 451–452
        graphlets, 448–450
        treelets, 452–453
        tree patterns, 450–451
        Weisfeiler–Lehman, 451
    pattern weighting, 458–459
        branching cardinality, 459
        generalized MKL, 460–461
        infinite MKL, 461

    multiple kernel learning, 460
    a priori methods, 459
    ratio depth/size, 459–460
    simple MKL, 460
Graphlets, 448–450, 452
Graph measures
    correlation of, 18–19
    uniqueness of, 13–15
Graph models, in image analysis,
        357–358
Graphs, 143–146, 182, 427
    of bounded tree, 182
    cloning, 246
    cut methods, 357–358
    databases, query evaluation in,
        223–224
    discrimination of, 357
    with girth, 288–290
    line, 113
    mathematical preliminaries, 112
    matrix representations of, 113–114
    metric in, 112–113
    metric invariants of, 114–115
    operations, 242–243
    parameters, 237–242
    path in, 112
    preliminaries, 305–306
    quantifying, 111–112
    simple/schlicht, 112
    Wiener index of, 286–288
Graph similarity measures
    frequent graphs, 432
    graph edit distance, 432–433
        approximation, 433
        and machine learning methods,
            434–435
    maximum common subgraphs, 431
Graph-theoretical branch, 3
Graph theory, 39–40, 185–186, 279
Graph transformation, 96–97
Gray-tone spatial-dependence
        matrix, 370
Greenpeace, 408
Gutman index, 117, 268, 281

H

Hamiltonian graph, 243
Hamming distance, 9–10
Haralick features, 370–371
Harary index, 118, 365
Hardy–Ramanujan formula, 246
Havel–Hakimi algorithm, 333–334
Havens, in digraphs, 197–198, 201
Hexagonal chains, 83–85
    arbitrary, 99, 104–105
    complete families, 87
    with extremal Wiener index, 89–90
    growing, 98
    linked, 86–89
    representation of, 85–86
    segment in, 84–86
    self-linked, 86–87
    symmetrically linked families, 89
    zigzag, 85–86
Hexagonal network, 82–83
Hexagonal ring, 83
Higher line graph iterations, 293–296
Homeomorphism, 294–296
Homogeneity, 183, 214
Hosoya index, 281
Hosoya polynomial, 118–119, 286
Hydrogen-completed graphs, 264
Hydrogen-suppressed molecular
    graph, 260
Hypergraph, 306
    definition, 429
    object-specific overlap
        hypergraphs, 315
        bounding theorem, 318–319
        MIS measure, 316–318
        OGSM MIS relaxation, 319
Hyperplane splitting data, 442
Hyper-Wiener index, 117

I

Image analysis, 10, 357–358
Implicit information inequalities, 15–17
Inert invisible reachability game, 209
Inert robber, 189

Inexact graph matching, 357
Infinite MKL, 461
Information inequalities, 15–18
Information polynomial, 15
Information-theoretic approach, 12, 102
Information-theoretic indices, 365
    Bonchev–Trinajstić information
        indices, 366
    Dehmer entropies, 367–369
Institutional Revolution Party (PRI),
    390–393
Integral values, flow with, 146–148
Invisible cops and robber game, 188
Invisible node searching, 188
Isometric embedding, 436–437
Isomorphism-based measures, 5–6
Isomorphisms, preliminaries, 306–307
Isoperimetric number, 253
Iterated line graphs, 282
Iterative methods, 8

K

Karush-Kuhn-Tucker (KKT) conditions,
    461
k-betweenness centrality, 236
k-colorable graph, 149
k-disjoint path problem, 219
KEGG COMPOUND
    ansamycins pathway map in,
        488, 490
    database, 479–481
    glycolysis pathway map in, 474
    MDL Molfile format, 481–482, 484
    molecular graph complexity
        algorithm, 482–484
        data set, 481–482
        treewidths analysis, 484–485,
            487–489
    for shikimic acid, 480
Kelly decompositions, 207–211
Kelly width, 206–210
Kenogram, 260, 264–265
Kernel between tree patterns, 450
Kernel combination, 439–440

Kernel ridge regression, 444–445
Kernel theory, 438
    chemoinformatics, 445
    definitions, 439
    kernel trick, 440–441
    ridge regression, 444–445
    and similarity measures, 441
    support vector machines,
        441–444
Kernel trick, 438, 440–441
Kirchhoff index
    lower bounds, 60–62
    numerical results, 71–72
    resistance-based indices, 57–60
    upper bounds, 62–63
Kleitman–Wang algorithm, 333–334
Königsberg bridge problem, 1

L

Labeled graph, 213, 428
Labeling information, 437
λ-variable Wiener indices, 263
Laplacian eigenvalues, 251–252
Laplacian Estrada index, 14
Laplacian index, 54–57
Laplacian matrix, 438
Laplacian of directed graph, 326–327
Length-scaled betweenness
    centrality, 236
Linearly scaled betweenness
    centrality, 236
Linear pattern kernels
    paths, 447–448
    random walks, 446–447
    tottering, 447
Linear representations, 477
Linear time temporal logic (LTL), 224
Line graph, 113
    iterated, 282
    Wiener index of, 286–288
Linkage problems, 219–223
Linked chains
    hexagonal, 86–87
    Wiener index of, 90–92

Linked families
    hexagonal chains, 87–89
    Wiener index of, 94–95
Local treewidth, chemical compound,
    478–479, 484, 486
Lone-pair electrons, 280
LTL, *see* Linear time temporal
    logic (LTL)

M

Machine learning methods,
    434–435
Majorization ordering, 37–38, 66
    extremal elements with,
        40–41
    maximal elements, 41–43
    minimal elements, 44–47
    notion of, 36
Maple programming language, 168
Marginalized kernel, 446
Mathematical preliminaries, 112
Matrix representations of graphs,
    113–114
Maximal common subgraph, 431
Maximal elements, 41–43
Maximal independent set (MIS) measure
    object-specific overlap hypergraphs,
        316–317
    antimonotonicity, 317–318
    necessary condition, 318
    sufficient condition, 318
    overlap-graph-based measures,
        312–314
Maximum centrality score, 394
Maximum common subgraph (MCS),
    7, 431
Maximum edge betweenness
    centrality, 236
Maximum vertex betweenness
    centrality, 235
MCS, *see* Maximum common
    subgraph (MCS)
Mean distance, 116
Menger's theorem, 200

Metabolic pathways, 472–473
Metric-extremal graphs, 120–122
   trees, 122–126
   unicyclic and bicyclic, 126–128
Metric in graphs, 112–113
Metric invariants
   of graphs, 114–115
   Harary index, 118
   Hosoya polynomial, 118–119
   Hyper-Wiener index, 117
   Szeged and terminal Wiener indices,
      119–120
   Wiener index, 115–117
Mexican network of power (MNP), 388
Mexican political system, 387–389
   centrality and centralization, 390
   color classes, 400
   Institutional Revolutionary Party,
      390–393
   network
      cohesion and centralization,
         393–396
      and decompositions, 399–403
   parameter values for, 389
   sensitivity and reliability of scores,
      397–399
Minimal elements, 44–47
Minimum clique partition (MCP)
      measure, 312–314
Minimum Randić index problem
   formulation and complexity,
      343–345
   heuristic for disconnected
      realization, 348
   solving, 345–348
Minimum spanning trees (MSTs), 10
Minimum weight perfect b-matching
      problem, 343
   formulation and complexity,
      343–345
   heuristic for disconnected
      realization, 348
   solving, 345–348
Min–Max kernel, 447
Model checking

foundations of, 224–225
   problem, 226
Modified Schultz index, 268
Molecular Design Limited (MDL)
      Molfile format
   KEGG COMPOUND, 481–482, 484
   linear representations, 477
   tabular representations, 475–476
Molecular graph, 112, 260, 265, 280, 430
   for chemical compound, 473–475
      complexity measures, 481
      KEGG COMPOUND (*see* KEGG
         COMPOUND)
      linear representations, 477
      range analysis, 484–487
      tabular representations, 475–476
   for shikimic acid, 475–476
Molecular similarity, 9–10
Molecular structure descriptors, 114
Molecular topological index, 117
Monadic second-order logic (MSO),
      182, 214
Monotonicity, in cops and robber games,
      189–191
Moore graph, 121
Morphological amoebas
   building graphs from, 363
   construction, 361
   Lerallut, 362
   metric, 362
   modified patch graphs, 363
   structuring element, 363
   texture representation by, 363–364
MSO, *see* Monadic second-order
      logic (MSO)
MSTs, *see* Minimum spanning
      trees (MSTs)
Multiple kernel learning (MKL), 460
   generalized, 460–461
   infinite, 461
   simple, 460
Multiplicative-degree-Kirchhoff index,
      63–64
Multiterminal network, 144

N

n-cycle graph, 72
Neighborhood matching set, 450
Neighborhood relationship, 427
Network(s), 143–146
    centralization index, 393
    cohesion and centralization,
       393–396
    and decompositions, 399–403
    definitions for, 389
    dynamics, 329–333
    paradigm, 387
Network sampling
    definitions, 326–328
    degree distributions, 335
    design of Internet router, 341–342
    directed graph, 326
    edge-degree correlation, 336–337
    Ford–Fulkerson maximum flow
       algorithm, 333–334
    Havel–Hakimi algorithm, 333–334
    Internet router and air transport, 342
    Kleitman–Wang algorithm, 333–334
    Laplacian of directed graph,
       326–327
    minimum Randić index
       problem, 343
       formulation and complexity,
          343–345
       heuristic for disconnected
          realization, 348
       solving, 345–348
    network dynamics, 329–333
    node-degree correlation, 336
    notations, 326–328
    optimization problem, 342–343
    perfect b-matching problem with
       minimum weight, 343
       formulation and complexity,
          343–345
       heuristic for disconnected
          realization, 348
       solving, 345–348
    quantile function, 335

Ramanujan graphs, 332–333
spectral properties, 329–333
structural properties of, 325
synchronization, 337–338
    description, 330
    neuronal networks, 338
    pulse-coupled oscillators,
       338–341
undirected graphs, 326
    degree sequence for, 327
    generalized Randić index, 327
    types of, 328
Neuronal networks, 338
Node-degree correlation, 336
Nonisomorphic cloned graphs, 246
Nonlinear pattern kernels
    graph fragments, 451–452
    graphlets, 448–450
    treelets, 452–453
    tree patterns, 450–451
    Weisfeiler–Lehman, 451
Nonmonotone strategies, 189
Nonoriented graph, 427
Nonoverlap-graph-based measures
    key-based support measures,
       307–308
    minimal image count support
       measure, 308
Nonterminal segment, 85
Nontrivial graph, 151, 175
Nonzigzag segments, 85
Normalized Laplacian index
    eigenvalue-based indices, 54–57
    graph theory, 39–40
Notations, 37–40
Nowhere crownful classes of digraphs,
    215–218
Nowhere-zero flows, 141
    arcs, 143
    dual graphs, 148–150
    flows, 143–146
    graphs, 143–146
    with integral values, 146–148
    networks, 143–146
    unsolved conjectures, 142

Nowhere-zero 5-flows, 158
   dihedral groups, reductions by,
      165–166
   discussion about computations,
      166–168
   forbidden networks, 162–163
   open problems, 169
   restrictions
   of cyclical edge connectivity,
      159–161
   of girths, 163–164
NP-complete, 6
Numerical graph invariants, 11–12
   correlation of graph measures,
      18–19
   information inequalities, 15–18
   uniqueness of graph measures,
      13–15

O

Object-specific overlap hypergraphs, 315
   bounding theorem, 318–319
   MIS measure, 316–317
      antimonotonicity, 317–318
      necessary condition, 318
      sufficient condition, 318
   OGSM MIS relaxation, 319
Optimal matching, 433
Ordinary least squares (OLS) multiple
      regression, 417
Oriented bipartite graph, 216
Oriented hypergraph, 429–430
Overlap graph
   node in, 304
   and overlap hypergraph, 316
   pairwise overlap graph, 309–310
Overlap-graph-based measures, 308
   antimonotonicity, 311–312
   bounding theorem, 312–314
   Lovász and Schrijver graph
      measures, 314–315
   MIS and MCP measure, 312–314
   pairwise overlap graph, 309–310
   support measure, 310–311

Overlap hypergraph, 305
   object-specific overlap
      hypergraphs, 315
   bounding theorem, 318–319
   MIS measure, 316–318
   OGSM MIS relaxation, 319
   overlap graph and, 316
Overlap-hypergraph-based support
      measures (OHSM)
   bounding theorem, 318–319
   conditions for antimonotonicity,
      317–318
   definition, 316

P

Pairwise dependency, 235
Pairwise overlap graph (POG), 309–310
   antimonotonicity, 311–312
   support measure, 310–311
Paraffins, 261
Parameterized problems, 186–187
Parent H-deleted graphs, 264
Parity games, 183, 224
Partial k-DAG, 209–210
Partial subgraph isomorphism, 428–429
Partitions, numbers of, 155–156
Patch graphs, texture representation
      by, 363–364
Paths, 427, 447–448
   width, 188
Pattern weighting, 458–459
   branching cardinality, 459
   multiple kernel learning, 460–461
   a priori methods, 459
   ratio depth/size, 459–460
Pearson correlation coefficient, 336–337
Pendant path, 294
Pendant vertex, 240
Perfect b-matching problem, with
      minimum weight, 343
   formulation and complexity,
      343–345
   heuristic for disconnected
      realization, 348
   solving, 345–348

Petersen graph, 142, 147
PI index, 262
Planar case, 154–155
Planar graph, 174
    with dual, 148
    orientation of, 149
Planar permutation group, 155
Plerogram, 260, 264
Polynomial time algorithm, 223
Positive definite kernel, 439
Power-law undirected graphs, 328
Predictive Toxicity Challenge (PTC)
    dataset, 464–465
Preliminaries, 37–40
    complexity theory, 186–187
    graph theory, 185–186
Probabilistic inequalities for graph
    measures, 17–18
Proper graph, 265
Pulse-coupled oscillators, 338–341

Q

Quadratic line graph, 291
Quantification of graphs,
    111–112
Quantitative graph descriptors
    distance-based indices
        Balaban index, 365
        Wiener and Harary index,
            364–365
    information-theoretic indices, 365
        Bonchev–Trinajstić information
            indices, 366
        Dehmer entropies, 367–369
Quantitative graph theory, 356–357
    classification of, 5
    comparative graph analysis (*see*
        Comparative graph analysis)
    concept formation, 1–3
    definition of, 2
    vs. descriptive graph theory, 3–4
    numerical graph invariants,
        11–12

    correlation of graph measures,
        18–19
    information inequalities, 15–18
    uniqueness of graph measures,
        13–15
    software
        conjecture engines, 19–20
        R-packages, 19
Quantitative structure–activity relation-
    ship (QSAR), 12, 261, 280
Quantitative structure–property relation-
    ship (QSPR), 260–261, 280
Quasicubic network, 162
Query evaluation in graph databases,
    223–224

R

Ramanujan graphs, 332–333
Randić index, 12, 281
Random walks, 446–447
Range analysis, 484–487
Rank decompositions, 213
Rank width, of undirected graphs, 213
Ratio depth/size, 459–460
Reachability problems, 226
Reachability searching, 189–190
Recurrent network, 225
REDMEX databases, 388, 397, 400, 404
Regression, support vector machines for,
    443–444
Regular path query (RPQ), 223
Relative angle inequality, 435
Relevant cycle graph, 456–457
Relevant cycle hypergraph, 457–458
Reliability of scores, 397–399
Resistance-based indices
    additive-degree-Kirchhoff index,
        64–68
    Kirchhoff index, 57–63
    multiplicative-degree-Kirchhoff
        index, 63–64
Revised edge-Szeged index, 263
Revised vertex-Szeged index, 263
ρ-index, 18

Robber monotone, 189
Robber monotonicity, 189
R-packages, 19
RPQ, *see* Regular path query (RPQ)

S

Schultz index, 117, 267, 281
Schur-convex functions, 36
    of degree sequence, 49
    of eigenvalues, 52
    energy index, 53
    extremal values of, 58
    Laplacian index, 54
    minimum value, 60
    order-preserving property of, 38
    resistance-based indices, 57
    topological indices, 48
Second line graph iteration, 291–293
Second-order logic, 214
Second reduction principle
    application, 172–174
    by superproper permutations,
        170–172
SED, *see* String edit distance (SED)
Segment, 84
Self-administered questionnaire, 415
Self-linked hexagonal chains, 86
Semiregular graph, 73
Sensitivity, 397–399
Separator graph, 186
Shape, texture and, 358–359
Shikimic acid
    KEGG COMPOUND for, 480
    molecular graph for, 475–476
Simple 5-cut graph, 174
Simple MKL, 460
Simple networks, 150–152
Simple regular path queries (SRPQ), 223
Simplified Molecular Input Line Entry
        System (SMILES), 477
Single-graph support measures
    antimonotonicity, 304
    description, 303–305
    nonoverlap-graph-based measures

key-based support measures,
        307–308
    minimal image count support
        measure, 308
    object-specific overlap
        hypergraphs, 315
    bounding theorem, 318–319
    MIS measure, 316–318
    relaxation of OGSM MIS, 319
    overlap-graph-based measures, 308
    antimonotonicity, 311–312
    bounding theorem, 312
    Lovász and Schrijver graph
        measures, 314–315
    MCP measure, 312–314
    pairwise overlap graph,
        309–310
    support measure, 310–311
    overlap notions, 321–322
    S-measure, 319–321
Snark graph, 147
Social movements
    collective action and, 411–412
    ENGOs, 411, 415–416
    identification, 412, 417–418
    one-mode network of ENGOs, 410
Social network centrality, 407
    activism level, 416–418
    concepts of, 410–411
    environmental movement, 408
    environmental organization
        memberships, 416
    weak ties to, 417
    Greenpeace, 408
    hypotheses, 413–414
    literature, 408–411
    network range and, 418–419
    OLS multiple regression, 417
    research questions, 412–413
    self-administered questionnaire, 415
    theoretical model, 414
    two-mode network, 409
    centrality measures, 410–411
    eigenvector centrality, 411
    visualizations, 408–409

Software, for quantitative graph analysis
conjecture engines, 19–20
R-packages, 19
Spanning subgraph, 143
Spectral embedding methods, 438
Spectral gap, 63
Spectral relations, 251–254
Spectral seriation method, 8
Sphere-regular graphs, 17
Spiral hexagonal chain, 86
SRPQ, *see* Simple regular path
queries (SRPQ)
Starlike tree, betweenness centrality,
239–240
Statistical graph matching, 10–11
Strictly convex, 294
String-based measures, 8
String edit distance (SED), 6
Strong component searching, 189
Strongly regular graphs, 245
Subgraph, 427, 429
Superproper permutations, 170–172
Support measures
key-based support measures,
307–308
Lovász graph measures, 314–315
minimal image count, 308
nonoverlap-graph-based measures,
307–308
overlap-graph-based measures,
310–311
pairwise overlap graph, 310–311
preliminaries, 307
Schrijver graph measures, 314–315
single-graph (*see* Single-graph
support measures)
Support vector machine (SVM), 427
kernel methods, 441–443
for regression, 443–444
Symmetrically linked families
hexagonal chains, 89
Wiener index of, 95–98
Synchronization
global, 340–341
network sampling, 337–338

description, 330
neuronal networks, 338
pulse-coupled oscillators,
338–341
Szeged index, 119–120, 263

T

Tabular representations, 475–476
Tanimoto index, 9–10
Tanimoto kernel, 447
Terminal hexagonal system, 83
Terminal segment, 84
Terminal Wiener indices, 119–120
Texture
descriptors, 359
segmentation, 359–361
and shape, 358–359
Texture discrimination, 369
evaluation procedure
graph indices, 371–373
Haralick features, 370–371
experimental results on,
373–377
mutual distances, 378–379
overlap of features, 377–381
test images, 370
Thorn graphs
notions, 259–266
preliminaries, 266–269
results, 269–275
3D pattern kernel, 453–454
Tonal distance, 362
Topological embedding, 437
Topological index, 2, 47–48, 114
chemical graph theory, 280
degree-based indices
over all vertices, 48–49
over edges, 50–52
distance-based, 81–82
eigenvalue-based indices
energy index, 53–54
Laplacian and normalized
Laplacian indices, 54–57
graph theory and, 260

resistance-based indices
  additive-degree-Kirchhoff index, 64–68
  Kirchhoff index, 57–63
  multiplicative-degree-Kirchhoff index, 63–64
Total Szeged index, 264
Tottering, 447
Trails, 427
Transition matrix, 39
Tratch–Stankevich–Zefirov (TSZ) index, 118–119
Tree, 260, 428
  of order, 239–240
  patterns, 450–451
  similarity, 9
Treelet kernel (TK), 452–453
Tree width, 182, 188
  chemical compound, 477–478
  molecular graph in KEGG COMPOUND, 484–485, 487–489
Trivial degeneracy class, 102
Trivial graph, 151
Two-mode network
  centrality measures, 410–411
  eigenvector centrality, 411
  social network analysis, 409

U

Underlying undirected graph, 185–186
Undirected graphs
  cops and robber games on, 188
  degree sequence for, 327
  generalized Randić index, 327
  network sampling, 326
  rank width of, 213
  types of, 328
Unicyclic graphs, 126–128, 287
Uniqueness of graph measures, 13–15

V

Valence, 280
Vertex cover problem, 187

Vertex-Szeged index, 262
Vertex-transitive graph, 245
Vertex transmission, 239
Visible cops and robber game, 188
Visible node-searching game, 188
Volkmann tree, 125–126

W

Wald–Wolfowitz test, 10
Walks, 427
Weighted digraph, 219, 221
Weighted graphs, 112
Weighted tree pattern kernel, 459
Weisfeiler–Lehman kernels, 451
Well-linked sets, 200–201
Width measures, 187
Wiener index, 82, 115–117, 282–286, 364
  of complete families, 92–93
  degeneracy of, 102–106
  degree analogue of, 267
  expansion of, 98–102
  of graphs and line graphs, 286–288
  hexagonal chains with extremal, 89–90
  of line graph, 279
  of linked chains, 90–92
  of linked families, 94–95
  molecular descriptor, 279
  of molecular graph, 261
  of symmetrically linked families, 95–98
  topological index, 280
Wiener polynomial, 286

Z

Zagreb index, 12, 117
  first general, 48–49
  first multiplicative, 49
Zigzag hexagonal chain, 85–86
Zigzag segment, 85
Zinc protoporphyrin (ZPP), 486